The Old Red Sandstone of Great Britain

THE GEOLOGICAL CONSERVATION REVIEW SERIES

The comparatively small land area of Great Britain contains an unrivalled sequence of rocks, mineral and fossil deposits, and a variety of landforms that provide a geological record of a large part of the Earth's long history. Well-documented ancient volcanic episodes, famous fossil sites, and sedimentary rock sections used internationally as comparative standards have given these islands an importance out of all proportion to their size. The long sequences of strata and their organic and inorganic contents have been studied by generations of leading geologists, thus giving Britain a unique status in the development of the science. Many of the divisions of geological time used throughout the world are named after British sites or areas; for instance, the Cambrian, Ordovician and Devonian systems, the Ludlow Series and the Kimmeridgian and Portlandian stages.

The Geological Conservation Review (GCR) was initiated by the Nature Conservancy Council in 1977 to assess and document the most important parts of this rich heritage. The GCR records the current state of knowledge of the key Earth science sites in Great Britain and provides a firm basis upon which site conservation can be founded in years to come. Each GCR title in the series of over 40 volumes describes networks of sites of national or international importance in the context of a portion of the geological column, or a geological, palaeontological or mineralogical topic.

Within each volume, the GCR sites are described in detail in self-contained accounts, consisting of an introduction (with a concise history of previous work), a description, an interpretation (providing geological analysis of the features of interest and assessing their fundamental scientific importance), and a conclusion (written in simpler terms for the non-specialist). Each site report is a justification of the particular scientific interest in a locality, of its importance in a British or international setting, and ultimately of its worthiness for conservation.

The aim of the Geological Conservation Review Series is to provide a public record of the features of interest in sites that have been notified, or are being considered for notification, as Sites of Special Scientific Interest (SSSIs). The volumes are written to the highest scientific standards but in such a way that the assessment and conservation value of each site is clear. The GCR Series is a public statement of the value placed on our geological and geomorphological heritage by the Earth science community and it will be used by the Joint Nature Conservation Committee, the Countryside Council for Wales, English Nature and Scottish Natural Heritage in carrying out their conservation functions. The three country agencies are also active in helping to establish sites of local and regional importance. Regionally Important Geological/Geomorphological Sites (RIGS) augment the SSSI coverage, with local groups identifying and conserving sites that have educational, historical, research or aesthetic value, enhancing the wider Earth heritage conservation perspective.

All the sites in this volume have been proposed for notification as SSSIs; the final decision to notify, or re-notify, sites lies with the governing councils of the appropriate country conservation agency.

Information about the GCR publication programme may be obtained from:

GCR Unit,
Joint Nature Conservation Committee,
Monkstone House,
City Road,
Peterborough PE1 1JY.

www.jncc.gov.uk

Copies of published volumes can be purchased from:

NHBS Ltd,
2–3 Wills Road,
Totnes,
Devon TQ9 5XN.

www.nhbs.com

Published titles in the GCR Series

1. **An Introduction to the Geological Conservation Review**
   N.V. Ellis (ed.), D.Q. Bowen, S. Campbell, J.L. Knill, A.P. McKirdy, C.D. Prosser, M.A. Vincent and R.C.L. Wilson

2. **Quaternary of Wales**
   S. Campbell and D.Q. Bowen

3. **Caledonian Structures in Britain South of the Midland Valley**
   Edited by J.E. Treagus

4. **British Tertiary Volcanic Province**
   C.H. Emeleus and M.C. Gyopari

5. **Igneous Rocks of South-West England**
   P.A. Floyd, C.S. Exley and M.T. Styles

6. **Quaternary of Scotland**
   Edited by J.E. Gordon and D.G. Sutherland

7. **Quaternary of the Thames**
   D.R. Bridgland

8. **Marine Permian of England**
   D.B. Smith

9. **Palaeozoic Palaeobotany of Great Britain**
   C.J. Cleal and B.A. Thomas

10. **Fossil Reptiles of Great Britain**
    M.J. Benton and P.S. Spencer

11. **British Upper Carboniferous Stratigraphy**
    C.J. Cleal and B.A. Thomas

12. **Karst and Caves of Great Britain**
    A.C. Waltham, M.J. Simms, A.R. Farrant and H.S. Goldie

13. **Fluvial Geomorphology of Great Britain**
    Edited by K.J. Gregory

14. **Quaternary of South-West England**
    S. Campbell, C.O. Hunt, J.D. Scourse, D.H. Keen and N. Stephens

15. **British Tertiary Stratigraphy**
    B. Daley and P. Balson

16. **Fossil Fishes of Great Britain**
    D.L. Dineley and S.J. Metcalf

17. **Caledonian Igneous Rocks of Great Britain**
    D. Stephenson, R.E. Bevins, D. Millward, A.J. Highton, I. Parsons, P. Stone and W.J. Wadsworth

18. **British Cambrian to Ordovician Stratigraphy**
    A.W.A. Rushton, A.W. Owen, R.M. Owens and J.K. Prigmore

19. **British Silurian Stratigraphy**
    R.J. Aldridge, David J. Siveter, Derek J. Siveter, P.D. Lane, D. Palmer and N.H. Woodcock

20. **Precambrian Rocks of England and Wales**
    J.N. Carney, J.M. Horák, T.C. Pharaoh, W. Gibbons, D. Wilson, W.J. Barclay, R.E. Bevins, J.C.W. Cope and T.D. Ford

Published titles in the GCR Series

21. **British Upper Jurassic Stratigraphy (Oxfordian to Kimmeridgian)**
    J.K. Wright and B.M. Cox

22. **Mesozoic and Tertiary Palaeobotany of Great Britain**
    C.J. Cleal, B.A. Thomas, D.J. Batten and M.E. Collinson

23. **British Upper Cretaceous Stratigraphy**
    R. Mortimore, C. Wood and R. Gallois

24. **Permian and Triassic Red Beds and the Penarth Group of Great Britain**
    M.J. Benton, E. Cook and P. Turner

25. **Quaternary of Northern England**
    D. Huddart and N.F. Glasser

26. **British Middle Jurassic Stratigraphy**
    B.M. Cox and M.G. Sumbler

27. **Carboniferous and Permian Igneous Rocks of Great Britain North of the Variscan Front**
    D. Stephenson, S.C. Loughlin, D. Millward, C.N. Waters and I.T. Williamson

28. **Coastal Geomorphology of Great Britain**
    V.J. May and J.D Hansom

29. **British Lower Carboniferous Stratigraphy**
    P.J. Cossey, A.E. Adams, M.A. Purnell, M.J. Whiteley, M.A. Whyte and V.P. Wright

30. **British Lower Jurassic Stratigraphy**
    M.J. Simms, N. Chidlaw, N. Morton and K.N. Page

31. **The Old Red Sandstone of Great Britain**
    W.J. Barclay, M.A.E. Browne, A.A. McMillan, E.A. Pickett, P. Stone and P.R. Wilby

The Old Red Sandstone of Great Britain

W.J. Barclay[1],

M.A.E. Browne[2],

A.A. McMillan[2],

E.A. Pickett[2],

P. Stone[2]

and

P.R. Wilby[1]

with contributions from

S.L.B Arkley[2]
J.R. Davies[1]
D.J. Hawley[3]
A.A. Monaghan[2]
R.A. Smith[2]
D. Stephenson[2]
N.H. Trewin[4]
B.P.J. Williams[4]

[1] British Geological Survey, Keyworth, Nottingham, UK
[2] British Geological Survey, Murchison House, Edinburgh, UK
[3] Department of Education, University of Wales, Swansea, UK
[4] Department of Geology and Petroleum Geology, University of Aberdeen

GCR Editor: L.P. Thomas

Published by the Joint Nature Conservation Committee, Monkstone House, City Road, Peterborough, PE1 1JY, UK

First edition 2005

© 2005 Joint Nature Conservation Committee

Typeset in 10/12pt Garamond ITC by JNCC
Printed in Great Britain by CLE Print Limited on Huntsman Velvet 100 gsm.

ISBN 1 86 107 543 X

A catalogue record for this book is available from the British Library.

Apart from any fair dealing for the purposes of research or private study, or criticism or review, as permitted under the UK Copyright Designs and Patents Act, 1988, this publication may not be reproduced, stored, or transmitted, in any form or by any means, without the prior permission in writing of the publishers, or in the case of reprographic reproduction only in accordance with the terms of the licences issued by the Copyright Licensing Agency in the UK, or in accordance with the terms and licences issued by the appropriate Reproduction Rights Organization outside the UK. Enquiries concerning reproduction outside the terms stated here should be sent to the GCR Team, JNCC.

The publisher makes no representation, express or implied, with regard to the accuracy of the information contained in this book and cannot accept any legal responsibility or liability for any errors or omissions that may be made.

British Geological Survey and Ordnance Survey copyright protected materials

1. The copyright of materials derived from the British Geological Survey's work is vested in the Natural Environment Research Council (NERC). No part of these materials (geological maps, charts, plans, diagrams, graphs, cross-sections, figures, sketch maps, tables, photographs) may be reproduced or transmitted in any form or by any means, or stored in a retrieval system of any nature, without the written permission of the copyright holder, in advance.

2. To ensure that copyright infringements do not arise, permission has to be obtained from the copyright owner. In the case of BGS maps this includes **both BGS and the Ordnance Survey.** Most BGS geological maps make use of Ordnance Survey topography (Crown Copyright), and this is acknowledged on BGS maps. Reproduction of Ordnance Survey materials may be independently permitted by the licences issued by Ordnance Survey to many users. Users who do not have an Ordnance Survey licence to reproduce the topography must make their own arrangements with the Ordnance Survey, Copyright Branch, Romsey Road, Southampton SO9 4DH (Tel. 023 8079 2913).

3. Permission to reproduce BGS materials must be sought in writing from the Intellectual Property Rights Manager, British Geological Survey, Kingsley Dunham Centre, Keyworth, Nottingham NG12 5GG (Tel. 0115 936 3331).

4. The National Grid is used on diagrams with the permission of the Controller of Her Majesty's Stationery Office, © Crown copyright licence no. GD 27254X/01/00.

Recommended example citations

Barclay, W.J., Browne, M.A.E., McMillan, A.A., Pickett, E.A., Stone, P. and Wilby, P.R. (2005) *The Old Red Sandstone of Great Britain*, Geological Conservation Review Series, No. 31, Joint Nature Conservation Committee, Peterborough, 393 pp.

Stone, P. and Barclay, W.J. (2005) Red Point. In *The Old Red Sandstone of Great Britain* (W.J. Barclay, M.A.E. Browne, A.A. McMillan, E.A. Pickett, P. Stone and P.R. Wilby), Geological Conservation Review Series, No. 31, Joint Nature Conservation Committee, Peterborough, pp. 74–8.

Contents

Acknowledgements	xi
Access to the countryside	xiii
Preface *N.V. Ellis*	xv

1 Introduction to the Old Red Sandstone of Great Britain — 1
W.J. Barclay

Introduction	3
GCR site selection	5
Scope	5
History of research	11
Old Red Sandstone palaeogeography	13
Stratigraphical framework for the Old Red Sandstone	16

2 The Orcadian Basin — 19

Introduction *P. Stone*	21
Shetland	25
Orkney	26
North-eastern Scottish mainland	28
Melby: Matta Taing to Lang Rigg, Shetland *P. Stone*	30
Footabrough to Wick of Watsness, Shetland *P. Stone*	37
Easter Rova Head, Shetland *P. Stone*	43
The Cletts, Exnaboe, Shetland *P. Stone*	47
South Stromness Coast Section, Orkney *E.A. Pickett*	52
Taracliff Bay to Newark Bay, Orkney *E.A. Pickett*	56
Greenan Nev Coast, Eday, Orkney *E.A. Pickett*	59
South Fersness Bay, Eday, Orkney *E.A. Pickett*	61
Yesnaby and Gaulton Coast Section, Orkney *E.A. Pickett*	64
Old Man of Hoy Coast, Orkney *E.A. Pickett*	68
Bay of Berstane, Orkney *E.A. Pickett*	71
Red Point, Caithness *P. Stone and W.J. Barclay*	74

Contents

Pennyland (Thurso–Scrabster), Caithness *P. Stone*	78
John o'Groats, Caithness *W.J. Barclay*	83
Wick Quarries, Caithness *P. Stone*	85
Achanarras Quarry, Caithness *P. Stone*	90
Sarclet, Caithness *P. Stone*	93
Tarbat Ness, Ross and Cromarty *P. Stone*	96
Dun Chia Hill (Loch Duntelchaig), Inverness-shire *D. Stephenson*	102
Tynet Burn, Moray *W.J. Barclay and N.H. Trewin*	107
Den of Findon, Gamrie Bay and New Aberdour, Aberdeenshire *W.J. Barclay*	112
Rhynie, Aberdeenshire *W.J. Barclay, P. Stone and N.H. Trewin*	117

3 The Midland Valley of Scotland and adjacent areas — **125**

Introduction *M.A.E. Browne and W.J. Barclay*	127
The Toutties, Aberdeenshire *M.A.E. Browne and W.J. Barclay*	140
Dunnottar Coast Section, Aberdeenshire *M.A.E. Browne and W.J. Barclay*	145
Crawton Bay, Aberdeenshire *R.A. Smith*	151
North Esk River, Aberdeenshire *R.A. Smith*	155
Milton Ness, Aberdeenshire *W.J. Barclay*	158
Aberlemno Quarry, Angus *M.A.E. Browne*	161
Tillywhandland Quarry, Angus *M.A.E. Browne*	163
Whiting Ness, Angus *M.A.E. Browne*	166
Tay Bank, Perth and Kinross *M.A.E. Browne and W.J. Barclay*	170
Glen Vale, Fife *M.A.E. Browne and W.J. Barclay*	172
Wolf's Hole Quarry, Stirlingshire *M.A.E. Browne and W.J. Barclay*	176
Auchensail Quarry, West Dunbartonshire *M.A.E. Browne and W.J. Barclay*	178
Siccar Point to Hawk's Heugh, Scottish Borders *M.A.E. Browne and W.J. Barclay*	181
Largs Coast, North Ayrshire *A.A. Monaghan*	187
North Newton Shore, Isle of Arran *S.L.B. Arkley*	192

4 Southern Scotland and the Lake District — **197**
A.A. McMillan

Introduction	199
Palmers Hill Rail Cutting, Scottish Borders	202
Pooley Bridge, Cumbria	204

5 The Anglo-Welsh Basin — **209**

Introduction *W.J. Barclay*	211
Porth-y-Mor, Anglesey *J.R. Davies*	221
Devil's Hole, Shropshire *W.J. Barclay*	228
Oak Dingle, Tugford, Shropshire *W.J. Barclay*	229
The Scar, Herefordshire *W.J. Barclay*	232

Contents

Cusop Dingle, Herefordshire–Powys *D.J. Hawley*	234
Sawdde Gorge, Powys *P.R. Wilby*	240
Pantymaes Quarry, Powys *W.J. Barclay*	246
Heol Senni Quarry, Powys *W.J. Barclay*	250
Caeras Quarry, Carmarthenshire *W.J. Barclay*	253
Craig-y-Fro Quarry, Powys *W.J. Barclay*	257
Abercriban Quarries, Powys *W.J. Barclay*	258
Afon y Waen, Powys *W.J. Barclay*	262
Duffryn Crawnon, Powys *W.J. Barclay*	265
Craig-y-cwm, Torfaen *W.J. Barclay*	268
Ross-on-Wye, Royal Hotel, Herefordshire *W.J. Barclay*	271
Wilderness (Land Grove) Quarry, Gloucestershire *W.J. Barclay*	274
Lydney, Gloucestershire *P.R. Wilby*	277
Albion Sands and Gateholm Island, Pembrokeshire *W.J. Barclay*	281
Little Castle Head, Pembrokeshire *W.J. Barclay*	284
West Angle Bay (North), Pembrokeshire *P.R. Wilby*	286
Freshwater West, Pembrokeshire *W.J. Barclay and B.P.J. Williams*	291
Freshwater East–Skrinkle Haven, Pembrokeshire *W.J. Barclay*	301
Llansteffan, Carmarthenshire *W.J. Barclay*	308
Portishead, North Somerset *P.R. Wilby*	312
Glenthorne, Devon *P.R. Wilby*	318

References	**325**
Glossary	**355**
Index	**373**

Acknowledgements

Compilation of this volume began in 2000, in a jointly funded contract between the British Geological Survey (BGS) and the Joint Nature Conservation Committee (JNCC). The volume describes sites in Old Red Sandstone strata originally selected for the Geological Conservation Review in the 1980s by the former Nature Conservancy Council, under the guidance of Dr W.A. Wimbledon, under the heading of the Non-marine Devonian GCR 'Block'.

This volume is the combined work of the 14 authors listed on the title page. W.J. Barclay was responsible for the collation and editing of the individual contributions, as well as the compilation of the volume. Since submission of the initial drafts to JNCC in 2002, a number of exciting new discoveries and newly published research in the Old Red Sandstone necessitated the updating of the original manuscripts. L.P. Thomas acted as the GCR editor and B.P.J. Williams reviewed the volume. We are grateful to both for their forbearance, support and encouragement. In the course of his review, Prof. Williams became an enthusiastic participant in the project, both contributing to site descriptions and greatly improving the volume as a result of his unique country-wide expertise in the Old Red Sandstone. He also recommended some additional sites for inclusion as potential candidates for protected status. In addition to the BGS contributors (Sarah L.B. Arkley, M.A.E. Browne, J.R. Davies, A.A. McMillan, Alison A. Monaghan, Elizabeth A. Pickett, R.A. Smith, D. Stephenson, P. Stone and P.R. Wilby), D.J. Hawley (University of Swansea) and N.H. Trewin (University of Aberdeen) provided important contributions.

The project was managed by N.V. Ellis for JNCC and M. Smith for BGS. Diagrams were drafted by J S Publications. Photographs were scanned by Caroline Adkin and Jayne Kmieciak (BGS Keyworth) and F. McTaggart (BGS Edinburgh). The volume was seen to press by the JNCC on behalf of the Countryside Council for Wales, English Nature and Scottish Natural Heritage. We thank the JNCC GCR Publications Editorial and Production Team of Neil Ellis (GCR Publications Manager), Emma Durham and Anita Carter (Production Editors).

Photographs from the BGS collection are reproduced by permission of the Director, BGS ©NERC; all rights reserved (PR/23–27). Other photographs are accredited accordingly. Where the content of illustrations has been replicated or modified from the work of others, appropriate acknowledgements are given in the captions. Figures derived from maps of the BGS are published by permission of the Director, BGS.

Access to the countryside

This volume is not intended for use as a field guide. The description or mention of any site should not be taken as an indication that access to a site is open. Most sites described are in private ownership, and their inclusion herein is solely for the purpose of justifying their conservation. Their description or appearance on a map in this work should not be construed as an invitation to visit. Prior consent for visits should always be obtained from the landowner and/or occupier.

Information on conservation matters, including site ownership, relating to Sites of Special Scientific Interest (SSSIs) or National Nature Reserves (NNRs) in particular counties or districts may be obtained from the relevant country conservation agency headquarters listed below:

Countryside Council for Wales,
Maes-y-Ffynnon,
Penrhosgarnedd,
Bangor,
Gwynedd LL57 2DW.

English Nature,
Northminster House,
Peterborough PE1 1UA.

Scottish Natural Heritage,
12 Hope Terrace,
Edinburgh EH9 2AS.

Preface

There is such a diversity of rocks, minerals, fossils and landforms packed into the piece of the Earth's crust we call 'Britain' that it is difficult not to be impressed by the long, complex history of geological change to which they are testimony. But if we are to improve our understanding of the nature of the geological forces that have shaped our islands, further unravel their history in 'deep time' and learn more of the history of life on Earth, we must ensure that the most scientifically important of Britain's geological localities are conserved for future generations to study and enjoy. Moreover, as an educational field resource and as training grounds for new generations of geologists on which to hone their skills, it is essential that such sites continue to remain available for study. The first step in achieving this goal is to identify key sites, both at national and local levels.

The GCR, launched in 1977, is a world-first in the systematic selection and documentation of a country's best Earth science sites. No other country has attempted such a comprehensive and systematic review of its Earth science sites on anything near the same scale. After over two decades of site evaluation and documentation, we now have an inventory of over 3000 GCR sites, selected for 100 categories ('Blocks') covering the entire range of the geological and geomorphological features of Britain. This volume is the 31st to be published in the intended 42-volume GCR series. It documents the results of a survey of Old Red Sandstone sites in Great Britain that was begun in the 1980s by the former Nature Conservancy Council, and revised between 2000 and 2002 by the present authors.

The rocks of the Old Red Sandstone were the sedimentary deposits of the Caledonian (Laurussian) continent, which formed by the amalgamation of the Lower Palaeozoic continents of East Avalonia, Laurentia and Baltica during the Caledonian Orogeny. The rocks are almost entirely of non-marine, terrestrial facies and as such are unfossiliferous, in contrast to their marine equivalents to the south in Devon. However, they locally preserve the fascinating remains and traces of the earliest animals to crawl onto dry land, breathe air and take wing, and the first fishes to appear in abundance, as well as the earliest plants. The rocks range from the Mid-Silurian (*c.* 428 million years) to Early Carboniferous (about 360 million years), but are predominantly Devonian in age. In Great Britain, they formed mainly in three areas, the Orcadian Basin, the Midland Valley of Scotland and the Anglo-Welsh Basin. Smaller basins include the Turriff Basin, the Rhynie Basin (internationally renowned for its early plants and insects), the Border Basin in southern Scotland and the Mell Fell Trough in the Lake District.

Preface

The Orcadian Basin is one of the great, classic lacustrine successions of the world, and of both immense academic interest and economic importance as a hydrocarbon source. The magnificent exposures in the cliffs and foreshore of Caithness, Orkney and Shetland are well represented in the GCR sites selected. The Old Red Sandstone of the Midland Valley of Scotland comprises a predominantly red-bed succession of Lower and Upper Devonian strata, much of it deposited in separate basins in a strike-slip regime. The GCR sites selected represent the range of sedimentary environments present. Whereas the Orcadian Basin and those in the Midland Valley of Scotland were rift basins within the Caledonian orogen, the Anglo-Welsh Basin was outside it for much of the Late Silurian and Early Devonian, the succession forming on alluvial plains on the margins of the marine basin to the south.

The volume describes sites in Old Red Sandstone strata originally selected for the Geological Conservation Review for the Non-marine Devonian GCR 'Block'. In addition to these, a number of sites were identified as being worthy of GCR status in the course of compilation of this volume, both at the start of the work and as a result of reviewing the draft text. These are referred to in the present volume as 'potential GCR sites'. Although these potential GCR sites have not yet been confirmed as GCR sites for the Non-marine Devonian (Old Red Sandstone) GCR 'Block', many have been selected independently for other GCR 'Blocks', and so are already established GCR sites, but for other reasons, such as palaeobotany or palaeoichthyology, and are referred to as 'potential ORS GCR sites'. The palaeontological credentials of these sites are to be found in volumes 9 and 16 of the GCR Series (Cleal and Thomas, 1995; Dineley and Metcalf, 1999). Furthermore, some sites selected for the Non-marine Devonian GCR Block are in Old Red Sandstone red-bed facies of Silurian age and are therefore included in the Silurian Stratigraphy GCR volume (volume number 19 of the GCR Series: Aldridge *et al.*, 2001). Old Red Sandstone igneous rocks are described in the volume on Caledonian igneous rocks (Stephenson *et al.*, 1999).

Most of the sites identified for the Geological Conservation Review are now designated as Sites of Special Scientific Interest (SSSIs) by the appropriate country conservation agencies (the Countryside Council for Wales, English Nature and Scottish Natural Heritage).

This volume records the scientific justification of conserving sites, but does not deal with site management details. The descriptions of the sites are necessarily technical, but are written to a format that allows access to the non-geologist, particularly in the 'conclusions' section of each account. A glossary of some of the technical terms is also aimed at improved ease of understanding to the non-specialist.

We hope that readers will appreciate that this volume presents an eclectic selection of Old Red Sandstone sites from many more potential candidates, with the aim of identifying the best GCR and potential GCR sites at the time of writing. As such, it brings together much widely scattered data and aims to ensure that the sites are documented in a systematic fashion and conserved for future examination and research. However, as new research is carried out, scientific discoveries at existing sites, as well as at new localities, will add to the rich treasure of knowledge that is described in this volume. We hope that this book will help to provide a stimulus to future research of these fascinating non-marine Devonian Old Red Sandstone rocks.

N.V. Ellis (GCR Publications Manager) and W.J. Barclay (British Geological Survey)
April 2004

Chapter 1

Introduction to the Old Red Sandstone of Great Britain

W.J. Barclay

Introduction

INTRODUCTION

The Old Red Sandstone is one of the two major 'red-bed' sequences of sedimentary rock in Great Britain, the other being the younger Permo–Triassic rocks that were formerly termed the 'New Red Sandstone' (see the companion GCR volume by Benton *et al.*, 2002) to distinguish them from the *Old* Red Sandstone, rocks that are about 150 million years (Ma) older. In the early days of geological research in the 1830s, the Old Red Sandstone was included in the Carboniferous System, but soon after was given separate status and accorded a Devonian age, in recognition of its equivalence to the marine Devonian rocks of south Devon and Cornwall.

The GCR sites described in this volume are representative of the continental Old Red Sandstone facies in Great Britain. The rocks are mainly of what is now formally defined as Devonian age (about 418 to 362 million years (Ma) old), but according to modern definitions extend back into the Silurian Period, perhaps locally into the Wenlock Series (424 Ma). They also extend upwards into what is now defined as the early Carboniferous at less than 362 Ma (see Figure 1.4, 'Stratigraphical framework for the Old Red Sandstone', this chapter).

The Old Red Sandstone crops out principally in five areas in Great Britain (Figure 1.1), which

Figure 1.1 Simplified sketch map showing the principal Devonian outcrops of Great Britain. Marine Devonian strata are confined to south-west England, the remainder being sedimentary rocks of Old Red Sandstone facies and volcanic rocks. Caledonian (Ordovician to Late Devonian) intrusive rocks are not shown.

Introduction to the Old Red Sandstone of Great Britain

broadly reflect the original sedimentary basins in which they were deposited. These are:

- the Orkney and Shetland islands and north-east Scotland (the Orcadian Basin);
- the Midland Valley of Scotland (in an amalgamation of several basins of which the largest was the Strathmore Basin);
- the Scottish Borders and Northumberland (the Scottish Border Basin);
- the southern Lake District (the Mell Fell Trough); and
- south Wales, the Welsh Borderland and Bristol (the Anglo-Welsh Basin).

Figure 1.2 shows the stratigraphical distribution of the main Old Red Sandstone sequences.

Traditionally, the base of the Old Red Sandstone in the Anglo-Welsh Basin was placed at the base of the Ludlow Bone Bed, a thin, lenticular, phosphatized 'lag deposit' marking the top of the Silurian Ludlow Series. However, the international agreement at the Montreal Devonian Symposium in 1972 to define the base of the Devonian System in the fully marine, graptolite-bearing succession exposed at Klonck in the Czech Republic, at the base of the *Monograptus ultimus* Biozone (e.g. House, 1977) now places the basal parts of the Old Red Sandstone in the modern Silurian System. The strata from the Ludlow Bone Bed up to the base of the modern Devonian System, which is as yet poorly defined in the Old Red Sandstone, belong to the Přídolí Series, the fourth, uppermost series of the Silurian System (White and Lawson, 1989). The age intervals (or stages) of the Devonian Period, also defined in the *marine* rocks of Europe, are applied to the *terrestrial* Old Red Sandstone succession with some difficulty because of its absence of marine fossils.

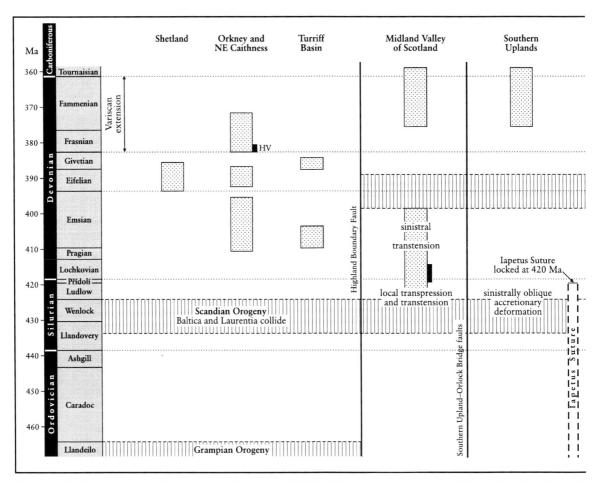

Figure 1.2 Stratigraphical distribution of the main Old Red Sandstone sequences of Great Britain. Tectonic events and their timing are from Soper and Woodcock (2003). Ages are from Williams *et al.* (2000). Small solid bars indicate the principal volcanic rocks. Individual chapter introductions provide more detailed stratigraphical distribution charts. (HV – Hoy Volcanic Member; MF – Mell Fell Conglomerate Formation; RC – Ridgeway Conglomerate Formation.)

Scope

GCR SITE SELECTION

The selection of Geological Conservation Review (GCR) sites described in this volume was carried out in the 1980s and 1990s, following the criteria set out in Ellis *et al.* (1996). The main reasons for qualification of a site for a particular GCR site selection category are:

- international importance – for example, the site may be important because it is a type locality for a geological time period, rock unit or fossil species, or is of historical importance in the development of geological science;
- possession of unique or exceptional geological features;
- national importance because a site is representative of a feature, event, process or rock body that is fundamental to the understanding of the geological history of Great Britain.

SCOPE

The GCR sites were selected according to thematic GCR 'Blocks', the present volume describing the 'Non-marine Devonian' GCR Block, which consists of 64 ratified GCR sites, together with a small number of potential GCR sites. The site descriptions are arranged geographically, from north to south, in areas that correspond to the original depositional basins. The sites are listed in Table 1.1, together with the principal criteria for their selection. Many of the sites have features that satisfy several selection criteria. Furthermore, there are numerous Old Red Sandstone sites that have been independently selected for other GCR palaeontological 'Blocks'. These sites are described in the companion GCR volumes on fossil fishes (Table 1.2; Dineley and Metcalf, 1999) and Palaeozoic palaeobotany (Table 1.3; Cleal and Thomas, 1995).

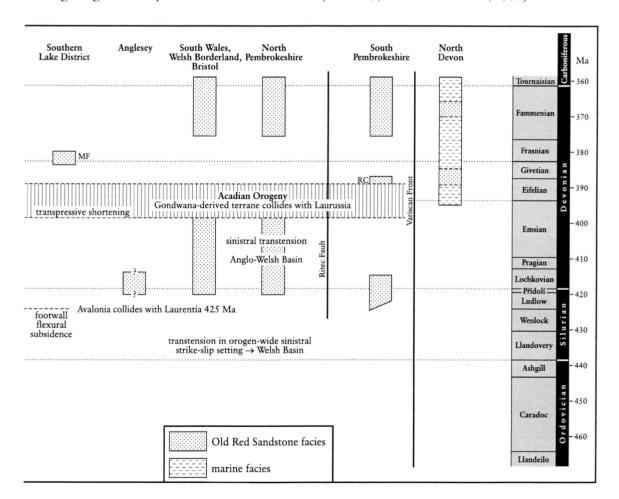

Introduction to the Old Red Sandstone of Great Britain

Table 1.1 GCR Old Red Sandstone sites and proposed sites, with main criteria for their selection. Continued on page 7.

Site	Age	Selection criteria
Orcadian Basin		
Easter Rova Head	Mid-Devonian	Spectacular sea-cliff exposures of conglomerates.
Footabrough to Wick of Watsness	Mid-Devonian	Continuous sea-cliff exposures of Walls Formation.
The Cletts, Exnaboe	Mid-Devonian	Continuous sea-cliff exposures of cyclic lacustrine, fluvial and aeolian facies of the Brindister Flagstone Formation. Also a fossil fish GCR site.
Melby: Matta Taing to Lang Rigg (P)	Mid-Devonian	Fossil fish GCR site. Spectacular sea-cliff and foreshore exposures of lacustrine Melby Formation.
South Stromness Coast Section	Mid-Devonian	Best section in Orkney through the Caithness Flagstone Group.
Taracliff Bay to Newark Bay	Mid-Devonian	Thickest, best-exposed section of the Eday Group.
Greenan Nev Coast, Eday	Mid-Devonian	Best section through the Eday Marl Formation in Orkney.
South Fersness Bay, Eday	Mid-Devonian	Well-exposed, accessible section of the Eday Group.
Yesnaby and Gaulton Coast Section	Early Devonian	Superb sea-cliff sections of the Yesnaby Sandstone Group, including unique aeolian facies. Also sections in the Lower Stromness Flagstone Formation containing the best stromatolites in the Orcadian Basin.
Old Man of Hoy Coast	Late Devonian	Spectacular sea cliffs of the Hoy Sandstone Formation, including the Hoy Volcanic Member.
Bay of Berstane (P)	Mid-Devonian	Unique onshore evidence of marine-influenced deposition in the Middle Devonian Eday Marl Formation.
Red Point	Mid-Devonian	Spectacular lake-margin deposits and basement-cover topography.
Pennyland (Thurso–Scrabster)	Mid-Devonian	Well-exposed lacustrine cycles of the Orcadian lake (Upper Caithness Flagstone Group). Also a fossil fish GCR site.
John o'Groats (P)	Mid-Devonian	Fossil fish GCR site. Type locality of John o'Groats Sandstone Group.
Wick Quarries	Mid-Devonian	Spectacular exposures of fish-bearing lake deposits and shrinkage cracks in the Lower Caithness Flagstone Group in an otherwise poorly exposed part of the Orcadian Basin.
Achanarras Quarry (P)	Mid-Devonian	Fossil fish GCR site, the richest locality in Great Britain. Type locality of Achanarras Limestone Member (Fish Bed), of importance regionally as a marker horizon between the Lower Caithness Flagstone Group and Upper Caithness Flagstone Group.
Sarclet (P)	Early Devonian	Sea-cliff sections in the Lower Devonian Sarclet Group.
Tarbat Ness	Mid-Devonian and Late Devonian	Complete section of the Balnagown Group and of the apparently conformable junction with the Strath Rory Group.
Loch Duntelchaig (Dun Chia Hill)	Mid-Devonian	Dramatic exposure of Middle Old Red Sandstone conglomerates resting unconformably on Dalradian metasedimentary rocks and late Caledonian granites in the south-west of the Orcadian Basin.
Tynet Burn (P)	Mid-Devonian	Fossil fish GCR site. Classic fossil fish locality yielding whole, well-preserved specimens, and important evidence on the nature of the southern margin of the Orcadian Basin at the time of maximum (Achanarras) lake extent.
Den of Findon, Gamrie Bay and New Aberdour (P)	Mid-Devonian (also Early Devonian)	Fossil fish GCR site, with superb coast sections at Gamrie Bay, Pennan (Lower ORS–Middle ORS unconformity), New Aberdour and Quarry Haven.
Rhynie (P)	Early Devonian	Fossil plant GCR site. World renowned floral and arthropod lagerstätte. Exceptional preservation in a hydrothermal spring deposit.
Midland Valley of Scotland and adjacent areas		
The Toutties (P)	Mid-Silurian	Fossil fish GCR site. Oldest (Mid-Silurian) Old Red Sandstone facies in Scotland containing fish and important arthropod fauna.
Dunnottar Coast Section (P)	Mid-Silurian–Early Devonian	Magnificent exposures in dramatic sea cliffs of conglomerates and sandstones.
Crawton Bay	Late Silurian–Early Devonian	Fine coastal sections of conglomerates and volcanics. Also a Caledonian igneous rocks GCR site.
North Esk River	late Early Devonian	Best sections of the Strathmore Group
Milton Ness (P)	Late Devonian–Early Carboniferous	Sea cliffs exposing one of the best sections of mature calcrete development in Scotland in the Kinnesswood Formation.
Aberlemno Quarry (P)	Early Devonian	Fossil fish and Palaeozoic palaeobotany GCR site. Also important for arthropod fossils. Important for Dundee Flagstone and Scone Sandstone formations.
Tillywhandland Quarry (P)	Early Devonian	Fossil fish GCR site. Unique lacustrine facies in the Midland Valley of Scotland in the Dundee Flagstone Formation containing fish, arthropods and trace fossils.
Whiting Ness	Early and Late Devonian	Sea cliffs exposing Early Devonian sandstones unconformably overlain by Late Devonian sandstones, the units being separated by a spectacular unconformity.

Scope

Table 1.1 – continued.

Site	Age	Selection criteria
Tay Bank	Early Devonian	Type locality of the Campsie Limestone Member, including the Stanley Limestone, representing mature calcrete development and providing an important stratigraphical marker horizon.
Glen Vale (P)	Late Devonian	Important sandstones (the Knox Pulpit Sandstone Formation) of aeolian origin.
Wolf's Hole Quarry (P)	Early Devonian	Fossil fish GCR site exposing the highest lava on the southern side of the Strathmore Basin.
Auchensail Quarry (P)	Early Devonian	Fossil plant GCR site yielding a well-preserved assemblage of land plants in the Teith Sandstone Formation.
Siccar Point to Hawk's Heugh (E)	Late Devonian–Early Carboniferous	Classic section at Siccar Point (Hutton's Unconformity). Fossil fish GCR site at Hawk's Heugh. Proposed extension to include the intervening superb sections of fluvial and ?aeolian sandstones at Pease Bay.
Largs Coast, Ayrshire	Late Devonian	Important coastal exposures in sandstones illustrating fluvial, braided river sandbody morphologies.
North Newton Shore, Arran	Late Devonian	One of the three classic Old Red Sandstone unconformities recognized by James Hutton.

Southern Scotland and the Lake District

Site	Age	Selection criteria
Palmers Hill Rail Cutting	Late Devonian	Exposures of calcrete in the Scottish Border Basin.
Pooley Bridge	Early Devonian	Best section of Old Red Sandstone facies conglomerates in north-west England.

Anglo-Welsh Basin

Site	Age	Selection criteria
Porth-y-Mor	Early Devonian	One of the best, most accessible Old Red Sandstone sections in the Anglo-Welsh Basin.
Devil's Hole (P)	Late Silurian–Early Devonian	Historically important fossil fish GCR site, with important 'Downtonian'–'Dittonian' boundary exposure.
Oak Dingle, Tugford (P)	Early Devonian	Fossil fish GCR site with well-documented sedimentological analysis.
The Scar	Late Silurian	Good inland exposure of the Raglan Mudstone Formation.
Cusop Dingle (P)	Late Silurian–Early Devonian	Best, most complete inland section through topmost Přídolí and lowermost Devonian strata.
Sawdde Gorge (E)	Late Silurian–Early Devonian	Proposed extension of GCR site to include higher Přídolí and basal Devonian strata.
Pantymaes Quarry (P)	Early Devonian	Excellent exposure of Lochkovian channel sandstones and floodplain mudstone facies, internationally known for its arthropod trackways.
Heol Senni Quarry	Early Devonian	GCR fossil fish site. Also representative of the Senni Formation.
Caeras Quarry	Early Devonian	Best exposure of local pebbly facies in the Brownstones Formation.
Craig-y-Fro Quarry (P)	Early Devonian	Classic fossil plant GCR site. Also important for exposure of the Senni Formation.
Abercriban Quarries	Late Devonian–Early Carboniferous	Type locality of the Grey Grits Formation.
Afon y Waen (P)	Late Devonian	Potential fossil fish GCR site. Also important for exposure of the Plateau Beds Formation.
Duffryn Crawnon (P)	Late Devonian	Type locality of the Plateau Beds Formation, including possible aeolian facies.
Craig-y-cwm (P)	Late Devonian–Early Carboniferous	Representative section of the Quartz Conglomerate Group.
Ross-on-Wye, Royal Hotel	Early Devonian	Excellent, well-documented, accessible section of the Brownstones Formation.
Wilderness (Land Grove) Quarry	Early Devonian	Fossil fish GCR site with superb exposure of the lowermost strata of the Brownstones Formation.
Lydney	Late Silurian–Early Devonian	Fossil fish GCR site with good section of the Psammosteus Limestone horizon.
Albion Sands and Gateholm Island	Late Silurian–Early Devonian	Magnificent sea-cliff and foreshore exposures of Wenlock marine strata and the overlying Old Red Sandstone.
Little Castle Head (P)	Late Silurian–Early Devonian	Reference section of the Přídolí Sandy Haven Formation and of the Townsend Tuff Bed.
West Angle Bay (North)	Late Silurian–Early Carboniferous	Continuous section of the entire Old Red Sandstone succession and of the underlying and overlying strata.
Freshwater West (P)	Late Silurian–Early Carboniferous	Superb, accessible dip section exposing the entire Old Red Sandstone succession.
Freshwater East–Skrinkle Haven ('Tenby Cliffs')	Late Silurian–Early Carboniferous	Excellent strike section of the entire Old Red Sandstone succession.
Llansteffan	Late Silurian–Early Devonian	Superb exposures of stacked carbonate palaeosols of the Chapel Point Calcretes Member (Psammosteus Limestone).
Portishead	Early and Late Devonian	Best section of the Old Red Sandstone succession east of Severn Estuary.
Glenthorne	Mid-Devonian	Best section of Old Red Sandstone facies south of the Bristol Channel.

P Potential site (most of these sites are confirmed GCR sites for their palaeontology)
E Proposed extension to site

7

Introduction to the Old Red Sandstone of Great Britain

Table 1.2 GCR sites in the Old Red Sandstone described in the fossil fishes GCR volume. After Dineley and Metcalf (1999). Continued on page 9.

Site	Stratigraphy	Criterion	Treatment in this volume Full description FD Summary description SD Not described ND
Orcadian Basin			
Westerdale Quarry	Mid-Devonian; Eifelian	One of oldest fish-bearing horizons in Orcadian Basin; complete specimens	ND
Achanarras Quarry	Mid-Devonian; Eifelian–Givetian boundary	Richest Old Red Sandstone fish site in Britain	FD
Cruaday Quarry	Mid-Devonian; Eifelian–Givetian boundary*	Best Old Red Sandstone fish site in Orkney	ND
Black Park, Edderton	Mid-Devonian; Eifelian–Givetian boundary*	Fish well preserved in three dimensions	ND
Den of Findon, Gamrie	Mid-Devonian; Eifelian–Givetian boundary*	Prolific fish fauna	SD
Tynet Burn, Elgin	Mid-Devonian*	Rich fish fauna and historically important	FD
Melby	Mid-Devonian; Eifelian–Givetian boundary*	Northernmost occurrence of Achanarras horizon	FD
Papa Stour	Mid-Devonian; Eifelian–Givetian boundary*	Fish in sedimentary rocks in predominantly volcanic sequence	ND
Dipple Brae	Mid-Devonian	Fish fauna younger than that of the Achanarras horizon	ND
Spittal Quarry	Mid-Devonian	Rare fish fauna, including only Mid-Devonian cephalaspid	ND
Banniskirk Quarry	Mid-Devonian	First ORS site to yield fishes	ND
Holborn Head Quarry	Mid-Devonian; mid-Givetian	10–11 fish species, including *Osteolepis panderi*	ND
Weydale Quarry	Mid-Devonian	Well-preserved *Osteolepis panderi* and *Dipterus valenciennesi*	ND
Pennyland	Mid-Devonian; Givetian	Many fish specimens from several fish-bearing horizons	FD
John o'Groats	Mid-Devonian; late Givetian	Youngest fish fauna in Caithness	SD
The Cletts, Exnaboe	Mid-Devonian; late Givetian	Northernmost late Givetian fish site	FD
Sumburgh Head	Late Mid-Devonian; late Givetian	Possibly youngest fish fauna of the Orcadian Basin	ND
Midland Valley of Scotland			
The Toutties	Late Wenlock	Oldest Old Red Sandstone facies rocks in Scotland; unique fish fauna	FD
Tillywhandland Quarry	Early Devonian	One of best Early Devonian fish sites in Scotland	FD
Aberlemno Quarry	Early Devonian	Best surviving of the famous Turin Hill fish sites; also a fossil plant GCR site (Table 1.3)	FD
Wolf's Hole Quarry	Early Devonian	Unique pteraspid fish fauna	FD
Whitehouse Den	Early Devonian	Fossil acanthodian fish	ND
Grampian Highlands			
Ardmore–Gallanach	Late Silurian–Early Devonian	Unique early fish fauna in sediments associated with Lorne lavas	ND
Bogmore, Muckle Burn	Earliest Late Devonian (Frasnian)	Diverse fish fauna with over 15 species	ND
Scaat Craig	Late Devonian	Diverse late Devonian fish fauna and a distinctive tetrapod	ND

* Achannaras Fish Bed horizon

Scope

Table 1.2 – *continued.*

Site	Stratigraphy	Criterion	Treatment in this volume Full description FD Summary description SD Not described ND
Southern Uplands			
Oxendean Burn	Late Devonian	Abundant fragments of *Bothriolepis*	ND
Hawk's Heugh	Late Devonian	Only British occurrence of *Remigolepis*	FD
Anglo-Welsh Basin			
Ludford Lane and Ludford Corner	Silurian; Přídolí	Internationally renowned for rich fish fauna; see also Table 1.4	ND
Ledbury cutting	Silurian; Přídolí	Historical site yielding complete specimens of *Auchenaspis* and *Hemicyclaspis*	ND
Temeside, Ludlow	Silurian; Přídolí	Historical site in Temeside Mudstone Formation yielding a rich fish fauna including *Hemicyclaspis murchisoni*	ND
Tite's Point (Purton Passage)	Silurian; Ludlow–Přídolí	*Thelodus parvidens* fish fauna, allowing correlation with Ludlow Bone Bed, and source of *Cyathaspis*	
Lydney	Late Přídolí–Early Devonian	Sequence of vertebrate faunas, including specimens of *Sabrinacanthus*	FD
Downton Castle area (network of 4 sites)	Early Přídolí	Several quarries in Downton Castle Sandstone yielding vertebrate remains	ND
Bradnor Hill Quarry	Late Přídolí	Late Přídolí thelodont fauna	ND
Devil's Hole	Přídolí–Lochkovian	Fish fauna straddling Downtonian–Dittonian boundary	SD
Oak Dingle, Tugford	Lochkovian (Dittonian)	Near-strike section of fish-bearing beds; earliest record of *Weigeltaspis*	SD
Cwm Mill	Lochkovian (Dittonian)	Unique preservation of complete cephalaspids, including three new species; also specimens of *Rhinopteraspis crouchi*	ND
Wayne Herbert Quarry	Lochkovian (Dittonian)	Well-preserved, diverse fish fauna	ND
Besom Farm Quarry	Lochkovian (Dittonian)	Rich, diverse fish fauna, including 7 type specimens and sole occurrence of 5 of them	ND
Heol Senni Quarry	Lochkovian–Pragian	Only occurrence of *Althaspis senniensis*	FD
Portishead	Late Devonian	Unique fish fauna, including only British occurrence of *Groenlandaspis*	FD
Prescott Corner	Late Devonian (Frasnian)	Extensive Late Devonian fish fauna	ND
Afon y Waen	Late Devonian	*Bothriolepis* and *Holoptychius* in Upper Old Red Sandstone	FD

Introduction to the Old Red Sandstone of Great Britain

Table 1.3 GCR sites in the Old Red Sandstone described in the Palaeozoic palaeobotany GCR volume. After Cleal and Thomas (1995).

Site	Stratigraphy	Criterion	Treatment in this volume Full description FD Summary description SD Not described ND
Orcadian Basin			
Sloagar	Mid-Devonian; Late Givetian	Only occurrence of *Svalbardia* in Britain	ND
Bay of Skaill	Mid-Devonian	Important floral assemblage in Sandwick Fish Bed; type locality of *Protopteridium thomsonii*	ND
Rhynie	Early Devonian	Renowned Devonian palaeobotanical site; 22 species unique to this site	FD
Midland Valley of Scotland			
Turin Hill	Early Devonian	Best example of *Zosterophyllum* Zone flora in world and type locality of *Cooksonia caledonica*	FD (as 'Aberlemno Quarry')
Ballanucater Farm	Early Devonian; Emsian	Best Emsian floral assemblage in Britain	ND
Auchensail Quarry	Early Devonian; Emsian	Well-preserved Emsian floral assemblage	FD
Anglo-Welsh Basin			
Targrove Quarry	Early Devonian; Gedinnian	Most diverse rhyniophytoid plant assemblage in world	ND
Capel Horeb Quarry	Late Silurian; Ludfordian–Přídolí	Oldest vascular plants in world in Ludlow Series; Long Quarry Formation yielded some rhyniophytoids including *Cooksonia*	ND
Perton Lane	Late Silurian; Přídolí	Classic locality and type locality of *Cooksonia*	ND
Freshwater East	Late Silurian; Přídolí	Most diverse Silurian flora in the world	FD
Llanover Quarry	Early Devonian; Siegenian	Classic locality yielding one of most diverse *Psilophyton* Zone flora in Britain	ND
Craig-y-Fro Quarry	Early Devonian	Some of best preserved Devonian plants in Britain; locality second only to the Rhynie site in Britain	SD

The GCR sites provide representative localities for the entire stratigraphical range of the Old Red Sandstone. The initial selection of GCR sites for the 'Non-marine Devonian' GCR Block included sites in the Anglo-Welsh Basin in strata that extended down from the base of the Devonian System to the Ludlow Bone Bed. These strata, comprising the Downton Group (the former Downtonian Stage), are of Silurian (Přídolí Series) age and the sites (Table 1.4) are described in the GCR volume on Silurian stratigraphy (Aldridge *et al.*, 2000). The GCR volume on Caledonian igneous rocks (Stephenson *et al.*, 1999) includes sites (Table 1.5) in which Old Red Sandstone strata are present in addition to the contemporaneous igneous rocks for which they are cited. All of the Old Red Sandstone sites described in the other GCR volumes are listed in Tables 1.2, 1.3, 1.4 and 1.5, along with the level of detail in which they are described in the present volume. Only some of these 'overlapping' sites are given full descriptions in the present volume, which emphasizes the sedimentological and lithostratigraphical features. The **Freshwater West** potential Old Red Sandstone GCR site, and part of the **Freshwater East–Skrinkle Haven** GCR site are also Variscan to Alpine structures GCR sites.

History of research

Table 1.4 GCR sites in the Old Red Sandstone described in the Silurian stratigraphy GCR volume. After Aldridge *et al.* (2000).

Site	Stratigraphy	Criterion	Treatment in this volume Full description FD Summary description SD Not described ND
Anglo-Welsh Basin			
Marloes	Wenlock–Přídolí	Classic site showing early transition from marine to Old Red Sandstone facies	ND (included with report for Albion Sands and Gateholm Island)
Albion Sands and Gateholm Island	Ludlow–Přídolí–Lochkovian	Complete, conformable succession from Ludlow into early Devonian	SD
Freshwater East (South)	Wenlock–Přídolí	Wenlock marine strata overlain by Old Red Sandstone; faulted/unconformable relationship	ND
Ludford Lane and Ludford Corner	Ludlow–Přídolí	Classic, internationally renowned site traditionally regarded as reference section for Silurian–Devonian boundary; earliest known land animals, early plants (see Table 1.3), unusual arthropods and fish remains in Ludlow Bone Bed	ND
Brewin's Bridge/Canal	Ludlow–Přídolí–Carboniferous	One of few sites in central England exposing marine Silurian–Old Red Sandstone junction, including Ludlow Bone Bed	ND
Capel Horeb Quarry	Ludlow–Přídolí	Good section of unconformity between Ludlow and Přídolí; internationally important plant site (Table 1.3)	ND
Little Castle Head	Přídolí	Old Red Sandstone facies rocks; Townsend Tuff Bed	SD
Lower Wallop Quarry	Ludlow–Přídolí	Marine to Old Red Sandstone transition later here, well into Přídolí	ND

Site selection is inevitably subjective, but the aim of the GCR is to identify the minimum number and area of sites needed to demonstrate the current understanding of the diversity and range of features within each GCR 'Block'. The preferred sites are generally those that are least vulnerable to the potential threat of destruction, are more accessible and are not duplicated elsewhere (Ellis *et al.*, 1996). The original selection of sites was made over 20 years ago, and all of these sites are included in this volume. In addition, a small number of sites were identified during the course of the compilation of the volume as representing stratigraphical units or unique features not included in the original GCR site selection, and are described as 'potential' GCR sites.

HISTORY OF RESEARCH

The name 'Old Red Sandstone' appears to have been first applied to the red rocks below the Mountain (Carboniferous) Limestone in the mistaken belief that they were the equivalents of the Permian Rotliegendes of Germany (Jameson, 1821; Simpson, 1959). It was initially mapped and named in southern Wales and the Welsh Borderland (Phillips, 1818; Conybeare and Phillips, 1822) and included as the lowermost part of the Carboniferous System. Murchison was the first to champion the Old Red Sandstone as a separate geological entity. According to Miller's account (1841), a visiting foreign geologist advised Murchison that 'you must inevitably give up the Old Red Sandstone:

Introduction to the Old Red Sandstone of Great Britain

Table 1.5 GCR sites with Old Red Sandstone sedimentary rocks described in the Caledonian igneous rocks volume. After Stephenson *et al.* (1999).

Site	Stratigraphy/ radiometric age	GCR selection criteria
Eshaness Coast	Mid-Devonian	Representative of Eifelian Eshaness volcanic succession, NW Shetland.
Ness of Clousta to the Brigs	Mid-Devonian	Representative of Givetian Clousta volcanic rocks, Walls, Shetland.
Point of Ayre	Mid-Devonian	Representative of Givetian Deerness Volcanic Member, mainland Orkney.
Too of the Head	Mid-Devonian	Representative of Givetian Hoy Volcanic Formation, Isle of Hoy, Orkney.
South Kerrera	Late Silurian to Early Devonian	Representative of Lorn Plateau Volcanic Formation. Exceptional examples of subaerial lava features and interaction of magma with wet sediment.
Ben Nevis and Allt a'Mhuilinn	Mid-Silurian 425 Ma	Representative of Ben Nevis Volcanic Formation. Exceptional intrusive tuffs. Internationally important as example of exhumed root of caldera, and historically for development of cauldron subsidence theory.
Stob Dearg and Cam Ghleann	Mid-Lochkovian 421 ± 4Ma	Representative of succession in eastern part of Glencoe caldera, including basal sedimentary rocks. Exceptional rhyolites, ignimbrites and intra-caldera sediments. Possible international importance for radiometric dating in conjunction with palaeontology close to Silurian–Devonian boundary.
Crawton Bay*	Late Silurian–Early Devonian	Representative of Crawton Volcanic Formation.
Scurdie Ness to Usan Harbour	Early Devonian	Representative of 'Ferrydean' lavas and 'Usan' lavas, comprising lower part of Montrose Volcanic Formation.
Black Rock to East Comb	Early Devonian	Representative of 'Ethie' lavas, comprising upper part of Montrose Volcanic Formation.
Balmerino to Wormit	Early Devonian (Lochkovian) 410.6 ± 5.6 Ma	Representative of eastern succession of Ochil Volcanic Formation. Possible international importance for radiometric dating in conjunction with palaeontology close to Silurian–Devonian boundary.
Sheriffmuir Road to Menstrie Burn	Early Devonian 416 ± 6.1 Ma	Representative of western succession of Ochil Volcanic Formation. Exceptional topographic expression of Ochil fault-scarp.
Tillycoultrie	Early Devonian 415–410 Ma	Representative of diorite stocks, intruded into Ochil Volcanic Formation, surrounded by thermal aureole and cut by radial dyke swarm. Exceptional examples of diffuse contacts due to metasomatism and contamination, with 'ghost' features inherited from country rock.
Port Schuchan to Dunure Castle	Early Devonian	Representative of Carrick Hills volcanic succession. Exceptional features resulting from interaction of magma with wet sediment are of international importance.
Culzean Harbour	Early Devonian	Representative of inlier of Carrick Hills volcanic succession. Exceptional features resulting from interaction of magma with wet sediment are of international importance.
Turnberry Lighthouse to Port Murray	Early Devonian	Representative of most southerly inlier of Carrick Hills volcanic succession. Exceptional features resulting from interaction of magma with wet sediment are of international importance.
Pettico Wick to St Abb's Harbour	Early Devonian *c.* 400+ Ma	Representative of volcanic rocks in the SE of the Southern Uplands. Exceptional vent agglomerates, block lavas, flow tops and interflow high-energy volcaniclastic sediments.

* described in this volume

it is a mere local deposit, a doubtful accumulation huddled up in a corner, and has no type or representative abroad.'

'I would willingly give it up if Nature would,' replied Murchison, 'but it assuredly exists, and I cannot'. Compared to the richly fossiliferous rocks of the Silurian System below and the Carboniferous System above, the Old Red Sandstone seemed relatively barren to the early Victorian workers, but as the remains of early fishes were discovered, first in Scotland by the young Swiss naturalist Louis Agassiz, and later in south Wales and the Welsh Borderland, interest gradually increased.

The Devonian System was established by Sedgwick and Murchison (1839) for the pre-Carboniferous marine rocks of Devon. These rocks were readily correlated with the Rhenish nearshore rocks and the Bohemian deep-water rocks of mainland Europe (House, 1977). With the recognition of large tracts of Old Red Sandstone in North America, Norway, Siberia, Poland and Russia (the last containing many of the same fish species as Great Britain), the strata assumed a new importance (Geikie, 1879). At the same time, Murchison (1839), impressed by the great thickness of Old Red Sandstone strata in the Welsh Borderland, and the difference between them and the overlying Carboniferous rocks, with which they had hitherto been merged, applied the status of 'system' to the Old Red Sandstone. This situation held, in the UK at least, for over 130 years, with the term used in a quasi-chronostratigraphical sense for rocks of continental facies and Devonian age. However, with the advent of more precise stratigraphical procedures and classification, and, in 1972, the new definition of the base of the Devonian System at a higher level, equivalent to a horizon within the Old Red Sandstone (see below), the term is now no longer used in a quasi-chronostratigraphical sense. Biostratigraphers tend not to use the term at all; the glossary in the companion GCR volume on Silurian stratigraphy, for example (Aldridge *et al.*, 2000), defines the Old Red Sandstone as 'a classic term still applied to the terrestrial, largely clastic facies of the late Silurian to earliest Carboniferous in Britain'. Sedimentologists retain the name as a facies (or magnafacies) term for all the terrestrial red beds and lacustrine deposits of Silurian to early Carboniferous (but predominantly Devonian) age (e.g. Friend and Williams, 2000). The term 'Old Red Sandstone' is also applied in an informal lithostratigraphical sense. The three subdivisions of the Old Red Sandstone recognized by Murchison – Lower, Middle and Upper – are similarly retained as informal, but long-established lithostratigraphical terms onshore in the United Kingdom and as formal groups offshore in the North Sea.

OLD RED SANDSTONE PALAEOGEOGRAPHY

The Old Red Sandstone represents a period when ocean closure and continental collisions resulted in a world geography vastly different to that of much of early Palaeozoic times. The drift of the early Palaeozoic continents and their relative positions can be estimated from the correlation of geological successions and their faunas, with palaeomagnetism providing data on palaeolatitudes. The Iapetus Ocean, which separated the northern (Laurentian) and southern (Gondwanan) continents, closed throughout the Ordovician and Silurian periods as the smaller continent of Avalonia fragmented from Gondwana and drifted northwards (Figure 1.3). As the Iapetus Ocean closed north of Avalonia, the Rheic Ocean opened behind it. To the east, the continent of Baltica also drifted northwards and eastwards and the Tornquist Sea, an arm of the Iapetus Ocean, slowly closed. The timing and nature of the convergence of the three components that were to make up the Old Red Sandstone continent remain matters of debate. Trench and Torsvik (1992) considered that Baltica and the eastern part of the Avalonia microcontinent collided first, in late Ordovician times, moving northwards together to make first contact with Laurentia by late Silurian time at about 420 Ma (Torsvik *et al.*, 1996; see Dewey and Strachan, 2003, fig. 1). However, Dewey and Strachan (2003) interpret the Scandian Orogeny as the result of collision, by sinistral transpression, of Baltica and Laurentia from about 435 Ma to 425 Ma, with a soft collision between Avalonia and Laurentia/Baltica (Laurussia) at about 425 Ma (Soper and Woodcock, 2003). By late Silurian (Ludlow) time, the continents had fully docked, with the Iapetus Ocean closed along the

Introduction to the Old Red Sandstone of Great Britain

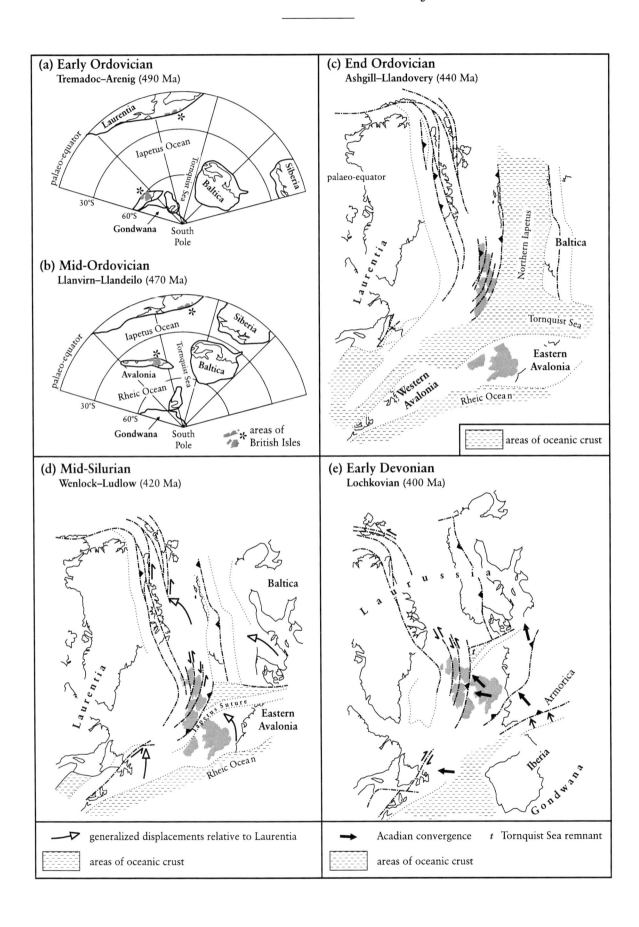

Old Red Sandstone palaeogeography

◀**Figure 1.3** Sketch maps showing the movements and amalgamation of the early Palaeozoic continents that produced the Old Red Sandstone (Laurussia) continent. (a) and (b) are global views to illustrate the fragmentation of Avalonia from Gondwana and its drift northwards as the Iapetus Ocean closed (adapted from Torsvik *et al.*, 1992, by Trench and Torsvik, 1992). (c), (d) and (e) show the later stages of the Caledonian Orogeny. Sinistral strike-slip movements in relation to the Laurentian margin culminated in the Acadian Orogeny in late Early Devonian (Emsian) times (after Stephenson *et al.*, 1999, adapted from Soper *et al.*, 1992).

line of subduction (the Iapetus Suture) under the Southern Uplands. Thus, the Caledonian–Appalachian Orogen (or North Atlantic Caledonides) and the newly amalgamated Old Red Sandstone continent (Laurussia or Euramerica) were formed. Continuing compression and shortening of the continental crust resulted in the filling to sea level of the Silurian marine basins, their inversion to upland areas and the establishment of terrestrial conditions in newly developing basins.

Palaeogeographical reconstructions (e.g. Scotese, 2001) suggest that the continent lay in tropical to sub-tropical latitudes from the equator to about 30°S, with the Anglo-Welsh Basin lying approximately 5°S to 15°S. Palaeomagnetic data from the Lower Old Red Sandstone in southern Wales suggest a latitude of 17 ± 5°S (Channell *et al.*, 1992). Sedimentological studies of the Old Red Sandstone, and particularly of its fossil carbonate soils (calcretes) confirm, by analogy with modern calcretes, a warm to hot, semi-arid tropical to sub-tropical setting (e.g. Allen, 1986) with rainfall confined to wet seasons (e.g. Marriott and Wright, 1993). Uplift of the orogen may have caused broad variations in the rainfall pattern, producing periods of wetter and drier climate.

Woodcock (2000a), Friend *et al.* (2000), Dewey and Strachan (2003) and Soper and Woodcock (2003) presented recent overviews of the tectonics and kinematics of Old Red Sandstone basin formation. Superimposed on the broadly compressive stresses associated with convergence of the Avalonian and Laurentian continental margins, the oblique angle of closure produced strike-slip transpressive and transtensional movements. The nature and extent of these movements remain the matter of debate, largely centred on whether there was a major, orogen-wide sinistral megashear or whether basins were controlled by strike-slip movements of different sense and at different times during the Caledonian orogenic cycle. Another debate concerns the possible role of gravitational collapse of the uplifted, granite-buoyed Caledonian Orogen in the formation of some at least of the internal basins (e.g. Woodcock, 2000a). Dewey and Strachan (2003) conclude that the diachronous closure of Iapetus, and subsequent deformation and basin formation were controlled by sinistrally dominated relative movement between the Laurentian and Avalonia–Baltica plates. The Old Red Sandstone basins probably formed as a result of sinistral transtension, with an estimated 1200 km of strike-slip movement between Laurentia and Baltica. Rheic convergence in the Emsian Age (late Early Devonian) from 400 Ma to 390 Ma resulted in the Acadian Orogeny, which affected the basins south of the Highland Boundary Fault (Soper and Woodcock, 2003).

The Old Red Sandstone basins were formerly divided into two main groups on the basis of their positions relative to the Caledonian Orogen (e.g. Allen, 1977; Woodcock, 2000a). Those within it (internal or intramontane basins) include the Orcadian Basin, the basins of the Midland Valley of Scotland (but see below) and the Scottish Border Basin. The Anglo-Welsh Basin was regarded as an external, or extra-montane basin, open to the sea to the south. However, the recent models, invoking major orogen-parallel, sinistral movement and three separate, temporally discrete collision events (Grampian, Scandian and Acadian) collectively making up the Caledonides have revised the former view of a continuously prograding Caledonian mountain front.

The recent models (Dewey and Strachan, 2003; Soper and Woodcock, 2003) envisage that the highly oblique, sinistral closure of the Iapetus Ocean resulted in, sequentially, transpression, strike-slip and transtension. The area of maximum uplift in the Scandian Orogen was to the north of Britain, in Scandinavia, in an orogen of Himalayan proportions (Dewey and Strachan, 2003). The compression in the Laurentian crust, of which the Scottish

Introduction to the Old Red Sandstone of Great Britain

Highlands were part, caused thrusting along major NE-trending faults, granitic intrusion, andesitic volcanicity and low-grade metamorphism in northern Britain (Stephenson *et al.*, 1999). The volcanic rocks were probably extensive, their eroded remnants being seen at Ben Nevis, Glen Coe, Lorn and just north of the Highland Boundary Fault. Volcanic rocks also occur extensively within the Midland Valley of Scotland, at Montrose, in the Sidlaw, Ochil and Pentland hills, and in Ayrshire. They also occur more locally in the Southern Uplands, where granitic intrusions such as the Cheviot were emplaced.

During the transcurrent and transtensional phases, much of the orogen-parallel, sinistral movement appears to have been taken up by the Great Glen Fault and its north-east continuation, with at least 700 km of displacement (Dewey and Strachan, 2003). The formation of the Late Silurian–Early Devonian Old Red Sandstone basins is also attributed to sinistral transtension (Dewey and Strachan, 2003; Soper and Woodcock, 2003). The Acadian Orogeny ended this phase of basin formation and caused transpressive shortening of the early Palaeozoic basins flanking the Midland Microcraton, as well as the inversion and erosion of the Old Red Sandstone rocks not underlain by the microcraton (Soper and Woodcock, 2003). The cause of the Acadian event was probably the collision of a Gondwana-derived continental fragment (Soper and Woodcock, 2003) with the Midland Microcraton segment of the amalgamated Laurussian continent. The evidence for the terrane boundary in the vicinity of the Bristol Channel is now confined to the Lizard mafic–ultramafic complex, interpreted as an ophiolite and a fragment of the Rheic suture (e.g. Soper and Woodcock, 2003).

The Orcadian Basin was a large Mid-Devonian intramontane lake basin, totally unconnected to the open sea, apart perhaps from a brief period. Its formation was probably due to a combination of both gravitational extension, and transtensional movements on basin-margin faults. The Midland Valley of Scotland was not a single discrete basin in the Devonian Period. Weakened by a long history of igneous activity, internal, transtensional fault movements opened pull-apart basins and transpressive movements subsequently inverted them, resulting in the recycling of the basin-fills and providing weak points for continuing volcanic extrusion (e.g. Bluck, 2000). The preserved sequences thus represent the deposits of separate pull-apart basins, formed and brought together in a strike-slip faulted collage. The Stonehaven Basin in the north-east is the earliest, its sedimentary fill dating perhaps from the Wenlock Epoch (Marshall, 1991). It and its larger successor basins, the Crawton and Strathmore basins, formed by sinistral strike-slip along the Highland Boundary Fault. The southerly Lanark Basin formed along the Southern Upland Fault. Large volumes of arc-related volcanic rocks were extruded along the central axis of the Midland Valley, on lines weakened by the transtensional stresses (e.g. Bluck, 2000). The late Devonian Scottish Border Basin formed after Acadian inversion in Mid-Devonian times and extended into the Midland Valley, Northumberland and the Solway Firth.

The Anglo-Welsh Basin was formerly interpreted primarily as the product of load-generated flexural subsidence of the Caledonian foreland (James, 1987; King, 1994; Friend *et al.*, 2000). Dewey and Strachan (2003) and Soper and Woodcock (2003) prefer a transtensional mechanism for its formation. Transtensional movement on faults produced variations in the basin-fill in Pembrokeshire (e.g. Marshall, 2000a,b) and introduced coarse, clastic, detritus farther north (Tunbridge, 1980a). The isolated succession in Anglesey was probably deposited contiguously with the Přídolí–Pragian sequences to the south, with which there are marked similarities, although the initial coarse conglomerates are unique and of local derivation, and lacustrine deposits suggest internal or impeded drainage.

STRATIGRAPHICAL FRAMEWORK FOR THE OLD RED SANDSTONE

Stratigraphical classification of the rocks in the geological record has traditionally fallen into two broad categories – lithostratigraphical and chronostratigraphical. Lithostratigraphical classification is based on the physical characteristics of a rock body, such as colour, rock type (lithofacies) and mode of formation. Chronostratigraphical classification is based on the relative age of a rock body, determined by its fossil content (biostratigraphy) as correlated with

Stratigraphical framework for the Old Red Sandstone

standard, defined and internationally agreed geological marker horizons (the 'golden spikes'), and in the case of igneous rocks, by radiometric age dating. Biostratigraphical classification is achieved by the study of component fossil and microfossil groups, with subdivisions based on marker species or assemblages of species. Thus, in the Devonian System, there are biostratigraphical zonal schemes for graptolites, ammonoids, brachiopods, fish, conodonts, microvertebrates and miospores.

The chronostratigraphical subdivisions of the Upper Silurian and Devonian (Figure 1.4) are internationally agreed and defined in fossiliferous marine strata in continental Europe. The Upper Silurian Přídolí Series (not yet divided into stages) and Lower Devonian Lochkovian and Pragian stages are defined in the deep-water, graptolite-bearing succession of the Prague Basin in the Czech Republic. The last two replace the previously used, but not completely equivalent Gedinnian and Siegenian stages defined in the nearshore succession of the German Rhenish Basin. The highest Lower Devonian stage is the Emsian, defined in Belgium. The Middle Devonian stages (Eifelian and Givetian) are defined in Germany, the Upper Devonian stages (Frasnian and Famennian) are named from the carbonate-bearing marine succession of southern Belgium.

The problems of classification and correlation of the Old Red Sandstone of Great Britain are inherent in its terrestrial origins and the patchy preservation of its non-marine fossils. The fossils that are present indicate that the Devonian Period was a time of profound changes in the evolutionary record, with the first significant colonization of terrestrial habitats by vascular plants (Edwards, 1979a), the rapid expansion of the first aquatic vertebrates, and their emergence onto land. However, no direct correlations can be made with the European marine successions and the internationally agreed stages. Because of this, a series of loosely defined local stages (Downtonian, Dittonian, Breconian and Farlovian) were erected for the Anglo-Welsh Basin, but now have been largely subsumed into the international stages as a result of increasing refinement in correlation. Figure 1.4 (based on House et al., 1977; and Marshall and House, 2000) shows the stages and their correlation.

		Period/System	Epoch	Series	Stage	Age (Ma)
Old Red Sandstone	Upper	Carboniferous		Tournaisian	Courceyan	362
		Devonian	Late	Upper	Famennian	376.5
					Frasnian	382.5
	Middle		Mid	Middle	Givetian	387.5
					Eifelian	394
	Lower		Early	Lower	Emsian	409.5
					Pragian	413.5
					Lochkovian	418
		Silurian		Přídolí		419
			Late	Ludlow	Ludfordian	
					Gorstian	424
			Mid	Wenlock	Homerian	

Figure 1.4 Major subdivisions of the Old Red Sandstone and its chronostratigraphical classification. Ages from Williams et al. (2000).

Introduction to the Old Red Sandstone of Great Britain

The principal macrofossils are fish fragments. A biozonal scheme was erected for the Old Red Sandstone in the Anglo-Welsh Basin (see Figure 5.3, Chapter 5) and was extended to continental Europe. Refinement of the scheme continues (Blieck and Janvier, 1989; Blieck and Cloutier, 2000), but the occurrence of fish remains is patchy and of limited use in high-resolution correlation. Miospore classifications (e.g. Richardson *et al.*, 2000; Streel *et al.*, 2000) and microvertebrate classification (Vergoossen, 2000) also aid correlation and stratigraphical resolution. However, the problem of detailed correlation of the terrestrial Old Red Sandstone succession with the Bohemian, German and Belgian marine stages, in which miospores are rare, remains. Progress is, however, being made by chains of correlation involving miospores that are common to the Old Red Sandstone and the Rhenish marine succession, the latter then being correlated with the Bohemian stages. For example, the recognition of the *Breconensis-zavallatus* Zone in the Ardennes allows correlation of the Anglo-Welsh and Rhenish Gedinnian–Siegenian successions (Richardson *et al.*, 2000) (Figure 5.3, Chapter 5). A widespread volcanic ash deposit (the Townsend Tuff Bed) and a basin-wide calcrete (the Psammosteus Limestone) are valuable marker horizons in the Anglo-Welsh Basin, providing lithostratigraphical correlation of the succession.

Chapter 2

The Orcadian Basin

Introduction

INTRODUCTION

P. Stone

The Devonian strata preserved in the north-east of mainland Scotland and the Orkney and Shetland islands are the remnants of deposits that accumulated in a major, intramontane lake rift-basin, the Orcadian Basin (Figure 2.1). Offshore well records and seismic data show a north-eastwards continuation of the basin beneath the North Sea, with probably equivalent strata also present to the west, in the West Orkney, Outer Isles and North Minches basins. The succession is over 4 km thick in Caithness and possibly twice that in parts of Shetland, where three distinct basins appear to have been juxtaposed by dextral strike-slip faulting (Mykura and Phemister, 1976). A restoration of the Shetland basins in relation to the main Orcadian depocentre is shown in Figure 2.2. Overviews of the Orcadian Basin successions and their regional context were provided by Anderton *et al.* (1979), Dineley (1999a), and

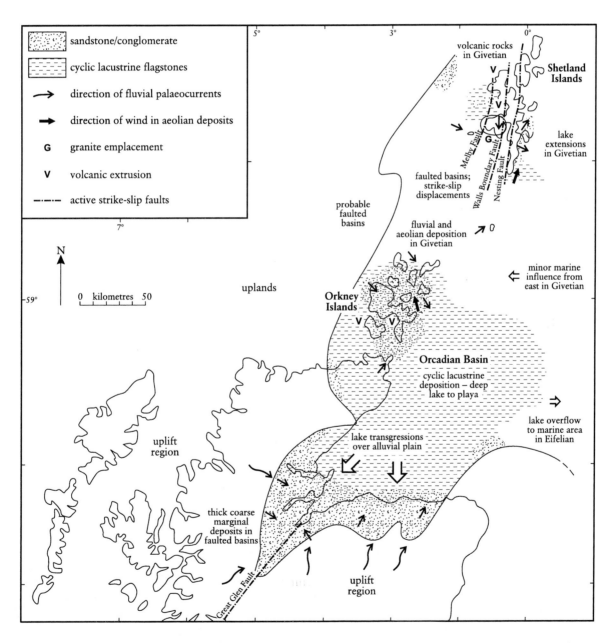

Figure 2.1 Generalized Mid-Devonian palaeogeography based on present geography of northern Scotland. After Trewin and Thirlwall (2002).

The Orcadian Basin

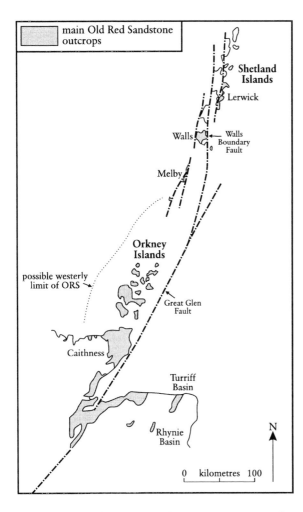

Figure 2.2 Old Red Sandstone outcrops in the Orcadian Basin and restoration of strike-slip displacements in the Shetland Islands. Based on Mykura and Phemister (1976) and Anderton *et al.* (1979).

Woodcock and Strachan (2000). A summary of the Caithness succession was provided by Trewin (1993) and palaeogeographical reconstructions are given by Bluck *et al.* (1992). Trewin and Thirlwall (2002) give a recent comprehensive account of the basin, and Marshall and Hewett (2003) provide a summary. The following introductory account is based on these sources.

Uplift and erosion dominated the northern Scottish region during Early Devonian times. Uplift was largely driven by the intrusion of an extensive suite of granitic plutons and was accompanied by strike-slip movement on major NE–SW-trending faults. Hence, the accumulation of Lower Devonian strata was localized and of variable lithology, lying unconformably on a surface of Moine, Dalradian and intrusive granitic rocks of the Highlands metamorphic complexes that had considerable local relief variations. A similar pattern continued into Mid-Devonian times, with the localized deposition of lake-margin facies, including carbonates, stromatolites and coarse alluvial breccias on the basement unconformity, but passing distally into and transgressively covered by fluvio-deltaic sandstones and lithologies of the developing deep-lake facies.

Lake transgression and regression occurred repeatedly through Mid-Devonian times and appear to have been relatively rapid events; sub-aerial, desiccation-cracked surfaces and deep-lake laminated mudstone facies occur locally in little more than 1 m of strata (Trewin, 1976). Elsewhere, a complex interplay of lacustrine and lake-shore environments gives intercalations of lake-floor, lake-delta, fluvial and aeolian deposits. During at least one episode, the lake waters extended across the whole basin area, resulting in deposition of the famous Achanarras fossil fish-bearing laminite and its correlatives. At other times, there were smaller and more localized lakes, separated by areas of fluvial or aeolian deposition. The lakes within the Orcadian Basin were fed by streams draining from surrounding high ground in a climate that was hot and varied between humid and semi-arid. The lakes were never very deep, perhaps up to about 80 m when the Achanarras Fish Bed was deposited (Hamilton and Trewin, 1988), at which time the lake covered some 50 000 km^2. However, basinal subsidence was rapid, since up to 5 km of strata accumulated in about only 10 million years. The evidence for decreasing alluvial-fan progradation and more extensive lakes suggests that regional thermal subsidence replaced crustal extension and normal faulting as the main subsidence mechanism in Mid-Devonian times.

The Middle Devonian fish-bearing laminites are a highly unusual lithofacies. Their regular, parallel, planar bedding facilitated their use as flagstones, examples of which can be seen in many British cities, and they are commonly known as 'Caithness Flags'. The GCR sites at **Achanarras Quarry** and **Wick Quarries** are disused flagstone quarries. The flagstones were deposited in the large, ephemeral lakes and have a cyclic character. Donovan (1980) recognized a basic sequence of four lithofacies associations

('A', 'B', 'C' and 'D'), representing a range of conditions from deep to shallow lake and exposed playa surface (Figure 2.3; Trewin, 1993; Trewin and Thirlwall, 2002). In practice, a great deal of facies variation arose through local differences in sediment supply, subsidence, carbonate production and water conditions. Donovan's 'A association' – the fish-bearing laminites – were deposited during the deep lake phase as intercalated silt, carbonate and organic laminae. They are interpreted as seasonal varves, controlled by annual climate variations. Clastic laminae represent input from rivers during the wet season, deposited from suspension. The carbonate laminae formed through increased photosynthesis in the dry season when algal phytoplankton blooms flourished in the lake waters, and the organic laminae formed by the subsequent annual decay of the phytoplankton. Shallower lake facies, the 'B & C associations' of Donovan, are thicker laminated and wave-rippled, with probable subaqueous shrinkage (synaeresis) cracks (although this interpretation is not universally accepted; see Astin and Rogers (1991) and subsequent discussions (Astin and Rogers, 1992, 1993; Trewin, 1992; Barclay *et al.*, 1993); see also Plummer and Gostin (1981) for a discussion on shrinkage cracks). Donovan's 'D association' formed in shallow, ephemeral lake conditions that produced abundant surfaces with polygonal arrays of desiccation cracks and shallow-water rippled surfaces; fluvial and aeolian influences may also have been present.

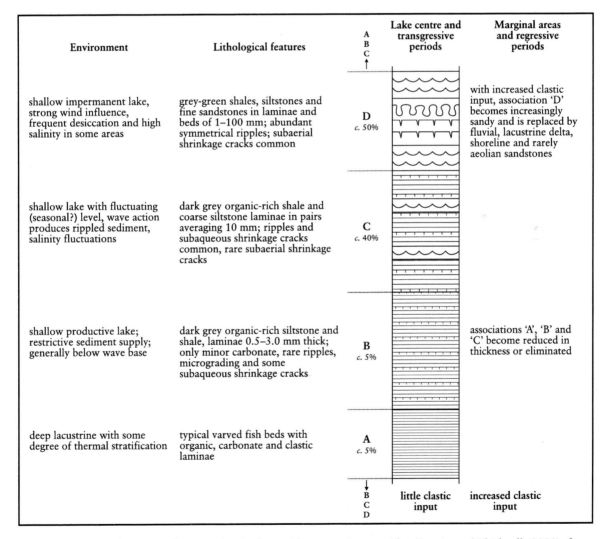

Figure 2.3 Cyclic lacustrine facies in the Caithness Flagstone Group. After Trewin and Thirlwall (2002), from Donovan (1980).

The Orcadian Basin

The Orcadian Basin lake(s) probably drained south-eastwards to the sea, the earliest evidence for possible marine influence within the basin being seen late in the Middle Devonian succession of Orkney (Marshall *et al.*, 1996). A marine connection became more important during Late Devonian times. Over most of the basin, the boundary between Middle and Upper Old Red Sandstone is conformable, but there is some evidence for localized tilting of the Middle Devonian strata, interpreted as a phase of limited basin inversion prior to Late Devonian sedimentation. More commonly, the base of the Upper Devonian Series is marked by a change in depositional environment. The Mid-Devonian lakes were replaced in the basin centre by sabkha plains that were periodically inundated by the sea, whilst meandering- and braided-river deposits accumulated on the basin margins.

Most of the preserved succession and the GCR sites in the Orcadian Basin are of Mid-Devonian age. The stratigraphical positions and ranges of the sites are shown in Figure 2.4. The lithostratigraphical scheme for the basin is not wholly satisfactory in modern terms and is currently under review. The version used in this account is largely that used by Dineley (1999a) in the *Fossil Fishes of Great Britain* GCR volume (Dineley and Metcalf, 1999), with small modifications to take account of recent developments.

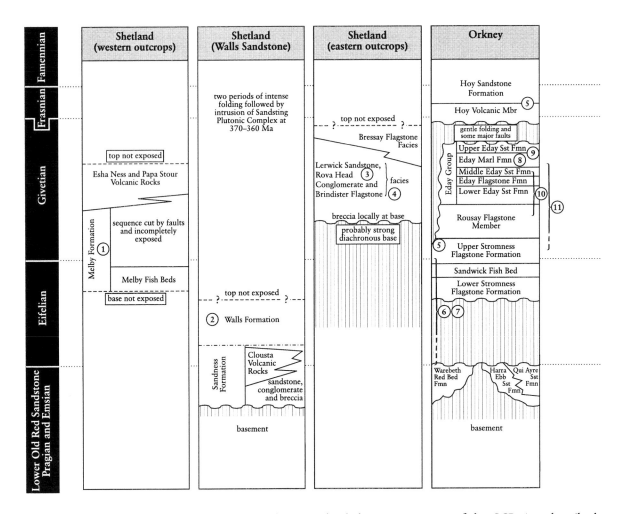

Figure 2.4 Stratigraphical successions in northern Scotland showing positions of the GCR sites described. (1 – Melby; 2 – Footabrough to Wick of Watsness; 3 – Easter Rova Head; 4 – The Cletts, Exnaboe; 5 – Old Man of Hoy Coast; 6 – South Stromness Coast Section; 7 – Yesnaby and Gaulton Coast Section; 8 – Bay of Berstane; 9 – Greenan Nev Coast, Eday; 10 – South Fersness Bay, Eday; 11 – Taracliff Bay to Newark Bay; 12 – Red Point; 13 – Pennyland (Thurso–Scrabster); 14 – Achanarras Quarry; 15 – John o'Groats; 16 – Wick Quarries; 17 – Sarclet; 18 – Tarbat Ness; 19 – Tynet Burn; 20 – Den of Findon; 21 – Dun Chia Hill (Loch Duntelchaig); 22 – Rhynie.)

Introduction

Shetland

The Devonian strata of Shetland are separated into discrete terranes by a series of major north–south faults (Figure 2.5). There are considerable lithological and structural differences between the sequences in each of the terranes, but all are of Eifelian to Givetian, Mid-Devonian age (Figure 2.4). They have been related to three separate depocentres, juxtaposed by dextral strike-slip fault movements (Mykura and Phemister, 1976). Their possible restoration in relation to the main Orcadian Basin is shown in Figure 2.2.

The westernmost terrane, to the west of the Melby Fault, is the smallest and may have originated closest to the main Orcadian Basin; its Old Red Sandstone lithofacies, of late Eifelian age, is certainly the most similar of the Shetland varieties to those of Caithness and Orkney. The **Melby** GCR site shows two wholly lacustrine intervals alternating with thick fluvial sandstones. The lacustrine beds are thinly laminated, silty mudstones with carbonate nodules that contain an important fossil fish fauna, described in the *Fossil Fishes of Great Britain* GCR volume (Dineley and Metcalf, 1999). This fauna is considered to be slightly older than the Achanarras (Caithness) fauna.

East of the Melby Fault, and bounded to the east by the Walls Boundary Fault, lies a terrane that contains the Devonian succession of the Walls Peninsula. This is represented by probable

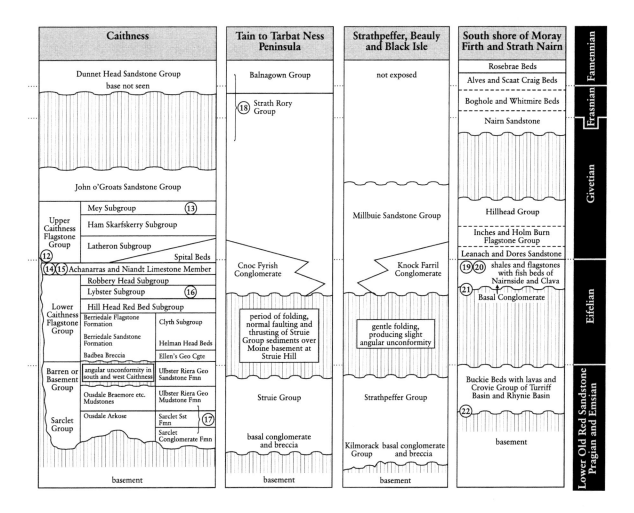

The Orcadian Basin

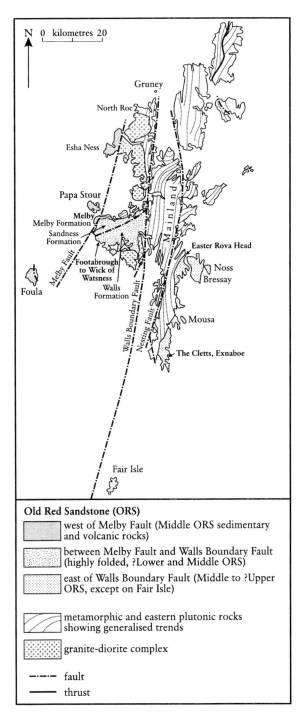

Figure 2.5 Geological sketch map of the outcrops of Old Red Sandstone rocks in Shetland and their structural relationships; with locations of GCR sites. After Mykura and Phemister (1976).

Givetian strata within the **Footabrough to Wick of Watsness** GCR site. The thick alternations of sandstone, siltstone and mudstone may have formed by turbidite deposition in a deep lake environment, but some combination of braided fluvial channel and overbank, littoral and shallow lacustrine lithofacies is perhaps more likely. The Devonian strata of the Walls Peninsula are more severely deformed than any other parts of the Orcadian Basin successions.

The Devonian succession of the eastern terrane, east of the Nesting Fault, is separated from those to the west by a basement terrane devoid of Devonian cover. There are two GCR sites, both with strata of Givetian age, at **Easter Rova Head** near Lerwick and at **The Cletts** near Exnaboe. The Easter Rova Head site shows a coarse, conglomeratic lake-margin alluvial fan comprising sheet-flood and mass-flow deposits. It probably rests unconformably on basement rocks, although the contact is not seen. Farther south, at The Cletts, a sequence of interbedded lacustrine mudstones and braided stream, alluvial-fan, deltaic and aeolian sandstones records the transgressive and regressive phases of the Orcadian Basin lake. The lacustrine mudstone contains a fossil fish fauna (described in the GCR fossil fishes volume, Dineley and Metcalf, 1999) of about the same age as that found at **Pennyland (Thurso–Scrabster)**, and slightly younger than the Achanarras fauna.

Orkney

The Orkney Islands consist principally of Devonian strata (Figure 2.6). The only significant exceptions are small basement inliers of granitic and schistose lithologies on the southwest coast of the island of Mainland and the adjacent small island of Graemsay. The pre-Devonian relief was of the order of 100 m, and locally steep and craggy. Two of the Orkney Devonian GCR sites, **South Stromness Coast Section** and **Yesnaby and Gaulton Coast Section**, show the basal unconformity. At Yesnaby, Lower Devonian sandstone, breccia and conglomerate rest on crystalline basement, whereas at South Stromness the breccias and conglomerate above the unconformity are believed to be of Mid-Devonian, Eifelian age. The likely age difference is emphasized by the tilting of the Lower Devonian strata at Yesnaby so that the basal sedimentary sequence is overlain unconformably by Middle Devonian rocks; at South Stromness the sedimentary relationships above the basal unconformity are all conformable. The Lower Devonian sequence represents alluvial-fan, lacustrine mudflat, aeolian dune and beach deposition.

Introduction

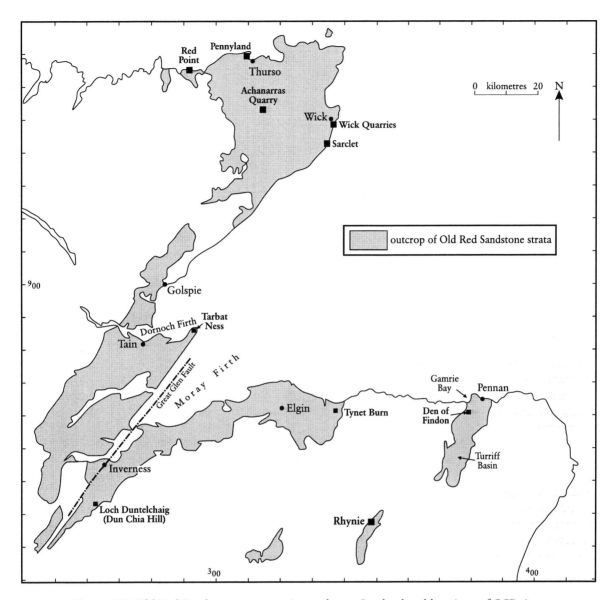

Figure 2.7 Old Red Sandstone outcrops in north-east Scotland and locations of GCR sites.

Lower Devonian rocks (the Crovie Group of the Lower Old Red Sandstone) are present in the largely fault-bounded Turriff and Rhynie basins. Middle Devonian (the Findon Group of the Middle Old Red Sandstone) rocks overlie the Crovie Group unconformably. The Turriff Basin extends about 30 km inland from the superb exposures of its fill on the Moray Firth coast at Gamrie Bay and New Aberdour. Coarse, basal conglomerates of the Crovie Group fill an irregular Dalradian basement topography and are succeeded by a 600 m-thick variable succession of alluvial-fan conglomerates and sandstones and floodplain/ playa mudstones. The Findon Group also comprises basin-margin alluvial-fan deposits, derived from the south, but fan progradation was halted at the time of maximum Achannaras lake extent, when the Gamrie Fish Bed was deposited, as seen at the **Den of Findon** GCR site in Banffshire. The description of this site includes a brief account of the magnificent cliff and foreshore sections of the Crovie and Findon groups nearby in Gamrie Bay, New Aberdour and Quarry Haven, which also merit protected status.

The main lacustrine lithofacies of the Orcadian Basin, spanning the Eifelian and Givetian stages, are represented by three GCR sites in Caithness (**Wick Quarries**, **Achanarras Quarry**, and **Pennyland (Thurso–Scrabster)**), in addition to two sites on the southern margin of the basin

(**Tynet Burn**, Moray, and the **Den of Findon**). The oldest strata, of Eifelian age, are seen in the old flagstone quarries and cliff sections at South Head, Wick. Two sequential lacustrine shallowing and deepening cycles are represented, with a splendid array of shrinkage cracks and soft-sediment deformation features. The disused flagstone quarry at Achanarras contains strata deposited at the acme of lacustrine development around the Eifelian–Givetian boundary. The extensive fossil fish fauna from this site is world-famous and has led to **Achanarras Quarry** being selected as a GCR site for fossil fishes (Dineley and Metcalf, 1999). Givetian strata at a level above that seen at Achanarras are exposed in the Pennyland cliff and foreshore to the west of Thurso. About 20 lithofacies cycles demonstrate the repeated, abrupt shallowing and deepening of the Orcadian Basin lake. In addition, there is good preservation of a spectacular array of shrinkage crack styles. Fossil fish have been recovered from several levels within the section and the site is separately selected for the GCR as a fossil fishes site (Dineley and Metcalf, 1999).

The highest Middle Devonian strata are seen at the **John o'Groats** GCR site, accorded protected status because of its fish fauna (Dineley, 1999a). A description of this site is extended here to include the fine exposures of the John o'Groats Sandstone Group to the east near Duncansby Head.

Upper Devonian strata are relatively rare within the mainland outcrop of the Orcadian Basin, but are represented in the GCR site at **Tarbat Ness** (Ross), close to the southern margin of the basin. They are also present at Dunnet Head near Thurso, where there are spectacular cliffs of the Dunnet Head Sandstone Group, of similar facies to the Hoy Sandstone Formation in Orkney. The cliffs are accessible at the north end of Dunnet Bay, where they provide a superb teaching section of bar and channel facies variations in a low-sinuosity sand-bed river system (B.P.J. Williams, pers. comm.); this section should be considered as a complementary site.

In the lower part of the succession exposed at Tarbat Ness, Givetian fluvial sandstones with sporadic lacustrine mudstones lie at the top of the Middle Devonian succession. They are conformably overlain by fluvial, aeolian and sabkha-like deposits, believed to be Frasnian to Famennian in age. The sabkha deposits contain striking examples of desiccation and evaporitic structures.

MELBY: MATTA TAING TO LANG RIGG, SHETLAND (HU 165 560–HU 177 575)

Potential ORS GCR site

P. Stone

Introduction

Fluvial and lacustrine strata of the Middle Devonian Melby Formation are spectacularly exposed within the sea cliffs and rocky foreshore between Matta Taing and Lang Rigg, on the north-west extremity of the Walls Peninsula, Shetland Mainland. The section is about 2–3 km west of Melby village, and to the north and west of Huxter. It lies to the west of the Melby Fault, in the westernmost of the tracts of Devonian strata of Shetland.

The lithofacies demonstrate the cyclical transgressions and regressions of a large lake within the Orcadian Basin. Highstands of the lake are marked by two units of laminated lacustrine mudstone with carbonate nodules. The nodules contain a regionally important fish fauna, which is described in the GCR fossil fishes volume (Dineley and Metcalf, 1999). During periods of lowstand, fluvial systems of braided channels prograded into the lake and deposited thick bodies of cross-stratified sands. Sandstone composition and palaeocurrent direction change through the succession, reflecting changes in the local palaeogeography.

The regional importance of the Melby site lies in the well-exposed evidence for the cyclicity of depositional environments in the northern extremity of the Orcadian Basin, and the insights provided into local palaeogeographical changes. The fish beds are also of great importance, being the northernmost correlatives of the Achannaras horizon of Caithness. All these factors allow better interpretation and characterization of the regional tectonic framework and development of the Orcadian Basin. Overviews of the geology are provided by Mykura (1976, 1991) and a detailed account of the section is given by Mykura and Phemister (1976). Details of the fish beds are given by Hall and Donovan (1978) and Dineley (1999a).

Melby: Matta Taing to Lang Rigg

Description

West and south-west from Melby and Huxter, high sea-cliffs and rocky foreshore areas provide excellent, continuous exposure through a sequence of fluvial and lacustrine Old Red Sandstone strata, the Melby Formation. The beds strike approximately north-east and dip consistently towards the south-east between 25° and 45°, so that a vertical succession of about 250 m is represented, with faulting causing some local repetition. The western section from Matta Taing (HU 166 560) to Quilva Taing (HU 171 572) has high, rugged sea-cliffs cutting across the strike, and the north-western coastline, from Quilva Taing to Lang Rigg (HU 177 577), is approximately a strike section (Figure 2.8). The fossil fish faunas recovered from two horizons and spore analysis establish a late Eifelian–early Givetian (Mid-Devonian) age.

The oldest strata are seen along the north-western coast between Lang Rigg and Ayre of Huxter (HU 173 574). There, over 50 m of reddish brown, feldspathic sandstones pass upwards into grey-brown, calcareous sandstones. Mykura and Phemister (1976) described the rocks as poorly graded arkoses, comprising up to 80% quartz, up to 40% feldspar and a small proportion of lithic grains. Conversely, sandstones described by Knudsen (2000) are quartz arenites with a modal content of over 90% quartz, less than 5% feldspar and less than 5% lithic fragments. There is very little detrital matrix, and carbonate occurs as a secondary cement in the higher part of the sequence. Large-scale cross-bedding is ubiquitous. The cross-bedded sets range up to 3 m in thickness and are tabular (Figure 2.9a) or show complex convolutions in their upper parts (Figure 2.9b). The convolute folds are overturned locally, invariably towards the east (Mykura and Phemister, 1976), in keeping with a dominant palaeocurrent flow from the west, as indicated by the dip of the foresets. Sporadic interbeds of siltstone reach about 50 cm in thickness, with the thickest example having an eroded top overlain by intraformational conglomerate. Locally, some of the thinner horizons of siltstone and shale contain sandstone-filled shrinkage cracks.

About 2 m of ripple cross-laminated sandstone mark the top of the sandstone-dominated sequence. These are overlain by approximately 12 m of thinly bedded, fine-grained lithologies that together form the lower of the two Melby

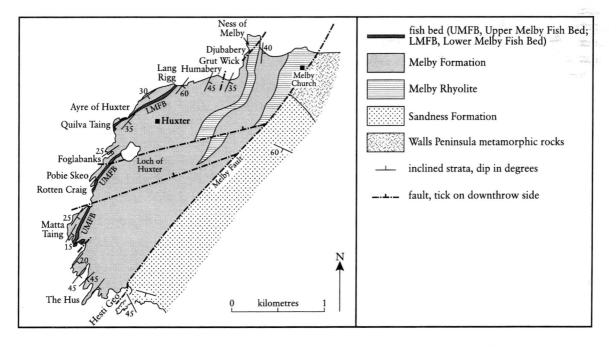

Figure 2.8 Geological sketch map of the Melby Formation on Shetland Mainland. After Mykura and Phemister (1976).

Figure 2.9 Cross-bedded sandstones between Ayre of Huxter and Lang Rigg. (Photos: P. Stone.)

fish beds (Figure 2.10, section a). There is considerable lateral variation within the Lower Melby Fish Bed, but it can be broadly subdivided into three units (Mykura and Phemister, 1976). The lowest unit, about 3 m thick, consists of pale grey, laminated siltstone and mudstone with desiccation cracks and sporadic plant remains. The middle unit, also 3 m thick, comprises grey to black, locally bituminous, laminated mudstone with irregular carbonate-rich ribs and nodules. A few thin interbeds of siltstone to fine sandstone have rippled upper surfaces. Plant remains are

Melby: Matta Taing to Lang Rigg

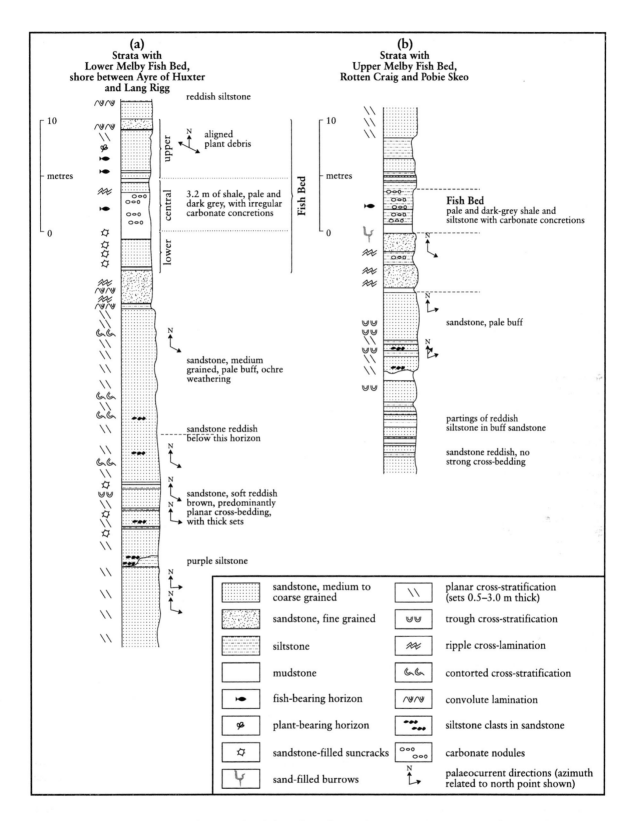

Figure 2.10 Stratal settings of the Melby fish beds. After Mykura and Phemister (1976), reproduced from Dineley (1999a).

abundant and fossil fish fragments have been recovered from the calcareous ribs and nodules, which are mostly up to about 25 mm by 75 mm. Disarticulated fragments of *Coccosteus* are the commonest fish fossil; Dineley (1999a) provides a full list of species recovered, including the index species *Pterichthyodes milleri*. The top 4.5 m unit consists of laminated siltstone and mudstone with sporadic lenses of fine-grained sandstone. Finely disseminated plant debris occurs on some bedding surfaces, the larger fragments having an ESE–WNW alignment.

The Lower Melby Fish Bed is overlain by about 100 m of predominantly fine- to medium-grained sandstones. Individual sandstone beds range up to about 1 m in thickness and are variably pale yellowish brown to red in colour. They are interbedded with reddish, lenticular siltstone layers up to about 1 m thick, which give a marked banded appearance to parts of the succession (Figure 2.11a). Many of the sandstone beds have large-scale, planar cross-bedding (Figure 2.11b) in which the foresets dip mainly towards the east. Trough cross-bedding is present locally on a small-scale (Figure 2.11c). Ripple cross-lamination is common, especially towards the top of the unit, and is locally convoluted, with pseudonodule structures. Compositionally, the sandstones are arkoses with approximately equal proportions of quartz and feldspar (Mykura and Phemister, 1976). Grains are angular to subrounded and there is virtually no detrital matrix. Small worm tubes occur in the highest of the sandstone beds, immediately below the abrupt junction with the Upper Melby Fish Bed.

The Upper Melby Fish Bed comprises about 4 m of thinly bedded, pale grey siltstone and darker, laminated shale with lenticular carbonate layers (Figure 2.12b) and nodular concretions of carbonate and sulphide; some calcareous layers are nearly black and slightly bituminous. The best-developed lamination is caused by alternating silt, carbonate and organic carbon laminae forming triplets 0.3 mm to 2 mm thick. The thin siltstone beds and rare fine-grained sandstone layers, the latter generally lenticular, are pervasively convoluted (Figure 2.12b) with rippled upper bedding surfaces (Figure 2.12c) and contain sporadic plant material. The carbonate layers and nodules contain a well-preserved but fragmentary fossil fish fauna listed by Dineley (1999a) that includes *Coccosteus cuspidatus*, *Homosteus milleri* and *Mesacanthus* sp.. The sulphide nodules in the upper part of the fish bed are complex, varved and tectonically deformed. They contain varying proportions of pyrite, chalcopyrite, covellite, bornite, sphalerite, galena and tennantite and have a complex, multi-phase origin (Hall and Donovan, 1978). The fish bed is overlain by brown, fine- to medium-grained, feldspathic sandstone with some large-scale cross-stratification. The sandstone has a quartz/feldspar ratio of 65:35, with a significant quantity of felsite grains (up to 10%), and relatively abundant, large grains of garnet (Mykura and Phemister, 1976).

Interpretation

The range of lithologies and sedimentary features seen in the Melby section between Matta Taing and Lang Rigg reflect the varying environment close to the margin of a large lake within the Orcadian Basin. Lake transgression and regression were the principal controlling influences. The fish beds, particularly the central part of the upper one, were deposited in reducing conditions during the acme of the transgression and maximum water depths, perhaps up to 50 m (Mykura and Phemister, 1976). Hall and Donovan (1978) described the laminated fish bed lithofacies as non-glacial varves deposited in a stratified tropical lake. Seasonal algal bloom resulted in increased photosynthesis and carbonate precipitation, the dead algae accumulating on the lake floor and being preserved as organic carbon. The clastic laminae formed either by continuous settlement from suspension or by small turbidity flow influx, perhaps with a seasonal or annual periodicity. Hall and Donovan (1978) emphasize one important difference between the Melby laminites and similar lithofacies elsewhere in the Orcadian Basin. At Melby, apparently uniquely, the clastic laminae consist largely of green clay, rather than mainly silt. The clay is a mixture of illite and chlorite and may have been derived by the weathering of a nearby volcanic massif. Hall and Donovan attributed the sulphide nodules in the Upper Melby Fish Bed to groundwater enriched by nearby volcanic activity.

The Lower Melby Fish Bed, and the top and bottom parts of the Upper Melby Fish Bed, are variable and record the irregular progress of lake transgressions and regressions. In these sections lacustrine sediments alternate with low-energy, fluvial sediments of channel, crevasse

Melby: Matta Taing to Lang Rigg

Figure 2.11 Sandstones between the two Melby fish beds. (a) Pobie Skeo; (b) large-scale planar cross-bedding; (c) small-scale trough cross-bedding. (Photos: P. Stone.)

Figure 2.12 The Upper Melby Fish Bed, Rotten Craig. (a) Laminated mudstones with dolomitic layers; (b) convolute lamination in thin sandstone bed; (c) rippled siltstone bedding surface. (Photos: P. Stone.)

and overbank origin. Full lake regression allowed relatively high-energy fluvial systems to prograde into the shallow lake margins, depositing thick, cross-stratified sands. Convoluted tops to many of the cross-bedding sets suggest rapid de-watering of the sediment. Below the Upper Melby Fish Bed, the foresets of most planar cross-bedded sandstones dip eastwards, suggesting a current flow predominantly from the west. This broad uniformity suggests deposition in braided channels, with the sporadic finer-grained beds representing the remains of overbank sediment accumulations. Thinly bedded, ripple cross-laminated sandstones below and between the two fish beds were probably laid down by relatively sluggish currents, probably in crevassed channels or as overbank deposits on an alluvial plain (Mykura and Phemister, 1976).

Above the Upper Melby Fish Bed, palaeocurrent flow direction shows a reversal, with consistent foreset dips to the west, indicating current flow from the east (Mykura and Phemister, 1976). This change coincides with the appearance in the sandstones of significant quantities of mainly felsic volcanic detritus, but also including some basalt. The volcanic rocks being eroded may have been the precursors to the Esha Ness and Papa Stour volcanic complexes, which are believed to overlie the Melby Formation. The identification by Hall and Donovan (1978) of volcaniclastic clay in the Upper Melby Fish Bed is further evidence for the active erosion of volcanic rocks during the deposition of the Melby Formation.

Conclusions

The cliffs and shoreline containing the Melby GCR site provide a well-exposed, representative section of the Middle Devonian (upper Eifelian–lower Givetian) strata in the westernmost of the Shetland structural tracts. The section demonstrates the varying depositional environments during lake transgression and regression, with lacustrine sediments interbedded within a dominantly fluvial sequence. The lacustrine strata are fine-grained, thinly bedded or laminated lithologies with carbonate (and locally sulphide) nodules. The carbonate nodules contain an important fossil fish fauna, allowing regional correlation with the Achanarras horizon of Caithness. Fluvial strata are mainly thick, cross-stratified, arkosic sandstones. Palaeoflow in the braided channel systems shows a marked change of direction within the succession exposed in the GCR site. The change coincides with a new volcanic source of detritus and is of regional palaeogeographical significance. Overall, the site provides important insights into lake evolution in the northernmost part of the Orcadian Basin and its fish fauna allows regional correlation with the Middle Devonian strata of mainland Scotland.

FOOTABROUGH TO WICK OF WATSNESS, SHETLAND (HU 179 502–HU 201 495)

P. Stone

Introduction

This section was originally named 'Fidlar Geo to Watsness' in the GCR archive. However, Fidlar Geo is in the middle of the section, and the south-east end of the site is better described as 'Voe of Footabrough', or simply as 'Footabrough'. At the north-west end of the site 'Watsness' should be either 'Wats Ness' (a location that is actually beyond the site boundary) or 'Wick of Watsness'.

The sea cliffs between Wick of Watsness and Voe of Footabrough on the west coast of the Walls Peninsula, Shetland Mainland, provide excellent, continuous exposure through folded strata of the Middle Devonian (Givetian) Walls Formation. Fidlar Geo (HU 190 493) is a prominent inlet in the central part of this section. The Walls Formation is part of the sequence lying between the N–S-trending Melby and Walls Boundary faults, and so forms part of the central Devonian structural tract within Shetland. Interpretation of the lithofacies present has proved difficult and controversial. The sequence is dominated by intercalated sandstone and shale that have been interpreted as turbidites deposited in a deep, lacustrine environment. However, a broad consensus from the most recent work (Melvin, 1985) is that a range of alluvial, braided stream, shoreline and shallow lacustrine environments are represented.

The regional importance of the site lies in the correct interpretation of its enigmatic lithofacies, thus allowing elucidation of depositional geometry within an otherwise poorly understood part of the Orcadian Basin. In this respect the site has an important role in characterizing the central north–south Devonian structural tract in Shetland. Detailed descriptions of the Walls Formation in the GCR site area are provided by Melvin (1976, 1985), Mykura and Phemister (1976) and Astin (1982). An overview of the geology is provided by Mykura (1976, 1991).

Description

Spectacular sea-cliffs along the coastline of the Walls Peninsula form the western extremity of Shetland Mainland. The cliffs expose an extensive section through part of the Walls Formation between Wick of Watsness and Voe of Footabrough (Figure 2.13). The sandstone-dominated formation was formerly thought to exceed 9000 m in thickness (Mykura, 1976), but may be much thinner (as discussed below). Only a part of this is exposed within the GCR site, but faulting and locally intense folding make thickness estimation very difficult.

A sparse and fragmentary fish fauna, together with some indeterminate plant detritus, has been recovered from the Walls Formation (including three localities within the GCR site) (Mykura and Phemister, 1976), but is indicative only of a broadly Mid-Devonian age. Highly carbonized palynomorphs collected by Melvin (1985) suggest a Devonian age no older than Emsian, and a more recent study of miospores has confirmed a Givetian age (Marshall, 2000). An approximate upper age limit is provided by the 360 ± 11 Ma date (K–Ar) from the Sandsting Plutonic Complex, which intrudes the Walls Formation in the south-east part of its outcrop (Mykura and Phemister, 1976).

The Walls Formation lies between the Melby and Walls Boundary faults, and so forms part of the central of the north–south structural tracts that contain the Shetland Devonian succession (Figure 2.2). The strata are folded about a series of upright, open to close synclines and anticlines (Figure 2.14) that trend approximately ENE and plunge gently towards the WSW.

Figure 2.13 The Walls Formation at Cotti Geo, Ram's Head. (Photo: P. Stone.)

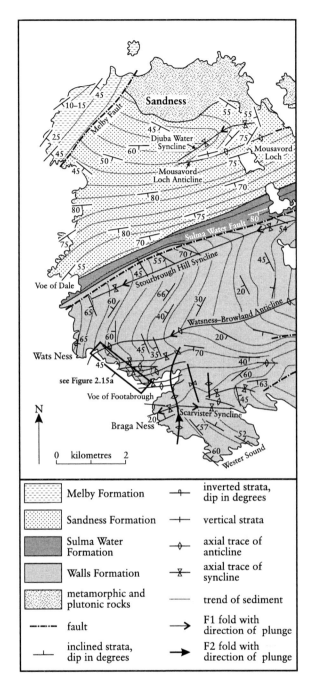

◄Figure 2.14 Major stratigraphical and structural features of the Walls Peninsula. After Mykura (1976).

Hence, bedding dips range from horizontal to sub-vertical and strikes are variable (Figure 2.15a,b). A locally strong cleavage is developed broadly axial planar to the folds. At least some of the deformation has been related to the intrusion of the Sandsting Plutonic Complex (Mykura, 1991).

Mykura and Phemister (1976) noted the vertical and lateral lithological uniformity of the Walls Formation. It comprises stacked cycles, each consisting of a basal fine-grained, dark grey sandstone that passes up into a finer-grained, thinly bedded unit consisting, in varying proportions, of shale, siltstone and, rarely, impure limestone. Complete cycles range in thickness from 0.75 m to 20 m, with individual sandstones up to 18 m thick. The ratio of sandstone to shale and siltstone varies between different parts of the formation, as is well illustrated southwards along the coast from Wats Ness, where shale-rich cycles are dominant for up to about 20 m but then alternate with sandstone-rich cycles 25–45 m thick. Melvin (1976, 1985) estimated that various types of sandstone-rich cycles form almost 80% of the succession, with the shale-rich cycles making up the remaining 20%.

The sandstone beds generally have a sharp base, with flute and groove marks developed locally, resting on an erosion surface. Rip-up clasts of shale and siltstone are common, but nowhere abundant, in the lowest few centimetres of the sandstone. Cross-bedding of various types is widespread, ranging from large-scale, planar forms to small-scale, trough cross-lamination. In some places, concentrations of heavy minerals line the cross-bedding foresets. Convolute lamination is also widespread, and some examples are truncated at the base of the overlying sandstone bed. Texturally, most of the sandstone is fine- or medium-grained and fairly well-sorted, with a matrix content up to about 25% (much of which is carbonate). The main detrital components are mono- and poly-crystalline quartz, potassium feldspar (mainly orthoclase) and some plagioclase; accessories include muscovite, biotite, granite and garnetiferous quartz-mica gneiss (Melvin, 1976, 1985; Knudsen, 2000). The approximate overall modal proportions are: 70% quartz, 25% feldspar and 5% lithic fragments, which classify the sandstone as a subarkose. Detrital heavy minerals are abundant in some of the sandstones and may comprise up to 17% of the grains. Zircon, tourmaline and epidote are the most abundant, with less common sphene, rutile, clinozoisite and apatite.

The shale-rich parts of the succession contain many beds with well-developed ripple cross-

The Orcadian Basin

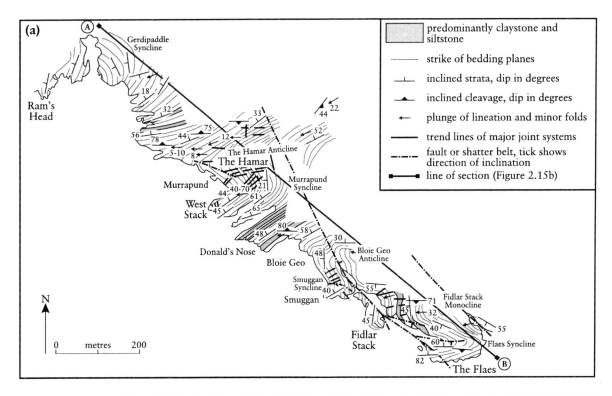

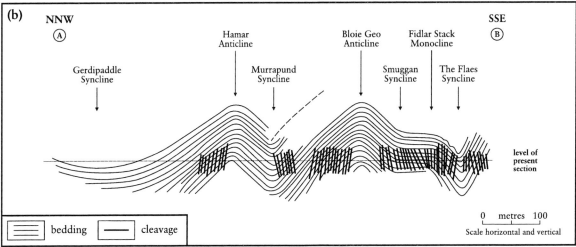

Figure 2.15 (a) Sketch map of the rocks of the Walls Formation exposed on the coast between Ram's Head and The Flaes. (b) Cross-section showing the structural pattern between Ram's Head and The Flaes. After Mykura and Phemister (1976).

lamination (Figure 2.16), which is locally accentuated by the coincidence of ripple crests with the bedding–cleavage intersection lineation. In some cases, successive beds show markedly different ripple orientations. Load casts are widespread and locally form pseudonodule layers. Desiccation polygons have been reported on some bedding surfaces by Melvin (1985), although Mykura and Phemister (1976) commented on their apparent absence from the Walls Formation. A typical section through the shale-rich lithofacies is shown in Figure 2.17a (Melvin, 1985). Thin beds of limestone and calcareous mudstone occur sporadically and are invariably finely laminated, with some included carbonaceous films. Most are disrupted by soft-sediment deformation, with subsequent preferential deformation by tectonic processes.

Figure 2.16 Rippled bedding surfaces in the Walls Formation. (Photo: P. Stone).

Interpretation

Various interpretations have been proposed for the depositional environment of the Walls Formation, and the topic remains controversial. Mykura and Phemister (1976) preferred an origin as turbidity flows within a deep, subsiding lake basin, although conceding the possibility of fluvial deposition. They were strongly influenced by the uniformity of the sandstone–shale succession over a considerable thickness (apparently up to 9000 m) and the apparent absence of diagnostic indicators of fluvial or subaerial environments. Examples of the latter, such as desiccation crack polygons, were subsequently discovered, and later workers (Melvin, 1976; Astin, 1982) preferred a shallow-water to fluvial depositional interpretation. Astin questioned the apparently great thickness of the succession and demonstrated sedimentological continuity northwards across the Sulma Water Fault (Figure 2.14) into the Sandness Formation, where fluvial depositional features are well developed. Further doubt was cast on the thickness estimate of 9000 m when Marshall (2000) showed from miospore evidence that both the Walls and Sandness formations were entirely Givetian in age.

The coast section forming the GCR site was shown by Melvin (1976, 1985) to consist largely of thick, multi-storey sequences of channelized, trough cross-stratified sandstone. Relationships between the successive channel units are complex, with cross-cutting scour surfaces underlying the bases of many sandstones. The assemblage of features suggests that in-channel deposition dominated, probably within a braided stream environment. A few of the sandstone-rich cycles show a particularly marked fining-upward trend and these were interpreted by Melvin (1985) to represent a progression of depositional environments from channel to sandbar, levee and floodplain (Figure 2.17b). The shale-rich cycles were interpreted by Melvin as shallow-water, floodplain deposits. Some thin, rippled sandstone beds within this overbank sequence were probably deposited from intermittent sheet floods, and the sporadic desiccation cracks suggest periods of subaerial exposure. More prolonged lacustrine deposition gave rise to the carbonate-rich beds, reflecting increased phytoplankton abundance.

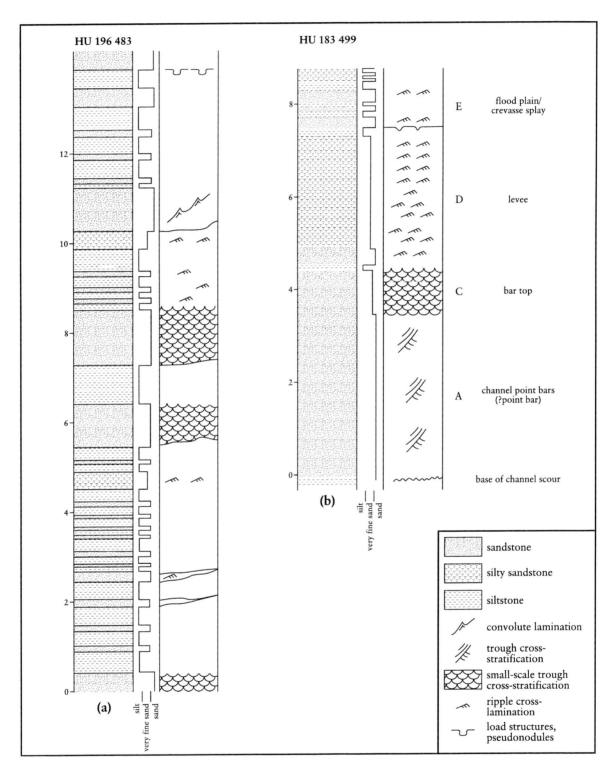

Figure 2.17 Examples of lithofacies associations in the Walls Formation. After Melvin (1985).

An important variation on a fluvial depositional interpretation was proposed by Astin (1982), who recognized a lateral facies transition from north-east to south-west. In his interpretation, the north-east part of the Walls Formation consists of an alluvial fan derived from the metamorphic rocks to the north (equivalent and probably contiguous with the Sandness

Formation). Towards the south-west the alluvial fan merges with more distal, lacustrine deposits and, most controversially, with beach ridges, the latter well represented within the GCR site as parts of the channel-bar and bar-top sandstone assemblages of Melvin (1976, 1985). The correct lithofacies interpretation is important for an accurate assessment of basin geometry and regional tectonics. The current consensus view has moved away from a deep lacustrine, turbidity-flow origin for the Walls Formation and now favours braided alluvial plain to shallow lacustrine and beach environments.

The tectonic deformation of the Walls Formation, involving two separate folding episodes (Mykura and Phemister, 1976), makes it difficult to deduce palaeocurrent flow directions from the sedimentary features. However, both Melvin (1976, 1985) and Astin (1982) deduced a broad flow regime from north to south, with sediment provenance to the north. The suite of detritus, together with its isotopic characteristics (Knudsen, 2000), suggests that the source was a composite granitic and metamorphic terrane with characteristics similar to those of the basement rocks now seen between the Melby and Walls Boundary faults. Thus, Devonian strata now occupying the central of the Shetland structural tracts could have been derived by erosion of the basement rocks from that same structural tract.

Conclusions

The Footabrough to Wick of Watsness GCR site provides a well-exposed section through part of the Walls Formation that is representative of the central of the Devonian structural tracts in Shetland. Sandstone, siltstone and shale combine in varying proportions to give broad alternations, over tens of metres, of sandstone-rich and shale-rich units with strong sedimentary cyclicity. The thick, sandstone-dominated units have a range of sedimentary features that have been interpreted in different ways. Turbidite deposition in a deep lake has been proposed, but the more recent consensus favours a combination of braided fluvial channel, littoral and shallow lacustrine environments. The interpretation of sedimentary depositional environments remains a subject of debate, but is of great importance in assessment of the regional basin geometry and tectonics.

EASTER ROVA HEAD, SHETLAND (HU 475 454)

P. Stone

Introduction

The sea cliffs and rocky foreshore around Easter Rova Head, about 4 km north of Lerwick on the east coast of Shetland Mainland, provide extensive exposures of spectacular, coarse cobble and boulder conglomerates. Together with subordinate sandstone interbeds, the conglomerates comprise the Rova Head Conglomerate Formation, of Mid-Devonian, probably Givetian, age. The formation is considered to comprise mainly sheet-flood deposits, with some mass-flow beds, that accumulated in an alluvial fan at the margin of a lake. It lies close to the base of the local Devonian sequence, unconformably overlying metamorphic basement.

The regional importance of the Easter Rova Head site lies in the unique insight it provides into Mid-Devonian alluvial environments and processes. Understanding of these allows a better interpretation of otherwise little-known aspects of the stratigraphy and tectonic framework of the Orcadian Basin. Overviews of the geology are provided by Mykura (1976) and Allen and Marshall (1981). A detailed sedimentological account of the site is given by Allen (1981a).

Description

At Easter Rova Head, the spectacularly coarse Rova Head Conglomerate Formation is well exposed in low sea-cliffs and across a rocky foreshore. The GCR site covers the island and the contiguous coastal outcrops on the mainland (Figure 2.18). Within the site area, the conglomerates and interbedded sandstones dip moderately to the south-east so that a vertical thickness of about 150 m is represented. The formation lies close to the base of the Devonian succession of eastern Shetland, within the structural tract east of the Nesting Fault (Figure 2.7) and is regarded as probably Givetian (Mid-Devonian) in age.

The clasts of the Rova Head Conglomerate Formation are rounded to well-rounded, generally with a moderate to high sphericity. They range up to about 75 cm in diameter and are poorly sorted. Quartzite is the most abundant clast lithology, but vein quartz and various granitic types are well represented,

The Orcadian Basin

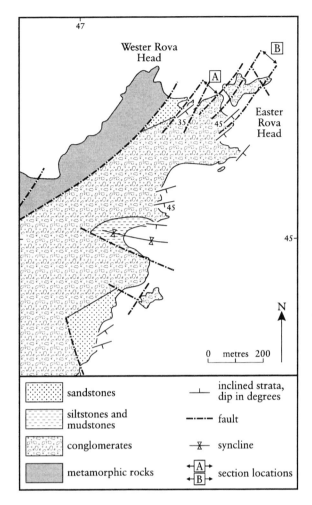

◄Figure 2.18 Outline geology of the Rova Head area. After Allen (1981a).

together with lesser amounts of psammite, felsite, schist and foliated granodiorite. Allen (1981a) recognized three types of conglomerate:

1. Clast- (or framework-) supported conglomerate, inversely graded at the base of the bed, ungraded and only crudely stratified in the centre of the bed, and normally graded at the top of the bed. This is the commonest type in the formation.
2. Matrix-rich or matrix-supported conglomerate, either ungraded or with inverse grading.
3. Clast- (or framework-) supported conglomerate with normal grading.

All three types occur at all levels in the succession, but there appears to be an up-sequence increase in the proportion of clast-supported types. However, many of the conglomerate beds have ambiguous characteristics and are not readily assigned to a specific category.

The lowest conglomerates within the site are seen at its north-west margin, along the shore of the North Bight of Rovahead. There, conglomerate beds range up to about 2 m, separated by thin sandstone beds and lenses up to about 20 cm thick (Figure 2.19). The conglomerates

Figure 2.19 Variably matrix- and clast-supported conglomerates with sandstone interbeds of the Rova Head Conglomerate Formation. (Photo: P. Stone.)

Easter Rova Head

are variably matrix- or clast-supported, with both textures commonly present in the same bed where pockets of pebbles are concentrated in an otherwise matrix-rich background. Some of the beds are normally graded and fine upwards, but others have a well-developed, inversely graded, upward-coarsening trend. There is some sporadic, but seemingly irregular pebble imbrication. Both the matrix and the sandstone interbeds consist of coarse-grained sand, and the interbeds show planar and cross-lamination. There are sporadic thin beds of fine-grained sandstone and red siltstone, the latter containing desiccation cracks (Allen, 1981a). A graphic log of this part of the succession by Allen (1981a) is shown in Figure 2.20A.

Higher in the succession, along strike from the lighthouse and towards the South Bight of Rovahead, there is a tendency for the conglomerates to be clast-supported rather than matrix-rich (Figure 2.21). There is also an overall up-sequence (SE-directed) trend to thinner beds (although still ranging up to about 2 m maximum thickness), a smaller proportion of conglomerate relative to sandstone (although

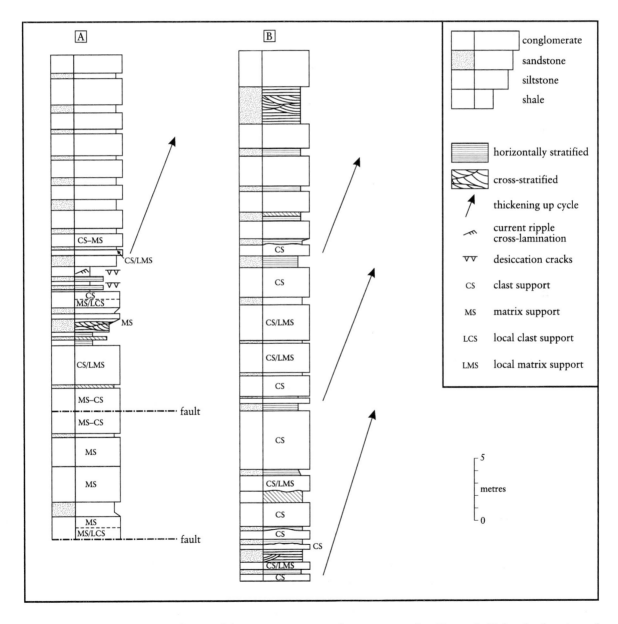

Figure 2.20 Graphic logs of part of the Easter Rova Head succession. See Figure 2.18 for the location of sections A and B. After Allen (1981a).

The Orcadian Basin

Figure 2.21 Clast-supported conglomerates of the Rova Head Conglomerate Formation forming cliffs at Rova Head. (Photo: P. Stone.)

conglomerate remains by far the major lithology) and a lower maximum clast size (Allen, 1981a). Some beds thin north-eastward, along strike. There are apparently no systematic changes, either with stratigraphical level or laterally, in clast compositional range, shape or modal size. Sandstone interbeds in the higher part of the sequence are rather more extensive than those lower down, but there are no compositional or textural changes, and both planar and cross-stratified beds are present throughout. A graphic log of the higher part of the succession within the GCR site, taken from Allen (1981a), is shown in Figure 2.20B.

Interpretation

The Rova Head Conglomerate Formation lies close to the base of the Devonian succession and interdigitates laterally towards the south-west with a breccia derived from, and believed to unconformably overlie, a basement of metamorphic rocks (Mykura, 1976). Palaeocurrent analysis shows the conglomerates and interbedded sandstones to have been derived from a broadly western source. Mykura (1976) recorded regional palaeocurrent flow towards the south-east, whereas Allen (1981a), in a more extensive and detailed study, demonstrated that palaeocurrent flow at Rova Head was broadly towards the north-east. From petrological and Sm–Nd isotope studies, Knudsen (2000) identified relatively local basement sources for the clast assemblage, all to the east of the Walls Boundary Fault and within the same structural tract as the Devonian strata.

Allen (1981a) concluded that the Rova Head conglomerates accumulated as successive sheet-flood deposits in a low-inclination alluvial fan at the margin of a lake. The inverse to normally graded, mostly clast-supported conglomerates were deposited as thick sheets and bars on a surface with little topographic relief. Finer material was rapidly transported during rising flood conditions and deposited as poorly sorted, unstratified gravels. As the flood developed, increased water flow transported and deposited coarser material, producing inverse grading. As the flood waned, normal grading and horizontal stratification were produced in progressively decreasing concentrations of sediment. The matrix-supported conglomerates, many of which are inversely graded, record high-concentration, mass-flow events.

Sand wedges were deposited over and against the coarse gravel bodies as the current flow waned. Thicker interbeds of horizontally or cross-laminated sandstone represent the migration of sand waves over the gravel bodies and are probably unrelated to the sheet-flood events. However, some of the most extensive sandstone units may have been deposited during small sheet-floods restricted to broad, shallow channels. Flood events may have also produced overbank floodplain deposition, represented by the sporadic red siltstone interbeds. Desiccation cracks in these siltstones show that the floodplain was subjected to periodic subaerial conditions in an arid climate (Allen, 1981a).

Conclusions

The Easter Rova Head GCR site provides a well-exposed and instructive section through part of an alluvial-fan sequence developed at the margins of a Mid-Devonian lake. The conglomerate sequence, with subordinate sandstone and siltstone, illustrates a range of depositional processes within an environment dominated by sheet floods. High-concentration, mass-flow conglomerates alternate with those recording rising and waning flood conditions. Between flood events, sand waves migrated across the conglomerate bodies in channelized fluvial deposition, and fine-grained overbank deposition and subaerial conditions were intermittently established. The conglomerates have a local source, and their presence provides an important control on the geometry of the depositional basin and its geotectonic evolution. The site is therefore of great importance in the regional understanding of the Orcadian Basin.

THE CLETTS, EXNABOE, SHETLAND (HU 409 137–HU 406 122)

P. Stone

Introduction

The sea cliffs between Vaakel Craigs and Millburn Geo, on the south-east coast of Shetland Mainland near Exnaboe provide excellent, continuous exposure through varied lacustrine, fluvial and aeolian strata of the Middle Devonian Brindister Flagstone Formation. The Cletts is a large and prominent bedding surface of pebbly sandstone within this section. The sequence lies to the east of the N–S-trending Nesting Fault, and so forms part of the easternmost of the three Devonian structural tracts of Shetland. The lithofacies present were produced during the cyclical deepening and shallowing of a large lake occupying part of the Orcadian Basin. Highstands of the lake are marked by two units of laminated mudstones, one including thin dolostone layers and lenses. The fossil fish fauna from this locality is of great importance, and is described in the GCR fossil fishes volume (Dineley and Metcalf, 1999). During lowstand periods of shallow-water, alluvial-fan gravels and deltaic sands prograded into the lake, and aeolian dunes prograded across the basin during periods of subaerial emergence.

The regional importance of The Cletts (Exnaboe) site lies in the well-preserved evidence for the cyclicity of depositional environments. This allows interpretation and characterization of otherwise little-known parts of the stratigraphy and tectonic framework of the Orcadian Basin. Overviews of the geology are provided by Mykura (1976, 1991) and Allen and Marshall (1981). A detailed sedimentological analysis for part of the site is given by Allen (1981b).

Description

The incised sea-cliffs forming the south-east coast of Shetland Mainland provide extensive exposure through a sequence of fluvial, lacustrine and aeolian Old Red Sandstone strata. The section spanning The Cletts (Figure 2.22), near the village of Exnaboe, lies within these cliffs, about 5 km north of Sumburgh Head. The strata are informally assigned to the Brindister Flagstone Formation and are of Givetian (Mid-Devonian) age. The beds dip consistently between 15° and 30° towards the south-east, resulting in several hundred metres of strata being exposed within the GCR site, although the continuity is locally disrupted by minor faulting, particularly at the northern end of the section. The section lies in the easternmost of three north–south, fault-bounded tracts of Devonian rocks in Shetland, and is characteristic of the strata in the tract.

The oldest strata present within the site crop out at its northern margin in the vicinity of

The Orcadian Basin

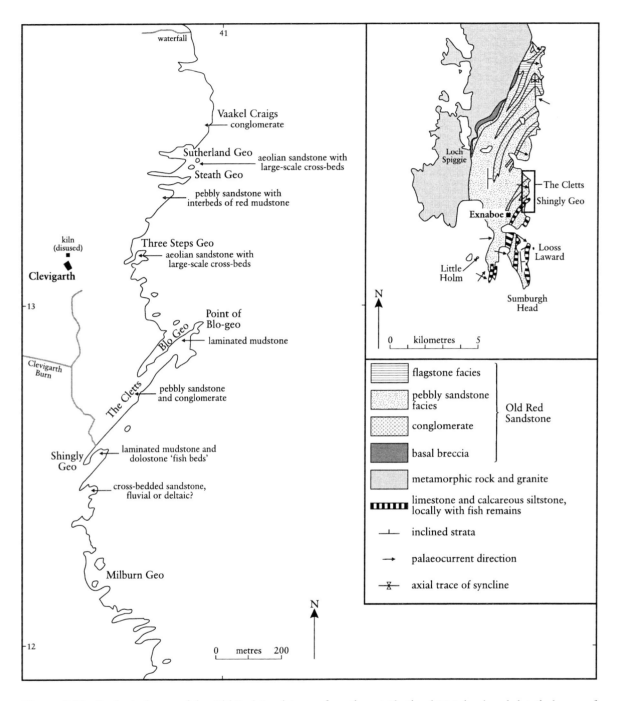

Figure 2.22 Geological map of the Old Red Sandstone of south-east Shetland Mainland and detailed map of the Cletts GCR site. After Mykura (1976).

Vaakel Craigs. There, coarse pebble and cobble conglomerate (Figure 2.23) is interbedded with thin, lenticular bodies of coarse-grained sandstone in a sequence at least 30 m thick. The conglomerate clasts range up to about 30 cm across, but most are 5–10 cm. The conglomerates occur in irregular bodies, either matrix- or clast-supported, in which clast imbrication is only weakly and locally developed. The clasts are predominantly of vein quartz and quartzite (both massive and laminated), but sporadic pebbles of granite (some of which are foliated) and gneiss are also present. Most clasts are subangular to well-rounded, but the degree of sphericity is variable. The interbedded sandstone lenses are commonly cross-bedded.

The Cletts

Figure 2.23 Conglomerates at Vaakel Craigs. (Photo: P. Stone.)

Southwards, and up the succession, the coarse conglomerate is succeeded conformably but quite abruptly by cross-bedded sandstone in thick sets up to about 2 m high. About 20 m of this cross-bedded sandstone form the cliffs around Sutherland and Steath Geo. The sandstone becomes more pebbly upwards, but the clasts are generally isolated and supported within the matrix. The clasts are generally less than 10 cm, with a few up to 25 cm, mostly rounded or well-rounded and have a high degree of sphericity. Most of them are of quartzite, commonly with a pinkish hue, but there are also some rare examples of granitic lithologies. Interbedded with the pebbly sandstone are thinly bedded units, up to about 50 cm thick, of red, laminated mudstone and siltstone.

South from Steath Geo, and up-sequence, the pebbles decrease in abundance and thick, cross-bedded sandstone bodies become dominant. These spectacular units are well exposed around Three Steps Geo (Figure 2.24), where individual sets range up to 2 m in thickness; sets up to 5 m thick (Mykura, 1976) and up to 3 m thick (Mykura, 1991) have been reported from elsewhere in this area. The sandstone is pale yellowish brown, variably quartzose or arkosic, and commonly with rounded grains. At least 50 m of the cross-bedded sandstone form the coastal cliffs between Steath Geo and Blo Geo, where they are abruptly overlain by about 4 m of very thinly bedded, laminated mudstone, siltstone and fine-grained, ripple-laminated sandstone, the last showing grading to siltstone over less than 5 mm. Preferential erosion of this unit created the deep inlet of Blo Geo.

Above the Blo Geo unit, there is an abrupt return to fairly massive pebbly sandstone. The concentration of pebbles increases irregularly upwards and most of them consist of quartz or quartzite; the pebbles range from subangular to well-rounded, sphericity is highly variable, and most are matrix-supported. Crude, large-scale cross-stratification is present locally. This lithology forms the prominent headland at Point of Blo-geo, and The Cletts is an extensive bedding surface running south-west from that point. Shingly Geo lies immediately south of The Cletts and is another example of the preferential erosion of a fine-grained, thinly bedded unit (Figure 2.25), this one abruptly overlying The Cletts pebbly sandstone.

About 4 m of laminated mudstone, siltstone and fine-grained sandstone with much small-

The Orcadian Basin

Figure 2.24 Planar cross-bedded, probably aeolian, sandstones, Three Steps Geo. (Photo: P. Stone.)

Figure 2.25 The Cletts (bedding plane of pebbly sandstone, left), and gully eroded through the fish bed. (Photo: P. Stone.)

scale cross-bedding and extensive rippled bedding planes are exposed at Shingly Geo. Within the top half of the unit are thin interbeds of dolomitic limestone that become more abundant and thicker upwards, to a maximum of about 5 cm. Fossil fish fragments have been recovered from this part of the sequence, the Exnaboe Fish Bed. Although not abundant, they include *Dipterus* sp., the commonest species, and *Stegotrachelus finlayi*, the second commonest. This is the only species of the genus and is found only in Shetland (Dineley, 1999a). The limestone layers are disrupted by soft-sediment structures, with evidence for both extensional and compressional deformation (Figure 2.26). There is also evidence for tectonic deformation, with an incipient cleavage cutting obliquely across the bedding and crenulating the more finely laminated lithologies. The exposure at Shingly Geo was modified by some limited, long-ceased quarrying, the limestone probably being burnt at the old kiln adjacent to the ruins of Clevigarth, about 500 m to the north.

The Shingly Geo unit is abruptly overlain by fine- to medium-grained, grey, micaceous, cross-laminated sandstone in beds generally 40–60 cm thick, with some up to about 1 m. This lithofacies continues to and beyond the southern margin of the GCR site.

Interpretation

The range of lithologies present within The Cletts GCR site represents deposition in the cyclically varying environments at the margin of a large lacustrine basin. The complexity of the facies relationships was stressed by Allen (1981b), who recognized the influence of both wave and current action in water depths of less than 5 m, and a deeper, offshore environment characterized by a lack of evidence for either wave- or current-action. The deep-water environment, during phases of maximum lake extent is represented by the thinly bedded units at Blo Geo, and, particularly, Shingly Geo. In both cases, mudstone/siltstone laminites were probably deposited from suspension of seasonal influx of clastic sediment, with the fine sandstone to siltstone gradations perhaps having a turbiditic origin. The carbonate laminae, lenses and thin beds may have formed during periods

Figure 2.26 Disrupted dolomitic layers in lacustrine laminites. Exnaboe Fish Bed at The Cletts. (Photo: P. Stone.)

of increased phytoplankton abundance, but some diagenetic re-distribution of carbonate seems likely. The primary bedding was further disrupted by soft-sediment deformation, facilitated by the earlier lithification of the dolostone, allowing it to behave in a brittle fashion while the enclosing clastic laminae were still ductile. The fish fossils at Shingly Geo are further evidence that this unit marks the maximum lake extent and probably the deepest water seen in the section. Water depths may have been up to 80 m (Hamilton and Trewin, 1988).

The pebbly sandstones that make up much of the lower part of the succession, and also separate the two thinly bedded lacustrine units, are probably braided stream and alluvial-fan deposits that prograded into the lake during periods of relatively low water-level. The matrix-supported nature of much of the pebbly sandstone suggests that slumping and mass-flow processes were also active. In their lower part, the pebbly sandstones are sporadically interbedded with thin red mudstone layers, representing overbank facies. The cross-bedded, fine-grained, micaceous sandstones at the top of the succession within the GCR site were probably deposited in prograding deltas during a relative fall in the lake water-level. The regional current flow pattern indicates dispersal of sediment in a general south-easterly direction (Mykura, 1991). The preponderance of quartz and quartzite clasts in the pebbly sandstones and conglomerates suggests that those lithologies were dominant in the sediment provenance area, although the high percentage of well-rounded quartzite pebbles may be due to recycling of older conglomerates. The granite and gneiss clasts are more positively ascribed to basement rocks that were probably exposed in a north–south mountain range that formed the western margin of a large lacustrine basin.

The lowest water-levels within the lake allowed aeolian sand dunes to advance across areas that had once been underwater. The large-scale, cross-bedded sandstones around Sutherland, Steath and Three Steps geos are good examples of such dune sandstones (Mykura, 1976, 1991). They contain rounded grains of quartz and feldspar, occur in sets up to 5 m thick, and appear to have been driven by winds from the south and south-west (Allen and Marshall, 1981).

Conclusions

The Cletts, Exnaboe GCR site provides a well-exposed section representative of the Devonian strata in the easternmost of the Shetland structural tracts. It demonstrates the interplay of depositional environments at the margin of a large lake within the Orcadian Basin. Fluctuation in lake water-level produced a vertical succession of stacked cycles, in which lacustrine, locally fish-bearing, fine-grained laminites alternate with braided stream, alluvial-fan and ?deltaic sandstones. Low lake water-levels allowed the encroachment of aeolian dunes. The succession exposed in this section provides important insights into the evolution of an otherwise little-known part of the Orcadian Basin. The importance of this site is further enhanced by the presence of the Exnaboe Fish Bed, which contains *Stegotrachelus finlayi*, the only species of this genus, which is unique to Shetland. The fish bed is younger than others in Shetland, and may be the only late Givetian site this far north.

SOUTH STROMNESS COAST SECTION, ORKNEY (HY 223 102–HY 254 078)

E.A. Pickett

Introduction

The South Stromness Coast Section GCR site lies on the south-west coast of the Orkney Mainland and extends along the coast from the Skerry of Ness, at the mouth of Stromness harbour, to Billia Croo, about 3 km WNW of Stromness. This stretch of coastline is excellently exposed and accessible, and provides the best section in Orkney through the Lower and Upper Stromness Flagstone formations. These formations make up the Middle Devonian Caithness Flagstone Group (approximately equivalent to the Middle Old Red Sandstone) in Orkney. At the eastern end of the site, the Lower Stromness Flagstone Formation rests unconformably, with a basal conglomerate, on Precambrian metamorphic rocks, exposed here in the largest of a few small inliers of basement in Orkney. Information from the overlying sequence has been critical in the recognition and palaeo-environmental interpretation of the rhythmic units or 'cycles' that characterize the Middle

South Stromness Coast Section

Devonian fluvio-lacustrine sequences of the Orcadian Basin (Fannin, 1970; Mykura, 1976). Features of particular sedimentary and palaeoenvironmental significance include impressive desiccation cracks, ripples, early diagenetic chert nodules, beds of algal debris, and large stromatolitic mounds.

The section also contains fault-repeated exposures of the Sandwick Fish Bed, the base of which is the junction between the Lower and Upper Stromness Flagstone formations (Astin, 1990). This horizon is important in regional correlation and justly famous for its fossil fishes (see also **Yesnaby and Gaulton Coast Section** GCR site report, this chapter).

Description

At the Skerry of Ness, which lies at the eastern end of the site (Figure 2.27), the Lower Stromness Flagstone Formation overlies a granite-gneiss basement surface that slopes gently to the south. This is one of a few small inliers of pre-Devonian basement which lie along an axis stretching from the island of Graemsay (directly south of Stromness) through Stromness to Yesnaby on the western coast of the Mainland (see also **Yesnaby and Gaulton Coast Section** GCR site report, this chapter). The basement is composed of pink, foliated feldspar-phyric granite with enclaves of foliated metamorphic rocks that resemble rocks of the Lewisian inliers within the outcrop of the Moine Supergroup of the Northern Highlands (Wilson *et al.*, 1935).

Unconformably overlying the basement are 5–20 m of coarse, poorly sorted, clast-supported breccia and conglomerate. These form the basal part of the Lower Stromness Flagstone Formation and consist of subrounded to angular clasts of gneiss, granite, schist and quartzite set in an arkosic, sandy matrix. The clasts are generally cobble-sized (up to 20 cm by 10 cm), although some are up to 50 cm. Fannin (1970) noted that some of the clasts show a poorly developed imbrication, dipping westward off the crystalline mass. Some of the clasts have thin algal coatings of interlaminated dolostone and siltstone. Above the conglomerates, there are further examples of algal coatings, as well as stromatolitic mounds, mats and ridges. The basal facies passes up into finer-grained conglomerates and arkosic, pebbly sandstones. The sequence fines markedly upwards, with pebble content decreasing and with very few pebbles in the beds higher than about 25 m above the base.

The BGS Warebeth Borehole (Mykura, 1976; Clarke, 1990) lies in the site, providing a further point of interest. The borehole proved 61 m of red-purple siltstones, sandstones and breccias (the 'Warebeth Red Bed Formation'), which are absent from the succession exposed on the coast. They lie above the metamorphic basement and below the basal conglomerate of the

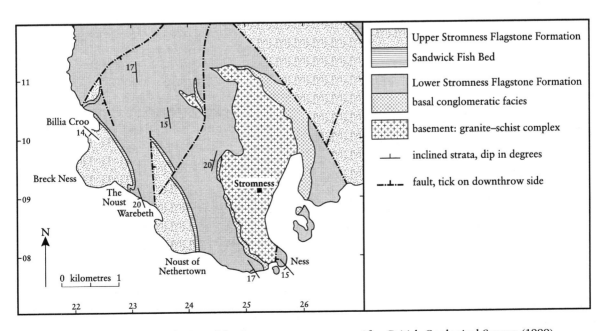

Figure 2.27 Geological map of the Stromness coast area. After British Geological Survey (1999).

Lower Stromness Flagstone Formation and are tentatively assigned to the Lower Old Red Sandstone (Lower Devonian).

The Lower Stromness Flagstone Formation is about 250 m thick and the Upper Stromness Flagstone Formation at least 500 m thick. Both formations, with the exception of the basal sequence of the Lower Stromness Flagstone Formation, comprise a series of cyclic units, each unit or 'cycle' ranging from 2 m to 15 m thick. Fannin (1970) studied the sequence in detail and subdivided the units into the Lower Stromness Flags (24 cycles), the 'Sandwick Fish Bed Cycle' (1 cycle), the 'Hoy Cycles' (4 cycles) and the Upper Stromness Flags (20+ cycles). The classification was subsequently simplified so that the informally defined 'Sandwick Fish Bed Cycle' and 'Hoy Cycles' of Fannin (1970) now form part of the Upper Stromness Flagstone Formation (Astin, 1990). The facies of a typical cycle were described by Mykura (1976) and are summarized below.

The base of each cycle is taken at the appearance of dark grey to black, calcareous, silty mudstone interlaminated with siltstone. This facies generally contains graded laminae that may have either a high bitumen and pyrite content or a high carbonate (mostly ferroan dolomite) content. Fossil fish remains may also be present. The dark mudstones and siltstones pass up into thinly interbedded bituminous siltstones and fine-grained sandstones, together with some discrete beds of massive siltstone and fine-grained sandstone, which fill small eroded channels. This part of the sequence contains evidence of soft-sediment slumping and is also characterized by numerous small lenticular cracks, which were infilled by sand or silt and then compacted and contorted. Algal stromatolite sheets and mounds are common. The upper parts of many cycles are characterized by ripple cross-laminated sandstones and massive siltstones with well-developed sand-filled desiccation polygons. Locally, stromatolite sheets and mounds cover the cracked surfaces and stromatolitic debris occurs in the cracks. The stromatolites are overlain by further ripple-laminated sandstones and siltstones with desiccation cracks. The top of the cycle is generally characterized by thinly laminated siltstones and mudstones with desiccation cracks that are overlain abruptly by black laminated mudstones that form the base of the next cycle.

Much of the sequence is rich in ferroan dolomite, causing the beds to weather to distinctive yellow, orange and buff colours. Fresh surfaces are commonly dark grey owing to their organic content. One of the most impressive features of the section is the presence of extensive, gently dipping bedding planes (Figure 2.28) covered with wave- and current-ripple marks or infilled desiccation cracks.

The algal stromatolites are an interesting and important feature of this site. About 800 m west of the Point of Ness are large (metre-wide) stromatolite-covered mounds (HY 2510 0765), described in detail by Fannin (1969). The mounds were built around coarse sand cores on an irregular erosion surface. The sandy cores are coated with linked hemispheroidal stromatolites and flanked by breccias of stromatolitic debris. The stromatolite-bearing sequence is locally silicified, with chert layers occurring at the top of some mounds. The succession above and below the mounds is rich in stromatolitic debris and smaller stromatolite developments are common in the beds above the mounds. Most of the stromatolitic rocks are mineralized by galena and sphalerite.

Small shows of mineralization, particularly galena-sphalerite-barite, are common elsewhere throughout the section. Barite veins also cut the metamorphic basement. Lead was mined at Warebeth Beach around 1775 and an infilled adit is still visible. Below the adit on the beach, a brecciated zone of normal faulting that contains veins of galena-calcite-barite is one of a series of 2 m-wide, parallel, brecciated zones. Blebs of galena and sphalerite also occur within the rock surrounding the veins. The sequence also exhibits pseudomorphs after gypsum, some of which appear to have modified sand-filled mud-cracks. Bedding planes on the shore by Stennigor exhibit pink barite pseudomorphs after gypsum and one west of The Noust contains calcite and quartz pseudomorphs after gypsum.

Towards the top of the Lower Stromness Flagstone Formation, about 50 m east of the east end of the Noust of Nethertown, an exposure cut by a minor fault contains numerous chert nodules (Parnell *et al.*, 1990). The relationship between the chert nodules and the lamination that passes through them shows that the nodules are replacive and developed before compaction, and that during compaction they were rotated and deformed. Rare sub-

South Stromness Coast Section

Figure 2.28 Upper Stromness Flagstone Formation at Warebeth. View south-west towards Hoy. (Photo: E.A. Pickett.)

millimetre sized sphalerite crystals in the chert are thought to be syndiagenetic (Parnell, 1987).

The Sandwick Fish Bed, the base of which marks the junction between the Lower and Upper Stromness Flagstone formations, crops out at three localities in the site owing to fault repetition – two localities at the Noust of Nethertown, and one at The Noust (near Breck Ness). It is a white-weathering carbonate laminite, above which are black lenses of chert and fossil fish fragments. At the Noust of Nethertown the section is deformed by N–S-trending box folds. The fish bed is rich in fossil fish specimens, Cruaday Quarry GCR site yielding 14 or 15 specimens, including common *Coccosteus cuspidatus*, *Mesacanthus* sp. and *Osteolepis macrolepidotus* (Dineley, 1999a).

The Caithness Flagstone Group contains many features of use in local correlation. The fish-bearing laminites, such as the Sandwick Fish Bed, are distinctive and laterally persistent. Other marker horizons include the large chert nodules (described above) in two thinly laminated iron-rich beds of silty dolomite 14 m and 59 m below the Sandwick Fish Bed. A further marker horizon, just above the base of the Sandwick Fish Bed, is a 25 cm-thick calcite mudstone that weathers to a distinctive blue-grey colour.

Interpretation

This site has the best section in Orkney of the rhythmic units or 'cycles' that form the Caithness Flagstone Group. These cycles are interpreted as reflecting fluctuations in water-level and sediment input in a major lacustrine rift basin, the Orcadian Basin, that extended across Orkney and Caithness during Mid-Devonian times (e.g. Fannin, 1970; Mykura, 1976; Astin, 1990). The cycles reflect fluctuations between wetter periods when there was a permanent lake in the Orcadian Basin, and drier periods when there was net evaporation and the lake was ephemeral (Astin, 1990).

The calcareous, fish-bearing, thinly laminated siltstone and mudstone at the base of each cycle represent a period when the water, although still shallow, was at its deepest. The sediments were laid down in relatively quiet and sometimes stagnant waters, undisturbed by wave action. The lake waters may have been thermally stratified at times and some of the graded laminae may have been deposited by turbidity currents (Mykura, 1976). The overlying sequence of siltstones and fine-grained sandstones contains algal stromatolites, suggesting very shallow water. The mineralization of the stromatolites by galena and sphalerite may also have environmental implications, Muir and Ridgway (1975) suggesting that the degradation of algal material produced organo-metallic complexes that precipitated sulphides in the presence of sulphide-reducing bacteria. Numerous small cracks, originally interpreted as subaqueous synaeresis cracks (e.g. Fannin, 1970; Donovan and Foster, 1972) have been re-interpreted as pseudomorphs of gypsum (Astin and Rogers, 1991; Rogers and Astin, 1991), but this remains controversial (Astin and Rogers, 1992, 1993; Trewin, 1992; Barclay et al., 1993; Trewin and Thirlwall, 2002).

Progressive shallowing of the lake produced mudflats that periodically dried to produce polygonal arrays of desiccation cracks. The presence of barite, quartz and calcite pseudomorphs after evaporite minerals such as gypsum suggest a playa-lake setting. The lake muds were covered by fine-grained, rippled sands and silts of alluvial fans that prograded over the shallow or dry lake floor. These fluvial and deltaic sediments were succeeded by thin-bedded silts and sands with mud-filled desiccation cracks, abruptly overlain by deeper-water fish-bed facies at the base of the next cycle.

The cyclicity of the sequences, together with the presence of wave-reworked surfaces, pseudomorphs after gypsum and mudcracks, reflects the frequent oscillation between shallow temporary lake and desiccated lake basin, the former estimated to have lasted in the range of 100–200 years to 5000–10 000 years (Astin, 1990). Palaeocurrent analysis of the sequence indicates that there was a dominant current flow towards the south. Fannin (1970) suggested that the Orcadian lake had a roughly E–W-trending shoreline and was fed by rivers entering from the north. Astin (1990) measured palaeocurrent directions from current ripples in sheet-flood sandstones in the Upper Stromness Flagstone Formation on the northern coast of the Mainland (north of this GCR site) and found the predominant sheet-flood drainage direction to be towards the SSW.

In a detailed study of the cyclicity in the Upper Stromness Flagstone Formation across Orkney, Astin (1990) recognized about 45 'first order' lake cycles between the Sandwich Fish Bed and the Lower Eday Sandstone Formation (which lies above the Upper Stromness Flagstone Formation). He interpreted these cycles as Milankovitch-type cycles controlled predominantly by orbital precession, causing long-term fluctuation in rainfall over about 1 million years. He also suggested that there is evidence for eccentric orbital cycles of longer timescale.

Conclusions

This site is of national importance as it is the best section in Orkney (and arguably the best anywhere) through the Caithness Flagstone Group, and its underlying metamorphic basement. The site provides a beautifully exposed section through a series of sedimentary 'cycles' that reflect the periodic oscillation between a shallow temporary lake and a desiccated lake basin. These cycles, together with the wide range of lithologies, sedimentary structures and biogenic structures that they contain, have been critical for the interpretation of sedimentary environments of the Orcadian Basin during Mid-Devonian times. The site also contains exposures of the well-known Sandwick Fish Bed, which separates the Lower and Upper Stromness Flagstone formations and is an important marker horizon in regional correlation.

TARACLIFF BAY TO NEWARK BAY, ORKNEY (HY 553 035–HY 568 043)

E.A. Pickett

Introduction

This site extends from Taracliff Bay to Newark Bay on the south-east coast of the Deerness Peninsula, in the south-eastern part of the Mainland. It exposes a long section through Middle Devonian (Middle Old Red Sandstone) strata comprising the uppermost part of the Caithness Flagstone Group and the overlying Eday Group (including the Lower Eday Sandstone, Eday Flagstone and Eday Marl formations). The section of the Eday Flagstone Formation is the thickest and best exposed in Orkney, and the site is important in containing evidence of contemporaneous volcanism. It is also particularly important for the interpretation of the depositional environments of the Eday Group and the palaeogeography of Orkney during late Mid-Devonian times, providing a comparison with the more proximal strata of the group in the Fersness Bay section on Eday (see **South Fersness Bay** GCR site report, this chapter).

Taracliff Bay to Newark Bay

Description

At the western end of the section, interbedded grey-green, calcareous siltstones and yellowish buff sandstones of the Caithness Flagstone Group are exposed in the vertical cliffs and on a wide shore platform (Figure 2.29). The sandstones are fine- to medium-grained, thin- to thick-bedded, planar and cross-laminated, and locally cariously weathered. The finer-grained, silty interbeds are rich in grey mudstone rip-up clasts and also display carious weathering. The Caithness Flagstone Group is overlain, with a transitional boundary, by the arenaceous Lower Eday Sandstone Formation. One of the best sections of the Lower Eday Sandstone Formation in Orkney lies west of the sea stack known as 'Muckle Castle', where over 175 m of quartzitic, well-bedded, cross-bedded and channelized sandstones with rare, scattered pebbles can be observed. A distinctive feature of these sandstones on Deerness and elsewhere is the presence of convolute or slumped bedding. The boundary between the Lower Eday Sandstone Formation and the Eday Flagstone Formation occurs just west of Muckle Castle. Muckle Castle itself (HY 562 032) is formed almost entirely of an intrusive plug of olivine dolerite that is fault-bounded on its western side (Kellock, 1969).

The thickest development and best exposures of the Eday Flagstone Formation in Orkney are found in and around the Deerness Peninsula (included in this site), where the formation is about 150 m thick. It thins northwards to 100 m in Shapinsay, 50 m in Stronsay and 10 m in Sanday. The formation comprises up to 12 cyclic sequences, which range in thickness from 2 m to several tens of metres. Most of these cycles comprise an upward-fining phase of buff, yellow, or more rarely, red sandstone and silty sandstone, and a phase of grey, black and locally purple, flaggy siltstones and mudstones (Mykura, 1976). Fish remains are common in the muddy facies at the base of some cycles, and two beds of flaggy siltstone near Muckle Camay have yielded remains of *Tristichopterus alatus* and *Pentlandia macroptera*. Desiccation cracks and small lenticular cracks, possibly pseudomorphs after gypsum, are common in the siltstones and fine-grained sandstones. Thin sandstones within the flaggy siltstones commonly have slump structures and load-casts at their bases.

Less than 200 m east of Muckle Castle, the Eday Flagstone Formation contains about 5 m of interbedded greenish tuffs and tuffaceous sandstones with sandstone and lava pebbles (HY 563 033). These form part of the Deerness Volcanic Member, which also includes thin basalt lava flows and basaltic breccias elsewhere on the Deerness Peninsula. Analcime, natrolite (a zeolite) and alkali feldspar occur in these rocks, but Thirlwall (1979) showed that the analcime and natrolite are secondary and that the basalts are calc-alkaline.

The Eday Flagstone Formation has a faulted contact with the Eday Marl Formation near Peerie Castle, a small sea stack on the west side of Newark Bay. The Eday Marl Formation forms cliffs on the western side of Newark Bay and

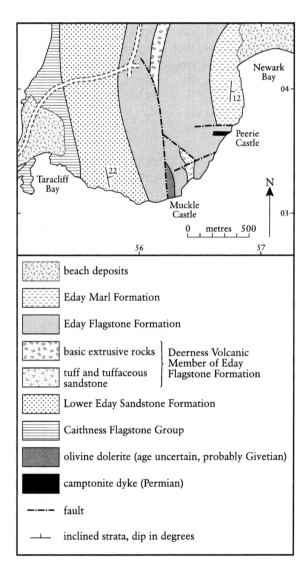

Figure 2.29 Geological map of the Taracliff Bay to Newark Bay area. Based on Kellock (1969) and British Geological Survey (1999).

comprises a series of fining-upward cycles, each made up of a sandstone unit 50 cm to 2 m thick overlain by reddish calcareous mudstones and siltstones with thin beds of convoluted sandstone (Figure 2.30). Grey-green, calcareous mudstones and siltstones are also common in the formation, some carious weathering in the mudstones testifying to the presence of carbonate. The sandstones at the base of the cycles are sharp-based, resting on erosion surfaces, and are commonly medium-grained, grading normally from coarse-grained bases. Other features include rip-up clasts of red siltstone, cross-bedding and desiccation cracks. These fining-upward sequences appear to grade upwards into more massive sandstones in the sequence exposed at Newark Bay.

Interpretation

The oldest part of the section is the Caithness Flagstone Group at the western end of the site. Although divided into the Lower and Upper Stromness Flagstone formations in the west Mainland, the group is undivided in east Mainland on the current published geological map (British Geological Survey, 1999). The Caithness Flagstone Group comprises a series of rhythmic units that record successive repeated changes in a lacustrine environment, from deep-water lake to ephemeral shallow lake with accompanying lake-beach and prograding alluvial-fan deposits. This cyclicity may have been driven by climate change (Astin, 1990).

The overlying Eday Group records the progradation of alluvial fans across the lake basin to form a regional braidplain, perhaps as a result of an episode of active extensional faulting along the basin margins (Astin, 1985). Braided river, aeolian-dune and lake-beach deposits are recognized in sequences of the Lower Eday Sandstone Formation across the east Mainland and the islands of South Ronaldsay, Shapinsay, Stronsay, Sanday and Eday (Astin, 1985).

On the basis of sedimentary structures and bed geometries, Astin (1985) interpreted the Lower Eday Sandstone Formation at Taracliff Bay as a sequence of braided river deposits. Palaeocurrent analysis of cross-bedding in the sandstones indicates that they were deposited from predominantly NE-flowing currents. Similar flow directions have been recorded in the Lower Eday Sandstone Formation on South Ronaldsay and in the John O'Groats Sandstone Group in Caithness (Astin, 1985). By contrast, palaeocurrents in the Lower Eday Sandstone Formation farther north, as exposed on Eday, Sanday, Stronsay and Shapinsay, record south-eastward flow directions. These patterns were interpreted by Astin (1985) as representing a SE-flowing river system in the north and a NE-flowing one in the south. The two systems overlap, and SE-directed currents are recorded at Taracliff Bay, in this GCR site, in an area where NE-flowing currents predominate (Astin, 1985). Aeolian sandstones have not been recognized in the Lower Eday Sandstone Formation at Taracliff Bay, but the foresets of aeolian dunes in the formation in other areas (including Eday, Sanday, Stronsay, Shapinsay and northern Deerness) show that wind directions varied from south-west to NNE, with an overall NNW dominance.

The Eday Flagstone Formation marks the re-establishment of cyclic, lacustrine sedimentation. However, in contrast to the cycles of the Caithness Flagstone Group, many of the cycles contain a buff or yellow, locally pebbly,

Figure 2.30 Load structures and convolute bedding in the Eday Flagstone Formation near Muckle Castle, south-west of Newark Bay. (Photo: BGS No. D1522, reproduced with the permission of the Director, British Geological Survey, © NERC.)

channelized sandstone phase. The formation exhibits great lateral changes in thickness and lithology, thinning northward and south-westward away from Deerness. The thinning is due to changes in facies within the individual cycles, with the sandstone phases of some cycles becoming thicker and coarser-grained away from Deerness, whereas the lacustrine phases become thinner and eventually wedge out. Contemporaneous, sporadic volcanic activity is represented by the olivine dolerite intrusion and tuffaceous sandstones at the site, which are typical of the volcanic rocks of Mid-Devonian age found in Orkney.

The Eday Marl Formation, at the eastern end of the section, comprises fluvial, fining-upward cycles, each interpreted by Mykura (1976) as a channel sandstone unit overlain by fine-grained overbank deposits. However, the range of sedimentary structures in the sandstones led Marshall *et al.* (1996) to interpret them as the products of fluvial channel and sheet-flood deposition on a muddy sabkha plain. The presence of pseudomorphs after halite and a marine microfauna at a locality in the **Bay of Berstane** (see GCR site report, this chapter), about 12 km north-west of this site, suggests that a marine incursion occurred during deposition of the formation (Marshall *et al.*, 1996).

The section, as a whole, records a series of transitions between lake basin and alluvial-plain environments, with fluvial sedimentation on a sabkha plain finally becoming predominant during deposition of the Eday Marl Formation.

Conclusions

This site displays a well-exposed section through the topmost Caithness Flagstone Group and a large part of the Eday Group. The Eday Group is characterized by great lateral thickness and lithological variations across its outcrop in Orkney. This succession is particularly important for comparison with other areas, such as Eday, and for regional palaeoenvironmental interpretations. The strata record the alternation between lake-basin and river braid-plain environments, with a trend towards open sabkha environments. An intrusive igneous plug and related tuffaceous sandstones within the Eday Flagstone Formation are also important features, indicating contemporaneous volcanism in late Mid-Devonian times.

GREENAN NEV COAST, EDAY, ORKNEY (HY 549 367)

E.A. Pickett

Introduction

This site extends for about 700 m around the point of Greenan Nev on the west coast of Eday and provides one of the best sections through the Eday Marl Formation in Orkney. It is the only accessible section of the formation in Eday, exposing the entire thickness, and it is the type section of the formation. The site lies on the eastern limb of the Eday Syncline, the axial trace of which passes just offshore from Greenan Nev. The site is especially important for palaeo-environmental interpretations of late Mid- to early Late Devonian times as it preserves well-exposed floodplain sandstones, calcareous mudstones and siltstones with fossil carbonate soil (calcrete) horizons, a facies association not observed elsewhere in the Orcadian Basin. The presence of abundant trace fossils including *Beaconites* and *Cornulatichnus* also adds to the importance of the site by providing further information on the depositional environments, and the habitats and types of animals that lived in the Orcadian Basin during Mid- to Late Devonian times.

Description

At Greenan Nev (Figure 2.31), the Eday Marl Formation comprises several fining-upward sequences consisting of buff, clean, fine- to medium-grained sandstone beds up to 2 m thick, interbedded with up to 15 m of bright red, calcareous mudstones and sandy siltstones (Figure 2.32). The sandstone beds are finely laminated and cross-bedded, with foresets picked out by red coloration. They commonly contain reddish wisps of calcareous mudstone, fine mudstone partings and red mudstone rip-up clasts. Some sandstone beds have undulose bases with erosion of, and slight loading into, the underlying calcareous mudstones. Sandstones containing convolute bedding and trough and tabular cross-bedding occur in an outcrop of Eday Marl Formation at the **Bay of Berstane** on the Mainland (Marshall *et al.*, 1996).

The red mudstones and siltstones are micaceous and contain ripple marks, desiccation cracks, small scours, hard calcareous layers and

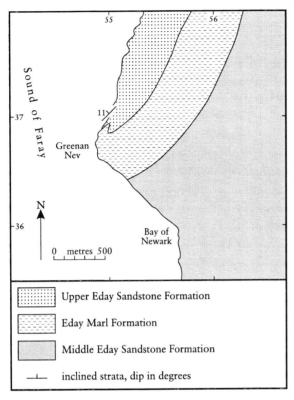

Figure 2.31 Geological sketch map of the Greenan Nev area. After British Geological Survey (1999).

abundant pale calcareous (calcrete) concretions. Bleached white reduction zones are especially common along joints and cracks. The mudstones and siltstones were bioturbated by burrowing organisms, bioturbation being most abundant in the more massive mudstones, as is seen particularly well on the gently dipping bedding planes at the base of the sea cliffs. Carroll (1991) recorded a diverse ichnofauna of 11 species within the Eday Marl Formation and the overlying Upper Eday Sandstone Formation. Carroll and Trewin (1995) described a newly recognized trace fossil *Cornulatichnus edayensis* in the Eday Marl Formation on Eday and the east Mainland, examples of which were observed at this site. The trace consists of a sub-vertical, downward-tapering conical burrow up to 40 cm long with a wide sub-circular cross-section up to 9 cm wide. Carroll and Trewin (1995) also recorded large back-filled burrows identified as *Beaconites* along with *C. edayensis* in the Eday Marl Formation.

The ratio of sandstone to mudstone increases northwards and up-sequence and there is a gradual passage into the overlying Upper Eday Sandstone Formation approximately at the northern boundary of the site.

Figure 2.32 Interbedded red marl and sandstone of the Eday Marl Formation at Greenan Nev. View towards the north-east. (Photo: E.A. Pickett.)

Interpretation

The Eday Marl Formation exposed at this site contains fining-upward cycles of sandstone overlain by calcareous mudstone and siltstone. The fining-upward cycles and the structures within the sandstones, including flat lamination and cross-lamination, as well as erosion surfaces at their bases, suggest that they represent fluvial-channel and sheet-flood (possibly channel overspill and crevasse-splay) deposition (Ridgway, 1974; Carroll, 1991; Marshall *et al.*, 1996). The cross-bedding suggests south-directed currents, similar to those determined by Astin (1985) for the Lower Eday Sandstone Formation.

The calcareous mudstones and siltstones were interpreted as floodplain overbank deposits by Mykura (1976) and as floodplain lake deposits by Ridgway (1974) and Carroll (1991). The trace fossils within the Eday Marl Formation were interpreted by Carroll and Trewin (1995) as the subaqueous shelter burrows of eel-like fish. Periodic and prolonged drying out of the floodplain is indicated by desiccation cracks and calcrete horizons. Sparite-filled vugs in the Eday Marl Formation on the Mainland have been interpreted as the moulds of evaporite nodules that formed in a sabkha-type environment (Marshall *et al.*, 1996). The marine microfauna, and pseudomorphs after halite, described by Marshall *et al.* (1996) in the Eday Marl Formation at the **Bay of Berstane** on the Mainland (see GCR site report, this chapter) suggest periodic marine inundation of the sabkhas and floodplains in mid- to late Givetian times.

Conclusions

This site is important as the type section of the Eday Marl Formation. It contains a well-exposed range of the rock types and sedimentary and biogenic structures that characterize this formation. The rocks are interpreted as floodplain sediments with associated river channel deposits, evaporitic sabkha deposits, pedogenic carbonate (calcrete) horizons and evidence of burrowing by eel-like fish. This represents an environment which is not found elsewhere in the rock record of the Orcadian Basin and is therefore of great significance in palaeoenvironmental reconstruction and the overall interpretation of the Mid- to Late Devonian evolution of the Orcadian Basin.

SOUTH FERSNESS BAY, EDAY, ORKNEY (HY 531 346–HY 543 332)

E.A. Pickett

Introduction

The South Fersness Bay GCR site extends for about 2 km along the south-west side of Fersness Bay on the west coast of Eday. It lies on the western limb of the Eday Syncline and provides a complete section through the Lower Eday Sandstone, Eday Flagstone and most of the Middle Eday Sandstone formations of the Middle Devonian Eday Group. The oldest part of the sequence is represented in the western part of the site by the Rousay Flagstone Member (forming the upper part of the Upper Stromness Flagstone Formation). The site is important because it contains an uninterrupted, well-exposed and accessible section through much of the lower part of the Eday Group. The rocks are not only representative of the stratigraphy of this part of the Orcadian Basin, but also contain a range of sedimentary structures that suggests the interaction of several depositional environments, including lake-beach, lake, braided river and aeolian dune-field. The site is therefore very important in palaeoenvironmental interpretations and for comparison with sites elsewhere in the Orcadian Basin.

Description

The rocks at the site young from west to east and form the western limb of the Eday Syncline, the axis of which runs approximately north–south through the centre of Fersness Bay. At the western end of the site, Fers Ness is composed of about 170 m of the Rousay Flagstone Member (formerly the Rousay Flags) at the top of the Upper Stromness Flagstone Formation (Figure 2.33). These comprise grey, flaggy rhythmic sequences of laminated, fish-bearing mudstone, siltstone and fine-grained sandstone. A distinctive pebbly sandstone at the top of the Rousay Flagstone Member (the Sacquoy Sandstone Member) is about 4 m thick and contains pebbles of quartzite, psammite and dolomitic limestone, with lesser amounts of quartz, granite and chert (Astin, 1990).

The Sacquoy Sandstone Member is overlain by the Lower Eday Sandstone Formation, which

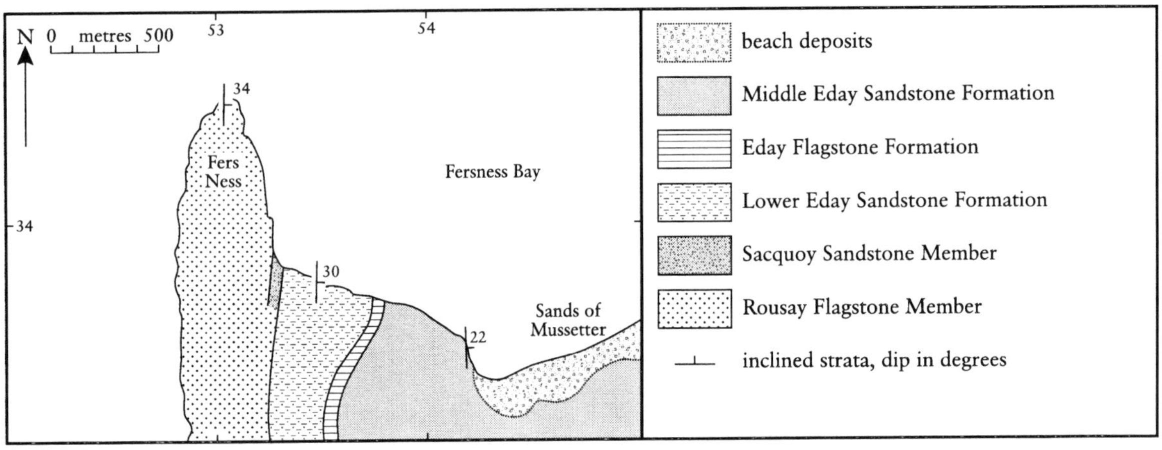

Figure 2.33 Geological sketch map of the south-west side of Fersness Bay. Based on Mykura (1976) and British Geological Survey (1999).

is about 200 m thick and consists of two main facies types (Mykura, 1976; Astin, 1985). The lower facies comprises reddish purple, medium- to coarse-grained, trough cross-bedded sandstone. Pebbly lenses and conglomeratic beds lie about 30 m above the base. The pebbles in the conglomerate are up to 7 cm across and include pink granite and pegmatite, granitic gneiss, quartzite, chert and vein quartz, with lesser amounts of sandstone (Mykura, 1976). The upper facies is a predominantly yellow, medium-grained, planar and trough cross-bedded sandstone with fewer pebbles.

The overlying Eday Flagstone Formation is relatively thin (less than 20 m) at this locality and consists of two cycles with fish beds. The cycles comprise a basal finely laminated fish-bearing facies, and overlying flaggy siltstone and mudstone with subordinate sandstone which pass up into buff, yellow or red sandstone and sandy siltstone.

A higher fish-bearing, calcareous siltstone (HY 530 335) lies about 110 m above the base of the overlying Middle Eday Sandstone Formation. The Middle Eday Sandstone Formation is an orange-yellow, pebbly and locally conglomeratic, trough cross-bedded sandstone containing scour structures, pebbly lags, mudstone rip-up clasts, convolute bedding and ripple-drift lamination (Figure 2.34). Some parts of the section contain abundant pebbles, up to 6 cm across, of quartz, porphyritic and spherulitic rhyolite, scoriaceous basic lava and granite/gneiss. The rocks locally display carious 'honeycomb' weathering and diagenetic red and yellow mottling.

Interpretation

The rhythmic units or 'cycles' that form the Rousay Flagstone Member of the Upper Stromness Flagstone Formation are interpreted as reflecting fluctuations in water-level and sediment input into a large, shallow lake that extended across Orkney and Caithness in Mid-Devonian times (e.g. Fannin, 1970; Mykura, 1976; Astin, 1990). The development of the Orcadian Basin was probably tectonically controlled, with the rhythmic sedimentation being caused by the interplay of tectonic and climatic changes. The dark, calcareous, fish-bearing siltstone/mudstone laminites at the base of each cycle represent deposition from suspension when the lake was at its deepest. Progressive shallowing culminated in mudflats that periodically dried up. Alluvial fans prograded across the shallow or dried-up lake floor producing fine-grained, rippled sandstones and siltstones. These fluvial sediments were deposited as sheet-flood and channel sands and show sediment transport to the south and south-west (Astin, 1990). The base of the following cycle is marked by an abrupt reversion to the deeper-water fish-bearing facies.

The Rousay Flagstone Member passes up into the Sacquoy Sandstone Member, which has distinctly different palaeocurrent directions, having been deposited from eastward-flowing rivers. This marks the first incursion of alluvial-fan sediments prograding from the north-westerly basin margin, probably sourced from metamorphic rocks or older Devonian conglomerates lying to the west and exposed as

Yesnaby and Gaulton Coast Section

Interpretation

The sequence at Yesnaby is interpreted as recording a changing sequence of fluvial, lacustrine and aeolian environments which characterized the Orcadian Basin in Early to Mid-Devonian times. The aeolian facies is very important in being one of only two occurrences recognized in the Lower Old Red Sandstone of Britain and Ireland. Trewin and Thirlwall (2002) suggest that a drier climate prevailed during its deposition. The basal breccias of the Harra Ebb Sandstone Formation may represent talus deposits which accumulated on the slopes of the basement palaeotopography. Fannin (1969) interpreted the coated basement clasts as algal oncolites, although there seems to be a gradation from these structures to the fracture-fillings in the basement and the pebble coatings may therefore be travertine drapes. The overlying sandstones, siltstones, mudstones and dolomicrites of the Harra Ebb Sandstone Formation are interpreted as representing alluvial-fan and sandflat deposits that prograded over a playa-lake system and its associated mudflats (Fannin, 1970; Clarke, 1990). Deposition was predominantly by sheet-flood, with some channelized flow, palaeocurrent directions suggesting south-westerly progradation of the alluvial-fan system (Fannin, 1970). The laminated carbonates probably had a biogenic origin as algal or microbial mats in the playa lake. The vertical breccia bodies have been interpreted as breccia pipes (cryptovents) produced by gas fluxion associated with Permian igneous activity (Mykura, 1976).

Clarke (1990) interpreted the basal part of the Qui Ayre Sandstone Formation as the deposits of a sandy braided river system that flowed south-westwards over the playa mudflats of the Harra Ebb Sandstone Formation. This was followed by a period of emergence and aeolian dune-field migration, when the predominant wind direction was towards the east or ESE (Fannin, 1970). Clarke (1990) interpreted the overlying facies as the result of renewed fluvial deposition by ephemeral streams and overbank sheet-floods, followed by development of sandy, low-sinuosity braided rivers flowing to the south-west and south-east. A different interpretation was presented by Fannin (1970) who tentatively suggested that the facies overlying the aeolian dunes represent a northward-encroaching, shallow, clastic shoreline that inundated and reworked the upper part of the dune deposits. He based this interpretation on analogy with modern marine shoreline environments, but pointed out that there is no evidence to show whether the advancing water body was fresh or marine.

Fannin (1970) distinguished three sources for the pebbles lying on the unconformity surface at the top of the Qui Ayre Sandstone Formation – the adjacent crystalline basement, the underlying Yesnaby Sandstone Group and an exotic source that lay outside the area and supplied a range of metamorphic and metasedimentary clasts.

The Caithness Flagstone Group, comprising the Lower and Upper Stromness Flagstone formations, marks a change to rhythmic, coarsening-upward lacustrine and alluvial cycles. Each cycle records a successive change in environment from deep-water lake to ephemeral shallow lake, lake-beach and alluvial fan. This cycle was repeated many times and may have been driven by a combination of cyclic climate change and tectonic activity (see **South Stromness Coast Section** GCR site report, this chapter, for a more detailed interpretation of the 'cycles' within the Caithness Flagstone Group).

Conclusions

This site is important in exposing a thick succession of Old Red Sandstone strata, including the Harra Ebb and Qui Ayre Sandstone formations of probable Early Devonian age, the oldest sedimentary rocks exposed in the Orcadian Basin. These represent a combination of alluvial-fan, aeolian dune and lake-beach deposits and are of great importance in the interpretation of sedimentary environments in the early development of the Orcadian Basin. The aeolian sandstones are particularly important in the uniqueness of their occurrence and its implications for early Devonian climate and the palaeogeography of Britain and Ireland. The well-exposed basement complex and its mantle of breccias and conglomerates also allow reconstruction of part of the pre-Devonian landscape. The Caithness Flagstone Group records the establishment of a lake environment and a cyclic pattern of sedimentation in which the lake deepened rapidly and shallowed slowly, with deep lake muds at the base of each cycle passing up into coarser-grained sediments containing evidence of shallowing and emergence. Rivers formed alluvial fans that built out over the dry lake-bed before becoming inundated as the cycle started again.

The Orcadian Basin

OLD MAN OF HOY COAST, ORKNEY (HY 211 052–HY 174 991)

E.A. Pickett

Introduction

This site includes about 8 km of the spectacular coastline of north-western Hoy. It extends from Rora Head to Hamar Hellia and includes the 137 m-high Old Man of Hoy sea stack and the 350 m-high cliffs at St John's Head (Figure 2.37). The cliff sections, which are largely inaccessible from inland, show the basal volcanic part (Hoy Volcanic Member) of the mainly Upper Devonian Hoy Sandstone Formation resting unconformably on the Middle Devonian Upper Stromness Flagstone Formation. The Hoy Volcanic Member has been dated as 379 ± 10 Ma (Halliday *et al.*, 1982) which places it close to the Mid–Late Devonian boundary and of probable Givetian age. The site is important because the Hoy Volcanic Member, seen only on Hoy, allows comparison with other Middle Devonian volcanic rocks of the Orcadian Basin. It is described in the GCR volume on the Caledonian igneous rocks (Stephenson *et al.*, 1999). Above the lavas lies a thick succession of sandstones of the Hoy Sandstone Formation, forming the impressive red sea-cliffs that dominate this site. These cliffs, where accessible, show excellent examples of the braided stream and aeolian facies that are typical of the Late Devonian of the Orcadian Basin. Sedimentary structures within the aeolian sandstones suggest wind directions different from those that prevailed in Early Devonian times (on the evidence of the unique aeolian sandstones at the **Yesnaby and Gaulton Coast Section** (see GCR site report, this chapter) and Mid-Devonian times (see **South Fersness Bay** GCR site report, this chapter). This site is therefore of great importance in the interpretation of the palaeoenvironments and palaeogeography of the Orcadian Basin.

Description

The oldest part of the exposed sequence is the cross-bedded, faulted and gently folded Upper Stromness Flagstone Formation, which is exposed at the base of the Old Man of Hoy sea stack (HY 176 008) and from the Geo of Hellia (HY 190 042) northwards (Figure 2.37). A particularly good section of the rhythmites of the formation is at the Bay of the Tongue (HY 207 047), near the northern end of the site. The cliffs at Kame of Hoy and the Bay of the

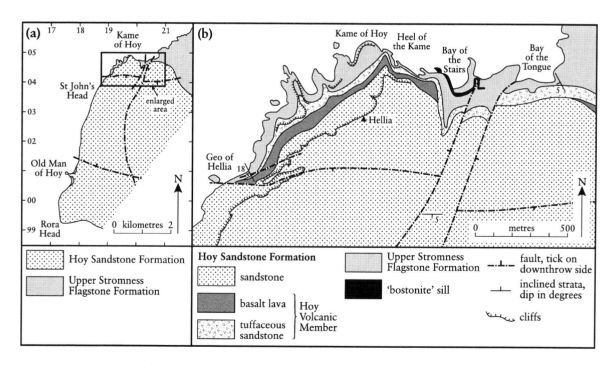

Figure 2.37 (a) Simplified geological map of north-west Hoy (after British Geological Survey, 1999). (b) Detailed geological map of the northern part of the GCR site (after Mykura, 1976).

Stairs consist mainly of cyclic sequences of mudstones, siltstones and sandstones of the formation. At Yelting Geo, about 300 m north-east of the eastern end of the site, Fannin (1970) noted thin (up to 4 cm) beds of tuffaceous siltstone in the formation.

The Upper Stromness Flagstone Formation is truncated by an undulating surface and unconformably overlain by the Hoy Volcanic Member of the Hoy Sandstone Formation, the unconformity being well exposed at the base of the Old Man of Hoy. The basal part of the Hoy Volcanic Member consists of tuffaceous sandstones and breccio-conglomerates. Between the Bay of the Tongue and the Geo of Hellia the tuffaceous sandstones are up to 15 m thick, brownish red, locally cross-bedded and contain angular blocks and pebbles of basalt and finely comminuted basaltic material (Mykura, 1976). At the Geo of Hellia and the Old Man of Hoy, these basal tuffaceous rocks are absent and the Upper Stromness Flagstone Formation is directly overlain by lavas.

The lavas of the Hoy Volcanic Member form five disconnected outcrops in the north-west of Hoy and one on the south coast of the island. McAlpine (1978) recognized five separate lava flows, although Mykura (1976) thought it impossible to determine whether the outcrops represent one or several flows. The lavas are alkali olivine basalts with porphyritic crystals of olivine and feldspar in a groundmass of iron oxides, augite and plagioclase (Wilson *et al.*, 1935). They are nepheline-normative (Thirlwall, 1979, 1981) and have been dated at 379 ± 10 Ma (Halliday *et al.*, 1982). A maximum exposed thickness of about 90 m is seen at Hellia, where the lavas form part of the impressive cliffs. They comprise a grey-weathering, vesicular lower part, a massive 60 m-thick columnar-jointed central portion and an upper, 15 m-thick, purplish slaggy zone (Mykura, 1976). At the Old Man of Hoy, the lavas are 3 m to over 7 m thick, and the lateral variations in the lavas can be seen particularly well at their base. Their upper surface at the Old Man of Hoy is a fairly even plane that appears to have been eroded before deposition of the overlying sandstones (Wilson *et al.*, 1935).

Above the Hoy Volcanic Member, the Hoy Sandstone Formation comprises over 1000 m of medium- to thick-bedded, medium- to coarse-grained, red and yellow sandstones with thin interbedded siltstone partings (Figure 2.38). The sandstones display planar and trough cross-bedding with individual cross-sets up to 1.2 m in height. Slumping and convolute bedding are locally common. Many cross-sets contain clasts of red and purple siltstone, and intraformational conglomerates at the bases of troughs are common. There are also rare small lenses of extraformational conglomerate containing pebbles of quartz, schist and gneiss.

McAlpine (1978) subdivided the sandstones into four members – the Lang Geo Sandstone, Lyre Geo Sandstone, Haist Pebbly Sandstone and Trowie Glen Sandstone. The Hoy Sandstone Formation in the northern part of the site is undivided, whereas the Lang Geo Sandstone Member and the overlying Lyre Geo Sandstone Member have been recognized in the southern part (British Geological Survey, 1999). The Lang Geo Sandstone Member, which forms the cliffs at the Old Man of Hoy, is distinguished from the Lyre Geo Sandstone Member mainly by its thinner-bedded and better-cemented sandstones. Both members contain fining-upward sandstone units and subordinate siltstones and mudstones.

A further feature of interest at the site is a microsyenitic ('bostonite') sill of uncertain age which cuts the Upper Stromness Flagstone Formation at the base of the cliff at Bay of the Stairs (HY 203 047). The sill consists predominantly of feldspar (mainly orthoclase) and some chlorite and carbonate (Mykura, 1976).

McAlpine (1978) correlated in detail the Hoy Sandstone Formation with the Dunnet Head Sandstone Group at Dunnet Head on the Caithness mainland, which can be seen from Hoy. The cliffs at Dunnet Head provide spectacular exposure of laterally continuous fluvial (braided river) sandstones (Trewin, 1993).

Interpretation

The coarsening-upward rhythmic cycles in the Upper Stromness Flagstone Formation at the base of the succession record changes in environment from relatively deep-water to ephemeral, shallow lake and prograding alluvial-fan. This cycle was repeated many times and may have been driven by a combination of cyclic climate change and tectonic activity (see **South Stromness Coast Section** GCR site report, this chapter). Thin bands of tuffaceous siltstone within the Upper Stromness Flagstone Formation near the eastern end of the site may represent

The Orcadian Basin

Figure 2.38 The Old Man of Hoy, a sea stack of Hoy Sandstone Formation lying just off the spectacular cliffs of north-west Hoy. The basal member, the Hoy Volcanic Member, lies at the base of the stack where it overlies the Upper Stromness Flagstone Formation. (Photo: BGS No. D1539, reproduced with the permission of the Director, British Geological Survey, © NERC.)

the start of Mid-Devonian volcanic activity that culminated in the extrusion of the volcanic rocks at Hoy and those in the Eday Flagstone Formation at Deerness on the Mainland (Fannin, 1970). The tuffaceous layers thin and wedge out northwards, suggesting that they emanated as small ash eruptions from a volcanic centre that lay farther south (Fannin, 1970).

The basal tuffaceous sandstones and conglomerates of the Hoy Volcanic Member were deposited on an undulating erosion surface. Mykura (1976) suggested that most of the volcaniclastic rocks were deposited as subaerial ash falls, since their bedding planes are parallel to the slopes of the underlying hummocky basement surface. However, McAlpine (1978) interpreted the tuffaceous beds as shallow-water deposits. Extrusion of basaltic lavas followed, probably as several flows. The planar upper surface of the lavas suggests that some erosion occurred before deposition of the overlying sandstones.

The Hoy Sandstone Formation records a period when over 1000 m of sands and subordinate silts were deposited in a predominantly fluvial, probably braided stream environment (Mykura, 1976). The great thickness of the formation has important implications for the subsidence history of the Orcadian Basin in Late Devonian times. The common fining-upward fluvial sequences are interbedded with sabkha and playa-lake siltstones and poorly cemented aeolian sandstones. McAlpine (1978) suggested a system of active wadi fans in which sediments were deposited as wadi-flood deposits. Beyond the distal margins of the wadi fans, sand accumulated as small barchan dunes while sand and silt accumulated in playa lakes and sabkhas in the interdune areas. Palaeocurrent analysis of the fluvial sandstones indicates north-east flow, and the aeolian sandstones appear to have formed by reworking of the fluvial sands by winds that blew predominantly towards the south-west (McAlpine, 1978).

Conclusions

The island of Hoy is the only locality in the Orcadian Basin where the Hoy Sandstone Formation and its basal member, the Hoy Volcanic Member, are exposed. For this reason Hoy, and more specifically this GCR site, are of great significance in the regional interpretation of the Devonian Orcadian Basin. Along a coastline of spectacular and commonly inaccessible cliffs, this large site exposes a series of important stratigraphical contacts and fine sections through the succession. At the base of the Old Man of Hoy, fluvial sandstones of the Upper Stromness Flagstone Formation are unconformably overlain by tuffaceous sandstones and basaltic lava of the Hoy Volcanic Member. Although a significant period of erosion probably occurred between deposition of the sandstones and eruption of the lavas, volcanic activity had already started during deposition of the Upper Stromness Flagstone Formation, as shown by the presence of thin tuffaceous sandstones. Volcanic activity ceased by Late Devonian times, when the Orcadian Basin was a wide plain with braided rivers, wadis, lakes, sabkhas and small dune-fields. The sediments laid down in these environments are preserved in the cliffs at this site. Sedimentary structures show that rivers flowed towards the north-east and dune-forming winds blew towards the south-west. These directions contrast with data from Lower Devonian and Middle Devonian rocks of the Orcadian Basin, thus making this site important for interpreting the palaeogeographical and palaeoenvironmental evolution of the basin through the Devonian Period.

BAY OF BERSTANE, ORKNEY (HY 476 111)

Potential GCR site

E.A. Pickett

Introduction

This potential GCR site lies in the north-west corner of the Bay of Berstane near Kirkwall on the Orkney Mainland. A short section of the Eday Marl Formation, including a mudstone containing a marine microfauna and pseudomorphs after halite, is exposed in a small, fault-bounded block. The mudstone is interpreted as representing a short marine incursion into the Orcadian Basin during mid–late Givetian (Mid-Devonian) times (Marshall *et al.*, 1996). If this interpretation is correct, this is the only record of a Devonian marine incursion onshore in Scotland, and the site is therefore of crucial importance for palaeoenvironmental and palaeogeographical interpretations of the Orcadian Basin. The site also provides the first onshore evidence to confirm the results of a BP/Chevron offshore well, drilled in the East Orkney Basin, which proved three intervals with marine microfossils in upper Givetian to lower Frasnian rocks (Marshall *et al.*, 1996).

Description

The following account is based wholly on a description by Marshall *et al.* (1996). The site (Figure 2.39) exposes about 7 m of red and green, calcareous mudstones and argillaceous siltstones and sandstones, punctuated by cleaner red and buff sandstones up to about 50 cm thick. The most important feature is a 1 m-thick bed of grey-green, parallel-laminated, mudstone and siltstone, with red mudstone layers spaced at about 10 cm intervals (Figure 2.40), which contains evidence suggesting a brief marine incursion into the Orcadian Basin. Many of the siltstone laminae are trains of starved wave ripples, the troughs of which contain common hopper-crystal pseudomorphs after halite up to 2 cm across. Only the corners of many of the larger pseudomorphs are present, or they are truncated below the overlying siltstone. The smaller pseudomorphs are generally complete. The lamination in the red mudstone layers is disrupted and compressed into tepee structures.

The Orcadian Basin

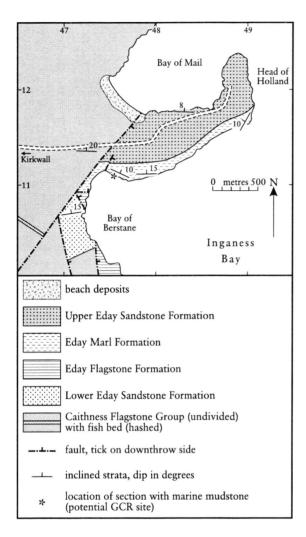

Figure 2.39 Geological map of the Bay of Berstane area. After British Geological Survey (1999).

The red mudstones above and below the grey-green unit are mostly apparently structureless and disrupted by successive polygonal arrays of desiccation cracks, or have a 'turbate', locally brecciated structure defined by contorted lenses and laminae of cleaner or coarser-grained sand. A diamict texture is present locally. Other less abundant structures include wave-rippled surfaces, deflation lags of coarse sand or granules (some with ripple form), gutter casts and water-escape structures. Traces of cross-lamination and flat lamination are preserved locally within the 'turbate' layers, and there are a few burrow-mottled horizons with tube-like, U-shaped, *Thalassinoides* and ?*Beaconites* burrows. Common calcite nodules are either small calcrete concretions or more commonly sparite-filled vugs, giving the mudstones a vesicular appearance. Some are filled with dog-tooth spar and others have geopetal fillings.

Samples of the grey-green mudstone contain a low-diversity palynomorph assemblage dominated by *Geminospora lemurata*, regarded as mid- to late Givetian in age. However, the main importance of this site lies in the discovery by Marshall *et al.* (1996) of two scolecodonts in the palynomorph assemblage. These are the chitinous jaw parts of polychaete worms, which are today restricted to marine environments, although all the sedimentological evidence at the site points to a non-marine origin.

This site is also important in providing an onshore link with offshore data obtained from the BP/Chevron offshore well 14/6-1, drilled in 1985 at the south-east margin of the East Orkney Basin (about 100 km ESE of Orkney). The well penetrated Devonian sedimentary rocks that are dated palynologically as late Givetian to early Frasnian in age (Marshall *et al.*, 1996), and three grey-green mudstone intervals, yielded marine microfossils, including an acritarch in the upper one and scolecodonts in the lower two.

Interpretation

The red beds of the Eday Marl Formation are interpreted by Marshall *et al.* (1996) as the deposits of a muddy sabkha plain, the 'turbate' structures being caused by growth of evaporite crusts and nodules, deposition and erosion around them by wind and water, repeated desiccation, remobilization of saturated sediment and bioturbation by burrowing organisms. The sparite-filled vugs are probably the moulds of dissolved evaporite nodules.

The presence of scolecodonts and pseudomorphs after halite in the grey-green mudstone in the Eday Marl Formation at the Bay of Berstane has been attributed to marine inundation of the sabkha plain (Marshall *et al.*, 1996). All living polychaete worms are found in marine environments and are accepted as indicators of similar environments in the geological past. Acritarchs would be expected to predominate over scolecodonts in a normal marine setting, and the dominance of scolecodonts in offshore well 14/6-1, as well as their restriction to simple forms such as *Micrhystridium* and *Veryhachium*, may be due to a 'stressed', marginal marine environment (Marshall *et al.*, 1996). This is supported by the fact that polychaete worms are common in the modern

Figure 2.40 Grey-green mudstone unit within the Eday Marl Formation at the Bay of Berstane. (Photo: J.E.A. Marshall.)

nearshore environment, where they can tolerate wide salinity variations.

The pseudomorphs after halite are also cited by Marshall *et al.* (1996) as important evidence for a marine incursion, although they also occur commonly in playa-lake deposits. Although evaporite pseudomorphs are widespread in the Orcadian Basin (e.g. Astin and Rogers, 1991) and the evaporitic crusts have been interpreted as causing the disrupted bedding and tepee structures in the Bay of Berstane section, the pseudomorphs dominantly replace gypsum, indicating the presence of sulphate-rich brines. The pseudomorphs after halite are thought to indicate a short episode when NaCl-rich brines flooded the basin, and their large size and high concentration suggest that marine waters were the source of the brines. The absence of desiccation cracks suggests that any periods of emergence were short. Marshall *et al.* (1996) envisaged an extensive, low-relief sabkha plain that was episodically flooded by marine waters, analogous to the Ranns of Kutch today, which is flooded during the monsoon season and is a dry, evaporitic sabka plain for the remainder of the year.

The evidence of the marine incursions in BP/Chevron Well 14/6-1 and in the Bay of Berstane section has led to a major revision of late Mid- and Late Devonian palaeogeography of the Orcadian Basin and North Sea. The discovery of marine Devonian rocks in the Argyll Field, farther south in the North Sea (Pennington, 1975) necessitated revision of the palaeogeographical model, incorporating a connection with the marine Devonian rocks of southern Britain and northern Europe along a narrow, probably fault-bounded zone. The discovery by Marshall *et al.* (1996) of the miospore *Archaeoperisaccus* in the marine beds in well 14/6-1, a genus restricted to the northern part of Laurasia to the north, indicates that the marine waters did not enter the Orcadian Basin through a southern connection. Marshall *et al.* (1996) suggest that the connection was from the east, along the Tornquist Zone, around the southern margin of the Fenno–Scandian High. When sea levels dropped, the connection was severed and continental sedimentation resumed in the Orcadian Basin.

The marine incursions represented at the Bay of Berstane and in well 14/6-1 were tentatively attributed by Marshall *et al.* (1996) and Marshall and Hewett (2003) to global sea-level rises, the Berstane occurrence being correlated with the Givetian Taghanic highstand of the 'standard' Devonian sea-level curve, as defined in New York

State (House and Kirchgasser, 1993). Marshall *et al.* (1996) further speculated that the sea level was controlled by two orders of orbital cyclicity, at periodicities of 39.5 ky and 413.9 ky.

Conclusions

This site's importance lies in providing what may be the first onshore record of a Devonian marine flooding event in Scotland. The marine interpretation is contentious, but if proved to be correct, there are important implications for the study of the palaeoenvironments and palaeogeography of the Orcadian Basin. A short section of the Eday Marl Formation contains a grey-green mudstone that has yielded marine microfossils and pseudomorphs of halite crystals. These, together with evidence from an offshore well drilled in the East Orkney Basin, may provide evidence that the sabkha plains of the Orcadian Basin were periodically inundated by marine waters in late Mid- and Late Devonian times.

RED POINT, CAITHNESS (NC 932 659)

P. Stone and W.J. Barclay

Introduction

The cliff sections around Red Point provide remarkable and unique examples of Middle Devonian lake margin deposits unconformably resting on granodiorite and gneiss basement. The lithofacies relationships are important in establishing fluctuations in playa-lake conditions and their interaction with the variable topography of the basin margin. Particularly noteworthy are the lake margin carbonate and coarse breccia deposits, and their lateral transition into braid-plain sandstones and deeper-water, lacustrine, laminated mudstones. A comprehensive description of the site is given by Donovan (1975), summarized by Trewin and Thirlwall (2002). Field guides are provided by Donovan (1978), Parnell *et al.* (1990) and Trewin (1993).

The regional importance of the Red Point site lies in the exceptionally well-preserved examples of marginal lacustrine deposits. The lithofacies are unique in their diversity and the clarity of their relationships with the basement rocks. The site, along with similar exposures to the west at Port Skerra and Baligill, allows interpretation and characterization of an otherwise unrepresented part of the stratigraphy of the Orcadian Basin, and apart from its intrinsic sedimentological value, it is important for the broader interpretation of the tectonics and palaeogeography of the basin.

The relationships between irregular basement topography and lowermost Old Red Sandstone conglomerates and sandstones are also well displayed at Port Skerra (HC 878 663) (Trewin, 1993). Here, cliff exposures show locally intense fracturing in the basement that was exploited and opened by Devonian weathering. In the bay, sheet-flood conglomerates and hard silica-cemented sandstones fill hollows, drape knolls and infill a 3 m-deep fissure in the irregular gneiss surface. Sedimentary structures include excellent convolute lamination in a sandstone exposed near low-tide mark and sand volcanoes (B.P.J. Williams, pers. comm.). West of Port Skerra, outcrops at Balligill (NC 855 659) and a section in Balligill Quarry (NC 852 657) provide excellent exposures of deeper lacustrine and lake-margin facies, the latter including a limestone mantling a basement hill (NC 857 661) (Trewin, 1993). The lacustrine laminites are fish-bearing at two horizons in the quarry. Fine examples of load structures are seen in the bases of fine-grained turbidite sandstones, as well as 'flow-roll' de-watering structures, diagnostic of lacustrine deposition. Trewin (1993) provided details of these excellent localities.

The dangers of the Red Point site should be stressed. The sea cliffs around Red Point are high and precipitous with much loose rock, and most of the critical exposures are perched at the top of the cliffs in precarious positions. Visitors must therefore take appropriate precautions to ensure their safety.

Description

Red Point lies about 18 km west of Thurso on the north coast of Caithness. It provides a remarkable section through marginal lithofacies of an Orcadian Basin lake, which rest unconformably on rusty brown, granodioritic and locally gneissose basement rocks. The basement relief is at least 30 m (Donovan, 1975), with the overlying sedimentary strata dipping variably away from the elevated basement areas. Local minor faulting complicates the unconformable relationship. The outline geology of the site and a sketch cross-section illustrating the relief are shown in Figures 2.41

Red Point

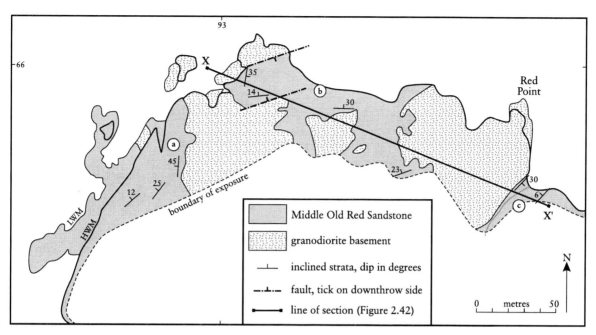

Figure 2.41 Locality map of the Red Point area. Based on Donovan (1975) and Trewin (1993).

and 2.42 respectively. There is no direct evidence for the age of the basin-margin strata, but they probably belong to the lower part of the Latheron Subgroup, the lowest division of the Upper Caithness Flagstone Group, and are therefore likely to be early Givetian (Mid-Devonian) in age. The deposition of the Red Point sediments was probably coincident with the maximum expansion of the Orcadian Basin lake.

Along the western side of the basement inlier ('a' on Figure 2.41) a remarkable limestone facies rests directly on the basement. The lowest limestone is a thin, calcite-cemented breccia of basement clasts, which is particularly concentrated in hollows in the erosion surface. Above this, up to 2 m of grey, fairly massive, limestone shows evidence for slumping, is irregularly silicified locally, and appears to merge laterally down-slope into laminated mudstone. Janaway and Parnell (1989) noted lower dolomitic and upper calcitic units and small laminated domes of possible algal origin. A coarse breccia of limestone and basement clasts in a sandy matrix overlies the limestone (Figure 2.43). Clasts are angular to subrounded and range up to about 40 cm across.

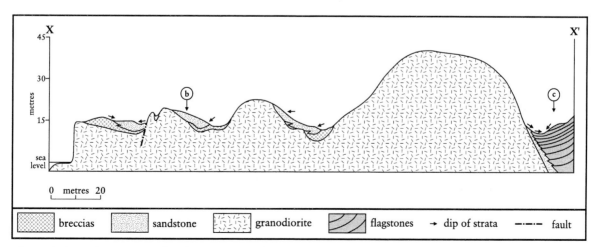

Figure 2.42 Cross-section of Red Point area to illustrate basement margin features. Based on Donovan (1980) and Trewin (1993).

The Orcadian Basin

Figure 2.43 Coarse breccia of basement and limestone clasts in a sandy matrix, overlying massive limestone, Red Point. (Photo: P. Stone.)

Across the central part of the basement inlier ('b' on Figure 2.41), breccias lie directly on the unconformity, although minor faulting affects the outcrop pattern. The breccias are cemented by calcite locally, and some beds are draped by very thin, carbonate-rich mudstone laminae. Northwards, and dipping away from the elevated basement area, the breccias are increasingly interbedded with coarse sandstone. Breccia tongues thin fairly abruptly down-dip (Figure 2.44) and pass into a sandstone-dominated sequence with lenses of breccia. Clast arrangement in the breccia tongues is mainly random, but weak clast imbrication in the lenses indicates a down-dip current flow. The sandstones are generally thinly bedded with internal, wavy and low-angle cross-bedding. Exposed bedding surfaces are rippled and have polygonal networks of desiccation cracks.

The eastern side of the basement inlier ('c' on Figure 2.41) is a spectacular exposure illustrating the relationship between a steep cliff of red gneiss and granodiorite, and breccias derived from it. Despite some local faulting, the sedimentary association is clearly preserved. The coarse breccia contains angular to subrounded clasts, mainly of the adjacent granodiorite, but there are also some clasts of limestone similar to that on the western side of the basement inlier. Locally, the breccia fills fissures in the steep basement cliff, and overall reaches a maximum thickness of about 10 m. The beds dip steeply away from the basement cliff, towards the east, but the dip decreases markedly in that direction and is sub-horizontal 50 m away, the decrease coincident with the breccias merging laterally with thinly interbedded sandstone and mudstone. These are variably laminated and cross-bedded, exposed bedding surfaces showing ripples and polygonal networks of desiccation cracks. Some of the sandstone beds contain sporadic, isolated clasts of red granodiorite.

Interpretation

The lithofacies assemblage at Red Point, and its unconformable relationship with the contemporary basement topography, illustrates the depositional pattern at the margin of an Orcadian Basin lake. The granodiorite–gneiss basement had significant relief in this area and so provided depositional environments that varied in response to fluctuating lake levels. The

Red Point

Figure 2.44 Breccia tongues interfingering with sandstone. (a) General view; (b) detail. (Photos: P. Stone.)

limestone to the west of the basement inlier represents lacustrine deposition at maximum lake water-level (a coincident lake margin of Donovan, 1975), when the water-level was high enough to transgress the margin of the basin. Carbonate deposition was controlled by photosynthesizing algae, with the upward transition from dolomitic to calcitic limestone probably reflecting reduced water salinity (Janaway and Parnell, 1989). Down-slope slumping of the limestone occurred prior to its complete lithification; in the same direction, it grades into the deeper-water mudstone–sandstone laminite facies.

The coarse breccias overlying the limestone were lake-margin beach deposits that prograded into the lake in response to a drop in water-level. Their angular limestone pebbles were clearly derived from fully lithified limestone. Again, there is a lateral transition, away from the elevated basement topography, as breccia tongues interfinger with thinly bedded sandstones. The sandstones may be a slightly deeper-water, nearshore lithofacies, but their sedimentary structures are equally compatible with an origin as high-energy, braided fluvial deposits. The latter interpretation is probably better suited to the current-imbricated breccia lenses within the sandstones. Wave-rippled bedding planes cut by desiccation polygons point to intermittent emergence of the lake floor.

The most dramatic example of basement relief is seen on the eastern side of the inlier. The steepness of the basement cliff against which the breccias are banked, and the abrupt transition away from the cliff into laminated sandstone and mudstone, suggests a scree-like accumulation at the lake margin. The laminated sequence contains many examples of wave-rippled bedding surfaces and polygonal desiccation crack networks, indicating alternating shallow-water and emergent conditions.

The range of marginal lacustrine lithofacies around Red Point is remarkable, but it is very hard to link the different exposures into a single, comprehensive model. Instead, it seems most likely that a range of stratigraphical levels is present, each reflecting slightly different conditions at different times in the evolution of the lake margin. The lacustrine transgressions and regressions recorded may possibly have had a climatic control. The local, minor faulting that cuts the site may also be significant in juxtaposing different parts of the sequence.

Conclusions

The Red Point GCR site exposes a unique assemblage of marginal lacustrine deposits and shows their relationships. It preserves a fragment of an Orcadian Basin lake margin, probably at the time of deposition of the earliest strata of the Upper Caithness Flagstone Group farther out in the lake and at the time of maximum lake extent in the basin. The lacustrine strata rest unconformably on a basement of red granodiorite with a visible topographic relief exceeding 30 m. Exceptional features of the site are the shallow-water limestones and coarse breccia tongues of possible beach origin. These pass laterally into high-energy, probably fluvial sandstones with clast-imbricated breccia lenses, and lacustrine mudstone–sandstone laminites. The latter display wave-generated ripples and polygonal desiccation crack networks, a combination that is strongly suggestive of a playa-lake environment. Overall, the Red Point site is critically important to an understanding of the palaeogeography of the Orcadian Basin and provides a unique insight into its lake margin environments.

PENNYLAND (THURSO–SCRABSTER), CAITHNESS (ND 102 696–ND 115 685)

P. Stone

Introduction

The cliff and foreshore section between Thurso and Scrabster provides excellent exposures of part of the Middle Devonian Upper Caithness Flagstone Group. The lithofacies present were produced during the cyclical deepening and shallowing of a large lake occupying part of the Orcadian Basin; about 20 cycles are represented, with abrupt transitions from laminated mudstones deposited in a deep lake, to cross-bedded sandstones deposited in an ephemeral, playa-lake environment. A remarkable range of shrinkage cracks have either a subaqueous or subaerial origin, and their correct interpretation is crucial for an understanding of lake history within the Orcadian Basin. Fossil fish faunas from the laminated mudstones are of great importance and have led to the site being separately selected for the GCR for its fossil fishes (Dineley and Metcalf, 1999). Of the 20

Pennyland (Thurso–Scrabster)

laminite horizons, 16 yield fish remains, most of them small fragments, but 7 or 8 laminites contain complete plates or complete fishes. Field guides to the site are provided by Donovan (1978), Parnell *et al.* (1990) and Trewin (1993).

The regional importance of the Pennyland (Thurso–Scrabster) site lies in the exceptionally well-preserved evidence for lacustrine depositional cyclicity, allowing interpretation and characterization of an otherwise poorly exposed part of the Orcadian Basin succession. The site is also important in the broader interpretation of the tectonics, lithostratigraphy and palaeogeography of the Orcadian Basin. The conservation value and importance of the site also rests in the good quality of its fossil fish remains.

Description

Along the south-west side of Thurso Bay, between Thurso and Scrabster, an extensive coastal section affords excellent exposure through part of the Mey Subgroup of the Upper Caithness Flagstone Group. The highest beds seen are probably transitional into the overlying John o'Groats Sandstone Group. A Givetian (Mid-Devonian) age is likely for the succession, which comprises lacustrine, fluvial and aeolian facies. Gentle, consistent north-west dips result in almost 200 m of strata being exposed along about 1.5 km of shoreline (Figure 2.45), but continuity is repeatedly disrupted by minor faulting. Exposure is in the cliffs and on a wide intertidal rock platform.

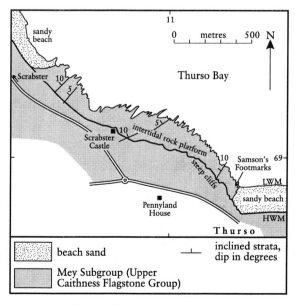

Figure 2.45 Locality map of Pennyland GCR site.

The oldest strata within the GCR site, at the eastern end of the section, are thinly bedded, laminated grey-green shales and paler, fine-grained, locally lenticular sandstones. They are affected by a range of styles of shrinkage cracks, all well exposed on bedding planes on the intertidal platform. Polygonal arrays (Figure 2.46a) are particularly well-developed (probably giving rise to the local name for this area – 'Samson's Footmarks') and grade into more orthogonal patterns of more widely spaced cracks (Figure 2.46b). Lenticular cracks tend to be smaller and to form complex, interlocking patterns (Figure 2.46c). Rippled surfaces commonly alternate with those affected by shrinkage cracks and some of them are also cracked. The sandstone layers commonly extend downwards into the small shrinkage cracks that penetrate the underlying shale or mudstone to create the so-called 'fang structures' of Donovan (1980). Differential compaction strongly deformed many of the crack infills so that in cross-section they are bulbous or sinuous (Figure 2.47). Originally mistaken for burrow-fills (Crampton and Carruthers, 1914), their origin remains a subject of controversy (see 'Interpretation', below). More likely examples of bioturbation are very small, *Skolithos*-like, circular structures less than 1 mm across on some bedding surfaces.

Finely interlaminated organic-rich mudstone and carbonate beds, some containing scattered, disarticulated fish remains, occur sporadically within the sandstone–mudstone laminites. There are 20 such units within the Thurso–Scrabster section, 16 of which contain fish. However, most of these units are disrupted by shrinkage cracks and contain only small fish fragments, and only seven or eight thicker calcareous laminites with no cracks contain complete plates or complete fishes (Dineley, 1999a). Complete specimens of *Dipterus* are fairly common and *Millerosteus minor* is very well-preserved as disarticulated and semi-articulated plates in the lower part of the section (ND 113 688–ND 110 692). Orange- or brown-weathering stromatolites also occur in the organic, fish-bearing strata, mainly as small, millimetre-scale hemispheroids (Parnell *et al.*, 1990) and as sheets (Trewin, 1993). The fish-bearing laminites occur on average at about 7 m intervals through the lower (eastern) part of the section (Parnell *et al.*, 1990). This cyclicity is superimposed on a general, but irregular, trend of upward-increasing sandstone proportion. Many of the sandstone beds have soft-sediment

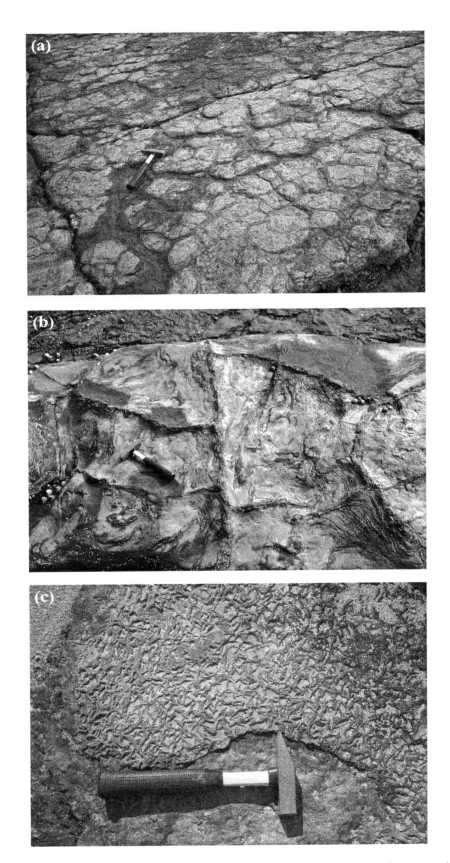

Figure 2.46 Styles of shrinkage cracks affecting thinly bedded mudstones and fine-grained sandstones. (a) Polygonal array of desiccation cracks; (b) more widely spaced orthogonal array of desiccation cracks; (c) lenticular shrinkage cracks forming complex interlocking pattern. (Photos: P. Stone.)

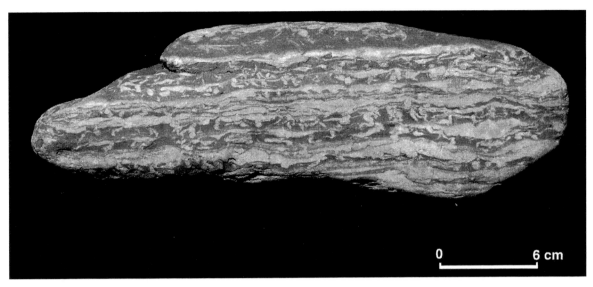

Figure 2.47 Compressed, deformed shrinkage cracks in cross-section. (Photo: BGS No. P547102, reproduced with the permission of the Director, British Geological Survey, © NERC.)

deformation structures, with widespread convolute lamination, load-induced 'pseudonodule' layers (Figure 2.48) and sandstone dykes, which cut up to 3 m of strata. Cross-bedded sandstone bodies up to 4 m thick occur towards the top (the western end) of the section. Some have trough cross-bedding, with locally abundant rip-up mudstone clasts, suggesting a fluvial origin. Others have low-angle cross-bedding, with lag accumulations of coarse, well-rounded quartz grains, and are probably aeolian. Small pyrite nodules are locally common in some of the more massive sandstone beds. Stratigraphically, this part of the Thurso section forms a link with the overlying John o'Groats Sandstone Group, in which high-energy, fluvial sandstones are

Figure 2.48 Load-induced 'pseudonodule' layer. (Photo: P. Stone.)

dominant. These can be examined near the type locality of **John o'Groats** south of Duncansby Head (ND 405 735; see GCR site report, this chapter). There, trough and planar cross-bedding, low-angle planar bedding and primary current lamination are seen in broadly cyclic, upward-fining channelized units, and mudstone clasts are common in the channel bases (Trewin, 1993).

Interpretation

The strata exposed in the coast section between Thurso and Scrabster record the interaction of lacustrine, fluvial and aeolian processes in and around an Orcadian Basin lake of fluctuating water-level. The lithofacies can be correlated with the facies associations recognized by Donovan (1980) (Figure 2.3), and record the repeated, abrupt variations in lake level from the deepest-water, fish-bearing laminites ('association A') to intermittently emergent playa-lake deposits ('association D'). The shallower lake deposits ('associations B and C') are poorly represented in the Thurso section.

The fish-bearing laminites ('association A') represent the slow accumulation from suspension of fine-grained sediment in the deeper part of a large lake over a period of hundreds, possibly thousands, of years. The sub-millimetre scale alternations of fine clastic siltstone, carbonate and organic laminae have been interpreted by Donovan (1980) as varved sediments deposited under annual, seasonal, climatic control. Trewin (1993) interpreted the clastic layers as representing input from rivers in the rainy season, the carbonate laminae as the deposits of the dry, warm summer season when the photosynthetic activity of phytoplankton was at a maximum, and the organic laminae as the product of autumnal, annual decay of the phytoplankton. Water depths may have been as great as 80 m (Hamilton and Trewin, 1988), but the presence of stromatolitic layers within the laminites necessitates shallower water in the photic zone, perhaps up to about 50 m, for at least part of the depositional cycle.

The laminites pass gradationally up through thinly interbedded fine-grained sandstones and mudstones into interbedded internally cross-laminated siltstones and sandstones. The latter lithofacies forms most of the succession between the fish-bearing laminites. It indicates a depositional environment in which the lake floor was periodically emergent and sand was transported by shallow streams and reworked by wave action (Trewin, 1993). The abundant polygonal desiccation cracks are evidence of the intermittent emergence and drying out of the lake-bed sediments, but the origin of the equally abundant, lenticular cracks has been the subject of debate. Donovan and Foster (1972), Trewin (1992) and Barclay *et al.* (1993) considered that the lenticular cracks were subaqueous in origin and developed by a synaeresis-like process perhaps linked with salinity changes in the lake. Astin and Rogers (1991, 1992, 1993) expressed a contrary view, suggesting that the formation of gypsum (and possibly halite) crystals during the drying-out of the lake (with commensurate increase in salinity) was an important precursor to the formation of the lenticular cracks. These were then initiated by the crystal pattern in a subaerial environment with the type of crack pattern developed being controlled by the extent of desiccation and the thickness of the sediment layer involved. During dry periods, wind-blown sand filled the open cracks and was deposited as thin lenses and laminae on the exposed playa-lake floor. The interpretation of Astin and Rogers requires many more and longer periods of subaerial conditions than the Donovan and Foster model.

Superimposed on the cycles of lake sedimentation is the overall upward increase in both the thickness and frequency of sandstone beds. These show a combination of sedimentary features that suggests aeolian reworking of fluvial sands into small dunes and rippled sheets. The environment was probably marginal to a receding lake, with fluvial sedimentation in broad, shallow channels and intermittent influxes of wind-blown sand. The thicker sandstones are commonly convoluted and disrupted by a combination of loading and water-escape structures.

Conclusions

The Pennyland (Thurso–Scrabster) GCR site provides an exceptionally well-exposed representative section for part of the Middle Devonian Upper Caithness Flagstone Group and its transition into the overlying John o'Groats Sandstone Group. The cyclical arrangement of the strata represents variations in the depth of the Orcadian Basin lake. The site is therefore of great importance in regional interpretation of the tectonics and palaeogeography of the Orcadian Basin. In addition, there is good preservation of a spectacular array of shrinkage cracks, the origins of which remain a matter of debate. The site is also of great importance because of its well-preserved late Givetian fossil fish fauna (Dineley, 1999a).

JOHN O'GROATS, CAITHNESS (ND 380 735–ND 407 735)

Potential ORS GCR site

W.J. Barclay

Introduction

The John o'Groats GCR site is already accorded protected status on account of its Mid-Devonian fish fauna, preserved in the John o'Groats Fish Bed. This is the best site for the highest Mid-Devonian fauna in Caithness and is described in the companion GCR volume on the *Fossil Fishes of Great Britain* (Dineley and Metcalf, 1999). The following account summarizes the account of Dineley (in Dineley and Metcalf, 1999) and incorporates important Old Red Sandstone sections to the east at Ness of Duncansby and Duncansby Head (Figure 2.49). The area provides the type locality of the John o'Groats Sandstone Group, the highest Middle Devonian unit of Caithness. The John o'Groats Fish Bed crops out on the foreshore 365 m north-east of the John o'Groats Hotel. Since the first reported discovery of *Microbrachius dicki* by Peach (1868), the bed has yielded complete fish remains, although in less rich concentrations than the earlier fish beds in the Caithness Flagstone Group (Westoll, 1948). Trewin's (1993) field guide to the section and to the outcrops at Ness of Duncanby and Duncansby Head provides the basis of this account. Donovan (1978) gave a summary.

Description

Red sandstones of the Last House Formation (Foster, 1972; Donovan *et al.*, 1974) dominate the lowest outcrops, seen immediately east of the harbour (ND 380 735). They occur in beds up to 50 cm thick, in which trough and planar cross-bedding, parallel lamination with primary current lineation and ripple cross-lamination are seen. Soft-sediment deformation de-watering structures are common (Astin, 1985). Green to grey, thin-bedded sediments 100 m east of the harbour contain arrays of polygonal desiccation cracks, as well as wave- and current-rippled beds and lenticular cracks of subaqueous origin. Dineley (1999a) gives a detailed section. Three thin, dark grey, calcareous, varved siltstone laminites with pale carbonate concretions contain scattered fish fragments. The lowest of these is the 0.25 m-thick John o'Groats Fish Bed, which has yielded *Tristichopterus alatus*, *Pentlandia macroptera*, *Microbrachius dicki*, *Watsonosteus fletti* and *Dipterus* sp.. The fauna of this fish bed is typical of the Eday Group of Orkney, and although fish fragments occur elsewhere, this is the only known bed to yield whole specimens in the 610 m-thick John o'Groats Sandstone Group. The beds are displaced by a small NE-trending fault in a notch on the foreshore, and are only visible at low tide. Red sandstones and grey-green laminites with fish scales crop out to the east.

Permian volcanic rocks intrude the John o'Groats Sandstone Group in the axial area of a syncline at Ness of Duncansby (ND 390 739). There are two exposures, but it is not clear if they represent a single vent or two separate ones. Nepheline basalt dykes cutting the vent agglomerate have been K–Ar whole-rock dated at around 270 Ma (Macintyre *et al.*, 1981; Stephenson *et al.*, 2003).

To the east of Ness of Duncansby, sandstones and laminites lie on the eastern limb of the syncline, dips increasing to 40° in the Bay of Sannick. A dark grey laminite visible at low tide here may be the John o'Groats Fish Bed. To the east, the John o'Groats Sandstone Group is faulted against the Mey Subgroup (Upper Caithness Flagstone Group), the latter forming Duncansby Head (ND 405 735). The fault is marked by a gully, and, on the foreshore, a zone of fractured rocks. The John o'Groats Sandstone Group here comprises high-energy, channelized fluvial sandstones with trough and planar cross-bedding and low-angle planar bedding, and lacks the lacustrine facies seen to the west. Fining-upward cycles commence with flat-bedded, fine- to medium-grained sandstones with large-scale planar and trough cross-bedding. Intraformational mudstone clasts line the bases of units, which overlie erosion surfaces. The cycles end with thinner-bedded parallel- and ripple-laminated sandstones passing up into red and pink mudstones. The trough cross-bedding indicates consistent north-east palaeocurrents. Individual sandbodies are wedge-shaped and up to 1 m thick and 10 m

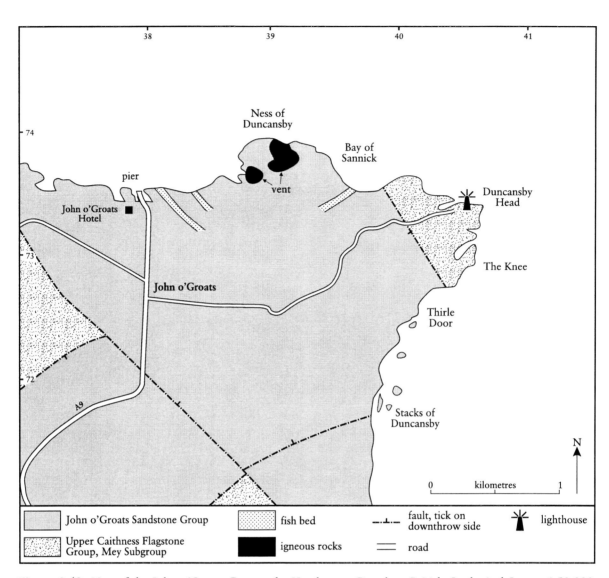

Figure 2.49 Map of the John o'Groats–Duncansby Head area. Based on British Geological Survey 1:50 000 Sheet 116 (Scotland), Wick (1985) and Trewin (1993).

wide. The Stacks of Duncansby offshore preserve more continuous sheet-like bodies up to 1 m thick. The Mey Subgroup comprises the typical grey-green, thinly bedded flagstones.

Interpretation

The John o'Groats Sandstone Group is interpreted as the deposits of shallow, braided or low-sinuosity streams on a broad, low-angle alluvial fan (Foster, 1972) that was subject to periodic inundation by lake waters. The John o'Groats Fish Bed has yielded fewer species than the earlier fish beds of the Orcadian Basin, suggesting to Dineley (1999a) that the fluvial-dominated environments were less suitable for some fish groups and that a long period of climatic stability may have given way to more aridity and only punctuated lacustrine lake development and fish habitation during cooler and/or wetter periods. The dark grey, fish-bearing laminites record periods of more permanent, deeper lacustrine conditions.

There appears to be a substantial amount of strata cut out by the fault separating the Mey Subgroup from the John o'Groats Sandstone Group east of the Bay of Sannick. The John o'Groats Sandstone Group is correlated with the Eday Group of Orkney (Astin, 1985), and on the basis of their respective fish faunas, it appears that all of the *Asterolepis orcadensis* Zone (Watson, 1935) is absent in Caithness (Dineley, 1999a).

Wick Quarries

Conclusions

The John o'Groats site is protected because of the presence of the John o'Groats Fish Bed, the best occurrence of a latest Mid-Devonian fish fauna in Caithness. It allows comparison with the richer faunas of Orkney and shows that lake development here was sporadic, with alluvial deposition being predominant. The outcrops around and to the east of the site are also important in providing the type locality of the John o'Groats Sandstone Group and the most easterly outcrops of the Mey Subgroup on the Scottish mainland.

WICK QUARRIES, CAITHNESS (ND 377 498–ND 374 491)

P. Stone

Introduction

At South Head, Wick, good exposure through part of the Middle Devonian Lower Caithness Flagstone Group is provided by a series of disused flagstone quarries cut into the sea cliffs. The quarries were last worked almost 100 years ago, since when differential weathering of the cliff faces has revealed an extraordinary array of sedimentary features. These and the lithologies present comprise a series of lithofacies produced during the sequential shallowing and deepening of a large lake occupying part of the Orcadian Basin. Of particular note are the laminated lacustrine deposits, including a fish-bearing bed; the remarkable range of shrinkage cracks, which have been variously interpreted as subaqueous and/or subaerial in origin; and the soft-sediment deformation structures showing both compressional and extensional bed-parallel movement. Field guides to the site are provided by Donovan (1978), Parnell *et al.* (1990) and Trewin (1993).

The importance of the Wick Quarries site lies in its exceptionally well-displayed sedimentary features. This allows interpretation and characterization of an otherwise poorly exposed part of the stratigraphy of the Orcadian Basin. The site is also important for the broader tectonic, lithostratigraphical and palaeogeographical interpretation of the Orcadian Basin, as well as providing exposure of rocks that have a high total organic carbon content (TOC) and are a source for oil.

Description

The South Head promontory forms the south side of Wick Bay. Between the coastguard station and the Castle of Old Wick (Figure 2.50), a sea cliff, modified by extensive quarrying, provides an outstanding section through lacustrine strata of the Lybster Subgroup of the Lower Caithness Flagstone Group. The strata are probably of Eifelian (Mid-Devonian) age. The beds dip consistently northwards by about 10°, so that just over 100 m of strata are present within the GCR site, although minor faulting at the northern end of the section disrupts continuity.

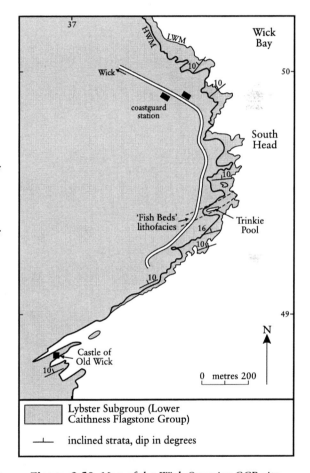

Figure 2.50 Map of the Wick Quarries GCR site.

The southernmost of the old flagstone quarries within the site area, adjacent to the road-end (Figure 2.50), affords a fine section through thinly laminated mudstones and fine-grained sandstones (Figures 2.51a,b). This transitionally overlies a sequence dominated by thicker sandstone beds, ranging up to about

Figure 2.51 (a) Laminated mudstones and thin sandstones of the Lower Caithness Flagstone Group, South Head. (b) Detail of laminites, Lower Caithness Flagstone Group, South Head. (Photos: P. Stone.)

10 cm, which are commonly disrupted by extensive desiccation polygons (Figure 2.52). The quarry face exposes hundreds of couplets of mudstone–fine sandstone, each between a few millimetres and several centimetres thick; it is an excellent example of 'B to C association' flagstones as defined by Donovan (1980) (see 'Introduction', Figure 2.3 for further details). The sandstone layers commonly extend downwards into small shrinkage cracks that penetrate the underlying mudstone, the so-called 'fang structures' of Donovan (1980). Differential compaction strongly deformed many of the crack infills so that they are bulbous or sinuous in cross-section. It is understandable that they were originally interpreted as bioturbation (Crampton and Carruthers, 1914) and their origin remains controversial (see 'Interpretation' below).

Interbedded with the mudstone–sandstone couplets are abundant, thin dolomitic layers up to about 2 cm thick that weather to a distinctive orange-brown colour. During diagenesis, the dolostones were lithified earlier than the adjacent mudstone–fine sandstone couplets and so behaved in a more competent fashion during early deformation. As a result, the dolostones preserve a remarkable array of small-scale compressional and extensional, brittle and ductile

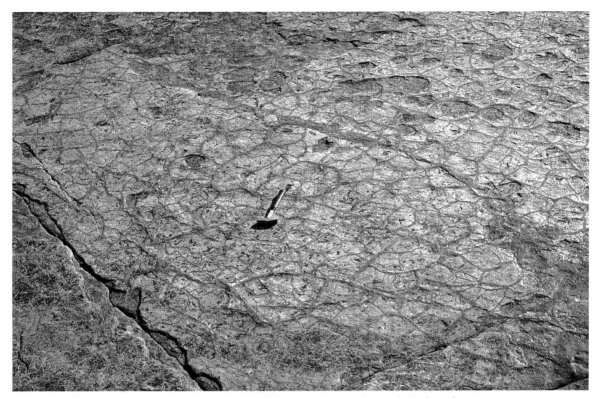

Figure 2.52 Desiccation polygons, Wick Quarries, South Head, Wick. (Photo: P. Stone.)

structures (Figure 2.53). Recumbent folds and thrusts demonstrate considerable local shortening (Figure 2.53a,b), but other beds in close proximity are boudinaged or offset by small arrays of en-echelon normal faults (Figure 2.53c).

The main face of the quarry exposes a sandstone body above the laminites. About 2 m thick, in cross-bedded sets up to about 30 cm, the sandstone is the 'D association' of Donovan (1980). North from here, the central part of the GCR site spans a quarried area around Trinkie Pool, which is slightly higher in the sequence than that exposed in the southernmost quarry. The distinctive mudstone–fine sandstone laminites again dominate, but sporadic thicker (up to about 10 cm) sandstone beds are also present. There is one thicker development of laminated mudstone ('A association' of Donovan, 1980) from which fish remains have been recovered and which is reported to be cut by hydrocarbon-bearing veins (Trewin, 1993). To the north, and higher in the succession, the laminites are interbedded with increasing amounts of thin sandstone beds, which are generally internally cross-laminated and, in some cases, extensively and chaotically convoluted.

Thin sandstone dykes cutting obliquely across 1 m or more of strata are further evidence of wet-sediment mobility. Exposed bedding surfaces of the sandstones are extensively rippled (Figure 2.54) and carry a range of cracks. Some are polygonal, but most are commonly lenticular and either randomly arranged or aligned. Locally, the crack patterns interfere to give complex, mixed arrays that are probably polygenetic.

Interpretation

The strata exposed in the series of old quarries at South Head, Wick were originally laid down in a large lake. The lithologies reflect changes in lake conditions, primarily water depth, and these have been grouped together into four general lithofacies associations by Donovan (1980) (see 'Introduction', and Figure 2.3 for further details). The South Head sequence shows two major deepening–shallowing cycles, on which bed-by-bed variations are superimposed.

The southernmost, oldest strata are sandstones, which were probably shallow-water deposits, with some polygonal desiccation cracks indicating sporadic emergence as the lake

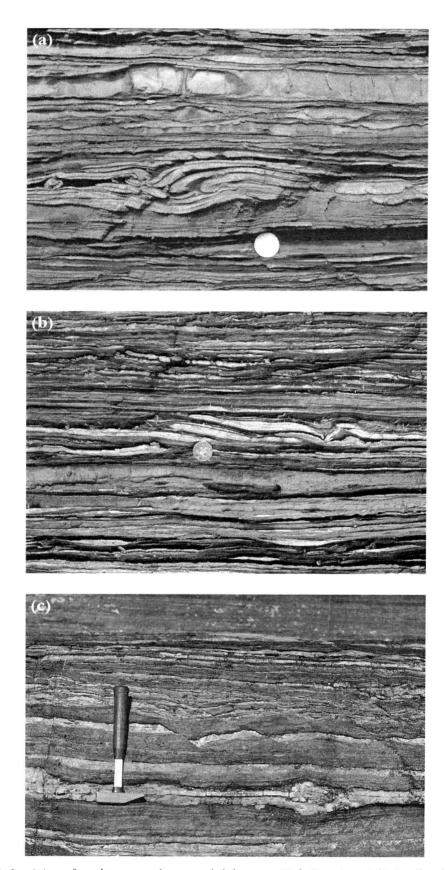

Figure 2.53 Laminites of mudstone, sandstone and dolostone, Wick Quarries. (a,b) Small-scale recumbent shortening structures; (c) small-scale brittle extensional faults. (Photos: P. Stone.)

Figure 2.54 Rippled sandstone surface, Wick Quarries. (Photo: P. Stone.)

dried up. Stratigraphically above these sandstones, the laminated mudstone–fine-grained sandstone couplets ('B to C association') seen in the southernmost quarry are the deposits of a shallow lake with a seasonally fluctuating water-level. Sporadic, thicker sandstone beds are probably turbidite units. Higher in the sequence, the significantly thicker and more massive sandstone body (the 'D association') is evidence for a return to very shallow water depths. It represents an increased clastic input to what had previously been a relatively distal lacustrine area, the shallower water allowing wave or fluvial reworking of the sediment in cross-bedded sets. Parnell *et al.* (1990) go further and suggest a partial aeolian origin for the cross-bedding, implying temporary emergence.

Subsequent deepening of the lake is shown by the re-appearance of the 'B to C association' laminites, a trend that reaches its fullest development with the fish-bearing mudstone laminite close to Trinkie Pool. In detail, this bed comprises sub-0.5 mm, alternating layers of fine, clastic sediment, carbonate and organic carbon probably derived from seasonal algal blooms. The whole assemblage is interpreted as 'a non-glacial varve recording sequential deposition in a tropical, eutrophic lake whose waters were subject to some degree of thermal stratification' (Donovan, 1980). The 'A association' unit is evidence for a deep lake environment remaining relatively stable for at least hundreds, and possibly thousands, of years. Thereafter, renewed shallowing of the lake is shown by the up-sequence (northward) return to 'B to C association' laminites and the increasing proportion of sandstone beds with polygonal desiccation cracks and rippled surfaces.

Polygonal crack patterns on bedding surfaces have been cited above as evidence for shallowing of the lake, with periodic emergence and desiccation of the sediments. However, most of the shrinkage cracks seen in the South Head section are isolated, curved and lenticular, and their interpretation is more controversial. Donovan and Foster (1972) considered them to be subaqueous in origin, and formed by a synaeresis-like process, perhaps linked to salinity changes in the lake. This interpretation has been supported by Trewin (1992, 1993), Barclay *et al.* (1993) and Trewin and Thirlwall (2002), and precludes subaerial conditions. Rogers and Astin (1991) and Astin and Rogers (1992, 1993) expressed a contrary view, suggesting that the formation of gypsum, and possibly halite, crystals during the drying out of the lake (with commensurate increase in salinity) was an important precursor to the formation of the lenticular cracks. The crack patterns and geometry were controlled by that of the crystals, the extent of desiccation and the thickness of sediment layer. During dry periods, wind-blown sand filled the open cracks and was deposited as thin lenses and laminae on the exposed playa-lake floor. The Astin and Rogers interpretation requires longer periods of subaerial conditions than the interpretation of Donovan and Trewin.

One of the most striking features of the 'B to C association' laminites is the widespread presence of small-scale, but intense compressional and extensional structures (Figure 2.43). The structures are picked out by the dolomitic layers that appear to have lithified before the enclosing mudstone, and so responded differently to early deformation. Compressional structures are more common than extensional features, but both are extensively developed and may occur in close proximity to each other, both vertically and horizontally. A complex, polyphase history of bed-parallel movement seems likely. This may have partly occurred during burial and loading, but may also be a result of adjustment during extensional faulting and subsidence of the Orcadian Basin.

Conclusions

The Wick Quarries GCR site provides an exceptionally well-exposed representative section of part of the Middle Devonian Lower Caithness Flagstone Group. Two lithofacies cycles are seen, representing the sequential shallowing and deepening of the Orcadian Basin lake. The site is of great importance in regional interpretation of the tectonics and palaeogeography of the Orcadian Basin. In addition, the excellent preservation of a spectacular array of shrinkage-crack styles provides pertinent evidence to current scientific debate concerning their origin. Polygonal crack patterns have traditionally been associated with desiccation, whereas lenticular cracks have been regarded as subaqueous. A recent re-interpretation of the lenticular cracks in the Wick Quarries section as the result of evaporite mineral formation and subsequent dissolution, subaerial desiccation and infilling by aeolian sand presents a radically different interpretation of lake chemistry and sedimentation in this part of the Orcadian Basin. A further aspect of the importance of the site lies in it being one of the few localities where potential hydrocarbon source rocks can be examined.

ACHANARRAS QUARRY, CAITHNESS (ND 150 544)

Potential ORS GCR site

P. Stone

Introduction

The disused quarry on Achanarras Hill provides a rare exposure of the Middle Devonian Achanarras Limestone Member. This distinctive unit is a lithostratigraphical marker bed separating the Upper and Lower Caithness Flagstone groups and allowing correlation of Orcadian Basin sequences from Shetland and Orkney south to the Moray Firth (Trewin and Thirlwall, 2002; Marshall and Hewett, 2003). The richest Old Red Sandstone fish site in Great Britain, the site is already a palaeontological GCR site, of international importance and renowned for its abundant and varied, well-preserved fossil fish fauna, including many whole specimens. The following account supplements that in the fossil fishes GCR volume (Dineley, 1999a). The seminal modern work on the Achanarras Limestone Member and its remarkable fish fauna is by Trewin (1986), who has also provided a field guide to the site (Trewin, 1993).

Achanarras Quarry was intermittently worked for flagstone and roofing slate from about 1870 to 1961. Since 1980, the quarry has been managed by the Nature Conservancy Council (and subsequently by Scottish Natural Heritage), and strict access and fossil collecting conditions apply. The limestone member represents the fullest development of a deep-water, lacustrine lithofacies seen in the Orcadian Basin. In addition to its unique sedimentological features and fish fauna, the site has important implications for an overview of basin palaeogeography and tectonics. The broad geological setting is described by Johnstone and Mykura (1989) and Mykura (1991).

Description

The GCR site is centred on the disused quarry excavated on the north side of Achanarras Hill, about 2 km west of the village of Spittal (Figure 2.55), where (at the time of writing) flagstones are being quarried from a stratigraphical level slightly above that seen at Achanarras. At Achanarras Quarry, the worked rock faces provide exposure through the Achanarras Limestone Member, a distinctive, 3.6 m-thick unit of fish-bearing, carbonate-rich laminites taken to mark the top of the Lower Caithness Flagstone Group (Donovan *et al.*, 1974; Trewin, 1986). The fish bed is at or slightly below the Eifelian–Givetian boundary (Paton, 1981). In the exposed quarry section (Figure 2.56), the laminites strike approximately north–south and dip a few degrees towards the east.

The section exposed is summarized in Figure 2.57 (after Trewin, 1986, 1993). At the base, thinly bedded, grey siltstone contains paler, silty laminae and small, isolated, rippled lenses of

Achanarras Quarry

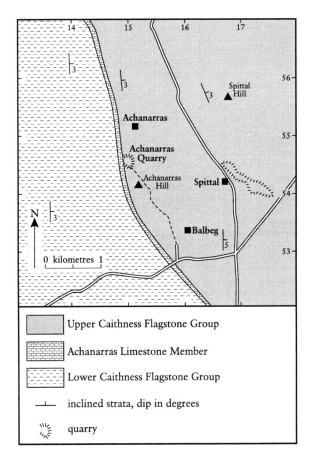

Figure 2.55 Geology of the Achanarras area and location of the Achanarras Quarry GCR site.

fine-grained sandstone; plant detritus is fairly common. These beds are the topmost of the Robbery Head Subgroup of the Lower Caithness Flagstone Group. Above, there is an abrupt transition to the dark grey, finely laminated fish bed, which forms the lowermost 1.95 m of the Achanarras Limestone Member (Trewin, 1986). The lamination is caused by a fine alternation of clastic, organic and carbonate (calcitic or dolomitic) laminae in varying relative proportions. The laminae are sub-millimetre in thickness, with an average clastic–carbonate pair about 0.7 mm thick; the organic laminae are very thin, and mostly less than 0.1 mm (Trewin, 1986). Trewin found six horizons of different fish diversity and relative abundance of species, and full details of the fauna are given by Dineley (1999a). The site is the type locality of *Rhamphodopsis threiplandi* Watson and *Palaeospondylus gunni* Traquair. Trewin (1993) noted that *Gyroptychius* is absent at the site, although common in the stratigraphically equivalent Sandwick Fish Bed of Orkney.

The upper part of the Achanarras Limestone Member, from which few fish fossils have been recovered, consists largely of clastic and dolomitic laminae. The fish bed and part of the overlying laminite unit are below water-level in the flooded quarry, but the topmost 50 cm of the

Figure 2.56 Part of the disused Achanarras Quarry, now flooded, showing the upper 50 cm of the Achanarras Limestone Member overlain by the basal beds of the Upper Caithness Flagstone Group. (Photo: P. Stone.)

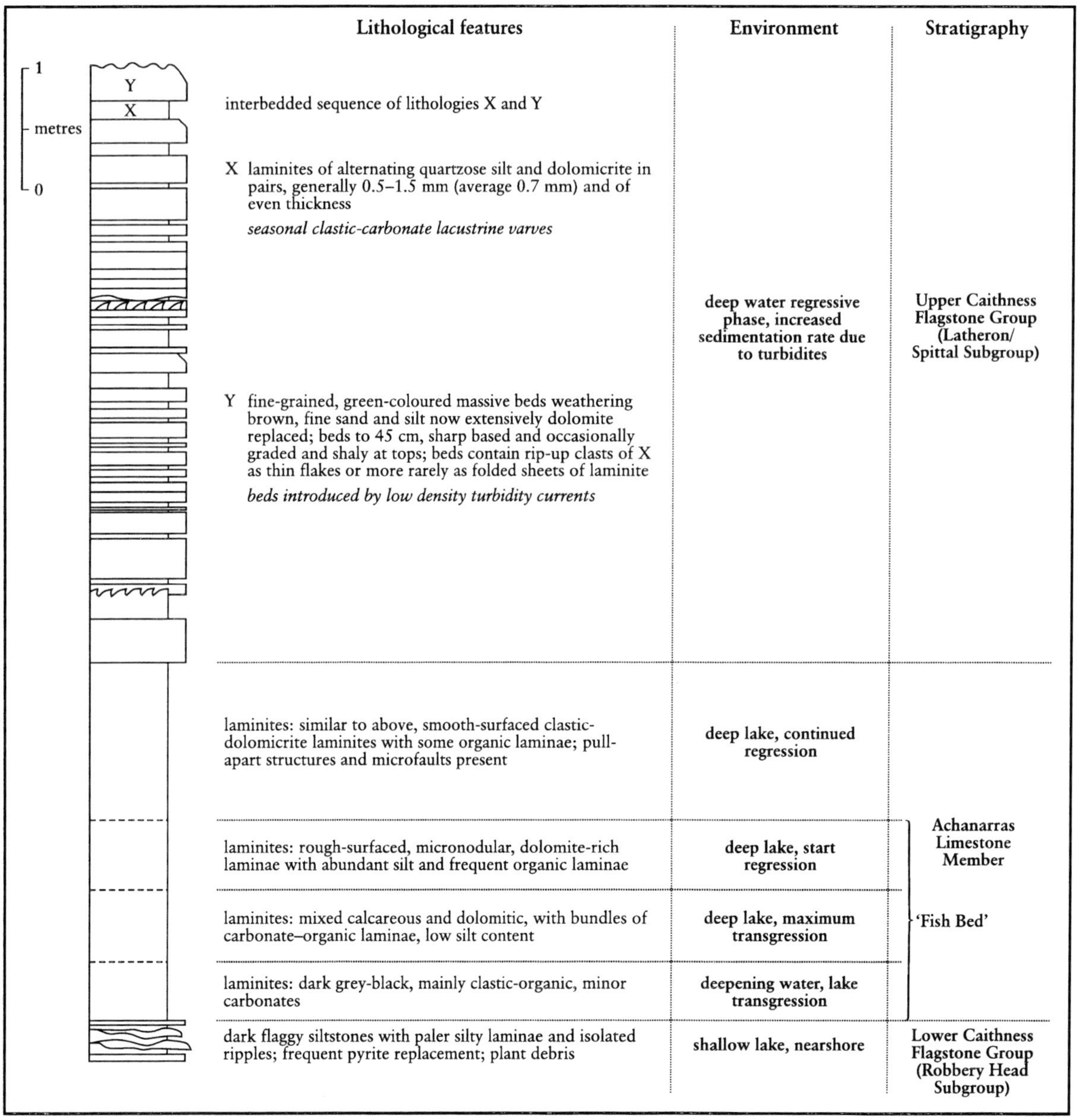

The Orcadian Basin

	Lithological features	Environment	Stratigraphy
	interbedded sequence of lithologies X and Y		
	X laminites of alternating quartzose silt and dolomicrite in pairs, generally 0.5–1.5 mm (average 0.7 mm) and of even thickness *seasonal clastic-carbonate lacustrine varves*	deep water regressive phase, increased sedimentation rate due to turbidites	Upper Caithness Flagstone Group (Latheron/ Spittal Subgroup)
	Y fine-grained, green-coloured massive beds weathering brown, fine sand and silt now extensively dolomite replaced; beds to 45 cm, sharp based and occasionally graded and shaly at tops; beds contain rip-up clasts of X as thin flakes or more rarely as folded sheets of laminite *beds introduced by low density turbidity currents*		
	laminites: similar to above, smooth-surfaced clastic-dolomicrite laminites with some organic laminae; pull-apart structures and microfaults present	deep lake, continued regression	Achanarras Limestone Member
	laminites: rough-surfaced, micronodular, dolomite-rich laminae with abundant silt and frequent organic laminae	deep lake, start regression	
	laminites: mixed calcareous and dolomitic, with bundles of carbonate–organic laminae, low silt content	deep lake, maximum transgression	'Fish Bed'
	laminites: dark grey-black, mainly clastic-organic, minor carbonates	deepening water, lake transgression	
	dark flaggy siltstones with paler silty laminae and isolated ripples; frequent pyrite replacement; plant debris	shallow lake, nearshore	Lower Caithness Flagstone Group (Robbery Head Subgroup)

Figure 2.57 Section at Achanarras Quarry. After Trewin (1986, 1993).

limestone member form part of the rock face immediately above water-level (Figure 2.56). Most of the strata seen are the lowest beds of the Upper Caithness Flagstone Group (the Latheron Subgroup, locally known as the 'Spittal Beds'). They consist of thin units of clastic–dolomitic laminites, lithologically similar to those in the upper part of the Achanarras Limestone Member, alternating with beds of fine-grained sandstone ranging up to 45 cm thick. The sandstone beds are fairly massive and greenish grey where fresh, but weather brown owing to a high proportion of secondary dolomite. The bed bases are mostly sharp and some grading is common, sporadically fining upwards to a thin mudstone at the top of the bed. Rip-up clasts of clastic–dolomitic laminite are fairly common in the lower parts of the sandstone beds.

Interpretation

The varying lithofacies in the Achanarras section were ascribed by Trewin (1986) to the effects of lake transgression and regression. The fine laminations of the fish bed were interpreted by Rayner (1963) as lacustrine varves with an annual periodicity. Donovan (1980) refined Rayner's model, proposing that the clastic–carbonate–organic triplets resulted from deposition in a thermally stratified, tropical lake, with cold, anoxic bottom conditions in the deeper parts of the lake. Rayner and Donovan agreed that increased seasonal algal growth and photosynthesis would have caused a rise in water pH, with resulting carbonate precipitation. The overlying organic laminae were the accumulated remains of the dead phytoplankton as the algal bloom decayed. The latter phenomenon might also have been responsible for periodic mass mortality in the fish population. The clastic laminae may have been seasonally influenced by rainfall and increased run-off into the lake, or may have arisen from repeated microturbidite flow unrelated to seasonal climatic events.

The lowest strata seen, below the Achannaras Limestone Member, were deposited in a near-shore, shallow lake environment. The overlying fish bed forming the base of the Achannaras Limestone Member demonstrates the effects of lake transgression and deepening water. The upper part of the limestone member, above the fish bed (*sensu stricto*), contains relatively few organic laminae, reflecting a decline in plankton productivity, and a higher proportion of clastic input. This marks the beginning of lake regression, but the fully developed, deep-water regressive phase is represented by the abrupt incoming of substantial, low-density turbidite sands at the base of the Upper Caithness Flagstone Group. Between turbidity flows, deposition of the clastic–carbonate laminites continued. Assuming a broad seasonal control on laminite deposition, the Achanarras Limestone Member is estimated to represent accumulation over about 4000 years (Rayner, 1963; Trewin, 1986).

Conclusions

The GCR site at Achanarras Quarry provides a rare section through the Achanarras Limestone Member and its contacts with the overlying and underlying strata of the Upper and Lower Caithness Flagstone groups. The limestone member is a valuable stratigraphical marker bed for correlation within the Orcadian Basin, and contains a remarkable fossil fish fauna of international importance. The lithofacies record the transgression and regression of a lacustrine environment, possibly recording the maximum transgression by the Orcadian Basin lake. Lacustrine deposition of the fine laminites was controlled by seasonal increased algal productivity and the resulting changes in lake water chemistry. The organic laminae are the product of decay of seasonal algal blooms, the carbonate laminae were deposited during periods when increased photosynthesis raised the pH of the lake waters. The fine-grained clastic laminae were deposited from suspension of material introduced to the lake, probably mainly during periods of seasonal rainfall and increased run-off.

SARCLET, CAITHNESS (ND 353 431–ND 354 435)

Potential GCR site

P. Stone

Introduction

The cliff sections on either side of The Haven, Sarclet provide good exposure of conglomerates and sandstones of the Lower Devonian Sarclet Group. Fluvial and probable aeolian lithofacies are overlain to the north by mudstone marking a major lacustrine transgression. A spectacular aspect of the section is the presence of deformed sheets overlying bed-parallel basal detachments, with other dislocation planes cutting bedding at a low angle. The deformation is likely to be of tectonic origin, but its age is uncertain. Field guides to the site are provided by Armstrong *et al.* (1978a) and Trewin (1993), and much detail is contained in a PhD thesis by Donovan (1970).

Good exposure of Lower Devonian strata is rare in the northern part of the Orcadian Basin. The Sarclet site is therefore of regional importance as an example of an otherwise poorly represented part of the stratigraphy, providing data necessary for an overview of the tectonics and palaeo-geography of the basin. The broad geological setting of the site is given by Johnstone and Mykura (1989) and Mykura (1991).

Description

From the village of Sarclet a steep track leads down into The Haven, a long-abandoned base for herring fishing. Sarclet Head is a rugged promontory on the south side of The Haven. The cliffs around the headland and those extending north from The Haven provide extensive exposure through the Sarclet Conglomerate and Sarclet Sandstone formations of the Sarclet Group. These occupy the broad hinge zone of a large, open anticline that plunges gently towards the north-west and which is broken up locally by intersecting N–S- and ENE–WSW-trending fault sets (Figure 2.58). To the north and west, the Sarclet Sandstone Formation is overlain by mudstone and sandstone of the Ulbster/Riera Geo and Ulbster/Ires Geo formations. A spore assemblage from the latter was assigned a late Emsian (Early Devonian) age (Collins and Donovan, 1977). The strata in the Sarclet section lie on the northern limb of the regional anticline and dip consistently north-west between 20° and 30°, so that 150 m of beds are present.

The lowest strata seen are polymictic conglomerate lenses and pebbly sandstones of the Sarclet Conglomerate Formation that form Sarclet Head, the southern margin of the site. The conglomerates are poorly sorted and contain clasts of granite, schist, quartzite and basalt, the last generally abundant and dominant in some lenses, together with sandstone intraclasts. The clasts are mostly subrounded pebbles, but range up to 30 cm boulders. The junction with the overlying Sarclet Sandstone Formation is transitional, with the proportion of pebbly beds diminishing over a few tens of metres.

The cliffs surrounding The Haven and extending to the north are of reddish brown, medium-grained sandstone in parallel-sided beds up to 50 cm thick. Many of the parallel beds are internally cross-laminated, others contain wispy mud laminae or parallel lamination defined by size variation in the rounded sand grains (Trewin, 1993). The sandstones are quartz-cemented and extensively fractured in some parts. They are also affected by dislocation planes cutting across the sequence at low or moderate angles, and locally by chaotically deformed units overlying basal, bed-parallel slide planes (Figure 2.59). In the latter examples, the degree of deformation decreases upwards from the basal detachment, immediately above which the sandstone is highly fractured with abundant small quartz veins.

The coastal outcrop of the Sarclet Sandstone Formation is terminated in the north by faulting at Riera Geo (ND 354 439), where thinly bedded, greenish mudstones and grey, calcareous siltstones of the Ulbster/Riera Geo Mudstone Formation are thrown down to the north.

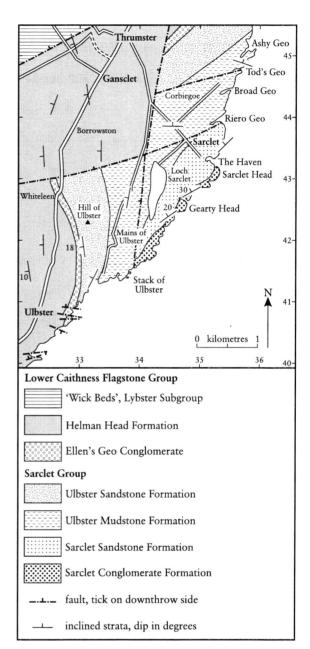

Figure 2.58 Geological map of the Sarclet area. After British Geological Survey 1:50 000 Sheet 110 (Scotland), Latheron (1985).

Interpretation

The lowest strata seen in the site area, the conglomerates and pebbly sandstones of the Sarclet Conglomerate Formation, are alluvial-fan deposits laid down by braided streams. The abundance of basalt detritus is of particular significance, suggesting a nearby area of contemporaneous volcanism (Trewin, 1993), which, from the evidence of the palaeocurrent indicators (Johnstone

Sarclet

Figure 2.59 Sarclet Sandstone Formation. Deformed sandstones rest on a bedding-parallel detachment; a low-angle dislocation cuts the regularly bedded sandstone below the detachment. (Photo: P. Stone.)

and Mykura, 1989; Mykura, 1991), lay to the southeast. Some of the overlying Sarclet Sandstone Formation may also have a fluvial origin, Mykura (1991) noting evidence for increased meandering of the rivers with time. However, Trewin (1993) noted that the wispy lamination defined by thin, irregular mud laminae resembles aeolian adhesion ripples. The lamination produced by size sorting of rounded sand grains provides further evidence for deposition by aeolian processes. Higher in the succession, the thinly bedded mudstones and calcareous siltstones of the Ulbster/Riera Geo Formation record deposition in a shallow lacustrine environment, and represent one of the earliest, significant lake transgressions in the Orcadian Basin.

Of particular interest at this site is the localized deformation of beds in the Sarclet Sandstone Formation above bed-parallel slide surfaces (Figure 2.59). The overall appearance is suggestive of synsedimentary sheet slumping, but the most deformed sandstone is fractured and quartz-veined, and was clearly lithified when deformed (Trewin, 1993). The age of the deformation at Sarclet therefore remains uncertain.

Conclusions

The Sarclet site provides good exposure through a section representative of the Lower Devonian Sarclet Group. The site is of regional importance as outcrop of Lower Devonian strata is rare within Caithness and the northern part of the Orcadian Basin. Alluvial-fan conglomerates form the base of the exposed succession, with an upwards transition to sandstones that may be either fluvial or aeolian in origin, or formed by a combination of both processes. The Sarclet Sandstone Formation is cut by numerous low-angle dislocation planes and contains chaotically deformed sheets resting on bed-parallel detachment planes. The origin of this deformation is unclear, with both synsedimentary and post-lithification tectonic indicators, and its timing also remains uncertain. Overlying the Sarclet Group, mudstones and calcareous siltstones of the Ulbster/Riera Geo Mudstone Formation represent a major lacustrine transgression, one of the earliest significant lake developments in the Orcadian Basin.

The Orcadian Basin

TARBAT NESS, ROSS AND CROMARTY (NH 929 873–NH 939 851)

P. Stone

Introduction

The cliff and foreshore sections on either side of Tarbat Ness afford excellent exposure through parts of the Strath Rory Group (Middle Devonian) and the Balnagown Group (Upper Devonian). The site also provides evidence pertinent to the debate over the relationship between the Middle and Upper Devonian successions. A regional unconformity seems likely, but there appears to be a conformable succession at this site. A range of lithofacies is present, including fluvial pebble conglomerates and pebbly sandstones, aeolian sandstones, and a sabkha-type facies of sandstone and mudstone containing evidence of evaporitic deposition, desiccation, intermittent rainfall and subaerial animal activity. The most comprehensive description and interpretation of the site are by Rogers (1987), whose results were incorporated into a regional palaeogeographical analysis by Marshall *et al.* (1996).

The importance of the Tarbat Ness site is twofold. Firstly, it provides a complete section through the Upper Devonian succession, allowing interpretation and characterization of an otherwise poorly represented part of the stratigraphy of the Orcadian Basin. It also provides one of the few apparently conformable transitions between the Middle and Upper Devonian strata. The site thus has sedimentological value, and is important to the broader interpretation of the palaeogeography, development and tectonics of the Orcadian Basin.

Description

The Tarbat Ness GCR site consists of the cliff and foreshore exposures to the west and south of the lighthouse. It forms the extreme north-eastern promontory of Easter Ross between the Moray and Dornoch firths. The exposed strata span the Middle Devonian Strath Rory Group and Upper Devonian Balnagown Group, the contact between them being variously interpreted as faulted or conformable. The dip is fairly uniform towards the north-west, steepening in that direction from 10° to nearly 40°, within the south-east limb of the Black Isle Syncline. About 800 m of beds are seen within the site, although continuity is disrupted by faulting towards the top and the bottom of the section.

The oldest strata are preserved on the east coast of the promontory, as a strike section along the shore southwards from a point about 1.5 km south of Tarbat Ness (Figure 2.60). They are thickly bedded, yellow and red, fluvial sandstones with subordinate (but fairly common) thin interbeds of calcareous mudstone. The beds were assigned by Armstrong (1977) to the Strath Rory Group. The sandstones are generally cross-bedded on a fairly large scale and internal lamination is highly convoluted locally. The mudstone interbeds contain sporadic carbonate concretions, some of which have yielded fish fragments that support an early Givetian (Mid-Devonian) age. Elsewhere, the Strath Rory Group ranges down into the Eifelian Stage (Dineley and Metcalf, 1999).

The top of the Strath Rory Group was placed by Armstrong (1977; see also Institute of Geological Sciences, 1973) at a fault on the south side of Port Tarsuinn, probably the same one identified by Rogers (1987) as the South Wilkhaven Fault (Figure 2.60). This interpretation allows for an unconformable relationship between the Strath Rory Group and the overlying strata, the unconformity itself being farther inland and unexposed. By contrast, Rogers (1987) proposed a complete and conformable succession, compiled in various fault blocks around Port Tarsuinn. He noted a thick sandstone body, probably of aeolian origin (the Port Tarsuinn Member of the Rockfield Formation in his informal stratigraphy), at the top of the Strath Rory Group conformably overlain by coarse, pebbly sandstones of the Balnagown Group.

Rocks of the Balnagown Group (Armstrong, 1977) form the headland of Tarbat Ness and are exposed from there along the north coast of the promontory. Red, pebbly sandstones with conglomeratic lenses and interbeds predominate at the base of the group, and are exposed northwards from Port Tarsuinn and around the headland itself. Rogers (1987) informally named these beds the 'Tarbat Ness Formation'. The pebbles are invariably well-rounded and mainly quartzose. They form conglomeratic beds up to about 20 cm thick, but more commonly occur as lenticular bodies (Figure 2.61a) or as isolated clasts strewn along bedding surfaces (Figure 2.61b). The

Tarbat Ness

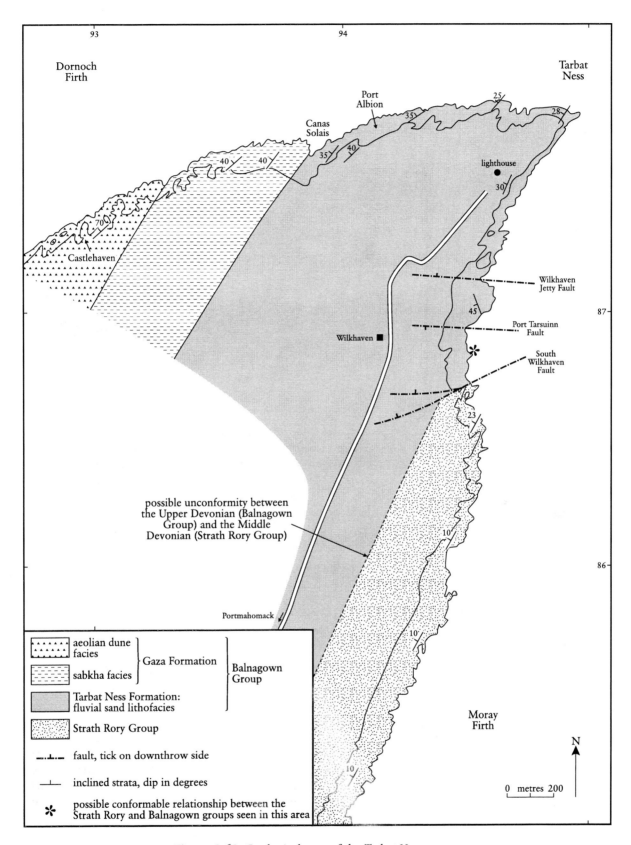

Figure 2.60 Geological map of the Tarbat Ness area.

The Orcadian Basin

Figure 2.61 Tarbat Ness Formation. (a) Conglomerate lenses; (b,c) pebbly layers and isolated pebbles in cross-bedded sandstones. (Photos: P. Stone.)

formation is cross-bedded on a large scale, with conglomeratic layers forming individual cross-sets and isolated pebbles lining foresets (Figure 2.61c). The concentration of pebbles decreases up-sequence towards Tarbat Ness, but thereafter appears to increase again towards the top of the Tarbat Ness Formation. Around the lighthouse and headland, many of the thick, red sandstone beds contain only sporadic, isolated pebbles and many are pebble-free. The sandstones are cross-bedded on a large scale (Figure 2.62a), and there are many examples of convolute bedding and post-depositional deformation, probably caused by slumping and de-watering, the latter phenomenon having produced sand volcanoes in places (Figure 2.62b).

The top of the Tarbat Ness Formation, to the west of the headland, shows an irregular increase in the proportion of pebbles in the thick, cross-bedded red sandstones towards Port Albion and Canas Solais. There, pebble content and bed thickness abruptly decrease, above which the higher part of the Balnagown Group is dominated by thinly interbedded sandstone and mudstone. Rogers (1987) named these beds the Gaza Formation.

Tarbat Ness

The characteristic lithologies of the Gaza Formation are very thinly bedded, argillaceous, fine-grained sandstones and rare mudstones (Figure 2.63a). Colour ranges from red to yellow-green, with many beds having a mottled appearance. An irregular, wavy lamination, defined by colour variation or silty partings, is fairly ubiquitous in the thin sandstone beds, and very low-amplitude ripples are apparent on some bedding surfaces, particularly where they are picked out by a colour variation. Red mudstone flakes are common on some bedding planes and sporadic, thicker, sandstone beds also contain isolated mudstone clasts. The thicker sandstones may be either fluvial or aeolian, but higher in the Gaza Formation, towards the north-western boundary of the GCR site at Castlehaven, Rogers (1987) identified compound aeolian dunes (draas) forming sandstone bodies up to 14 m thick. These sandstones show large-scale, internal cross-bedding, described by Rogers in terms of three orders of bounding surface. He also noted the presence of small (0.5–3 mm) 'adhesion warts' within the cross-sets (Figure 2.63b) and interpreted them as clusters of grains held together by moisture from sparse rainfall. Rare examples of rain impact pits were also noted elsewhere in the sequence.

Figure 2.62 Tarbat Ness Formation. (a) Tabular cross-bedding; (b) sand volcano. (Photos: P. Stone.)

The Orcadian Basin

Figure 2.63 (a) Thinly bedded sandstones of the Gaza Formation west of Canas Solais. (b) 'Adhesion warts' in the Gaza Formation near Castlehaven. (Photos: P. Stone.)

Rogers (1987) described a range of trace fossils from the Gaza Formation, mostly in the Castlehaven area. The forms include tetrapod tracks attributed to early amphibians, arthropod tracks and burrows, and zones of strata homogenized by bioturbation. Although compatible with a Late Devonian age, the trace fossils do not provide a more exact date for the formation, which is generally presumed to be Frasnian to Famennian in age (Mykura, 1991; Dineley and Metcalf, 1999).

Interpretation

The lower part of the sequence exposed within the GCR site comprises thick, cross-bedded, fluvial sandstones of the Strath Rory Group, deposited in response to the basin rejuvenation that accompanied early Mid-Devonian deformation and uplift (Armstrong, 1977). Thick conglomerate beds in the lower part of the Strath Rory Group, not represented within the GCR site, indicate derivation from uplifted areas of Moine and Dalradian rocks to the west and south; a similar provenance direction is indicated by the cross-bedding in the sandstones.

Within the fluvial sandstones of the Strath Rory Group, interbeds of red mudstone with calcareous laminae and concretions probably represent lacustrine episodes and indicate the

Tarbat Ness

intermittent extension, into the Easter Ross region, of the main Orcadian Basin lake that was centred on Caithness. The thick aeolian sandstones (Port Tarsuinn Member) identified by Rogers (1987) at the top of the Strath Rory Group, provide evidence of temporary subaerial conditions prior to the re-establishment of a fluvial regime at the base of the Balnagown Group. This is marked by the appearance of conglomeratic lenses, pebbly sandstones and thick, cross-bedded fluvial sandstones, which, together with the absence of red mudstone interbeds, are characteristic of the Tarbat Ness Formation. The conglomerate pebbles are invariably well-rounded (Figure 2.61), suggesting some degree of reworking of the underlying Strath Rory Group. The fluvial sandstones show widespread internal convolute lamination and liquefaction phenomena such as sand volcanoes (Figure 2.62b), indicating dewatering during rapid burial of water-saturated sand, probably in a large, braided river system.

The top of the Tarbat Ness Formation marks a major environmental change. The overlying non-pebbly, fluvial sandstones and thinly interbedded sandstones and mudstones (the Gaza Formation) were interpreted by Rogers (1987) as having formed in a sabkha-like environment. The presence of the fragmented remains of desiccated mud layers as mudstone flakes in the sandstones, and 'adhesion warts' produced by differential wetting of a dry sand surface support this interpretation. Rippled surfaces may have formed either by adhesion of blown sand to an irregularly damp surface or by sheet flooding, but Rogers (1987) speculates that, in either case, wind scour may have accentuated the features. Roughly polygonal to irregularly sinuous ridges rising up to 1 cm above some bedding surfaces caused localized disruption of the underlying sandstone laminae. Rogers (1987) interpreted them as a result of surface buckling during expansion due to evaporite formation. This interpretation supports a sabkha origin. Rain pitting, and the tetrapod and arthropod tracks add further evidence of predominantly subaerial environments during deposition of the Gaza Formation.

At the north-western extremity of the GCR site, the Gaza Formation is overlain by a thick, compound aeolian dune sequence identified as a draa by Rogers (1987). From measurements of the internal set and co-set relationships, Rogers calculated dune heights up to about 6 m, with interdune distances of up to 80 m. Most of the animal tracks were recorded in the dune sandstones, on surfaces strewn with 'adhesion warts'.

A palaeoenvironmental model for the Upper Old Red Sandstone of the Tarbat Ness area, after Marshall *et al.* (1996) is shown in Figure 2.64. This envisages fluvially dominated marginal facies passing distally to sandy sabkha with migrating dunes.

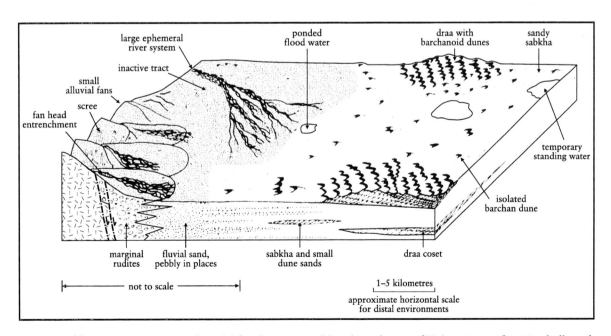

Figure 2.64 Palaeoenvironmental model for the Upper Old Red Sandstone of Tarbat Ness. After Marshall *et al.* (1996).

The Orcadian Basin

Conclusions

The Tarbat Ness GCR site provides an exceptionally well-exposed representative section of part of the Middle Devonian Strath Rory Group and its boundary with the overlying Upper Devonian Balnagown Group. This is one of the few localities where the Middle–Upper Devonian boundary can be examined, and here it appears to be conformable, in contradiction to the generally accepted model of a regional unconformity. The range of lithofacies exposed is unusually large. Fluvial sandstones with sporadic lacustrine mudstone interbeds are typical of the upper part of the Strath Rory Group. Two markedly different facies associations are seen in the Balnagown Group – a fluvial association of pebble conglomerate and cross-bedded sandstone forms the Tarbat Ness Formation at the base of the group; above this, a fluvial–aeolian–sabkha association forms the Gaza Formation. The latter contains striking examples of desiccation and evaporitic structures, features associated with sporadic rainfall in an arid environment, and evidence of subaerial animal activity. Overall, Tarbat Ness is a very important site, providing a rare insight into Mid- and Late Devonian environments and habitats at the margin of the Orcadian Basin.

DUN CHIA HILL (LOCH DUNTELCHAIG), INVERNESS-SHIRE (NH 600 285–NH 600 291)

D. Stephenson

Introduction

At the south-west limit of the Orcadian Basin, the outcrop of Middle Old Red Sandstone (MORS) rocks extends down the south-east side of the Great Glen as far as Foyers (Stephenson, 1972, 1977; Mykura, 1982). Within this outcrop 'extension', the basal beds are characterized by coarse breccias, breccio-conglomerates and conglomerates that rest on a highly irregular palaeo-landsurface of Grampian Group (Dalradian) metasedimentary rocks and late Caledonian granitic rocks. The unconformity is well exposed in several places, the most dramatic and instructive being on the south-east side of Dun Chia hill, between Loch Ruthven and Loch Duntelchaig. There, the unconformable junction is exposed continuously for a distance of some 1300 m and the overlying breccio-conglomerates form cliffs from Tom Mor (NH 591 280) in the south-west, through Craig na h-Iolaire (NH 597 282), to the spectacular, 50 m-high Creag nan Clag (Crag of the Bells; NH 601 289) in the north-east (Figure 2.65). The effects of pre-MORS weathering and erosion on the underlying Grampian Group rocks are well seen and the breccio-conglomerates yield much information about sedimentary processes and palaeogeography of the region in Mid-Devonian time. Similar basal beds are also well exposed in cliffs between Creag Dhearg (NH 618 294) and Carn Mor (NH 627 303), 2–3 km to the ENE of Creag nan Clag, but the unconformity is less well-exposed.

The Old Red Sandstone rocks of the area were first described by Wallace (1880), who described the unconformity and the overlying beds at Dun Chia and Creag Dhearg in some detail. Further descriptions and interpretations of these localities were given by Mould (1946) in her account of the nearby Foyers granitic pluton. The most detailed investigation was that of Mykura (1982), which was used in the compilation of the 1:50 000 geological map, Sheet 73E (Foyers) (British Geological Survey, 1996) and is the basis for much of the following account.

Description

Between Loch Duntelchaig and Loch Ness in the Great Glen, there is a cross-strike outcrop width of about 5 km of Middle Old Red Sandstone strata. This represents a succession over 2 km thick that includes coarse and fine breccias, conglomerates, gritty, pebbly feldspathic and fine-grained sandstones, with subordinate mudstones, siltstones and calcareous beds. In general, the lower parts of the succession (up to 900 m) are of breccio-conglomerate and coarse bimodal conglomerate overlain by planar-bedded, medium- to fine-grained sandstones. The succession interdigitates along strike to the north-east with that of the Inverness and Nairnside districts, where fragmentary fish remains and plant spores of Mid-Devonian age have been found (Horne and Hinxman, 1914; Horne, 1923; Fletcher *et al.*, 1996). There is a regional dip of between 10° and 50° to the north-west.

Dun Chia Hill (Loch Duntelchaig)

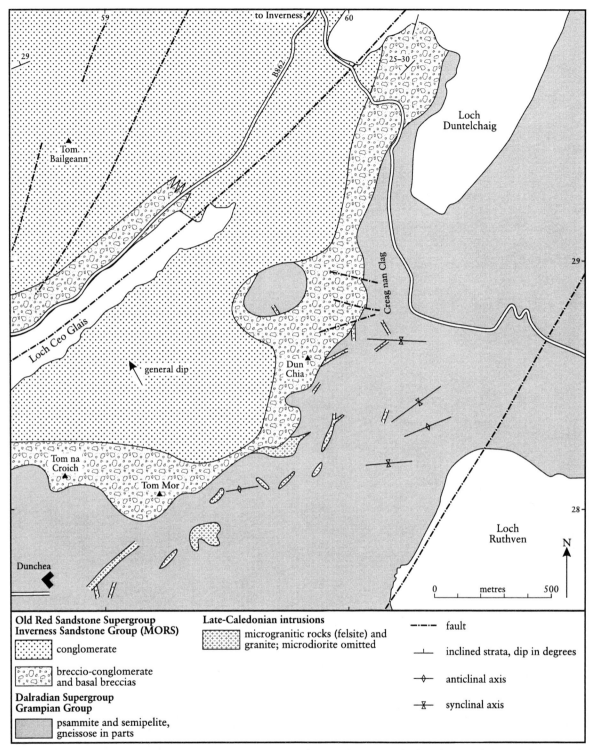

Figure 2.65 Map of the area around Dun Chia hill, at the south-west end of Loch Duntelchaig. After British Geological Survey 1:50 000 Sheet 73W (Scotland), Foyers (1996).

Throughout the area in general, the pre-MORS topography was undulating, with some steep slopes locally that probably reflected active fault scarps (Stephenson, 1972; Mykura, 1982).

Within the GCR site, between Tom Mor (NH 591 279) and Creag nan Clag (NH 600 287), the basal unconformity is exposed almost continuously for 1300 m at the foot of steep cliffs (Figure 2.66). It is undulating, but generally inclined in a NNW to westerly direction at angles ranging from 20° to 65°, with an average of 45°. Locally (NH 5985 2835), it is almost vertical on Craig na h-Iolaire.

The underlying rocks are platy psammites, gneissose in parts, of the Dalradian Grampian Group, which are generally vertical or steeply inclined, but exhibit some quite large-scale folding. Directly beneath the unconformity at the southern end of Creag nan Clag (NH 5988 2857–NH 6000 2870), a monocline, overturned towards the west, has a near-horizontal upper limb. Locally this limb is broken into small blocks, some rotated, but others still in their original position and enclosed by sandstone. Sandstone veins also penetrate downwards through the horizontal limb and axial plane of the monocline into the vertical limb beneath. A small inlier of psammite on the north-west slope of Dun Chia, some 40 m below the summit (Figure 2.65), probably represents the peak of a buried hillock or ridge crest. Here, just below the unconformity, is another monocline overturned to the south-west.

The psammites are intruded by dykes, up to 10 m thick, of microgranitic rock ('felsite') and microdiorite. Many of the felsites are intensely shattered at the unconformity and are overlain by a deposit composed almost entirely of felsite clasts.

The MORS rocks of the Moray Firth area are formally assigned to the Inverness Sandstone Group, and the basal beds at the GCR site are all part of the Bochruben Formation (Mykura, 1982). These are mainly breccio-conglomerates and conglomerates in which almost all of the clasts are locally derived from the Grampian Group or the nearby Foyers Pluton. Horizontal

Figure 2.66 The cliffs of Creag nan Clag, on the eastern flank of Dun Chia hill. The undulating unconformity between Middle Old Red Sandstone massive breccio-conglomerates and underlying flaggy psammites of the Grampian Group (Dalradian) can be traced for 1300 m along the base of the steep upper cliffs. (Photo: BGS No. D1813, reproduced with the permission of the Director, British Geological Survey, © NERC.)

and vertical variations in the basal beds of the formation are well displayed in the cliffs on the east side of Dun Chia.

At the south end of Creag nan Clag, the lowest 5–10 m are composed entirely of angular, locally derived clasts, with plates of psammite and felsite in roughly equal proportions; felsite clasts predominate in the vicinity of felsite dykes. Elsewhere, for example at the eastern end of the inlier on the north-west flank of Dun Chia, felsite clasts commonly constitute 40%, and in some restricted areas up to 90%, of the basal deposits.

The basal breccias of almost in-situ angular material are overlain by massive, very crudely bedded breccio-conglomerates composed of ungraded, subangular clasts up to 60 cm in diameter. The larger clasts are not in contact and are supported in a sparse red gritty matrix. Their long axes are randomly orientated, some being almost perpendicular to the bedding. Most are psammite and quartzite of local origin, with subordinate granite-gneiss and gneissose semipelite, but angular felsite clasts constitute up to 30% in places. Rounded boulders of biotite granite also occur sporadically. The breccio-conglomerates vary in thickness, being thickest in the cliffs of Creag nan Clag and thinning to the north-east and south-west, although they still form the greater part of all of the cliff sections.

The north-west flanks of Dun Chia and parts of the summit ridge are composed of bimodal conglomerates with a high proportion of leucogranite clasts. On Craig na h-Iolaire, coarse conglomerates with rounded granite boulders up to 60 cm in diameter overlie Dalradian 'basement' almost directly in places, although residual patches of local breccia only a few metres thick, and some breccio-conglomerates are also present. Sandstone lenses are intercalated locally with the conglomerates, such as those dipping at 15° to the WNW, between 6 m and 10 m above the unconformity at the south-west end of Creag nan Clag.

Interpretation

The prominent monocline in the Dalradian 'basement' at Creag nan Clag has an axial plane almost parallel to the unconformity and was interpreted by Mykura (1982) as flexuring due to hill creep prior to deposition of the MORS sediments. However, this coherent fold is on a remarkably large scale to have been formed in such a manner and subsequent visitors to the site have rejected the interpretation, observing that the geometry of the fold is of a tectonic style and conforms with others in the surrounding psammites. At one locality a dyke, truncated by the unconformity, is intruded along the vertical limb of the monocline but is not bent as it enters the horizontal limb. Mykura explained this by suggesting that the dyke was intruded after the hill creep but before MORS sedimentation. This somewhat implausible scenario does fit a further suggestion by Mykura (1982), concerning the origin of local concentrations of felsite in the basal MORS. Observing that, in several places on Dun Chia and Creag Dearg, the percentage of felsite clasts (40–90%) is much higher than the percentage of felsite dykes seen in the Dalradian beneath, Mykura suggested that the dykes may have been feeders to small extrusive domes on the pre-MORS land surface. Such domes are usually unstable and would have crumbled to breccia shortly after extrusion; hence no definite domes have been identified. However, in the absence of any other evidence for volcanism in the Grampian Highlands in Mid-Devonian time, this scenario is also regarded as highly tenuous.

The local origin of most of the clasts in the basal MORS and the low degree of sorting indicate that many of the deposits were formed quite near to the source of their constituents. Much of the angular material in the basal breccias is virtually in-situ and is in effect a regolith, or, if on a steep original slope, a 'fossil scree'. The lack of movement is particularly noticeable above some felsite dykes where it is difficult to separate the in-situ felsite from 'felsite scree'. This effect is even more characteristic of basal MORS deposits to the south-west of the GCR site, where 'granitic screes' are indistinguishable in places from underlying in-situ granitic rocks (Stephenson, 1972; Mykura, 1982). The freshness of many of the clasts indicates rapid accumulation in a climate in which chemical weathering was minimal. However, oxidation is commonly intense, imparting a red or purple hue to the matrix of the rocks, so climatic conditions during deposition were probably arid or semi-arid.

The breccio-conglomerates that dominate the GCR site merge with the basal breccias and have no marked basal erosion surfaces. They are unstratified and poorly sorted, with mainly subangular clasts of local material 'floating' in a matrix, usually of mud or silt grade. These are the characteristics of debris flows that moved as dense, viscous masses in which waterlogged matrix was able to support and transport clasts up to boulder size. Such deposits form typically in areas of high relief and rapid erosion, in which deposition on steeply sloping alluvial-fans results from occasional sheet floods during sudden violent rainstorms in an otherwise dry area. The thickest development of breccio-conglomerate, in the 50 m-high cliffs of Creag nan Clag, has no horizontal breaks and it is possible that the entire section represents a single debris flow, possibly filling a palaeovalley. Lenses of conglomerate and sandstone within the sequence probably originated as channel-fill deposits within the alluvial fans and some of these have cut down to rest directly on basement in places.

Higher in the sequence, reworking by water becomes apparent as the breccias become slightly better sorted, some bedding is evident and clasts become more rounded. Debris of a less local nature appears, including leucogranite clasts that must have been transported for some distance. Bimodal clast-supported conglomerates, which occur locally as channel-fills within the breccio-conglomerates, become dominant higher in the succession, from the summit of Dun Chia north-westwards. These higher conglomerates are of fluvial origin and probably originated as braided river deposits on the distal parts of fans and on the piedmont plain. Beyond the GCR site, still higher parts of the succession record the eventual burial of the fans by finer-grained sediments that accumulated on a broad floodplain (Mykura, 1982).

By the time of deposition of the MORS sediments, the Grampian Highlands had already experienced considerable rapid uplift at the end of the Caledonian Orogeny. An estimate of the scale of uplift in the northern Grampians can be obtained from petrological evidence in the aureole of the Foyers Pluton, which Marston (1971) suggested had a possible cover of 7 km at its time of emplacement in Early Devonian times. The cover was removed in a maximum of about 20 Ma to expose the pluton in Mid-Devonian time. This represents an erosion rate of at least 3.5 cm per 100 years, comparable with rates in present-day newly formed mountain areas.

The Great Glen Fault was already in existence in Mid-Devonian times and may have been the locus of a major valley that drained north-east and was bounded to the south-east by high mountains (Stephenson, 1972, 1977). The MORS basal sediments were deposited predominantly in coalescing alluvial-fans along the foot of these mountains and in the valley bottom, which widened out north-eastwards into a piedmont plain on the edge of the main Orcadian Basin. The plain probably extended across the Great Glen Fault into the area now occupied by the Black Isle and Tarbat Ness (Mykura, 1982, fig. 11). Close to the Great Glen Fault, in the Foyers–Inverfarigaig area, the highlands to the south-east were probably bounded by major active NE–SW fault scarps, related to the Great Glen Fault and with a component of downthrow to the north-west. However, around Loch Duntelchaig, the edge of the hills probably swung to the ENE; here major fault scarps were probably absent and the boundary zone between plain and mountain may have been a belt of foothills with several small active local fault scarps.

Conclusions

The spectacular cliffs on the south-east flank of Dun Chia hill, at the south-west end of Loch Duntelchaig, provide a representative section of the basal beds of the Middle Old Red Sandstone at the southern limit of the Orcadian Basin. The unconformable junction beneath these beds is very well-exposed almost continuously over a distance of 1300 m and is probably one of the best preserved 'buried landscape' unconformities in Great Britain. It is certainly of national importance and visitors cannot fail to be impressed by the scale of the exposure and by the many exceptional features that are exhibited. Beneath the unconformity, folded Neoproterozoic flaggy metasandstones, cut by felsitic dykes, are disaggregated and shattered close to an undulating former land surface. The overlying angular debris can be matched precisely to the immediately underlying rock in many places and probably accumulated as scree. Poorly sorted mixtures of angular fragments and rounded boulders (breccio-conglomerates) comprise the majority of the section and probably originated as debris flows or sheet floods associated with alluvial fans that debouched from mountainous terrane to the south-east and spread out over a broad piedmont plain to the north on the edge of the alluvial plain of the main Orcadian Basin.

TYNET BURN, MORAY (NJ 383 618)

Potential ORS GCR site

W.J. Barclay and N.H. Trewin

Introduction

Tynet Burn, 5 km north-east of Fochabers, Moray is a classic Middle Old Red Sandstone fossil fish site, discovered in 1838. Since then, it has yielded about 14 species of fish, most of which are found in the upper of two beds of calcareous nodules within laminated shales and limestones deposited in a lake. The fish fauna is correlated with the Middle Devonian (Eifelian) Achanarras Fish Bed in Caithness and other Orcadian Basin equivalents in Orkney and Shetland. The fish-bearing nodule beds alternate with conglomerates and sandstones deposited by rivers on an alluvial plain at the southern margin of the Orcadian Basin (Trewin and Thirlwall, 2002). Periodic transgression of the lake over the alluvial plain resulted in lacustrine sediments being deposited at times of high lake level; the contractions and expansions of the lake were controlled by climatic changes with Milankovitch periodicities. The palaeogeographical and stratigraphical history of the Orcadian Basin has been summarized by Trewin and Thirlwall (2002). Trewin and Davidson (1999) provided details of the sedimentary environments and palaeoecology at Tynet on the southern shore of the Orcadian lake about 387 million years ago.

Description

The geology of the site was described by Peacock *et al.* (1968). Gillen (1987) provided a brief field guide to the locality, and Wood and Norman (1991) described an excavation in the fish bed at Tynet in 1989–1990. Dineley (1999a) and Trewin and Davidson (1999) gave historical details of research at the site, and the latter gave an account of excavations made at the site in 1996, providing the main source for this account. The attractive pink, red and purple colours of the fossil fish, in contrast to the green or beige matrix of the nodules in which they occur, made the fossils very popular, and specimens from Tynet are widely dispersed in museums, university departments and private collections throughout the world.

The site (Figure 2.67) lies in the Tynet Burn below Lower Mills of Tynet (NJ 388 618), where the stream is incised into Old Red Sandstone through a cover of glacial till. The beds belong to the Middle Old Red Sandstone, which comprises about 200 m of red sandstones, conglomerates and mudstones in this area. The fish-bearing succession in the Tynet Burn consists of 28 m of conglomerates, sandstones, mudstones and limestones (Figure 2.68), fault-bounded at the top and bottom. The lowest bed (NJ 3837 6196) is a 0.9 m-thick conglomerate with numerous angular pebbles. The highest beds are faulted against a fining-upward succession of about 120 m of gently dipping beds that are exposed downstream. These comprise a basal red boulder conglomerate 39 m thick passing up through interbedded red sandstones and conglomerates into cross-bedded sandstones with subordinate conglomerate. The Lower Nodule Bed is exposed in the river (NJ 3837 197) just downstream of the 0.9 m-thick conglomerate.

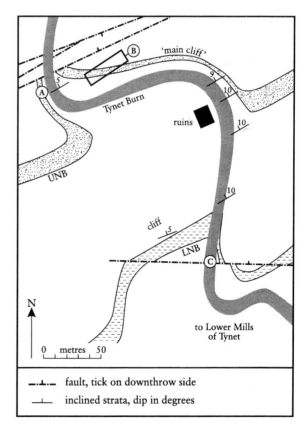

Figure 2.67 Sketch map of Tynet Burn: (A) section cleared in 1996 (Trewin and Davidson, 1999); (B) section excavated in 1989–1990 (Wood and Norman, 1991). Based on Peacock *et al.* (1968) and Davidson and Trewin (1999).

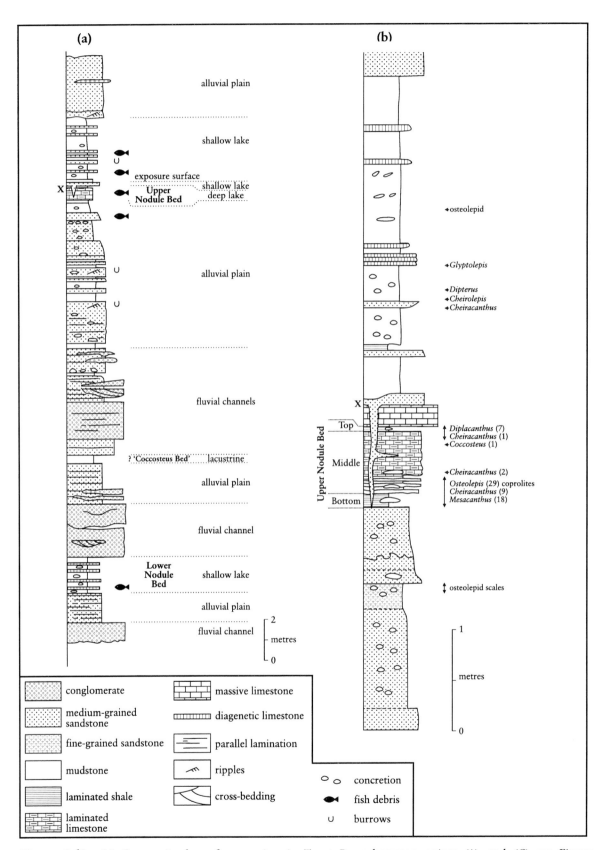

Figure 2.68 (a) Composite log of succession in Tynet Burn between points (A) and (C) on Figure 2.67. The position of the Coccosteus Bed is inferred. (b) Summary log of section excavated at Point (A). After Trewin and Davidson (1999).

Tynet Burn

The best section was formerly in a 12 m-high cliff (the 'Main Cliff') on the east bank of the stream (NJ 384 620). The higher nodule bed (the Upper Nodule Bed) crops out along the top of this cliff, but the disused quarry is covered by talus and the exposure is obscured. The bed is exposed and more accessible farther downstream in a meander scar on the west bank (NJ 3823 6203) close to the fault that forms the northern limit of the block. Excavations (NJ 3828 6205) near the 'Main Cliff' in 1989 and 1990 (Wood and Norman, 1991) and clearance of the exposure on the west bank (NJ 3823 6203) in 1996 (Trewin and Davidson, 1999) provided detailed sections of the strata here, although soft-sediment extensional and compressional structures, sandstone intrusions and minor tectonization of the beds in proximity to the main fault to the north disrupt the continuity of the beds.

The succession contains two main fish-bearing horizons (the Upper Nodule Bed and Lower Nodule Bed) and a third, intermittent horizon between them (the Coccosteus Bed), the exact position of which is uncertain (Figure 2.68). These horizons represent lake deposition within the fining-upward fluvial to lacustrine cycles. The cycles are truncated by erosion surfaces and commence with conglomerates that contain rounded to subrounded pebbles up to 15 cm in diameter, predominantly of quartzite, but with some of metamorphic and igneous lithologies. The pebbles are set in a red, medium- to coarse-grained sandstone matrix and show sorting and imbrication. Lenses of laminated and cross-bedded sandstones also occur, and there are some cross-bedded conglomerate bodies up to 0.75 m thick.

The sandstones are red with minor patchy reduction zones. They are mainly fine- to medium-grained and micaceous, and occur in beds ranging from 5 cm to 40 cm thick. Parallel lamination is the dominant structure, although most beds are only weakly laminated. A few beds are trough cross-bedded and the tops of some are linguoid current-rippled. The sandstones pass up into brick-red, massive mudstones, with small (up to 10 cm) irregular to subspherical unfossiliferous diagenetic carbonate concretions (or nodules). The mudstones containing the Upper and Lower Nodule beds are red to chocolate-brown and drab grey-green. They are mainly massive to weakly laminated on a millimetre to centimetre scale, but the fish-bearing horizons are well laminated on a sub-millimetre scale. The carbonate concretions occur in several forms within the laminites and are commonly fish-bearing.

There are three types of limestone in the section. A massive white to pale pink limestone with red, hematite-stained limestone intraclasts and small spar-filled cavities forms a prominent bed up to 0.15 m thick immediately above the Upper Nodule Bed in the main exposure (Figures 2.68, 2.69). Secondly, white to dark red, irregularly stained, laminated, platy limestones form the middle part of the Upper Nodule Bed and are interlaminated with siliciclastic laminites, particularly at the top and base of the Upper Nodule Bed. These are typical carbonate, fish-bearing laminites, the fish being preserved in flat concretions as well as in the non-concretionary carbonate laminae. Thirdly, concretions and thin, laterally persistent non-laminated beds of green limestone occur in the mudstone-dominated beds above the Upper Nodule Bed and contain rare fish.

The laminated limestones in the Upper Nodule Bed and the massive limestone that caps it are extensively cut and disrupted by irregular, green to purple, medium- to coarse-grained, micaceous sandstone veins (Figures 2.68, 2.69). Their irregular nature is probably due to several factors, including deformation of the bed subsequent to the formation of the veins. Some containing sandstone intraclasts were probably sand-filled desiccation cracks that were subsequently deformed by differential compaction. Some sub-horizontal veins lacking intraclasts were probably intruded into weakly compacted muds as liquefied wet sand during movement of the sediment pile.

Calcite veins up to 5 mm wide cut the limestone layers and carbonate concretions, and fill cracks that post-date the initial compaction of the sediments. The calcite encloses filamentous and possible coccoid bacteria that invaded the sediment during lowstand events when the sediment was in the vadose zone, prior to crystallization of the calcite (Trewin and Knoll, 1999).

The Lower Nodule Bed has yielded unidentifiable fish scales (Peacock *et al.*, 1968), an articulated osteolepid, scattered osteolepid scales and a *Cheiracanthus* fin spine (Trewin and Davidson, 1999). The Coccosteus Bed has not been traced recently, is probably lenticular,

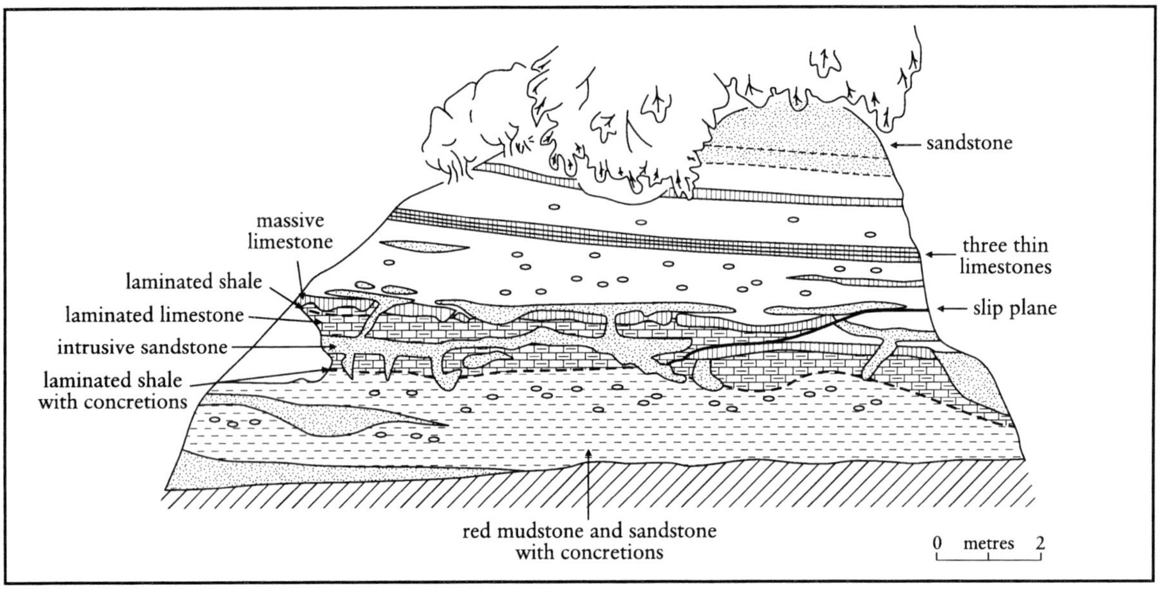

Figure 2.69 Field sketch of exposure at Point A in Tynet Burn showing the disrupted nature of the Upper Nodule Bed. After Trewin and Davidson (1999). This exposure is now largely obscured by talus and vegetation.

and apparently yielded only *Coccosteus* and some scales. *Dipterus* (Malcolmson, 1859) and *Osteolepis* scales (Trewin and Davidson, 1999) have been recorded from the beds below the Upper Nodule Bed, both in sandstones, the latter in small nodules 1 m below the base of the nodule bed.

The Upper Nodule Bed (Figure 2.68b) is the main fish-bearing horizon. Many of the carbonate nodules are shaped like the fish they contain. Detailed sections from excavations were recorded by Wood and Norman (1991) and Trewin and Davidson (1999). Although only 50 m apart, the sections show a marked variation in thickness of the bed. Wood and Norman recorded a thickness of over 2 m, compared to 0.75 m recorded by Trewin and Davidson. The difference may be reconciled by invoking deposition of the bed on a channelled surface of the underlying sands, the irregular topography probably contributing to instability, soft-sediment slumping and disruption of the bed (Figure 2.69).

In detail, the Upper Nodule Bed comprises three horizons (Top, Middle and Bottom). The top and bottom units are chocolate brown, clastic laminites with sporadic fish-bearing concretions. The lower unit ranges from 0.05 m to 0.9 m in thickness and has yielded a fauna dominated by osteolepids but including the acanthodians *Cheiracanthus* and *Mesacanthus*. Acanthodians (*Diplacanthus* and *Cheiracanthus*) dominate in the upper unit, which shows a similar thickness range. The intervening middle laminated limestone unit is 0.5 m thick and has yielded mainly acanthodians.

The complete faunal assemblage from the Upper Nodule Bed is listed by Dineley (1999a) and Trewin and Davidson (1999) as:

Placodermi:
Coccosteus cuspidatus Miller (1841)
Rhamphodopsis trispinatus Watson 1938
Pterichthyodes milleri (Miller 1841)
Acanthodii:
Cheiracanthus murchisoni Agassiz
Cheiracanthus latus Egerton 1861
Diplacanthus crassisimus Duff 1842
 (= *D. striatus* Agassiz 1835)
Mesacanthus pusillus Agassiz 1844
 (?= *M. peachi* Egerton 1861)
Rhadinacanthus longispinus (Agassiz 1844)
Osteichthyes:
Cheirolepis trailli Agassiz 1844
Glyptolepis leptopterus Agassiz 1844
Glyptolepis paucidens Agassiz 1844 ? (not seen in any collection by R.G. Davidson)
Dipterus valenciennesi Sedgwick & Murchison 1828
Osteolepis macrolepidotus Agassiz 1835
Gyroptychius spp.

The assemblage includes type specimens of *Diplacanthus striatus* and *Rhamphodopsis trispinatus* Watson 1938, and possibly that of *Cheiracanthus latus* Egerton 1861. The material is sufficiently well-preserved to distinguish the scale morphology of several acanthodians (Young, 1995). Wood (in Trewin and Davidson, 1999) carried out a statistical analysis of the assemblage of the Upper Nodule Bed, calculating that acanthodians comprise 82% of the assemblage of the Top Bed and sarcopterygians (mainly *Osteolepis*) 72% of the Bottom Bed. *Osteolepis?*, *Glyptolepis*, *Cheirolepis* and *Cheiracanthus* were collected between 1 m and 2 m above the Upper Nodule Bed by R.G. Davidson.

The preservation of the fish ranges from complete carcasses with full articulation to isolated scales and bones. The best material comes from the nodule beds and limestone laminites of the Upper Nodule Bed, in which the osteolepids at the base and the acanthodians in the Middle and Top beds are commonly articulated or suffered only minor disarticulation before burial. Isolated scales and coprolites also occur in the concretions alongside the whole fishes.

The striking pink, red and purple colour of the fish is the result of oxidation, probably by chemotrophic bacteria (Trewin and Davidson, 1999; Trewin and Knoll, 1999). Zones of iron oxide surround the fish, commonly permeating the scale and bone structure and infilling original canals in the bone. The phosphatic scales are commonly dissolved, leaving calcite-filled cavities. Relics of internal organs and eyes have been recognized in some specimens, represented by dark red to black stains (Davidson and Trewin, 1999, 2002). Gut contents are preserved in one specimen of *Coccosteus*.

Interpretation

Hamilton and Trewin (1988) and Trewin and Davidson (1999) provided a detailed analysis of the sedimentary environments and of the diagenetic processes by which the fish were preserved. Figure 2.68 shows a summary of the range of sedimentary environments in which the succession was deposited.

The conglomerates at the bases of the cycles were the deposits of high-energy, gravel-bed (?braided) stream channels. The overlying mainly parallel-bedded laminated sandstones were probably deposited rapidly as unconfined sheet floods. Some of the finer, thinly bedded sandstones contain *Skolithos* and *Diplocraterion* burrows and may have been deposited in shallow floodplain lakes. Red mudstones with carbonate nodules record formation of incipient caliche soil profiles on the floodplain. On three, or perhaps four occasions, the alluvial plains were rapidly inundated from the north by lake waters, resulting in the deposition of the fish-bearing beds. These lake transgressions record two major lake highstand events, when the Orcadian lake reached its greatest extent and was deep enough for lacustrine laminites to form at Tynet during the deposition of the Upper Nodule Bed.

The massive limestone at the top of the Upper Nodule Bed is interpreted as a shallow-water lake-margin deposit, similar to that seen at the **Red Point** GCR site. Exposure followed its deposition, sand-filled desiccation cracks cutting the limestone.

The Lower Nodule Bed may represent a shallower water and/or shorter-lived lacustrine transgressive event, as may the problematic Coccosteus Bed (discussion in Trewin and Davidson, 1999). Lacustrine conditions were established several times at Tynet for long enough to allow carbonate deposition in low-energy environments in which articulated fish were preserved. The presence of channelized sandstones above some horizons may explain their lateral discontinuity, with fluvial incision removing all or part of the lacustrine interval.

The correlation of the Tynet Burn succession with the Achanarras horizon of Caithness and the Sandwick horizon on Orkney is based on the presence of at least seven species common to both, minor differences in the faunas being explained by ecological factors in their distribution or a slight age difference (Trewin, 1986). If the whole of the Tynet Burn fish-bearing succession is equivalent to the Achanarras Limestone Member, the three cycles here point to a situation more complex than at Achanarras, which was situated in a deeper area of the lake. The marginal position of the Tynet Burn locality, combined with greater clastic influence, made it more sensitive to lake level fluctuation.

The preservation of the fish is summarized by Trewin and Davidson (1999, fig. 12). The differentiation of their post-mortal and pre-burial disarticulation is made difficult because of soft-sediment deformation, sand mobilization, synsedimentary faulting and calcite veining. Also, the carbonate nodules were not entirely nucleated on the fish, and where they are only partially enclosed, post-burial differential compaction and movement resulted in the in-situ breakage of the fish fossils.

Covering of fish carcasses by lake-bed sediment and subsequent compaction was followed by wet sediment deformation and disruption, probably aided by its position overlying an irregular lake bottom. A drop in lake level exposed the lake bed, and desiccation resulted in cracking of the semi-consolidated sediment and infill of the cracks by sand. Further compaction and formation of the carbonate concretions took place in the vadose zone, followed by colonization of fractures in the nodules by chemotrophic bacteria. The selective deposition of iron oxide by the bacteria at haemoglobin-rich organ sites may be the cause of preservation of these soft parts as dark-stained traces.

Conclusions

Tynet Burn is a classic Middle Old Red Sandstone fossil fish site, having yielded a rich fauna of acanthodians, placoderms and bony fishes since its discovery in 1838. The importance of the site lies in the remarkable preservation of the fish in carbonate nodules. The pink, red and purple colours of the fish, in contrast to the green or beige matrix of the nodules, have made the fossils very popular and specimens are widely dispersed throughout the world.

The fish beds were deposited as part of a cyclic fluvial and lacustrine succession on the southern margins of the Mid-Devonian Orcadian lake. The cyclicity was controlled by climatic fluctuations that influenced expansion and contraction of the lake. At times of lake expansion the alluvial plain on the shore of the lake was rapidly inundated by lake waters, which deposited laminated muds and carbonates. Bacterial oxidization in the subsurface preserved the fish in fine detail, including in some cases, traces of soft tissue. The material collected continues to provide new insights into the morphology and diet of Devonian freshwater fish and their lake habitat.

DEN OF FINDON, GAMRIE BAY AND NEW ABERDOUR, ABERDEENSHIRE (NJ 796 635–NJ 882 650)

Potential ORS GCR site

W.J. Barclay

Introduction

The Den of Findon near Gamrie, Banffshire has been independently selected for the GCR for fossil fishes (Dineley and Metcalf, 1999). It is the easternmost occurrence of the Mid-Devonian Achanarras fauna and has yielded 12 species of fossil fishes. The fish bed (the Gamrie Fish Bed) lies within the upper part of the Findon Group of the Middle Old Red Sandstone and the site is important historically as the first prolific Scottish Old Red Sandstone fish site to be exploited. The section is now rather inaccessible and poorly exposed in a steep, overgrown ravine and no collection or excavation is permitted, but the opportunity is taken to describe briefly the important Old Red Sandstone coast sections nearby at Gamrie, and farther east at New Aberdour (Figure 2.70). The following account summarizes that of Dineley (in Dineley and Metcalf, 1999), Sweet (1985), Trewin and Kneller (1987a–c) and Trewin (1987a).

Description

Dineley (1999a and references therein) gave a detailed account of the discovery and history of fish discoveries at the Den of Findon. The fish bed lies up to about 17 m above the base of the Findon Group. The basal unit of the group is a conglomerate comprising rounded clasts of quartzite, felsite and local Dalradian lithologies. Within 0.15 m, the conglomerate fines up into red clay, above which are 1.25 m of grey, laminated mudstones containing fish-bearing calcareous concretions. The parts of the fish within the nodules are preserved complete or with only slight disturbance. The species recovered here include the acanthodians *Diplacanthus striatus*, *D. tenuistriatus*, *D. (Rhadinacanthus) longispinus*, *Cheiracanthus murchisoni*, and *C. latus*; the placoderms *Pterichthyodes milleri* and *Coccosteus cuspidatus*; the actinopterygian *Cheirolepis trailli*; and the osteolepids *Glyptolepis leptopterus*, *Osteolepis*

Den of Findon, Gamrie Bay and New Aberdour

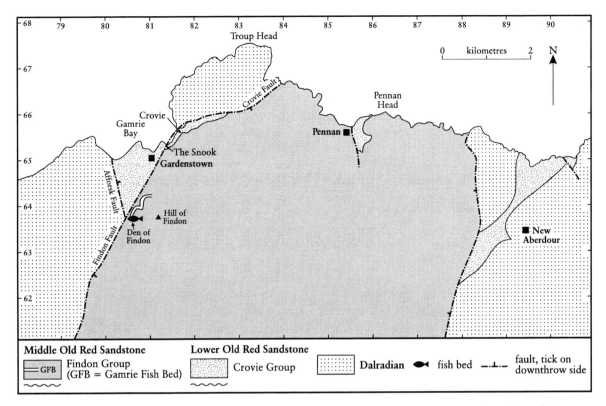

Figure 2.70 Geological map of the Gamrie–New Aberdour area. Based on Institute of Geological Sciences 1:50 000 Sheet 96 (Scotland), Banff (1955) and British Geological Survey 1:50 000 Sheet 97 (Scotland), Fraserburgh (1987).

macrolepidotus and *Gyroptychius* n. spp.. Red clays and shales (0.6 m thick) overlying the laminites are truncated by an erosion surface at the base of a breccia/conglomerate of local Dalradian clasts.

The site lies at the northern end of the Turriff Basin, the fill of which comprises Lower Old Red Sandstone strata (the Crovie Group) resting unconformably on Dalradian basement and unconformably overlain by Middle Old Red Sandstone (the Findon Group). In Gamrie Bay to the north of the Den of Findon, the Afforsk Fault forms the western margin of the basin (Figure 2.70). Conglomerates in the Crovie Group to the east of the fault (NJ 7922 6448) contain locally derived clasts of Dalradian Southern Highland Group lithologies, but also further-travelled clasts of hornblende schist and cleaved greywackes.

A succession of sandstones and conglomerates of the Crovie Group is exposed along the foreshore and in the sea cliffs of Gardenstown between the Afforsk Fault and the Findon Fault to the east. Cut by numerous small faults, the succession is a fine example of a coarsening-upward fluvial sequence, with mudstones and thin sandstones at the base and conglomeratic sandstones at the top (Trewin and Kneller, 1987a). The sandstones are markedly porous and permeable; their red colour is due to the breakdown of large quantities of detrital biotite (Archer, 1978; Donovan *et al.*, 1978). The lowest part of the succession (NJ 802 651), seen on the wave-cut platform between the Findon Fault and the breakwater at the east side of the harbour, comprises thinly bedded grey-green sandstones and mudstones. The sandstones are generally less than 5 cm thick, with parallel lamination and ripple-lamination, as well as small load structures and disrupted beds with detached sandstone balls. Calcrete nodules occur in the mudstones and polygonal desiccation cracks are common at the lowest level exposed, and also near the harbour wall, where the sandstones are thicker and contain more current ripples and ripple-drift lamination.

Immediately west of the harbour wall (NJ 799 648), the succession continues with interbedded micaceous sandstones, mudstones and siltstones. The beds display a range of colours, the sandstones being brown and green and the mudstones red, grey and purple. The sandstones are in beds 5–30 cm thick, with current ripple-, continuous ripple-drift-, parallel- and convolute lamination present. Desiccation cracks are present at the tops of the mudstones, and rip-up clasts occur in the bases of some sandstones.

To the west, the beds flatten into a small syncline, in which there are fine examples of arthropod-produced trace fossils in red, micaceous, rippled sandstones. The traces are *Diplichnites*, *Beaconites*, *Isopodichnus*, *Diplocraterion* and *Rusophycus* (Carroll, 1991; Trewin and Thirlwall, 2002). Continuing up-sequence (NJ 798 647), the succession becomes coarser and the sandstones thicker bedded and redder. Cross-bedding, in sets up to 50 cm high, appears as current ripple-lamination rapidly dies out up-section. Red mudstone occurs in sporadic, thin, desiccation-cracked interbeds and as burrow-fills in the sandstones.

Where small pebbles appear first, they lie mainly in the bases of small channelized, medium-grained sandstone bodies (NJ 796 646), with mudstone interbeds virtually absent. The succession continues to coarsen upwards from the last house on the harbour wall to the cliffs (NJ 7940 6445), with conglomerate lenses lining the bottoms of channels up to 2 m deep. Cross-bedded and parallel-bedded sandstones completing the channel-fills are variously bright red and green as a result of a complex diagenetic history of oxidation and reduction. Dark brown to black zones in some paler sandstones are rich in vanadium. A small fault in the cliff east of the burn mouth (NJ 7925 6445) brings down a conglomeratic sandstone containing angular blocks of red sandstone.

East of the Findon Fault, the Crovie Group outcrops on the foreshore to Crovie and is unconformably overlain by basal conglomerates of the Middle Old Red Sandstone Findon Group. The outcrop of the Crovie Group is much faulted, and Trewin and Kneller (1987a) made no attempt at a correlation of the succession with that to the west of the Findon Fault, as had been proposed by Read (1923) and Donovan *et al.* (1978) (Trewin and Thirlwall, 2002). There is a complex relationship between the conglomerates of the Findon Group east of the Findon Fault and the red conglomeratic sandstones on the foreshore. To the east of The Snook (NJ 803 651) the contact is sharp and probably unconformable, although disturbed by faults. West of The Snook, the contact is a fault in the foreshore exposures.

From The Snook north-eastwards to Crovie, a broadly coarsening-upward succession is traversed down-sequence to Crovie. There, the Crovie Group is faulted against Dalradian basement of Troup Head by the Crovie Fault, which is probably a continuation of the Findon Fault. The Crovie Group comprises red, cross-bedded, mainly channelized sandstones and red mudstones with calcrete nodules. The basal beds of the Findon Group are conglomerates containing a large proportion of rounded pebbles and some boulders of quartzite, vein quartz and felsite. Around The Snook, these are overlain with low-angle discordance by locally derived breccias of Dalradian slate clasts.

The unconformity between the Lower Old Red Sandstone and Middle Old Red Sandstone is best seen in the impressive cliffs to the west of Pennan (NJ 846 655) (Trewin, 1987a). Deposition of the Crovie Group appears to have been controlled by synsedimentary faulting. Re-activation of these faults and erosion before deposition of the Findon Group produced a complex unconformity, the basal conglomerate of the group draping and filling hollows in the underlying eroded surface (Figure 2.71).

At New Aberdour, the Crovie Group is represented by a westward-dipping succession of conglomerates, sandstone and siltstone over 400 m thick and divisible into a lower sandstone–conglomerate unit and an upper mudstone-dominated unit with sandstone lenses in its upper part (Sweet, 1985; Trewin and Kneller, 1987b). The unconformity at the base of the succession is magnificently exposed at the east end of the section (NJ 898 652), where coarse breccia at the base of the Crovie Group rests on an eroded surface of Dalradian andalusite schists and psammites. The unconformity surface dips from 45° to near-vertical, the breccia mantling a highly irregular topography in which resistant greywacke ridges protruded. The breccia is up to about 10 m thick and consists mainly of locally derived psammitic clasts along with some vein quartz and felsite. Westwards, the breccia is absent (NJ 897 651)

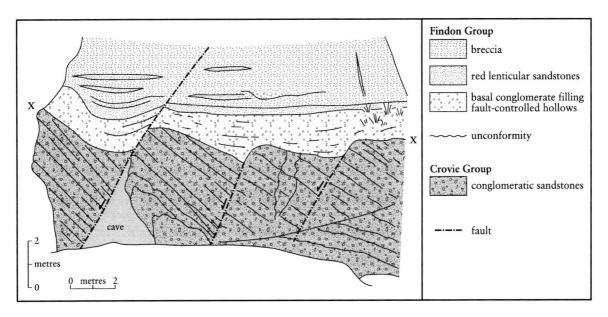

Figure 2.71 Sketch of the lower part of the cliff near Pennan (NJ 842 658) showing the unconformity between the Lower Old Red Sandstone (Crovie Group) and overlying Middle Old Red Sandstone ((Findon Group). The faults in the Crovie Group do not all appear to affect the Findon Group, the basal conglomerates of which fill hollows eroded along the faults. The Crovie Group was therefore faulted and eroded prior to the deposition of the conglomerates. After Trewin (1987a).

and sandstones with conglomerate lenses and granitic debris of angular quartz and feldspar grains overlie the unconformity. Conglomeratic red sandstones rich in granitic debris dominate the succession up-sequence until a marked, but punctuated change into fine-grained sandstones (NJ 894 649). These fine upwards and pass up into red and green mudstones with calcrete nodules, which locally coalesce to form thin, continuous beds (Stage III calcrete) with pseudo-anticlinal structures. Polygonal arrays of desiccation cracks appear to have controlled the development of the calcrete nodules. Thin sheets of laminated and ripple-laminated sandstone interrupt the mud-prone succession.

The fine-grained succession is seen to a point on the wave-cut platform (NJ 839 649) where thin, coarse-grained sandstones appear again and persist upwards. A remarkable feature is the presence of pink felsite pebbles and boulders up to 60 cm in a sandstone or mudstone matrix. They were clearly derived from nearby outcrop and may have slid gently onto the mud- or sand-flat. Another striking feature of this part of the succession is the abundance of *Beaconites* burrows. The succession is terminated by the Dundarg Fault, which throws down to the east, and a similar succession of conglomeratic sandstones (NJ 889 647) and mudstone-dominated beds (NJ 886 648–NJ 882 650) is seen to its west.

Close to the east of New Aberdour, exposures on the east side of Quarry Haven (NJ 9081 6578) reveal a small outlier of Old Red Sandstone rocks, faulted against Dalradian schists and psammites on their western margin and unconformably overlying them to the east (Trewin and Kneller, 1987c). In contrast to New Aberdour, there is no thick development of basal conglomerate facies, the mudstone facies lying much closer to the unconformity.

Interpretation

The Turriff Basin developed in Early Devonian times, its Lower Old Red Sandstone alluvial fill (the Crovie Group) comprising conglomerates and sandstones. Basal breccias mantled the irregular land surface of Dalradian meta-sedimentary basement, infilling valleys and hollows and forming accumulations at the faulted basin margins. Sweet (1985) gave a detailed interpretation of the facies at New Aberdour. The basal conglomerates there filled two SSW-draining palaeovalleys. They represent debris-flow and flash-flood deposits, passing upwards and distally into alluvial-fan braided stream, sheet-flood and fluvial channel sands and gravels. The basal conglomerates are absent at Gamrie Bay, the Crovie Group being faulted against basement and comprising a coarsening-upward fluvial sequence of mud-dominated floodplain deposition succeeded by alluvial-fan progradation. The basal conglomerate is thin to absent at Quarry Haven, its restricted clast suite confirming that there was only limited fan progradation at this point on the basin margin (Trewin and Kneller, 1987c).

The floodplain deposits comprise mudstones with thin sheets of sandstone, the latter probably representing unconfined sheet-flood deposition. Sweet (1985) and Trewin and Kneller (1987a,b) favoured a playa-lake environment for the mudstone deposition, impermanent, shallow lakes forming after flood events. Periods of exposure and low rates of aggradation produced desiccation cracking of the mudflats, with prolonged carbonate soil formation producing calcrete nodules, and in some cases, mature calcrete horizons. A semi-arid, hot climate with seasonal rainfall is inferred. The impressive range of burrows and trails at Gamrie Bay were probably made mainly by arthropods. The mudflats were succeeded by sandy alluvial plains which were combed by rapidly migrating stream channels.

Spores from the Crovie Group at Gamrie Bay are of late Early Devonian (Pragian–Emsian) age (Westoll, 1977). Correlation of the sequences on both sides of the Findon–Crovie Fault in Gamrie Bay was attempted by Read (1923) and Archer (in Donovan *et al.*, 1978). Similarly Sweet (1985) correlated the New Aberdour and Quarry Haven sections. Trewin and Kneller (1987a) adopted a more cautious approach, pointing out that rapid facies variations in the locally derived sediments in the basal part of the succession (probably controlled by synsedimentary faulting), and numerous faults between The Snook and Crovie at New Aberdour make correlation tentative.

Inversion and erosion of the basin-fill followed deposition of the Lower Old Red Sandstone (Crovie Group), represented by a major unconformity overlain by the Middle Old Red Sandstone (Findon Group), spectacularly seen at Pennan. The basal conglomerates and breccias seen at The Snook in Gamrie Bay represent the progradation of two discrete alluvial-fans from different valleys, with two sediment sources involved, although the fans interfingered at times. The earlier one deposited rounded, far-travelled quartzite boulders, although these may have been reworked from older conglomerates. The later fan was of much more local derivation, from perhaps no more than 2 km, depositing angular clasts of Dalradian slate (Trewin and Kneller, 1987a).

The Gamrie Fish Bed at the Den of Findon represents an interruption to alluvial deposition, when the Orcadian lake reached maximum development (at the end of the Eifelian Age) and transgressed over the alluvial plain on the southern margin of the basin (Trewin and Kneller, 1987a). Lake deposition occurred over a period of at least about 4000 years, based on the interpretation of the fine laminations as annual, seasonally controlled varves. The bed is correlated on its fish fauna with the Tynet Burn fish bed (see **Tynet Burn** GCR site report this chapter), the Achanarras Fish Bed of Caithness and the Sandwick Fish Bed of Orkney. There are differences, however, between Gamrie and the other sites, with, for example, *Dipterus*, which is common elsewhere, being absent. Subsequent uplift and erosion took place prior to deposition of the locally derived fan breccia seen at the Den of Findon. The fish bed was probably removed by erosion in Gamrie Bay and Pennan, where the basal Findon Group breccia rests with angular discordance on the conglomerates at the top of the Crovie Group (Dineley, 1999a).

Conclusions

The Gamrie Fish Bed is historically important as one of the first prolific Scottish Old Red Sandstone fossil fish sites. The Den of Findon GCR site is overgrown, but magnificent cliff and foreshore sections nearby in Gamrie Bay, and to the east at Pennan and New Aberdour provide the best sections available of the Turriff Basin. Lower Old Red Sandstone and Middle Old Red Sandstone successions, and the intervening unconformity, are completely exposed, allowing detailed analysis of the sedimentary rocks and the environments in which they were deposited. Of particular importance are the trace fossil assemblages, the variations in facies and the bounding faults with the Dalradian, all spectacularly displayed in these sections.

RHYNIE, ABERDEENSHIRE (NJ 494 277)

Potential ORS GCR site

W.J. Barclay, P. Stone and N.H. Trewin

Introduction

The Rhynie outlier of Old Red Sandstone strata lies about 50 km WNW of Aberdeen. It has an elongate outcrop measuring about 21 km from north to south and up to 3 km from east to west. Structurally, the inlier has been interpreted as a half-graben, with a major, low-angle extensional fault zone at its western margin and strata generally dipping moderately towards the west from an unconformity at the eastern margin. However, recent re-mapping of the northern part of the basin by Rice and Ashcroft (2004) has demonstrated that faulting and folding can be related to basin formation in an Early Devonian regional strike-slip system. The site lies close to the north of the village of Rhynie (Figure 2.72) on the western margin of the basin. Natural exposure within the site is confined to some tuffaceous sandstone, but the site is an established palaeontological GCR site for the Rhynie cherts and its biota, which comprises the best-preserved and most diverse early terrestrial/freshwater ecosystem in the world. The diverse biota was silicified and exceptionally well-preserved by hot spring activity. It includes plants, together with algae, fungi and cyanobacteria, as well as a number of species of terrestrial and freshwater arthropods. Many of the plants and arthropods are unique to this site. It is a world-renowned lagerstätte that has been of crucial importance in providing insights into plant and arthropod evolution. The plants include *Rhynia*, the type-genus of the Rhyniophytina and widely regarded as the archetypal early land plant, and *Asteroxylon*, the earliest well-documented lycopsid. The arthropods include terrestrial opilionid (harvestman) spiders, trigonotarbids, mites, hexapods, euthycarcinoids, centipedes, freshwater crustacea, and several other arthropods of uncertain affinities. A recent re-examination of *Rhyniognatha hirsti*, a fragmentary fossil from Rhynie, suggests that it may have been not only the earliest true insect, but also the first winged species (Engel and Grimaldi, 2004). *Leverhulmia* has also been re-interpreted as an insect (Fayers and Trewin, in press).

The Rhynie cherts originated as siliceous sinters produced by hot springs. Chert was first recorded as loose blocks at the surface, but the site has since been extensively investigated by trenching and drilling. The international palaeobotanical importance of the site led to it being also selected as a GCR site for its Palaeozoic palaeobotany (see Cleal and Thomas, 1995 and references therein); independently it was selected for the GCR for its fossil arthropods. It is also of great importance as one of the earliest preserved surface expressions of a hydrothermal hot spring system.

Recent studies on the general geology of the area are by Rice and Trewin (1988), Trewin and Rice (1992), Trewin (1994, 1996), Rice *et al.* (1995, 2002), Gould (1997), Trewin *et al.* (2003), Rice and Ashcroft (2004) and Trewin and Wilson (2004). A general account of the Old Red Sandstone is provided by Trewin and Thirlwall (2002). Freshwater and terrestrial arthropod faunas recovered from the nearby Windyfield Chert are described by Anderson and Trewin (2003) and Fayers and Trewin (2004). Recent palaeobotanical discoveries include the recognition of gametophytes of some plants (Remy *et al.*, 1993; Kerp *et al.*, 2004), the Zosterophyll plants *Trichopherophyton* (Lyon and Edwards, 1991) and *Ventarura* (Powell *et al.*, 2000a), a parasitic relationship between fungi and green alga (Taylor *et al.*, 1992a,b), ascomycete fungi (Taylor *et al.*, 1999), and the earliest known lichen (Taylor *et al.*, 1997).

The fossiliferous horizons are placed into a context of the evolving geological environments at the margins of a small, active half-graben. The importance of the site lies in the in-situ biota that provides a unique 'snapshot in time' of an Early Devonian terrestrial ecosystem.

Description

The GCR site spans an area of land adjacent to the road from Rhynie to Cabrach (Figure 2.72). Several patches of the distinctive chert have been recorded as float, but there are no natural chert exposures, and the stratigraphical evidence for the position and relationships of the cherts has been obtained from geophysical studies, trenching and drilling. The drilled core and trench samples are housed at the University of Aberdeen. The chert occurs in the highest part of the succession (the Rhynie Chert Member of Gould, 1997; Rhynie Cherts Unit of Rice *et al.*,

The Orcadian Basin

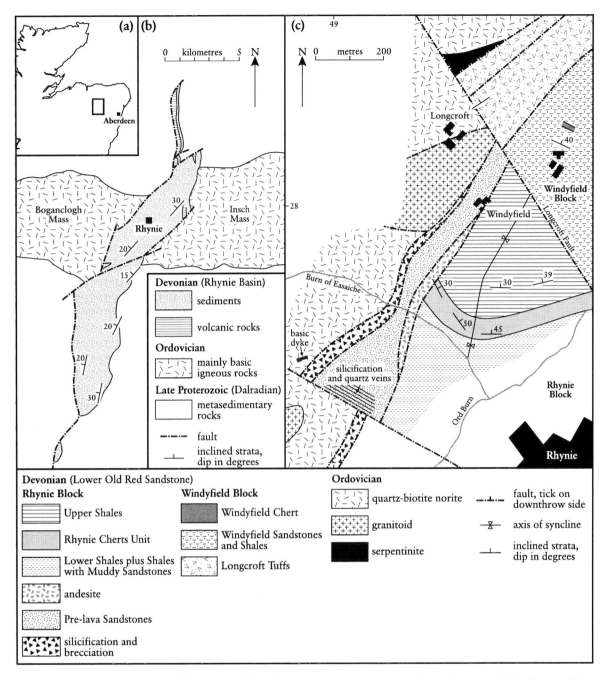

Figure 2.72 Geological map of the area of the Rhynie GCR site. Inset maps show location of the figure. After Rice *et al.* (2002). See Rice and Ashcroft (2004) for a new structural interpretation of the northern part of the basin.

2002) in the Rhynie outlier, in a sequence of shales and tuffaceous sandstones comprising the Dryden Flags Formation (Gould, 1997; Rice *et al.*, 2002). The GCR site lies in the Rhynie Block, the southern of two chert-bearing blocks that are separated by the Longcroft Fault (Figure 2.72). The northern (Windyfield) Block lies beyond the boundary of the palaeobotanical GCR site. In this GCR site however, a basin-margin faulted succession older than the cherts comprises about 20 m of lava and 30 m of sandstone and conglomerate. Most of the Old Red Sandstone succession seen elsewhere in the outlier is absent in the palaeobotanical GCR site area. Recently discovered altered lapilli tuffs east of the Longcroft Fault (the Longcroft Tuffs) are

Rhynie

unique to the Rhynie Basin and other Devonian basins in north-east Scotland (Rice *et al.*, 2002). Palynological evidence (Richardson, 1967) suggested a Pragian age for the cherts; Wellman (2004), using borehole samples, has placed the Rhynie succession within the *polygonalis–emsiensis* Spore Assemblage Biozone, giving an early (but not earliest) Pragian to earliest Emsian age. Radiometric (Ar–Ar) dating has given an age of 396 ± 12 Ma (Rice *et al.*, 1995); the Milton of Noth andesite (Figure 2.73) has yielded an ID-TIMS U–Pb zircon-titanite age of 409.6 ± 1.1 Ma that is probably close to the age of the Rhynie Cherts Unit (Parry, 2004), and consistent with the Pragian age based on palynology.

In the Rhynie area the basement to the Old Red Sandstone succession consists mainly of basic igneous rocks (quartz-biotite norite and minor serpentinite) and some granitoids belonging to the Ordovician Boganclogh intrusion. These are separated by an extensional, basin-margin fault zone from the overlying Old Red Sandstone sedimentary and volcanic rocks. The basin-margin fault zones in the Rhynie and Windyfield blocks comprise faulted slices of the basin-fill, as yet uncorrelated between blocks, although Rice *et al.* (2002) estimate that the Longcroft Fault throws down to the north-east by about 160 m, with chert outcrops in the Windyfield Block lying at a higher level than the Rhynie cherts.

The basal part of the succession in the basin-margin fault zone in the Rhynie Block comprises 30 m of mostly lithic sandstone, pebbly sandstone and conglomerate (Figure 2.73), named the 'Pre-lava Sandstones' by Trewin and Rice (1992) and correlated by Gould (1997) and Rice *et al.* (2002) with the Tillybrachty Sandstone Formation of the southern and eastern parts of the Rhynie Basin. The succession fines upwards, and the sandstones are mostly massive, with some parallel lamination and cross-bedding. The abundant pebbles are locally derived from the underlying igneous rocks, some being up to 10 cm in diameter. A calcareous cement is widespread, some sandstones are silicified, and rare calcrete nodules are present (Trewin and Rice, 1992).

A faulted sliver of purple vesicular, andesitic lava separates the Pre-lava Sandstones from the overlying Dryden Flags Formation. This formation is subdivided informally by Rice *et al.* (2002) into five units (Figure 2.73). It is at least 200 m thick and consists mainly of shales and sandstones with minor cherts. Over 40 m of clean, white, parallel- and ripple-laminated sandstones with some cross-bedding (White Sandstones unit) lie at the base of the formation. Beds have sharp bases resting on erosion surfaces, contain large mudstone rip-up clasts and some pebble beds. Mudstone interbeds become commoner upwards and the unit passes into a thick succession of argillaceous sandstones and

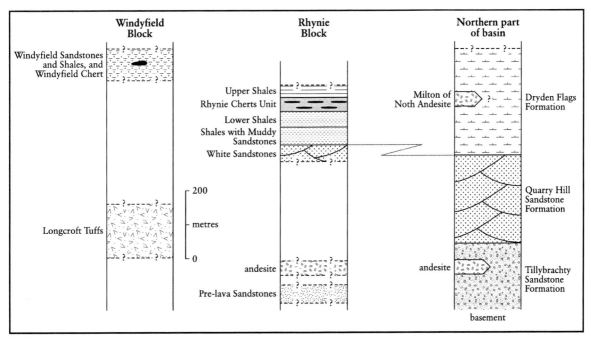

Figure 2.73 Generalized stratigraphy and basin correlations of the Rhynie Basin. After Rice *et al.* (2002).

mudstones (Shales with Muddy Sandstones unit). The argillaceous sandstones are massive, dark, up to 1 m thick and contain numerous small rip-up clasts. There are also some clean, pale sandstones with sharp bases, parallel- and ripple-lamination, and some pebble beds. Interbeds of laminated mudstone are commonly disrupted and locally burrowed. This unit is overlain by laminated mudstones with thin sandstones, subdivided into lower (Lower Shales) and upper (Upper Shales) units that are separated by the Rhynie Cherts Unit. The Lower Shales unit comprises green to blue-black, laminated, locally burrowed mudstone and fine-grained, pale sandstones. The latter occur in beds up to 0.15 m thick that have basal erosion surfaces and are parallel- and ripple-laminated. They contain rip-up clasts, and there are also some beds of pebbles of granitic basement and volcanic rocks. The Upper Shales unit comprises graded green mudstone and siltstone laminae in couplets up to 10 mm, interbedded with sandstones up to 0.1 m thick. The sandstones are similar to those in the Lower Shales, but in addition contain a few desiccation cracks, carbonaceous plant debris and patchy calcite cement.

The Rhynie Cherts Unit (Rhynie Chert Member of British Geological Survey, 1993 and Gould, 1997) is 35 m thick and contains the Rhynie cherts in a succession of interbedded mudstones, carbonaceous sandstones and minor tuffs. Individual chert beds range up to about 0.5 m thick and contain the remarkably well-preserved plant and arthropod material. Powell *et al.* (2000b) described 53 chert beds (totalling 4.2 m) in 35.41 m of core. One composite bed of 6 cherts is 0.76 m thick. The cherts show a range of laminated, brecciated, vuggy and geopetal textures typical of siliceous sinters (Trewin and Rice, 1992; Trewin *et al.*, 2003). In some of the beds, the plants are preserved partly in growth position, above a substrate of fine-grained sandstone that contains plant rhizomes. Trewin and Wilson (2004) have shown that whereas the chert-bearing unit can be correlated over 45 m to 65 m between three boreholes, individual chert beds are laterally impersistent and there is strong lateral variation in the flora. The published fauna and flora of the site are listed below. Several recent finds, including a nematode worm, a large trigonotarbid, an eoarthropleurid and possible spider remains, await publication (H. Kerp, S.F. Fayers, pers. comm.).

PLANTS

Trachyophytes (Sporophytes)
Rhynia gwynne-vaughanii Kidston & Lang 1917, 1920a; Edwards 1986
Horneophyton lignieri Kidston & Lang 1920a; El-Saadawy & Lacey 1979a
Aglaophyton major (Kidston & Lang 1920a); Edwards 1986
Nothia aphylla Lyon 1964; El Saadawy & Lacey 1979b; Kerp *et al.* 2001
Asteroxylon mackiei Kidston & Lang 1920b; Lyon 1964
Trichopherophyton teuchansii Lyon & Edwards 1991
Ventarura lyonii Powell *et al.* 2000a

Trachyophytes (Gametophytes)
Remyophyton delicatum (♀ ♂) (of *Rhynia*) Kerp *et al.* (2004)
Langiophyton mackiei (♀ ♂) (of *Horneophyton*) Remy & Hass 1991a; Kerp *et al.* (2004)
Lyonophyton rhyniensis (♀ ♂) (of *Aglaophyton*) Remy and Remy 1980; Remy & Hass 1991b; Kerp *et al.* (2004)
Kidstonophyton discoides (♀) (of Nothia) Remy & Hass 1991c

Nematophytes
Nematophyton taiti Kidston & Lang 1921
Nematoplexus rhyniensis Lyon 1962

Algae *sensu lato*
Mackiella rotunda Edwards & Lyon 1983
Rhynchertia punctata Edwards & Lyon 1983
Palaeonitella cranii Kidston & Lang 1921; Edwards & Lyon 1983; Kelman *et al.* 2004

Lichen
Winfrenatia reticulata Taylor *et al.* 1997

Cyanobacteria
Archaeothrix contexta Kidston & Lang 1921
Archaeothrix oscillatoriformis Kidston & Lang 1921
Kidstoniella fritschii Croft & George 1959
Langiella scourfieldii Croft & George 1959
Rhyniella vermiformis Croft & George 1959 (generic name pre-occupied by the springtail *Rhyniella praecursor*)
Rhynicoccus uniformis Edwards & Lyon 1983

Fungi
Palaeomyces gordonii (No. 2) var *major* (No. 3) Kidston & Lang 1921
Palaeomyces asteroxyli (No. 7) Kidston & Lang 1921
Palaeomyces horneae (No. 8) Kidston & Lang 1921
Palaeomyces vestita (No. 9) Kidston & Lang 1921
Palaeomyces agglomerata (No. 10) Kidston & Lang 1921
Palaeomyces simpsonii (No. 13) Kidston & Lang 1921
(Kidston & Lang described 15 types of fungi though only 7 were named, 2 being variants of the same species, the others were only given numbers and are not included here)
Glomites rhyniensis Taylor *et al.* 1995
Palaeoblastocladia milleri Remy *et al.* 1994
Milleromyces rhyniensis Taylor *et al.* 1992a
Lyonomyces pyriformis Taylor *et al.* 1992a
Krispiromyces discoides Taylor *et al.* 1992a
Ascomycetes Taylor *et al.* 1999

ANIMALS

'Worms'
Nematoda
   Nematode (undescribed) noted in Dunlop *et al.* (2004)
Annelida
   Polychaete (undescribed)

Crustaceans
Lepidocaris rhyniensis Scourfield 1926, 1940
Castracollis wilsonae Fayers & Trewin, 2003
Ebullitiocaris oviformis Anderson *et al.* 2004
Nauplii (of *Lepidocaris*?) (Fayers *et al.* in prep.)
   Noted in Fayers & Trewin 2004

Euthycarcinoid
Heterocrania rhyniensis Hirst & Maulik, 1926a,b; Anderson & Trewin, 2003

Trigonotarbids
Palaeocharinus rhyniensis Hirst 1923; Shear *et al.* 1987
Palaeocharinus hornei (Hirst 1923); Shear *et al.* 1987
Palaeocharinus tuberculatus Fayers *et al.* (in press)
(*Palaeocharinus hornei* (previously *Palaecharinoides hornei*) and *Palaeocharinus rhyniensis* are probably the only valid species of the five originally described by Hirst)
Unnamed large trigonotarbid Noted in Fayers & Trewin 2004

Araneae (Spiders)
Palaeocteniza crassipes* Hirst 1923 (*this specimen is now regarded as a juvenile trigonotarbid (Selden *et al.* 1991))

Opilionids
Harvestman spider Dunlop *et al.* 2003
= *Eophalangium sheari* Dunlop *et al.* 2004

Arcari (Mites)
Protacarus crani Hirst 1923
Protospeleorchestes pseudoprotacarus (Hirst 1923); Dubinin 1962
Pseudoprotacarus scoticus (Hirst 1923); Dubinin 1962
Palaeotydeus devonicus (Hirst 1923); Dubinin 1962
Paraprotacarus hirsti (Hirst 1923); Dubinin 1962

Arachnida?
Unnamed arachnid(?)

Eoarthropleurids
Eoarthropleura sp. Noted in Fayers & Trewin 2004

Chilopods
Crussolum sp. Shear *et al.* 1998; Anderson & Trewin 2003
Unnamed scutigeromorph Fayers & Trewin 2004
Unnamed centipede Fayers & Trewin 2004

Hexapods
Rhyniella praecursor Hirst & Maulik 1926
Rhyniognatha hirsti Hirst & Maulik 1926; Engel & Grimaldi 2004
Leverhulmia mariae Anderson & Trewin 2003; Fayers & Trewin (in press)

Arthropoda *incertae sedis*
Rhynimonstrum dunlopi Anderson & Trewin 2003

In the basin-margin fault zone of the Windyfield Block, the Longcroft Tuffs comprises at least 140 m of intensely altered andesitic lapilli tuffs with subordinate sandstones and minor andesite. The tuffs are up to 1 m thick in normally graded beds with lithic, vesicular and non-vesicular clasts. The sandstones are laminated, in fining-upward beds, with convolute lamination and possible bioturbation. The Windyfield Sandstones and Shales comprise at least 65 m of green sandstone and shale. The sandstone beds have sharp basal erosion surfaces and are parallel- and ripple-laminated. There are a few coarse beds of granitic debris, andesite, sandstone and metamorphic rock. The Windyfield Chert (Fayers and Trewin, 2004) comprises chert lenses up to 1 m across and 0.3 m thick within blue claystones with sandstone and pebble beds. The chert contains freshwater and terrestrial plants and arthropods (Anderson and Trewin, 2003; Fayers and Trewin, 2004) and vent sinter occurs as float (Trewin, 1994).

There is widespread hydrothermal alteration of the Devonian rocks, particularly in the vicinity of the basin-margin faults (Rice *et al.*, 1995, 2002). The altered rocks and the chert sinters contain high (but non-commercial) concentrations of gold and arsenic.

Interpretation

The interpretative model of the Rhynie inlier continues to be refined as more data become available. An earlier structural model invoking a sub-vertical basin-margin fault zone and a gently dipping basal unconformity (Rice *et al.*, 1995) was revised (Rice *et al.*, 2002) in the light of 1997 drilling results. In the 2002 model (Figure 2.74), a zone of extensional faulting along the western basin margin dips 35° eastwards, flattening to a 15° listric basement/cover fault under the basin. The basin-fill succession dips north-westwards, except close to the basin margin, where the strata are folded into a syncline. Further mapping supported by excavation and geophysical (magnetic) survey (Rice and Ashcroft, 2004) has revealed more extensive faulting and folding consistent with a dextral strike-slip origin for the northern part of the basin.

In the Rhynie Block, the Pre-lava Sandstone is interpreted as a small basin-margin alluvial-fan produced by a combination of flash-flood and channel deposition (Trewin and Rice, 1992; Rice

The Orcadian Basin

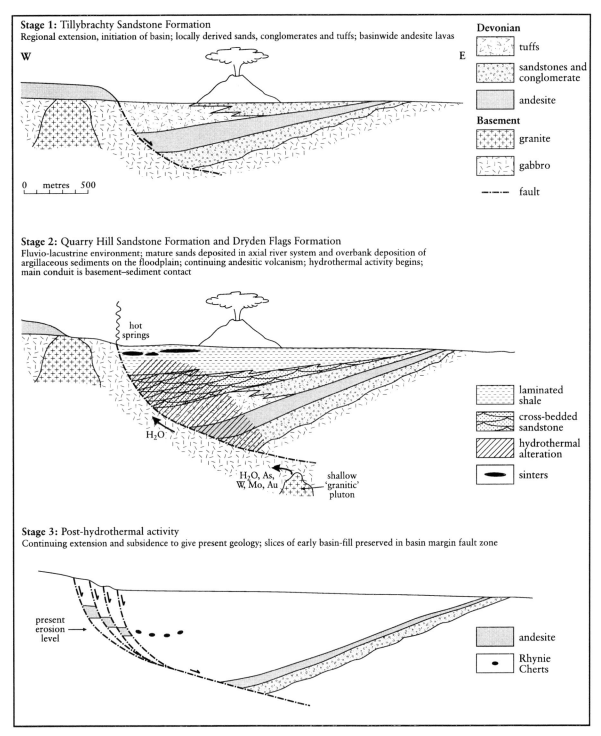

Figure 2.74 Evolution of the Rhynie Basin and hot spring development. After Rice *et al.* (2002).

et al., 2002). The rare calcrete nodules indicate soil formation in a subaerial, seasonally wet environment and semi-arid climate. The clastic rocks are derived from the local igneous basement and there is no evidence for volcanicity prior to the eruption of the overlying andesitic lava flow. The White Sandstones are interpreted as the deposits of shallow, fluvial channels and sheet-flood events. The Shales with Muddy Sandstones, Lower Shales and Upper Shales are predominantly lacustrine and floodplain deposits. Trewin and Rice (1992) and Rice *et al.* (1995,

2002) interpreted the depositional environment as ephemeral, shallow lakes on an alluvial plain that periodically dried to desiccated mudflats. The fine lamination of the mudstones is attributed to minor seasonal changes in sediment supply, the thin, rippled sandstones representing clastic input to the lakes during flooding events. The tuffaceous sandstones at the base of the Shales with Muddy Sandstones (the Tuffaceous Sandstones of Trewin and Rice, 1992) are largely fine grained and quartzose, but also contain angular, vesicular andesitic clasts. Some of the more tuffaceous beds may be volcanic ashfall deposits, implying contemporaneous volcanicity, but most of the volcanic material was transported by water from a local source. Trewin and Rice (1992) placed these beds stratigraphically immediately above the andesitic lava, but the contact between the units is now interpreted as a fault (Rice et al., 2002). Archer (1978) and Trewin and Rice (1992) interpreted the beds as alluvial-plain deposits, with regional axial current flow from the south (Archer, 1978). Trewin and Rice suggested that a nearby tuff cone undergoing erosion provided the volcanic debris, and that the laminated shales were overbank deposits of small ephemeral pools. The rare calcrete nodules are evidence of carbonate soil formation in a climate that continued to be semi-arid and seasonally wet.

The Rhynie Cherts Unit has received much attention on account of its plant-bearing cherts. For a historical summary see Cleal and Thomas (1995). Interpretations are given by Trewin and Rice (1992), Trewin (1994), Rice et al. (1995, 2002) and Trewin and Wilson (2004). The revised model (Rice et al., 2002) invokes silica deposition from mineral-rich hydrothermal fluids that emanated from hot (about 100°C) springs along basin-margin faults during the waning of local volcanism. Hot spring activity may have occurred at intervals along at least 2 km of the fault zone as pulses of fluid were released by subsidence movement along the fault zone. The silica-rich fluids permeated an alluvial plain with scattered small ponds, depositing surface sinter. Trewin and Wilson (2004) conclude that sinter deposition took place on a low-angle outwash apron from a hot spring, and that overbank flooding from a river system deposited the sand and mud interbeds (Figure 2.75). The silicification of standing plants and plant debris resulted from flooding of the land surface, but the aquatic arthropods and algae preserved in the chert probably inhabited cooler pools within areas of sinter. The wide range of preservation of the plants is probably due to silicification at different times in the cycle of plant growth and decay, combined with the variable efficiency of the silicification process. An outline model for the chert formation proposed by Rice et al. (2002) is shown in Figure 2.74. The sinter textures in the chert compare with those in modern subaerial and

Figure 2.75 Cartoon view looking north to illustrate the Rhynie hot spring system and the low-angle sinter apron crossed by streams emanating from hot spring vents. Plants colonized the apron along stream banks and ponds, and on alluvial floodplain areas. The sinter apron was periodically flooded by waters from the axial river system. Some coarse detritus and reworked volcanic ejecta were sourced from the west. After Trewin and Wilson (2004).

subaqueous sinters in the cooler outwash areas of hot springs in Yellowstone National Park USA, and indicate a depositional temperature below 30°C for the majority of the plant-bearing cherts (Trewin *et al.*, 2003).

In the Windyfield Block, the Longcroft Tuffs are interpreted by Rice *et al.* (2002) as the product of volcanic activity from a nearby centre, producing lava and airfall tuffs, as well as the product of the erosion of the volcanic source. This introduced coarse debris during flood events into an otherwise low-energy fluvio-lacustrine floodplain. The Windyfield Chert was the product of hydrothermal silicification of the contents of freshwater pools in the vicinity of a hot spring vent.

Rice *et al.* (2002) suggest correlation of the strata in the Rhynie and Windyfield blocks with the succession recognized in the northern part of the Rhynie Basin (Figure 2.73). The Pre-lava Sandstones are correlated with the Tillybrachty Sandstone Formation (British Geological Survey, 1993; Gould, 1997) and the White Sandstones may correlate with the topmost part of the Quarry Hill Sandstone Formation. The succeeding beds of both the Rhynie and Windyfield blocks, including the Rhynie and Windyfield cherts, are correlated as part of the Dryden Flags Formation.

Conclusions

The GCR site at Rhynie is arguably the most important Old Red Sandstone site in Great Britain and one of the most important in the world. In addition to its status as a unique Early Devonian faunal and floral lagerstätte, it is also one of the earliest-known occurrences of the surface expression of a hydrothermal spring system. The strata include cherts that contain an exceptionally well-preserved, silicified, internationally important early land-plant and arthropod assemblage, the whole in-situ biota providing an insight into plant and arthropod evolution and the development of terrestrial ecosystems. The fossiliferous horizons are placed into a context of evolving geological environments at the margins of a small, subsiding basin. Initial alluvial-fan deposition was followed by the eruption of andesitic lavas, marking the local onset of volcanicity. Subsequent sedimentation was on an alluvial plain with ephemeral lakes and ponds. The climate throughout was semi-arid and seasonally wet. As volcanicity waned, the basin-margin faults acted as conduits for silica-rich, mineralized fluids. These emerged on the alluvial plain as hot springs and preserved the biota in siliceous sinters, now represented by the Rhynie and Windyfield cherts.

Chapter 3

The Midland Valley of Scotland and adjacent areas

Introduction

INTRODUCTION

M.A.E. Browne and W.J. Barclay

The Midland Valley of Scotland (Figure 3.1) is an elongate NE-trending tectonic feature lying between the Highland Boundary Fault to the north and the Southern Upland Fault to the south (Cameron and Stephenson, 1985). It extends from under the North Sea (Forth Approaches Basin) south-westwards across Scotland and under the Irish Sea into Ireland. Usually described as a rift (or graben), the 'valley' contains substantial hilly and upland areas. Reflecting its origin as part of a collage of tectonically juxtaposed crustal fragments, it is also referred to as the 'Midland Valley Terrane' (Bluck, 2001, 2002).

The strata exposed today in the Midland Valley range from Ordovician to Palaeocene in age, but are mainly of Devonian and Carboniferous age. The Devonian strata are represented by volcanic and Old Red Sandstone sedimentary rocks, laid down in semi-isolated basins. The succession is divided into informal lithostratigraphical units (Lower Old Red Sandstone (LORS) and Upper Old Red Sandstone (UORS)) that have loosely defined age connotations and correlation with the Lower (Early) and Upper (Late) Devonian respectively. The Lower Old Red Sandstone

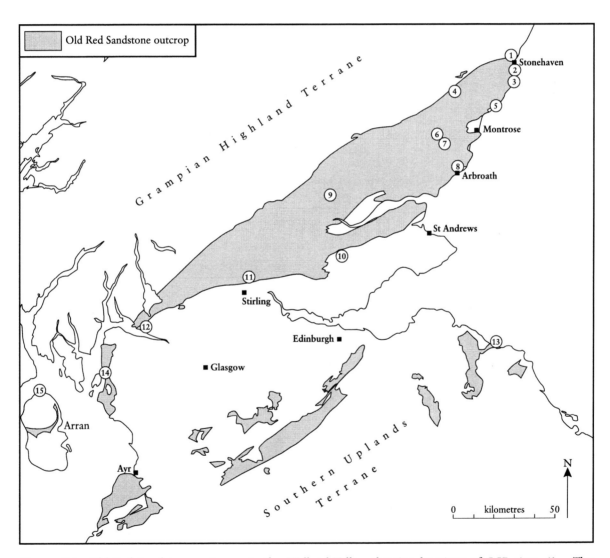

Figure 3.1 Old Red Sandstone outcrops in the Midland Valley, showing locations of GCR sites: (1 – The Toutties; 2 – Dunnottar Coast Section; 3 – Crawton Bay; 4 – North Esk River; 5 – Milton Ness; 6 – Aberlemno Quarry; 7 – Tillywhandland Quarry; 8 – Whiting Ness; 9 – Tay Bank; 10 – Glen Vale; 11 – Wolf's Hole Quarry; 12 – Auchensail Quarry; 13 – Siccar Point to Hawk's Heugh; 14 – Largs Coast, Ayrshire; 15 – North Newton Shore, Arran). After British Geological Survey 1:625 000 Solid Geology Map, UK North Sheet, 4th edn (2001).

spans the Wenlock–Ludlow to Emsian (middle Silurian to lower Devonian) stages and the Upper Old Red Sandstone is Famennian (Late Devonian) to Courceyan (early Carboniferous) in age. The lithostratigraphy of the Old Red Sandstone used in this account is based on a recent review by the British Geological Survey (Browne *et al.*, 2002) (Figure 3.2). Bluck (2002) provides a recent, comprehensive summary of the Old Red Sandstone of the Midland Valley.

Devonian palaeogeography of the British Isles is illustrated and summarized in Cope *et al.* (1992). Recent models of the palaeogeography of the Midland Valley were given by Bluck (2000, 2001), Woodcock (2000a) and Trewin and Thirlwall (2002). Palaeomagnetic data suggest that the Midland Valley was situated in the equatorial zone, ranging from about 25° south in early Devonian times to about 10° south in late Devonian times (Trench and Haughton, 1990). The palaeoclimate appears to have been hot and, at times, semi-arid to arid, with seasonal rainfall, as indicated by pedogenic limestones in the Scone Sandstone Formation (Arbuthnott–Garvock Group) and the Kinnesswood Formation (Inverclyde Group) and the aeolian sediments in the Auchtitench Sandstone Formation (Lanark Group), Cromlix Mudstone Formation (Strathmore Group) and Knox Pulpit Sandstone Formation (Stratheden Group). In general, global sea-level rose from a lowstand at about the beginning of the Devonian Period to a highstand in Frasnian (Late Devonian) times, and falling and oscillating somewhat before rising again during the major Carboniferous transgression (e.g. Woodcock, 2000a).

The Midland Valley had a complex Siluro–Devonian geological history involving multiple phases of transcurrent strike-slip faulting, sediment deposition, igneous activity, uplift and erosion (e.g. Smith, 1995). This followed its collision with the Laurentian continental margin in Ordovician times (e.g. McKerrow *et al.*, 1991), with deposition of the Old Red Sandstone succession taking place after the amalgamation of terranes that were previously widely separated (see 'Old Red Sandstone palaeogeography', Chapter 1). The closure of the Iapetus Ocean resulted in sinistral strike-slip fault re-activation, crustal compression, uplift, granitic intrusion, andesitic volcanicity and low-grade metamorphism. The Midland Valley and Grampian Highland terranes amalgamated by late Silurian time, after which strike-slip movements between them appear to have been of the order of only tens of kilometres (Trench and Haughton, 1990). The compression in the Laurentian crust, of which the Scottish Highlands were part, was accompanied by the intrusion of large volumes of granitic rocks and the extrusion of predominantly andesitic volcanic rocks. The latter were probably extensive, their eroded remnants being seen at Ben Nevis, Glen Coe, Lorn and immediately north of the Highland Boundary Fault. Volcanic rocks also crop out extensively within the Midland Valley, at Montrose, in the Sidlaw, Ochil and Pentland hills and in Ayrshire, in a central axial belt separating the Old Red Sandstone sedimentary basins (Figure 3.3). They also occur more locally in the Southern Uplands, where granitic intrusions such as the Cheviot were emplaced. The amalgamation of the Midland Valley, Grampian Highland and Southern Uplands terranes was probably accompanied by strike-slip movements on the Highland Boundary and Southern Upland faults.

The cycles, initiated by sinistral, strike-slip faulting, fine upwards from basal coarse gravels and become petrologically more mature upwards, reflecting the decline in tectonic influence and reduction in the relief of the source area. Calc-alkaline volcanism punctuated the clastic infilling of the basins. The basins are thus interpreted as mainly pull-apart structures, formed by sinistral strike-slip re-activation of pre-existing faults that were weakened by a long history of igneous activity (e.g. Phillips *et al.*, 1998; Bluck, 2000). However, when regional crustal shortening and intrusion of buoyant granitic magma had caused uplift of the Caledonian mountain chain, gravitational collapse of the orogen and extensional spreading may also have been a factor (e.g. Woodcock, 2000a). Transpressive movements subsequently inverted the basins, resulting in the recycling of the basin-fills and providing weak points for continuing volcanic extrusion (Bluck, 2000).

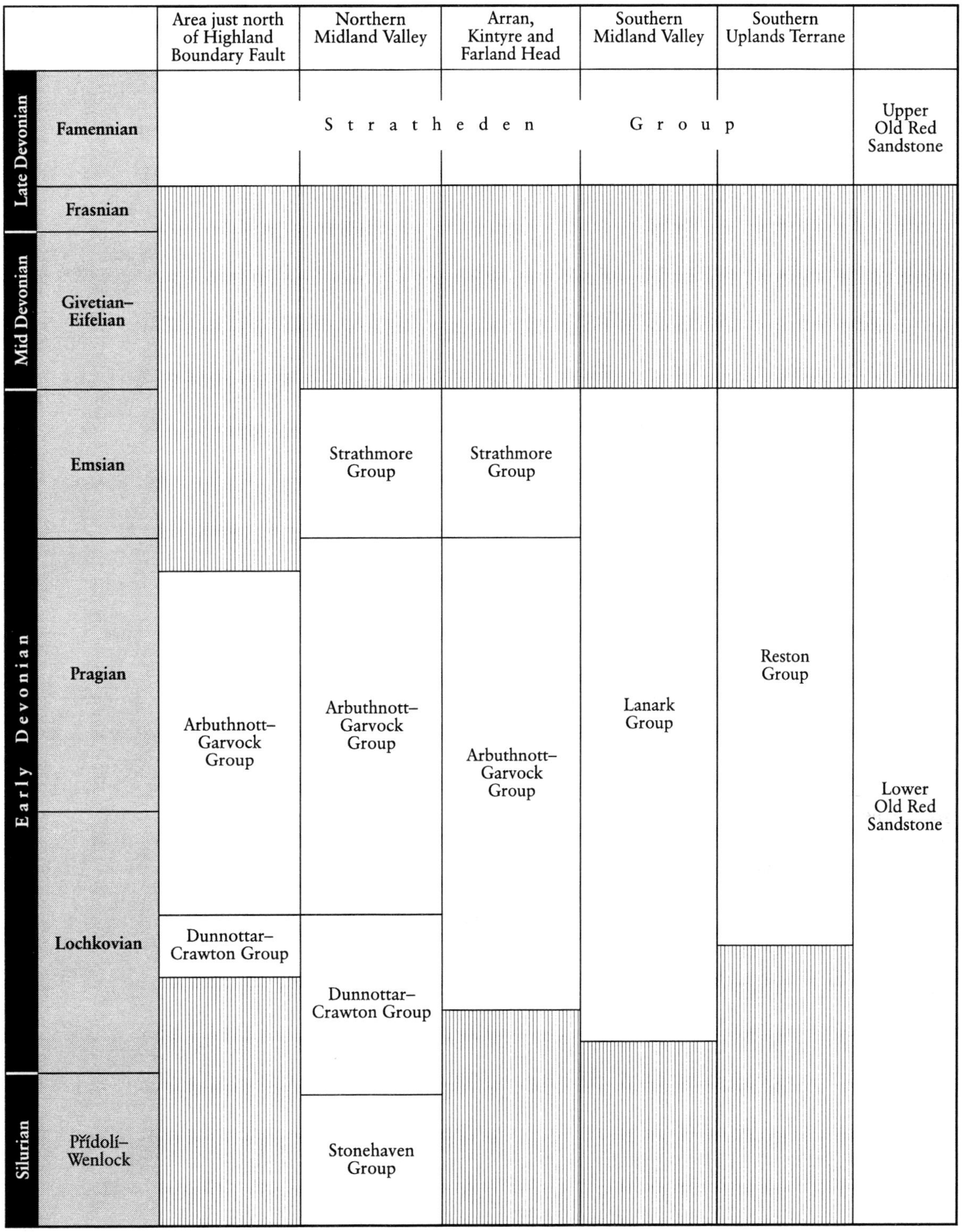

Introduction

	Area just north of Highland Boundary Fault	Northern Midland Valley	Arran, Kintyre and Farland Head	Southern Midland Valley	Southern Uplands Terrane	
Late Devonian — Famennian		Stratheden		Group		Upper Old Red Sandstone
Frasnian						
Mid Devonian — Givetian–Eifelian						
Early Devonian — Emsian		Strathmore Group	Strathmore Group			
Pragian	Arbuthnott–Garvock Group	Arbuthnott–Garvock Group	Arbuthnott–Garvock Group	Lanark Group	Reston Group	Lower Old Red Sandstone
Lochkovian	Dunnottar–Crawton Group	Dunnottar–Crawton Group				
Silurian — Přídolí–Wenlock		Stonehaven Group				

Figure 3.2 Lithostratigraphical groups of the Old Red Sandstone of the Midland Valley. After Browne *et al.* (2002).

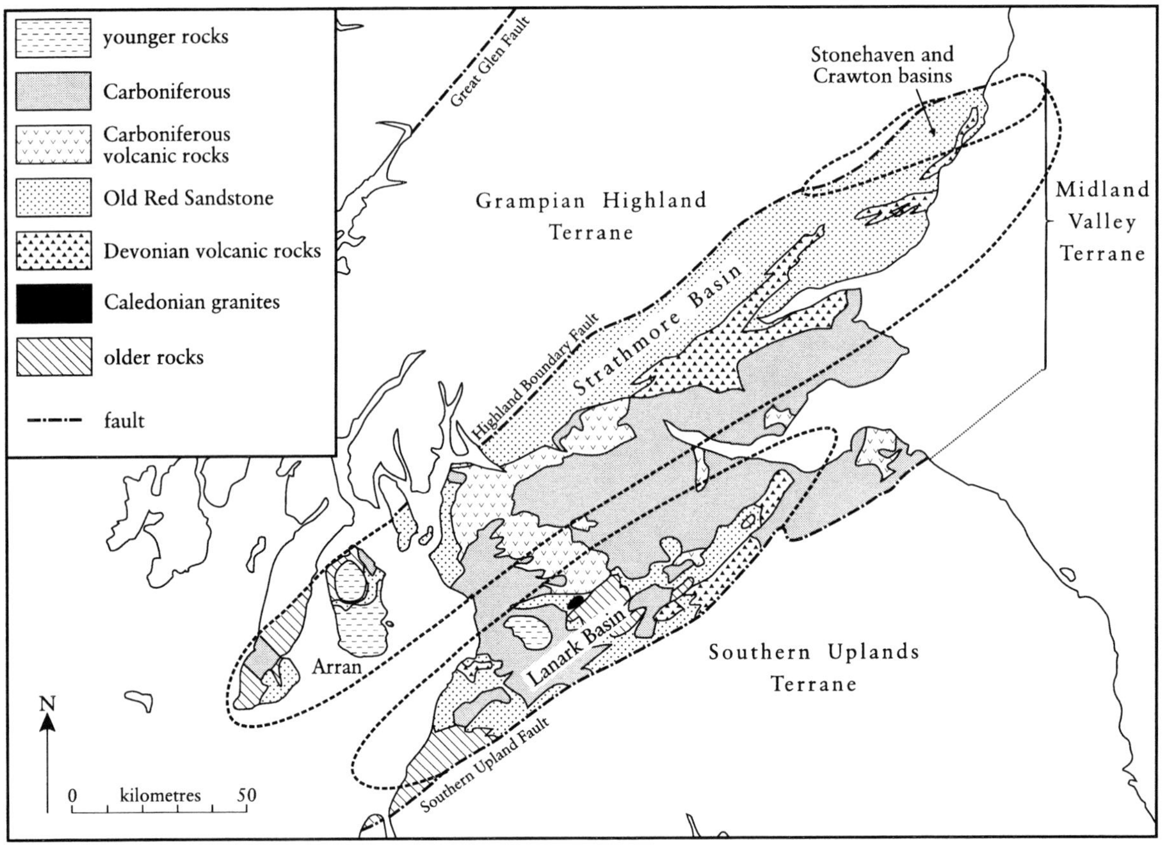

The Midland Valley of Scotland and adjacent areas

Figure 3.3 Old Red Sandstone basins in the Midland Valley.

The roles of the Highland Boundary Fault and Southern Upland Fault throughout much of the Devonian Period remain uncertain (Bluck, 2000, 2001, 2002). Over 8 km of coarse conglomerates and immature sandstones were deposited in the footwall of the Highland Boundary Fault, but the hanging-wall terranes of this and the Southern Upland Fault (Grampian Highland to the north and Southern Uplands to the south) do not appear to have contributed significant volumes of first-cycle detritus to the fill of the basins, having apparently been eroded to low relief before the Devonian Period (Bluck, 1984, 2000, 2001, 2002). Thus, earlier broad interpretations of the Old Red Sandstone as the molasse deposits of the Caledonian mountain chain have been refined (e.g. by Haughton, 1988; Phillips *et al.*, 1998; Bluck, 2000). However, the source of much of the sediment is enigmatic; a now-concealed flysch terrane under the Midland Valley (Haughton, 1988; Bluck, 1992), and a limestone-covered metamorphic–volcanic arc terrane between the Midland Valley and Southern Uplands (Armstrong and Owen, 2000) appear to have supplied sediment initially (Figure 3.4a). Later in early Devonian times, major SW-flowing axial drainage was sourced from the Scandian Orogen to the north-east (e.g. Bluck, 2000). Following Acadian uplift, basin inversion and erosion in Mid-Devonian times, the late Devonian (Famennian) to early Carboniferous Upper Old Red Sandstone succession was deposited unconformably on the Lower Old Red Sandstone. The LORS and UORS broadly represent two major cycles of sedimentary basin-fill, separated by the Acadian unconformity. Fluvial dispersal in late Devonian times appears to have been mainly from the west, towards the north-east (Figure 3.4f).

The main graben development started in Wenlock times in the north-east, with the small Stonehaven and Crawton basins (see **The Toutties** GCR site report, this chapter) forming as a result of sinistral shear along the Highland Boundary Fault Zone. The sediments of the Stonehaven Group were sourced to the south-west and include medium-grade metamorphic clasts from south of the Highland Boundary

Introduction

Fault (Robinson *et al.*, 1998). The Crawton Basin (Haughton, 1988; Haughton and Bluck, 1988; Marshall *et al.*, 1994; Phillips *et al.*, 1998; Bluck, 2000) formed later, extending south-westwards. It was filled with large volumes of recycled conglomerates and sandstones mostly derived from the north (the Dunnottar–Crawton Group) (see **Dunnottar Coast Section** and **Crawton Bay** GCR site reports, this chapter). These contain clasts of Ordovician muscovite-biotite granite, late Silurian granitoids and staurolite-grade metamorphic rocks (Haughton *et al.*, 1990). Some conglomerates were southerly derived, from outcrops of deep-water sedimentary rocks and high-level granite within the Midland Valley. The succession in the Crawton Basin appears to be linked to that of the Stonehaven Basin (Phillips *et al.*, 1998), but the development of the former was accompanied by large-scale synsedimentary faulting. Conglomerates, derived from the south-east, containing recycled quartzite clasts occur in the Callander–Loch Lomond area (Bluck, 1984). Bluck (2000) suggests an intricate arrangement of stacked, superimposed basins (the Stonehaven, Crieff–Callander and Arran basins) in Strathmore, based on the recognition of culminations in thickness at Edzell and Balmaha.

Subsequent Lower Old Red Sandstone alluvial-fan, fluvial and lacustrine deposition appears to have been focused in the Strathmore Basin (Arbuthnott–Garvock and Strathmore groups) in the north (Figure 3.5) and the Lanark Basin (Lanark Group) in the south (Figure 3.6). These basins were separated by an ill-defined high of little-known geology, but possibly comprising an Ordovician volcanic arc intruded by micro-granitic bodies such as that at the base of the Salsburgh No. 1 Well in the central Midland Valley. The cryptic flysch terrane proposed by Haughton (1988) may have been part of this block. Farther west, it is possible that the Strathmore and Lanark basins merged and continued south-westwards into Ireland. The calc-alkaline lavas and associated intrusions in the Lower Old Red Sandstone have been dated at about 410 Ma (Thirlwall, 1988). However, volcanism may have started earlier in the north, as minor amounts of acid/intermediate lava occur in the Stonehaven Group. Thick piles of lavas accumulated in composite centres and strato-volcanoes in the Dunnottar–Crawton and Arbuthnott–Garvock groups, as well as thinner, but quite widespread ignimbrites and other volcaniclastic deposits.

The large size of the coarse conglomerate-filled channels, complex palaeoflow patterns and evidence of high rates of sediment flux and water discharge in the Crawton Basin suggest deposition from antecedent rivers in large 'wet-type' fans (Haughton, 1989). The later, larger Strathmore Basin was filled axially with fluvial sediments deposited by a major SW-flowing river system extending along the north of the Midland Valley (Figure 3.4c,d) and interstratified calc-alkaline lavas and volcaniclastic rocks. The river system may have been sourced in the mountains of Norway and Greenland, where there is evidence of considerable Scandian uplift from 410 Ma to 380 Ma. The sandstones of the Arbuthnott–Garvock Group are characterized by palaeocurrents directed to the south-west in a mainly braided river system. Cross-bedding and palaeocurrent directions in the Arbuthnott–Garvock Group suggest a SW-directed braided fluvial system and a basin margin east of the present Angus/Kincardine coastline (Armstrong *et al.*, 1985). Localized lacustrine deposits in the Arbuthnott–Garvock Group, such as those of Lake Forfar (see **Tillywhandland Quarry** GCR site report, this chapter), are thought to be the result of impeded drainage caused by local volcanic activity or synsedimentary faulting (Trewin and Davidson, 1996). A widely developed concentration of calcrete profiles at the top of the Arbuthnott–Garvock Group represents a prolonged period of non-deposition about the Pragian–Emsian boundary (see **Tay Bank** GCR site report, this chapter).

The overlying thick succession of siltstones and mudstones of the Cromlix Mudstone Formation (basal Strathmore Group) may have been alluvial floodplain deposits of the distal parts of alluvial fans, sourced from the Grampian Highland Terrane or farther afield, which escaped fluvial reworking (Armstrong *et al.*, 1985). Alternatively, an aeolian origin (cf. Dare-Edwards, 1984; Yang, 1997) or fluvial reworking of pedogenic mud aggregates (e.g. Ékes, 1993) may have been responsible for at least some of the argillaceous rocks. Cross-bedded and ripple-laminated sandstones up to 2 m thick are interpreted as the overbank deposits of large river systems (Haughton and Bluck, 1988). Thick arenaceous fluvial deposits (the Teith Sandstone Formation) occur widely in the Strathmore Basin above the Cromlix Mudstone Formation. Local, thick fanglomerates were still being deposited close to the Highland Boundary Fault at that time,

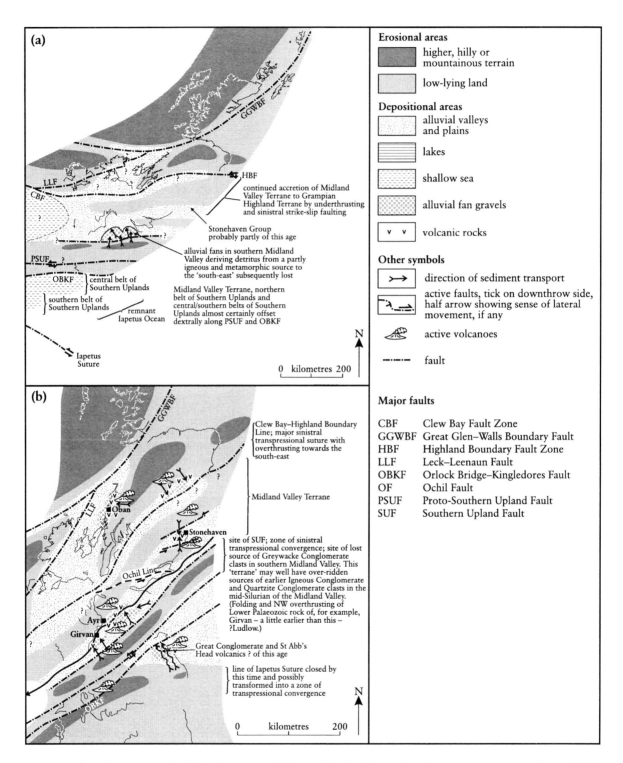

Figure 3.4 Palaeogeographical development of the Midland Valley and adjacent areas. (a) Early Wenlock; (b) Mid-Přídolí. (a) after Bassett *et al.* (1992); (b) after Bluck *et al.* (1992). Continued on page 133.

Introduction

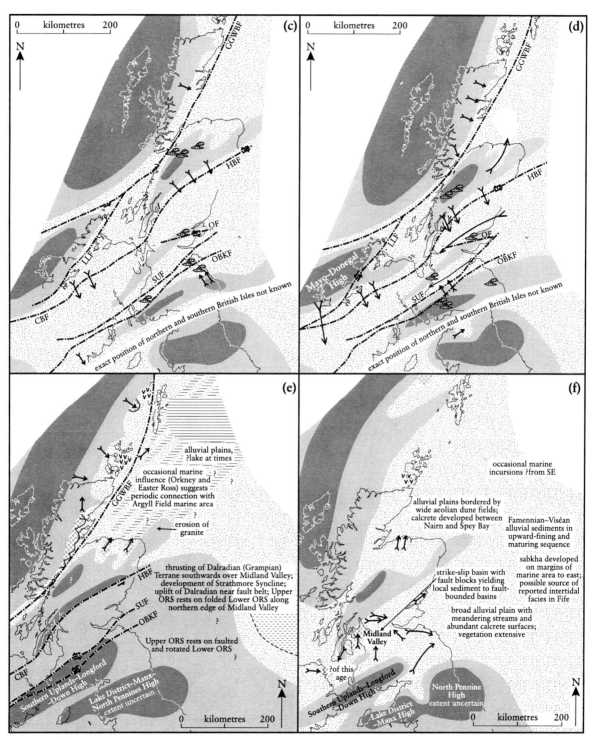

Figure 3.4 – *continued*. Palaeogeographical development of the Midland Valley and adjacent areas. (c) Early Devonian (Lochkovian); (d) Early Devonian (late Pragian–early Emsian); (e) Mid-Devonian (Givetian); (f) Late Devonian (Frasnian–early Famennian). After Bluck *et al.* (1992).

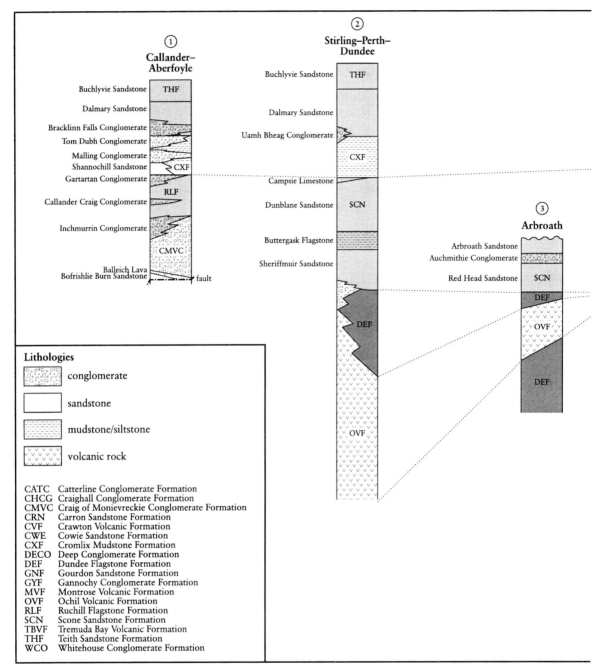

Figure 3.5 Sections of the Late Silurian–Early Devonian rocks in the northern Midland Valley. After Browne *et al.* (2002).

sourced from the Grampian Highland Terrane to the north (see **North Esk River** GCR site report, this chapter). On Arran, the Strathmore Group comprises argillaceous floodplain deposits intercalated with thin, sheeted fanglomerates.

In the Lanark Basin, alluvial-fan gravels generally fine upwards into fluvial sandstones. The detritus is thought to have been derived from horsts of flysch to the east (Syba, 1989) created by strike-slip fault movements within the Midland Valley Terrane. A period of calc-alkaline volcanic eruptions from at least four centres produced substantial piles of subaerial volcanic rocks. These were locally eroded and the detritus reworked by rivers which also carried quartzose metamorphic detritus from a more distant source to the north-east. Lithological similarities between the upper part of the Arbuthnott–Garvock

Introduction

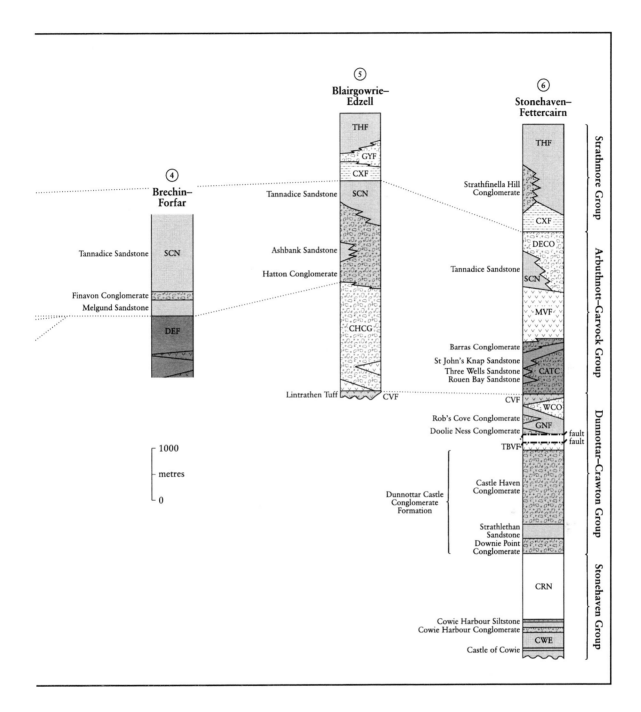

Group in the north of the Midland Valley and the Auchtitench Sandstone Formation of the Lanark Group in the south (Phillips *et al.*, 1998) suggest that by Pragian–Emsian times, there may have been a link between the Strathmore and Lanark basins, at least in the west. Along the southern margin of the Midland Valley, the Lower Palaeozoic marine basin, now represented by rocks of the Pentland Hills, Hagshaw Hills and Lesmahagow inliers, was inverted, with terrestrial fluvial and lacustrine sedimentation continuing into early Devonian times before erosion began.

Within the Grampian Highland Terrane, in the Lorn–Oban area, calc-alkaline volcanism preceded the deposition of fossiliferous sedimentary rocks of Přídolí to Lochkovian age on peneplained Dalradian basement (Marshall, 1991; Durant, 1999a,b; Bluck, 2000). The area of preserved

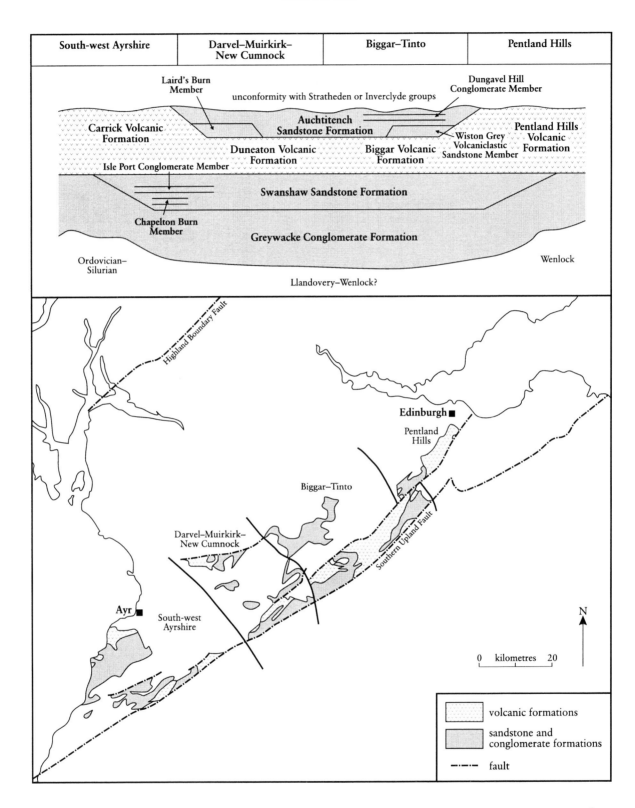

Figure 3.6 Outcrops of the Lanark Group in the southern Midland Valley and schematic cross-section. After Browne *et al.* (2002).

Introduction

sub-horizontal lavas and ignimbrites (the Lorn Plateau Volcanic Formation) may be only a small remnant of their original extent. The sedimentary rocks accumulated in small fault-bound basins, as scree breccias and fanglomerates passing basinward into fluvial sands and then lacustrine muds with thin carbonate beds (Stephenson and Gould, 1995).

Early Devonian sinistral strike-slip on the Southern Upland Fault appears to have brought together unrelated parts of the Midland Valley and the basins in the Southern Uplands Terrane to the south, as indicated by the different provenance of the basal conglomerates and sandstones. Also, the calc-alkaline volcanic rocks in the Southern Uplands have slightly different geochemistry to the volcanic rocks of the Midland Valley, and their original position in relation to the Midland Valley Terrane is therefore unclear. The Lower Devonian sandstones and conglomerates (the Reston Group) rest on an eroded surface of Lower Palaeozoic turbidites, suggesting that the north-east part of the Southern Uplands Terrane was almost peneplained (Bluck *et al.*, 1992; Bluck, 2000). Lava fields may have had a limited extent in the Southern Uplands Terrane, as the overlying fluvial deposits overlap on to the Lower Palaeozoic rocks around Eyemouth. The Cheviot volcanic field, intruded by granite at about 395.9 Ma (Thirlwall, 1989), was extensively eroded, and appears to be younger than the volcanic formations in the Midland Valley.

Upper Devonian rocks (the Stratheden Group of the Upper Old Red Sandstone; Figure 3.7) were deposited unconformably on the Lower Old Red Sandstone and older rocks after uplift and erosion during the Acadian Orogeny. The Acadian unconformity is seen at the **Siccar Point to Hawk's Heugh**, **Whiting Ness** and **North Newton Shore** GCR sites. Clastic deposits of fluvial and aeolian origin predominate, tending to be generally finer grained than the Lower Old Red Sandstone. A single, major ENE-trending basin has been modelled towards the north of the Midland Valley (e.g. Hall *et al.*, 1998). The deposits in the western and northern parts of this basin are thicker and coarser than those in the south and east, braided fans of relatively mature detritus prograding from the west and north (see **Largs Coast** GCR site report, this chapter). Vein quartz and schist clasts are thought to have been derived from Dalradian rocks, whereas rounded quartzite, lava and granite boulders may have been reworked from Lower Devonian strata. Penecontemporaneous strike-slip faulting has been postulated (Bluck, 1980a, 2000) and movements on the Largs Fault occurred before deposition of the overlying Carboniferous Inverclyde Group (Paterson *et al.*, 1990).

In the northern part of the Midland Valley, the predominant palaeocurrent directions were towards the east and south-east, indicating an axially-draining braided river system. Bluck *et al.* (1992) postulated substantial rivers over 10 m deep in the west, possibly sourced from the north. The presence of substantial volumes of aeolian deposits (see **Glen Vale** GCR site report, this chapter) and calcrete (see **Milton Ness** GCR site, this chapter) in the younger formations indicates that rainfall may have diminished by late Devonian times. The prevailing easterly and north-easterly winds caused the migration of dune sands to the west and south-west (Hall and Chisholm, 1987). In places to the west, dune sands were blown on to the floodplains or reworked by the rivers. Famennian fish beds occur in finer-grained shallow-water and floodplain facies.

On the south side of the Midland Valley, exposure of the Stratheden Group is limited to the Ayrshire coast in the west, where rudaceous to arenaceous fluvial deposits may have formed part of a separate fan sourced from the remnant high between the Lower Devonian Lanark and Strathmore basins. To the east, in the present North Sea, a gulf of the Mid- to Late Devonian sea briefly connected to the Orcadian Basin (Marshall *et al.*, 1996). The thin successions of Upper Old Red Sandstone fluvial and lacustrine red beds in the eastern part of the Southern Uplands (including the Cockburnpath–Pease Bay outcrops) may have been deposited in a basin separate from the Midland Valley. However, in both the Midland Valley and Southern Uplands, alluvial gravels and fluvial/lacustrine beds pass up into fluvial sandstones with pedogenic carbonate of the Carboniferous Kinnesswood Formation. Leeder (1973) postulated internal drainage in a Scottish Border

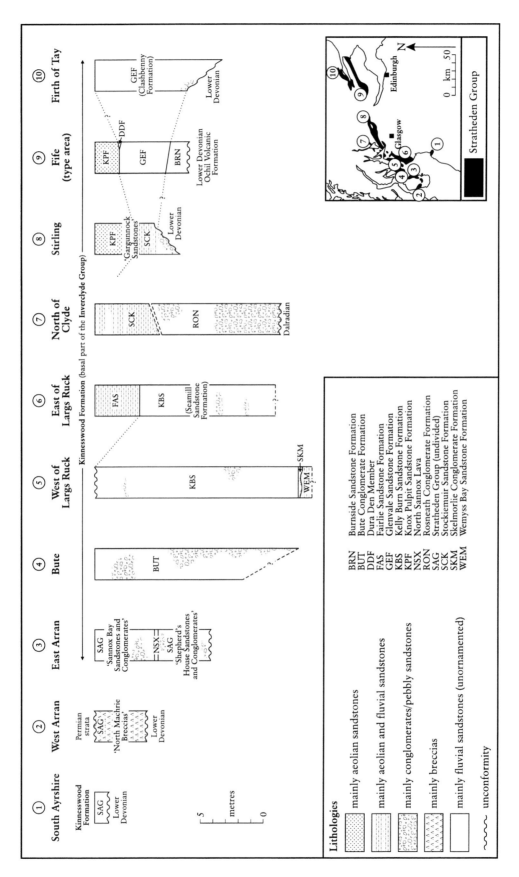

Figure 3.7 Sections of the Late Devonian Stratheden Group in the Midland Valley showing stratigraphical positions of GCR sites. Inset shows outcrop of Stratheden Group and locations of sections. After Browne et al. (2002).

Introduction

basin in the Jedburgh area, but the overall pattern can be explained by an easterly flowing river system (Paterson *et al.*, 1976) which may have connected with the Northumberland Basin farther east. In the central North Sea, red sandstones of relatively mature fluvial systems are assigned a Frasnian age, but may extend into the early Carboniferous (Gatliff *et al.*, 1994).

Some of the Caledonian igneous rock GCR sites in and adjacent to the Midland Valley, described by Stephenson *et al.* (1999), include Old Red Sandstone sedimentary rocks (Table 1.5, Chapter 1). These are not described in this volume. The sedimentary rocks of the Montrose Volcanic Formation at the Scurdie Ness to Usan Harbour and Black Rock to East Comb sites (Smith, 1999a) are clast-supported conglomerates and coarse-grained, locally cross-bedded sandstones consisting of volcanic detritus with varying amounts of non-igneous components. The Balmerino to Wormit, Sheriffmuir Road to Menstrie Burn, and Tillicoultrie sites (Browne, 1999) are in the Ochil Volcanic Formation. The main sedimentary rocks at these sites are volcaniclastic conglomerates, commonly matrix-supported and of debris-flow origin. At Balmerino to Wormit, the lavas and volcaniclastic conglomerates are interbedded with fluvio-lacustrine sandstones and minor mudstones of the Dundee Flagstone Formation. The mudstones have yielded fish, arthropod and plant fossils, providing palaeontological evidence of an early Devonian (Lochkovian) age to complement local radiometric ages of 410.5 ± 5.6 Ma. GCR sites in Ayrshire with Old Red Sandstone rocks are Port Schuchan to Dunure Castle, Culzean Harbour and Turnberry Lighthouse to Port Murray (Durant, 1999a). The principal feature of these sites is the interaction of unconsolidated, finely laminated, sandy sediments and andesitic igneous rocks formerly regarded as lava flows, but re-interpreted by Kokelaar (1982) as high-level andesitic sill intrusions. In the Scottish Borders, volcaniclastic conglomerates with some sandstones are interbedded with lava flows of the St Abb's Volcanic Formation at the Pettico Wick to St Abb's Harbour GCR Site (Stephenson, 1999). In the Grampians, sedimentary rocks are present on the island of Kerrera, around Oban, in Allt a'Mhuilinn (Ben Nevis) and Glencoe.

At Kerrera (Durant, 1999b), Lower Old Red Sandstone conglomerates and sandstones (the Kerrera Sandstone Formation of Browne *et al.*, 2002) underlie the Lorn Plateau Volcanic Formation (dated as 424–415 Ma) locally. They have yielded fossil fish and arthropod remains, as well as spores of late Silurian to early Devonian age (*tripapillatus-spicula-micrornatus-newportensis* biozones) (Marshall, 1991). In the central down-faulted block of the Ben Nevis igneous complex, mudstone/siltstone laminites with non-volcaniclastic interbeds (the Allt a'Mhuilinn Mudstone Formation) rest unconformably on Dalradian metasedimentary rocks and are overlain by volcaniclastic sediments (McGarvie, 1999a). The deposits are interpreted as lacustrine, playa-lake deposits. In Glencoe, sedimentary rocks underlie the volcanic complex and are sandwiched locally between volcanic rocks (McGarvie, 1999b). Basal conglomerates locally infill channels and canyons incised into the underlying Dalradian metasedimentary rocks. Within the central caldera, 20 m of finely laminated, greenish grey mudstones and sandstones (Group 6 of the caldera succession) are interpreted as caldera lake deposits. At the base of Stob Dearg (NN 224 547), a sequence of breccias, conglomerates, well-bedded, quartzose sandstones, red, laminated sandstones and fissile mudstones underlies the volcanic complex. *Psilophyton* and *Pachytheca* were obtained from a mudstone slab, and spore assemblages indicate a mid-Lochkovian (*micrornatus-newportensis* Biozone) age (Wellman, 1994; McGarvie, 1999b). The overlying lavas are dated at 421 ± 4 Ma (Thirlwall, 1988).

Several Old Red Sandstone GCR sites are described in the companion GCR volume on fossil fishes (Dineley and Metcalf, 1999; Table 1.2, Chapter 1). Of these, the Whitehouse Den GCR site (NO 426 397) near Dundee, is not described in this volume. It is one of three localities at which the Tealing Fish Bed has been identified. Poorly exposed sandstones and black lacustrine mudstone laminites of the Dundee Flagstone Formation (Arbuthnott–Garvock Group) have yielded fish and plant fossils, and arthropod remains and tracks.

GCR Old Red Sandstone fossil plant sites are described in the companion GCR volume on Palaeozoic palaeobotany (Cleal and Thomas, 1995; Table 1.3, Chapter 1). Of these, Ballanucater Farm, Callander is not described in this volume. There, plant remains in grey, coarse-grained sandstones and blue-green mudstones are typical of the Lower Devonian (Emsian) Teith Sandstone Formation (Strathmore Group).

THE TOUTTIES, ABERDEENSHIRE (NO 881 866)

Potential ORS GCR site

M.A.E. Browne and W.J. Barclay

Introduction

The Toutties GCR site on the foreshore at Cowie Harbour north of Stonehaven, Aberdeenshire (Figure 3.8) has been long known for the fossil fish specimens from the Cowie Harbour Fish Bed and the arthropods from the overlying mudstones. The fish are described in the companion GCR volume on fossil fishes (Dineley and Metcalf, 1999). The site has recently gained international prominence with the discovery of fossil terrestrial millipedes, including a new air-breathing species, *Pneumodesmus newmani* (Wilson and Anderson, 2004). Traditionally referred to the 'Downtonian' (Přídolí), recent palynological evidence (Marshall, 1991; Wellman, 1993) suggests a mid-Silurian (Late Wenlock–Early Ludlow) age for the beds. Although controversial (Dineley, 1999b), this dating makes the strata among the oldest Old Red Sandstone lithofacies in Britain and the terrestrial diplopod the earliest air-breathing animal of any kind recorded in the geological column. The strata are near-vertical to overturned, lying on the north-west limb of the asymmetric Strathmore Syncline. They belong to the Cowie Sandstone Formation and Carron Sandstone Formation of the Stonehaven Group (Figure 3.9).

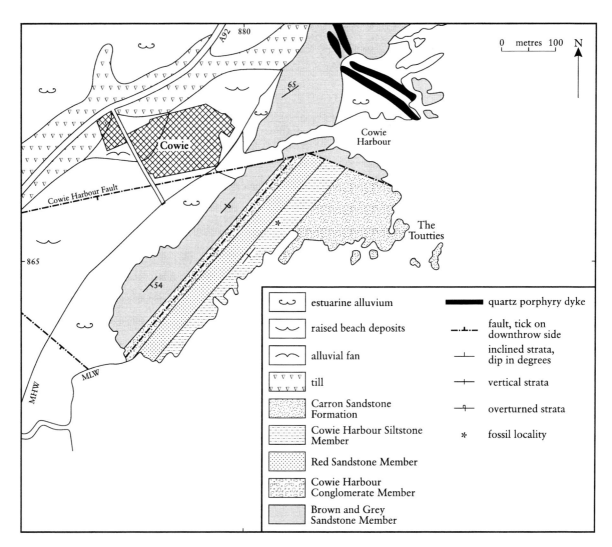

Figure 3.8 Geological sketch map of the area of The Toutties GCR site. After British Geological Survey, 1: 10 000 Manuscript Map NO 88NE (1996).

The Toutties

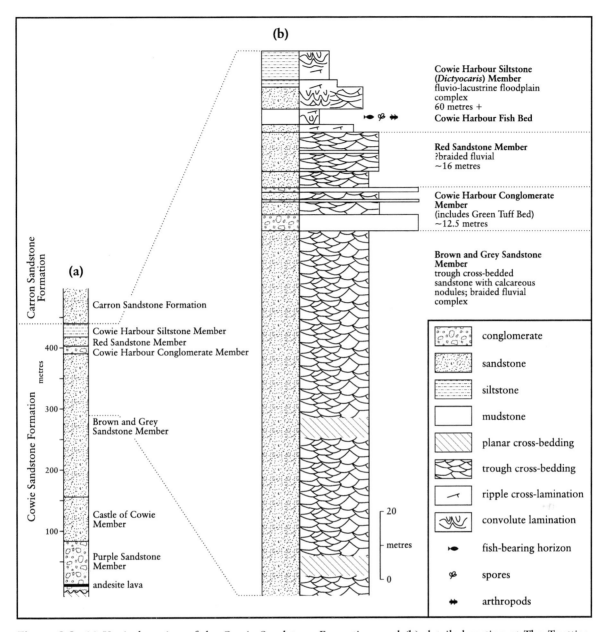

Figure 3.9 (a) Vertical section of the Cowie Sandstone Formation; and (b) detailed section at The Toutties. Based on Armstrong *et al.* (1978b) and Dineley (1999b).

Description

The geology of the site has been described by Armstrong *et al.* (1978b), Gillen and Trewin (1987) and Carroll (1995a). Trewin and Thirlwall (2002) provided a summary. Dineley (1999b) described the fossil fishes and summarized the early work on the fauna. The Cowie Sandstone and Carron Sandstone formations consist mainly of cross-bedded and horizontally laminated, quartzo-feldspathic sandstones. There are numerous intercalations of silty mudstone and siltstone in the Cowie Sandstone Formation and abundant fragments of volcanic rock in the overlying Carron Sandstone Formation. Robinson *et al.* (1998) reported that the sandstones in both formations show a dominant palaeocurrent direction to the north-west and contain detrital garnets indicating a metamorphic source terrane to the south-east. In contrast, Gillen and Trewin (1987) reported palaeocurrent indicators in the Cowie Sandstone Formation showing transport to the ESE near the base and to the south-west in the remainder of the formation. The Cowie Sandstone Formation is late Wenlock to early Ludlow in age (Marshall, 1991; Wellman, 1993), but the Carron Sandstone Formation may be significantly younger, Marshall *et al.* (1994) arguing that there is a break in sedimentation between them, with differences in mineralogical composition (Armstrong *et al.*, 1978b), palaeomagnetism (Sallomey and Piper, 1973) and burial history.

The Cowie Sandstone Formation consists of dull red, purple, grey, buff and yellow, cross-bedded and flat-bedded, medium-grained, quartzo-feldspathic sandstones with intercalations of red, purple and grey siltstone and mudstone. It is exposed at The Toutties and northwards from Cowie Harbour to the unconformity with the Highland Border Complex at Ruthery Head close to the Highland Boundary Fault. About 450 m of beds are present at The Toutties, where faulting may attenuate the succession (Figure 3.9). The formation is subdivided into six members, the names of Donovan (in Armstrong *et al.*, 1978b) as amended by Carroll (1995a) being used in the following description.

The two lowermost members (Purple Sandstone and Castle of Cowie members) crop out to the north of The Toutties site. The Purple Sandstone Member (as re-defined by Carroll, 1995a) consists of about 93 m of purple, medium-grained, trough cross-bedded, quartzo-feldspathic sandstones. The basal 60 m of the member are pebbly and conglomeratic (the Basal Breccia Member of Armstrong *et al.*, 1978b) and contain an andesite lava flow. The overlying Castle of Cowie Member comprises 75 m of interbedded medium-grained, quartzo-feldspathic sandstones, red siltstones and sandy siltstones.

Within and just to the north of the site, over 240 m of soft, brown, grey and green, locally pebbly, trough cross-bedded, quartzo-feldspathic sandstones with common calcareous nodules constitute the Brown and Grey Sandstone Member. The sandstones contain a greater proportion of lithic clasts in comparison with the sandstones of the underlying members. Lenses of breccia containing rip-up siltstone clasts up to cobble size are locally abundant and volcaniclastic material makes its first appearance in the succession.

The Brown and Grey Sandstone Member is overlain by 12.5 m of conglomerates and interbedded green tuffaceous sandstone (the Cowie Harbour Conglomerate Member of Carroll, 1995a and the Volcanic Conglomerate Member of Armstrong *et al.*, 1978b). The conglomerates consist of well-rounded pebbles and cobbles generally up to 10 cm, but with some boulders, of acid volcanic rocks. The unit is variable along strike (Gillen and Trewin, 1987), Carroll (1995a) noting that it consists of three structural blocks bounded by strike parallel (?reverse) faults at Cowie Harbour. The lower block contains two broadly upward-fining sequences in which an irregular erosion surface is overlain by poorly sorted pebbly, coarse- to very coarse-grained volcaniclastic sandstone, which passes upwards over about 7 m into medium-grained volcaniclastic sandstone. The contact between the lowermost conglomerate and the underlying Brown and Grey Sandstone Member is repeatedly displaced eastwards (dextrally) by small syndepositional faults. Beds in the lower part of the member are overstepped by the overlying massive and trough-cross-bedded medium-grained volcaniclastic sandstone. A short break in exposure marks a line of truncation of bedding, the presence of fragments of fractured, soft, sandy fault breccia in this gap suggesting that a near-strike-parallel fault cuts the section here, possibly causing some tectonic attenuation. The upper contact is a gradual transition from medium-grained, green volcaniclastic sandstone into the overlying red sandstone.

The Red Sandstone Member comprises about 16 m of red, lithic, medium-grained, cross-bedded sandstones and tuffaceous sandstones with common convolute bedding. Trough cross-bedding indicates palaeocurrents towards the south-west. In the western part of the Stonehaven area, westwards from Tewel (NO 826 855), a thicker succession of massive and thinly planar-bedded, medium-grained, lithic sandstone may be an indication of varying rates of subsidence across the basin (Carroll, 1995a).

The Cowie Harbour Siltstone Member of Carroll (1995a) (the *Dictyocaris* Member of Armstrong *et al.*, 1978b) consists of interbedded sandy siltstone and planar-bedded, fine-grained sandstone. Both lithologies are predominantly planar laminated, but small-scale cross-lamination is also present, mainly in asymmetrical, current-generated ripples. Symmetrical, wave-generated ripples and convolute de-watering structures occur on the upper surfaces of sandstone beds directly overlain by siltstone.

The Toutties

The Cowie Harbour Fish Bed (Figure 3.10) lies in a succession of intercalated grey sandstones and fissile siltstones and mudstones at the base of the Cowie Harbour Siltstone Member. The other biota occur in grey beds intercalated with sandy siltstones and red sandstones with convolute de-watering structures. Planar lamination predominates, but small-scale ripple lamination occurs locally (Figures 3.9, 3.10). The fish bed and adjacent strata are truncated by a NW-trending fault at the northern end of the outcrop, where they are displaced eastwards to below low-tide level in Cowie Harbour. Abundant remains of *Dictyocaris slimoni* and other arthropods including *Archidesmus* sp., *Kampecaris*?, *Ceratiocaris* sp., *Hughmilleria norvegica* and *Pterygotus* sp. (Campbell, 1913; Westoll, 1951, 1977) have been recovered from a bed of grey mudstone. Cephalaspids were found in a reddish sandy mudstone (the Cowie Harbour Fish Bed) below (see Dineley (1999b) for references). The fish fauna includes the anaspid *Birkenia* sp., the holotype of the cephalaspid *Hemiteleaspis heintzi* and the holotype of *Traquairaspis campbelli*, a heterostracan of which this is the only species.

Three new species of fossil millipedes (*Albadesmus almondi*, *Cowiedesmus eroticopodus* and *Pneumodesmus newmani*) discovered recently at the site are of international importance, particularly in view of the fact that *Pneumodesmus newmani* is the earliest-known record of a fully terrestrial, air-breathing species (Wilson and Anderson, 2004).

The junction between the Cowie Harbour Siltstone Member and the overlying Carron Sandstone Formation is exposed at low tide at The Toutties (Carroll, 1995a). The Carron Sandstone Formation consists predominantly of

Figure 3.10 Cowie Harbour Fish Bed (NO 8813 8667). Fissile mudstones and siltstones with thin ripple-laminated sandstone. (Photo: BGS No. D2455, reproduced with the permission of the Director, British Geological Survey, © NERC.)

brown, dull reddish brown and grey, locally pebbly, medium-grained, lithic sandstones with a substantial volcanic content locally. The sandstones are thinly bedded and weakly planar laminated in the lower part of the formation in and adjacent to the GCR site, and trough cross-bedded in the upper part near Downie Point to the south of Stonehaven Harbour.

Interpretation

The Stonehaven Group is interpreted as the fill of a pull-apart basin (the Stonehaven Basin) controlled by strike-slip faulting (Bluck, 2000, 2001). In contrast to the Crawton and Strathmore basins, there is no broad upward-fining trend to the sedimentary basin-fill and conglomerates are rare. Vitrinite reflectivities of the Stonehaven Group (represented here by the Cowie Sandstone Formation) are compatible with a maximum burial of 3–5 km (Marshall *et al.*, 1994). This, along with the distinctive structure, sediment dispersal pattern, petrography and palaeomagnetism, supports the suggestion of Marshall *et al.* (1994) that the Cowie Sandstone Formation was the fill of a separate strike-slip sub-basin, and not part of the general fining-upward Lower Old Red Sandstone megacycle of which the Carron Sandstone Formation is a part. In this interpretation, the junction between the Cowie Sandstone and Carron Sandstone formations is a major structural and stratigraphical discontinuity (Trewin and Thirlwall, 2002). However, Carroll (1995a) and Phillips and Carroll (1995) record a transitional boundary, invoking the development of forced folds by continuous sedimentation and deposition in a strike-slip basin (cf. Christie-Blick and Biddle, 1985; Serrane, 1992). In this model, the bedding in the older deposits adjacent to the basin margin is progressively rotated, resulting in them retaining a structurally high position. Hence the Cowie Sandstone Formation could have been marginal to the supposed Crawton Basin deposits, as indicated by present outcrop distribution, and have been buried at shallow depth, as indicated by the vitrinite reflectivity data of Marshall *et al.* (1994).

The sandstones of the Purple Sandstone Member and Brown and Grey Sandstone Member are interpreted as the deposits of a braided river complex (Armstrong *et al.*, 1978b). Phillips and Carroll (1995) interpreted the sediments to have been laid down by small, bedload-dominated, braided streams on the lower part of alluvial fans, with some sheet-flood deposits in the Purple Sandstone Member. Based on palaeocurrents, the drainage was lateral (towards the ENE) in the Purple Sandstone Member and longitudinal (south or SSW) in the Brown and Grey Sandstone Member. The interbedded sandstones and argillaceous rocks of the Castle of Cowie Member are interpreted as the channelized and floodplain deposits respectively of a sinuous (?meandering) river system with longitudinal flow to the south-west. The Cowie Harbour Conglomerate Member was deposited from bedload-dominated, braided streams either on a distal alluvial-fan or braidplain, with transport to the ENE. Marked scouring at channel bases suggests higher flow energies than those of the streams that deposited the other members of the formation. Phillips and Carroll (1995) suggested that deposition of this thin unit might be linked to the appearance of a small volcanic cone on the braidplain. A return to a bedload-dominated, braided stream environment is suggested by the Red Sandstone Member. Palaeocurrents indicate derivation from the north-east (Gillen and Trewin, 1987; Phillips and Carroll, 1995). Sedimentation by fluvial channel, floodplain and lacustrine processes were responsible for the deposition of the Cowie Harbour Siltstone Member. Phillips and Carroll (1995) interpreted the horizontally laminated and cross-bedded sandstones that form most of the member as distal deposits of turbidity currents that introduced fluvial sediment into the lake. The finer-grained, carbonaceous siltstones lack wave ripples and trace fossils, suggesting that the lake was deep and stratified periodically, with anoxic bottom conditions.

The fish and arthropod fauna of the Cowie Harbour Fish Bed were formerly considered to be of late Silurian–early Devonian ('Downtonian') age (now the Přídolí Series) (Campbell, 1913; Denison, 1956; Westoll, 1951, 1977). However, Lamont (1952) advised caution in the age dating and correlation of the fauna, on the basis of similar faunas of probable late Llandovery to early Wenlock age in the southern part of the Midland Valley. Hanken and Størmer (1975) later suggested an early Ludlow age for the eurypterid arthropods. Marshall (1991) and

Dunnottar Coast Section

Wellman (1993) provided palynological evidence of a late Wenlock or early Ludlow (mid-Silurian) age. On this basis, the site is only one of two, the other being in Pennsylvania, that has yielded late Wenlock palynomorphs from continental Old Red Sandstone facies.

The Toutties site is important because it is comparable with the red-bed successions of the Silurian inliers in the southern Midland Valley, in which the Silurian rocks are overlain unconformably by Lower Old Red Sandstone of latest Silurian to early Devonian age. A late Wenlock (Homerian) to Ludlow (Gorstian) age for the Cowie Harbour Siltstone (*Dictyocaris*) Member, as proposed by Marshall (1991) and Wellman (1993), implies that *Traquairaspis campbelli* is substantially older than had previously been thought, and of similar age to the heterostracans of the Cape Phillips Formation and Cape Storm Formation in the Canadian Arctic (Dineley, 1999b). It also implies that the site provides, in the discovery of the fully terrestrial millipede *Pneumodesmus newmani*, the earliest evidence of air-breathing in any animal of any kind (Wilson and Anderson, 2004).

Conclusions

The site's international importance lies in its freshwater fauna of fish, arthropods and newly discovered millipedes, and in its late Wenlock to early Ludlow palynomorphs, which date the rocks as some of the oldest Old Red Sandstone facies recorded. The unique heterostracan fish *Traquairaspis campbelli* is unknown elsewhere, with similar heterostracans providing a link with the Canadian and Baltic provinces. Recent discoveries of new species of fossil millipedes add a new dimension to the site's importance, particularly in the presence of the oldest record anywhere in the world of a fully land-based, air-breathing animal.

The section is also important in the evidence it provides into an understanding of the early development of the Midland Valley. The Stonehaven Basin appears to have been an early pull-apart basin formed before the development of the larger Strathmore Basin. The site offers opportunity for further research into its unique fauna and microflora and into the relationship between the Cowie Sandstone Formation and the overlying Carron Sandstone Formation.

DUNNOTTAR COAST SECTION, ABERDEENSHIRE (NO 883 853–NO 882 839)

Potential GCR site

M.A.E. Browne and W.J. Barclay

Introduction

The dramatic cliffs and foreshore of the coastline extending southwards from Downie Point, south of Stonehaven Harbour to Dunnottar Castle and beyond for 1 km to Tremuda Bay (Figure 3.11) expose a Lower Old Red Sandstone succession dominated by conglomerates. This potential GCR site includes the topmost beds of the Carron Sandstone Formation (of the middle Silurian Stonehaven Group) and the type section of the Dunnottar Castle Conglomerate Formation (of the upper Silurian to lower Devonian Dunnottar–Crawton Group) (Figure 3.5, Section 6). The northern limit of the outcrop of the Tremuda Bay Volcanic Formation marks the southern limit of the potential site. The strata lie on the steeply dipping to overturned north-west limb of the asymmetric, NE-trending Strathmore Syncline.

Description

Descriptions of the site were given by Gillen and Trewin (1987), Carroll (1995a,b) and MacGregor (1996a). Haughton and Bluck (1988) carried out a detailed analysis of the conglomerates in the section. Trewin and Thirlwall (2002) provided a summary of the succession and its depositional environments. The Stonehaven Group crops out northwards from Downie Point (Figure 3.11; see **The Toutties** GCR site report, this chapter) and comprises at least 1800 m (Carroll, 1995a) of cross-bedded and horizontally laminated, quartzo-feldspathic sandstones. The lower part of the group (the Cowie Sandstone Formation) has numerous mudstone/siltstone interbeds. The upper part (the Carron Sandstone Formation) is rich in volcanic detritus, has predominantly NW-directed palaeocurrents and contains detrital garnets indicating a metamorphic source to the south-east (Robinson *et al.*, 1998).

In this area, the sandstones of the Carron Sandstone Formation range from fine-grained to pebbly, gritty and lithic. They are predominantly

The Midland Valley of Scotland and adjacent areas

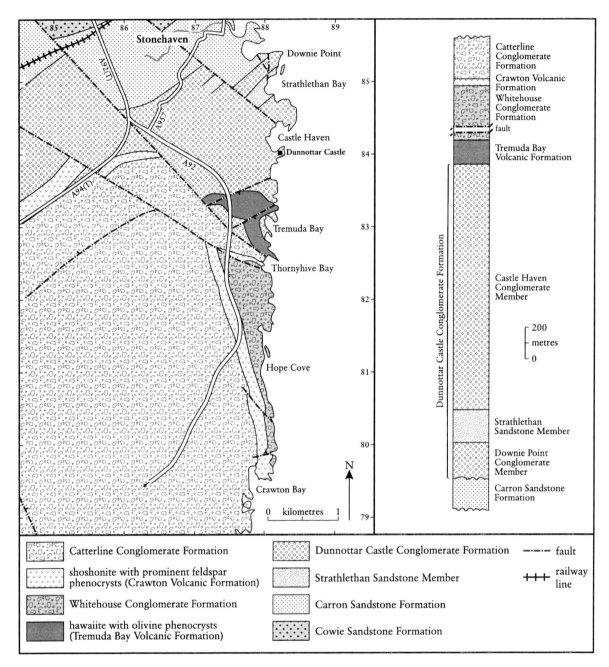

Figure 3.11 Geological map of the area of the Dunnottar Coast Section, from Downie Point to Crawton Bay, with generalized vertical section of the strata. After British Geological Survey 1:50 000 Sheet 67 (Scotland), Stonehaven (1999).

red to purple, with some pink, orange and yellow beds. Both tabular and trough cross-bedded sets (over 1 m thick) and planar laminated units occur, the former with convolute de-watering structures locally. Some of the sandstones are micaceous, others are tuffaceous, the latter being associated with conglomerates dominated by acid volcanic clasts. There are a few beds of red sandy mudstone up to 0.3 m thick, but little evidence of upward-fining cycles, with argillaceous overbank rocks being a minor component and largely confined to pale lilac rip-up mudstone clasts in some of the sandstones. The rare conglomeratic beds and the pebbly sandstones contain clasts of chert, jasper, schistose grit, quartzite, vein quartz and rhyolite.

The Carron Sandstone Formation is overlain unconformably by thick, clast-supported conglomerates (the Downie Point Conglomerate Member), which form the basal part of the Dunnottar Castle Conglomerate Formation (of the Dunnottar–Crawton Group; Browne *et al.*, 2002) at Downie Point. The Downie Point Conglomerate Member comprises 180–225 m of strongly jointed, bimodally sorted, clast-supported cobble to boulder (generally less than 0.6 m) conglomerate with a matrix of coarse-grained sandstone (Carroll, 1995a). Massive, thickly bedded and cross-bedded units occur. The largest, well-rounded boulders are up to 1 m and mainly of resistant 'Highland' quartzites and psammites. They account for 30–60% of the coarse fraction, the remainder being largely of igneous lithologies (mostly andesitic lava, with rare granite and feldspar-phyric microgranite). A few pebbles of Highland Border Complex rocks (metabasalt, serpentinite and red chert) are also present. Beds of sandstone less than 1 m thick occur locally. The basal unconformity is seen particularly well in the quarry north-west of Downie Point (Gillen and Trewin, 1987; MacGregor, 1996a), where there appears to be about 2 m of relief on the erosion surface.

The Strathlethan Sandstone Member overlies the Downie Point Conglomerate Member, its base resting on an irregular surface that is markedly oblique to bedding, cutting down into the Downie Point Conglomerate Member from west to east. The member crops out in Strathlethan Bay and consists mainly of grey and grey-green, medium-grained, cross-bedded, locally pebbly, lithic sandstones. A thin basal bed of red, massive, medium- to coarse-grained volcaniclastic sandstone contains angular clasts of felsite and andesitic lava with rarer, rounded clasts of quartzite. Thin planar beds with normal grading are present locally, but sedimentary structures are destroyed towards the north-east of the exposures by a 1 cm-spaced parting sub-parallel to the bedding. The top of the bed is transitional with the overlying green, locally pebbly, lithic sandstone in places, but in others there is a sharp, highly convolute junction. Foreshore exposures at the north end of Strathlethan Bay show spectacular disruption in a 100 m-thick zone. Disruption is most intense at the base of the Strathlethan Sandstone Member, with rotated and partly rounded blocks of grey-green sandstone in a volcaniclastic sandstone matrix with a streaky, foliated fabric (Gillen and Trewin, 1987; Robertson, 1987). The zone apparently thickens from west to east and some of the rotated sandstone blocks are up to 30 m long. At the top of the zone, 'flames' of massive sandstone extend up into the overlying undisturbed beds.

Trough cross-bedding in sets of 1–2 m are typical of the undisturbed upper beds and in coherent blocks within the disturbed zone. Lenses of clast-supported pebble conglomerate and very poorly sorted volcaniclastic sandstone and thin beds of mudstone with desiccation cracks are interbedded with the lithic sandstones in the topmost 20–30 m near the sea stack of Carlin Craig (NO 880 847) at the southern end of Strathlethan Bay, in a rapidly upward-coarsening transition into the Castle Haven Conglomerate Member.

The Castle Haven Conglomerate Member (Browne *et al.*, 2002) consists of massive and weakly bedded, clast-supported conglomerate with lenses of horizontally laminated, medium-grained sandstone. This member forms most of the Dunnottar Castle Conglomerate Formation and is almost continuously exposed in the cliffs and foreshore from Dunnicaer (NO 8825 8473) to Tremuda Bay (NO 880 830) 1 km south of Dunnottar Castle. The conglomerate is of essentially 'Highland' type, its pebble suite dominated by andesitic lavas, quartzite and psammite. There are more clasts of 'granite' and porphyritic microgranite than in the underlying conglomerates, as well as a diverse suite of lithologies including vein quartz, chert, metabasalt, gabbro, migmatitic 'gneiss' and flow-banded rhyolite (Haughton 1989). Gillen and Trewin (1987) noted a quartzite-dominated clast assemblage with jasper and jasper-rich grit pebbles typical of the Highland Border Complex from the base of the member near the southern end of Strathlethan Bay.

At the north end (NO 880 842) of its type locality of Castle Haven, exposures of the Castle Haven Conglomerate Member on the wave-cut platform show well-sorted conglomerates with current imbrication of the pebbles and thin sandstone interbeds, cross-bedding in the sandstones indicating mainly SSW flow (Gillen and Trewin, 1987).

A conglomerate at the south end of Castle Haven (NO 880 840) marks the first appearance of clasts typical of the Dalradian Southern Highland Group (Gillen and Trewin, 1987). They are less well-rounded and of cleaved, low-grade metamorphic rocks. An unusual displacive sparry calcite carbonate cement may be of pedogenic origin. Sparse cross-bedding data indicate ESE flow. The succeeding conglomerate (NO 8800 8404) is also distinctive in consisting mainly of matrix-supported red, weathered andesitic lava boulders and pebbles in a matrix of argillaceous siltstone/very fine-grained sandstone with convolute laminations. The conglomerates between this unit and Dunnottar Castle (Figure 3.12) contain clasts of apparent Dalradian lithologies.

Red, coarse-grained, medium-bedded, volcaniclastic sandstone up to 30 m thick in the upper part of the member are composed largely of coarse-grained, angular clasts of fine-grained 'felsitic' material. These are present around Old Hall Bay (NO 8799 8370) just south of Dunnottar Castle, where the dip of the strata decreases from 70° to 40° to the south-east.

The overlying Tremuda Bay Volcanic Formation crops out in the cliffs southwards from Maiden Kaim (NO 882 834) to Tremuda Bay. It is 170 m thick (Carroll, 1995a) and consists of a number of lava flows of microporphyritic, olivine-bearing hawaiite (Thirlwall, 1979; Le Maitre, 1989). The basal bed of the formation is a thin, fine-grained, sandstone-to-mudstone, fining-upward unit.

Interpretation

The cliff sections between Downie Point and Tremuda Bay expose superb examples of conglomerates of Old Red Sandstone lithofacies that are interpreted as the deposits of a large braided stream system and alluvial fans (Figure 3.13; Haughton and Bluck, 1988). The scale of

Figure 3.12 Sub-vertical conglomerates of the Castle Haven Conglomerate Member of the Dunnottar Castle Conglomerate Formation at Dunnottar Castle (NO 882 839). (Photo: BGS No. D5187, reproduced with the permission of the Director, British Geological Survey, © NERC.)

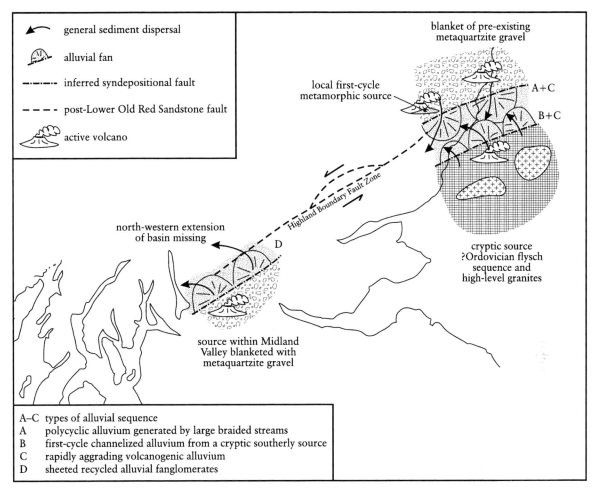

Figure 3.13 Model for Lower Old Red Sandstone sedimentation in the northern Midland Valley. After Haughton and Bluck (1988).

the river systems that deposited the detritus can be appreciated; both large braid bars of high-stage coarse gravels and low-stage and bar-tail cross-stratified sands occur.

At the north end of the section, exposures at Downie Point show the unconformable relationship between the sediments laid down during mid- to late Silurian times in the Stonehaven Basin (the Carron Sandstone Formation) and those in the later Crawton Basin (the Dunnottar Castle Conglomerate Formation). However, the time represented by the unconformity at Downie Point, and whether the Carron Sandstone Formation has more in common with the Dunnottar–Crawton Group than with the Cowie Sandstone Formation in the Stonehaven Group, remain matters of debate (Gillen and Trewin, 1987; Carroll, 1995a; Trewin and Thirlwall, 2002). The sources of some of the detritus are also unclear.

The Carron Sandstone Formation has been interpreted as the deposits of bedload-dominated, braided streams. Phillips and Carroll (1995) interpreted reversals of palaeoflow direction within the Stonehaven Group as a whole as reflecting the interaction of transverse and longitudinal drainage systems within a fault-controlled basin. Contemporaneous tectonic activity may also have changed the local palaeoslope. Convolute bedding structures also point to rapid deposition, liquefaction and de-watering in a tectonically active basin (Trewin and Thirlwall, 2002). The topmost beds of the formation were probably braidplain deposits laid down by moderate- to high-energy streams flowing to the south-west.

The Downie Point Conglomerate Member represents a sudden influx of coarse detritus and uplift, probably to the north-east, as palaeocurrents in a large braided river system are directed to the south-west (Gillen and Trewin, 1987; Haughton and Bluck, 1988; Trewin and Thirlwall, 2002). Robinson *et al*. (1998) report northward transport based on imbrication fabrics in the basal part of the Castle Haven Conglomerate Member and the Downie Point Conglomerate Member. There is no change in clast types in the higher conglomerates, in which imbrication indicates a southerly transport direction. Haughton and Bluck (1988) recognized six types of alluvial sequence (A–F) in the Strathmore Basin on the basis of structural, textural and compositional criteria and palaeocurrent directions. The Downie Point Conglomerate Member is interpreted as polycyclic alluvium generated by large braided rivers (Type A). Haughton (1989) considered the Downie Point Conglomerate Member to have been deposited by large, SW-flowing, multi-channelled, braided streams developed on 'wet' alluvial-fans (Phillips and Carroll, 1995). The drainage system that deposited this conglomerate and the higher Castle Haven Conglomerate Member was clearly much larger than those of the Stonehaven Group, as indicated by the channel sizes (Haughton, 1989) and maximum size of the clasts. Phillips and Carroll (1995) noted that the contact between the Stonehaven and Dunnottar–Crawton groups at Downie Point represents a rapid change in these factors, but assumed a period of continuous deposition, with only minor erosion preceding deposition of the Downie Point Conglomerate Member. These authors interpreted the coarse conglomerates as the deposits of a large drainage system antecedent to the Stonehaven Basin, captured by headward erosion of consequent drainage or by tectonic adjustments. This interpretation is in contrast to a more recent one by Bluck (2000, 2001), in which the Stonehaven Group is envisaged as the fill of an older, pull-apart basin separate from that in which the Dunnottar–Crawton Group was deposited. The presence of volcaniclastic sandstones in the upper part of the Castle Haven Conglomerate Member was interpreted by Phillips and Carroll (1995) to be due to the introduction of large volumes of volcanic detritus during periods when the more general coarse gravel deposition was overwhelmed. Although largely of braided stream origin, high-concentration flood deposits and debris flows are also present. Haughton and Bluck (1988) envisaged deposition of the conglomerates mainly as gravel braid bars that were continuously accreted on one bank as the channel migrated. The laterally extensive sheet geometry and limited number of thick cross-bedded bodies suggest that the channels were shallow and wide and that flow was unconfined periodically. The upward-coarsening units represent downstream migration of coarse bar-head deposits over the finer bar-tail ones.

Like the Carron Sandstone Formation, the Strathlethan Sandstone Member has been interpreted as the deposits of sandy bedload-dominated, braided streams, on the basis of the typical trough cross-bedding. The presence of mudstone with desiccation cracks and sandy mudstone rip-up clasts suggests distal, alluvial-fan, flood deposition or a decrease in stream discharges in a more proximal setting. Haughton and Bluck (1988) described part of the Strathlethan Sandstone Member as a sequence of rapidly aggrading, first-cycle alluvium (Type C), relating the diversity of the volcanic clasts to deposition of volcanic detritus in alluvial fans and aprons (Figure 3.13). The source volcanoes may have been of high relief, independent of tectonic processes that controlled the dispersal of non-volcanic sediment.

The zone of deformed sedimentary rocks at the base of the Strathlethan Sandstone Member was interpreted by Gillen and Trewin (1987), Carroll (1995a) and Trewin and Thirlwall (2002) as a major slide of partly lithified sands triggered by an earthquake of tectonic or volcanic origin. Robertson (1987) suggested sinistral, strike-parallel shear in weakly lithified sediments, rotated to the vertical by early tectonic movements. The restriction of the deformation to Strathlethan Bay favours a slide origin (Carroll, 1995a).

The rocks of the Castle Haven Conglomerate Member were interpreted by Haughton (1989) as the deposits of large braided streams flowing south or south-west on extensive alluvial-fans. Phillips and Carroll (1995) recognized debris flow and hyperconcentrated flood deposits in the volcaniclastic sandstones, introduced when large volumes of volcanic detritus, possibly of local origin, flooded the basin. One matter of debate concerns the change in clast composition recognized by Gillen and Trewin (1987), from conglomerates sourced from the Highland Border Complex at the base of the Castle Haven Conglomerate Member at the south end of Castle Haven to Dalradian-sourced conglomerates (not recognized by Bluck, 1984) to the south. This implies that the Grampian Highland Terrane was more-or-less in its current position at the time of deposition of the Castle Haven Conglomerate Member. However, the relative paucity of Dalradian clasts and the predominance of quartzite suggests that there was a thick cover of quartzite pebble conglomerate over the Grampian Highland Terrane, with little Dalradian outcrop available for erosion at any time during deposition of the Lower Old Red Sandstone (Gillen and Trewin, 1987; Trewin and Thirlwall, 2002).

There is a strong contrast in grain size between the sandstone-dominated Stonehaven Group and the conglomerate-rich Dunnottar–Crawton Group. Both groups are thought to have been deposited in two small, separate pull-apart basins – the Stonehaven and Crawton basins. Bluck (2000, 2001) related their opening and closure to transtension and transpression respectively, along the Highland Boundary Fault Zone. The closure of basins resulted in the recycling of their sediments, with volcanic activity a natural adjunct to the rifting of the weak crust of the Midland Valley. The major depositional cycles fine upwards and become more mature petrographically upwards, indicating that initial rifting was followed by decreasing tectonic influence and a reduction in the relief of the source area.

Bluck (2000) commented on the enigmatic sources of the sediments in the Midland Valley Terrane, with both the Grampian Highland and Southern Uplands terranes extensively eroded before latest Silurian times. The Midland Valley of Scotland was therefore not a post-orogenic molasse basin, with mountainous areas to the north and south. These areas were not undergoing sufficient uplift to provide the huge volume of sediment preserved in the Lower Old Red Sandstone succession, nor were there slopes of sufficient gradient to supply much of the coarse gravels. The thick coarse conglomerates, such as those at Dunnottar, may therefore have been the product of recycling of sediment from a locally inverted, pop-up basin, or from older sedimentary cover in the area of the Highlands to the north. A now-hidden flysch below the centre of the Midland Valley to the south may also have contributed to the fill of the Dunnottar–Crawton Basin (Haughton, 1988). However, whereas local drainage may have accounted for the coarse gravels, it could not have supplied the huge volume of sands deposited. Bluck (2000) suggested a source in the Greenland–Baltica collision zone to the north-east of Scotland where major Silurian to Carboniferous (Scandian) uplift took place. Major river systems draining these mountains would have been able to enter the Crawton and Strathmore basins when relief was lowered sufficiently to allow them access. Large river bars in the Scone Sandstone Formation (Arbuthnott–Garvock Group) west of Perth provide evidence to support Bluck's (2000) model of a distant source of these major distributaries.

Conclusions

The sea cliffs and foreshore from Downie Point south to Dunnottar Castle and Tremuda Bay expose a magnificent section of Lower Old Red Sandstone conglomerates and sandstones. The importance of the section lies in the evidence it provides towards an understanding of the development of the earliest Old Red Sandstone basins in the Midland Valley, including the sources of the coarse gravels and sands that filled them. However, the sources remain enigmatic and interpretations are a matter of debate, and there is scope for further sedimentological and petrographical study.

CRAWTON BAY, ABERDEENSHIRE (NO 897 797)

R.A. Smith

Introduction

The coastline of Crawton Bay (Figure 3.14) exposes a succession of Lower Old Red Sandstone conglomerates and is the type locality of the Crawton Volcanic Formation, the youngest formation in the Dunnottar–Crawton Group (Browne *et al.*, 2002). This formation is Late Silurian to Early Devonian in age and comprises olivine basalts, basaltic andesites and interbedded conglomerates, described in the GCR volume on the Caledonian igneous rocks (Smith, 1999b). The underlying alluvial conglomerates, intercalated volcaniclastic beds and overlying boulder conglomerates are the subjects of this description. The cliff sections, particularly at Crawton Ness and Trollochy, show superb examples of large braid bars with high-stage coarse gravels and low-stage cross-stratified sands. The braided stream character and the

The Midland Valley of Scotland and adjacent areas

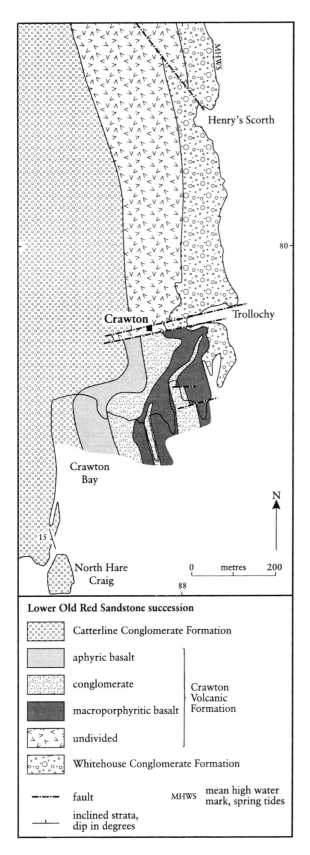

Figure 3.14 Geological sketch map of the Crawton Bay area. After Stephenson *et al.* (1999, fig. 9.19).

large scale of the river system that deposited the detritus in the Midland Valley of Scotland can be appreciated at this locality. The large proportion of quartzite and granite clasts in the conglomerates, in contrast to the clast types in the conglomerates elsewhere in the Dunnottar–Crawton Group (Haughton, 1988), suggests that the source was the Grampian Highland Terrane to the north. This evidence contributes to our knowledge of the unroofing history of the Grampian Highland Terrane and the development of the Midland Valley.

Description

The inter-relationships between the lava flows and the intercalated sedimentary rocks of 'Highland origin' were first noted by Geikie (1897). The locality has since been described in detail by Campbell (1913), Trewin (1987b), Carroll (1995b) and MacGregor (1996a). Carroll (1995b) divided the section, in ascending sequence, into the Whitehouse (Fowls Heugh) Conglomerate Formation (exposed at Crawton Ness), the Crawton Volcanic Formation (interbedded conglomerates and lavas) and the Catterline Conglomerate Formation, which is the basal formation of the Arbuthnott–Garvock Group. The provenance of the clasts in the conglomerates of the Dunnottar–Crawton Group conglomerates at the site has been studied by Haughton *et al.* (1990).

The oldest conglomerates at the site belong to the Whitehouse Conglomerate Formation of Browne *et al.* (2002) (the Fowls Heugh Conglomerate Formation of Carroll, 1995b), which dips gently to the WSW below the Crawton Volcanic Formation. The cliffs to the north and south of Trollochy (NO 881 798), the inlet east of Crawton, are eroded along ENE-trending faults cutting the Whitehouse Conglomerate Formation (Figure 3.14). The conglomerates of the formation are clast-supported and have weak bedding picked out by grain-size differences. Intercalated sandstone lenses are trough- and planar cross-bedded, moderately well-sorted and locally scoured below conglomerate beds. At the base of the cliff on the north side of Trollochy, sandstone rests on an inclined surface of conglomerate, and cross-stratification can be seen in the conglomerates nearby. A few metres up the cliff, a conglomerate that originally had open interstices, is now cemented by micritic calcite

152

(Trewin, 1987b). Coarse conglomerates immediately below the Crawton Volcanic Formation (NO 881 798) occur in units that coarsen upwards overall (Haughton et al., 1990) and have imbricate fabrics indicating palaeoflow from the north. Clast types depend on their sizes, which range up to 1 m. Volcanic rocks similar to those in the Crawton Volcanic Formation predominate, with lesser amounts of psammites, granites, granodiorites, hypabyssal rocks, quartzites and vein quartz. Of the granitic rocks, biotite-muscovite granite, porphyritic granite and equigranular red granite are the commonest in the boulder-size fraction, other granitoids and porphyries are common in the cobble to pebble fraction. Rb–Sr muscovite whole-rock ages for biotite-muscovite granite boulders from the top of the Dunnottar–Crawton Group are 473–457 Ma; a U–Pb age of 475 ± 5 Ma was obtained from a monazite in one of the clasts (Haughton et al., 1990). A Rb–Sr biotite age of 412 ± 4 Ma was obtained from a clast of porphyritic biotite granite. Mica whole-rock ages for the accompanying psammitic clasts suggest that their source included metamorphic rocks that were uplifted at the same time as the intrusion of the granites. Haughton et al. (1990) concluded that the source resembled the Grampian Highland Terrane of north-east Scotland, and that the north Midland Valley Terrane was juxtaposed within tens of kilometres of its present position with the Grampian Highland Terrane at the time of deposition. South of Trollochy (NO 8805 7970), laminated sandstone and mudstone immediately below the first lava flow are locally disrupted and baked by the lava (Trewin, 1987b).

The Crawton Volcanic Formation, which is up to 70 m thick (Carroll, 1995b), has conglomerates above the first and third lava flows. There is evidence of penecontemporaneous erosion of the third lava flow where it was partly eroded and potholed prior to deposition of the thick overlying conglomerates (Figure 3.15). These are clast- and matrix-supported, and consist of well-rounded clasts ranging from pebbles to boulders about 1 m set in a coarse-grained, poorly sorted, volcaniclastic sandstone matrix. The clasts comprise macroporphyritic andesite lava, psammite and quartzite, with lesser amounts of metabasalt, felsite, chert and greywacke. The coarse fraction of the clast suite is of broadly 'Highland' provenance (i.e. Highland Border Complex and probably some

Figure 3.15 Coarse conglomerates of the Crawton Volcanic Formation resting on the eroded top of the third lava. (Photo: N.H. Trewin.)

Grampian Highland Terrane), with a significant component of local lavas.

The overlying Catterline Conglomerate Formation is exposed in a steep cliff on the west side of Crawton Bay (NO 877 796). It comprises red-brown and grey, thickly bedded, pebble- to boulder-grade conglomerates that show weak grain-size stratification. They are mainly clast-supported, with well-rounded pebbles set in a matrix of less well-rounded, volcaniclastic, coarse sand grains and granules. Volcanic pebbles are more common than in the underlying conglomerates, and include massive basalt and vesicular and porphyritic andesites, probably of local derivation. Quartzite, vein quartz, greywacke sandstone, granitic rocks, jasper and metabasalt are also present. The last two rock types are typical of the Highland Border Complex and the greywacke sandstone has a spaced cleavage similar to that in the Dalradian Southern Highland Group north of the Highland Boundary Fault. Cross-bedded sandstones within the conglomerates with an apparent seaward dip are exposed towards the top of the cliff and form part of a southerly dipping fan (Trewin, 1987b).

Interpretation

Earlier models for Lower Old Red Sandstone deposition (Bluck, 1978) suggested that axial alluvial sedimentation in an elongate basin was dominant over lateral, coarse alluvial-fan deposition. Detailed work on the Dunnottar–Crawton Group (Haughton, 1988, 1989; Marshall et al., 1994) showed that other depositional packages, with clastic material derived from sources within the Midland Valley and the penecontemporaneous volcanic rocks, were present in the Crawton Basin, together with the polycyclic alluvium from a 'Highland' source (Haughton and Bluck, 1988).

The alluvial gravel overlying the top of the third lava flow (Figure 3.15) indicates that a short period of erosion removed any hematized bole, such as that developed between the second and third flows, but not all of the vesicular top to the lava. The gravels were carried by large, complex river systems from both local and distant 'Highland' sources. The large size of the cobbles and coarsening-upward units in the river gravels suggest that the sporadic, relatively thin and sheet-like Crawton lavas only temporarily diverted these powerful rivers.

All the conglomerates at this site contain varying proportions of locally derived volcanic rocks and northerly derived 'Highland' rocks. This contrasts with conglomerates derived from the south and east within the Midland Valley Terrane (Haughton, 1988), which contain first-cycle clasts of lithic arenite, granodioritic rocks and metagreywacke, as well as limestone clasts with Early Ordovician faunas. This cryptic Midland Valley source is distinct from the Southern Uplands Terrane and may have originally comprised displaced fore-arc slivers to the south of the Laurentian continental margin.

Carroll (1995b) considered the Crawton Volcanic Formation to be Late Silurian in age. This was based on an earliest Devonian age (on palynological evidence) for the overlying Arbuthnott–Garvock Group (Richardson et al., 1984) and a date of 415 ± 5.8 Ma for a dacitic ignimbrite (the Lintrathen Tuff Member) north of the Highland Boundary Fault (Thirlwall, 1988; Bluck, 2000). This is correlated with the Glenbervie 'Porphyry' at the top of the Crawton Volcanic Formation, and because it crops out north of the Highland Boundary Fault, Trench and Haughton (1990) considered that the relative lateral movement between the northern Midland Valley and the Grampian Highland terranes from Early Devonian time onwards could have been only of the order of tens of kilometres. This has an important bearing on the provenance of the intercalated sedimentary clasts. The Lower Old Red Sandstone lavas appear to have accumulated in a subsiding rift basin close to the Highland Boundary Fault, with the Grampian Highland Terrane to the north-west of the fault providing a source of coarse clastic detritus. The succeeding Arbuthnott–Garvock Group lacks any clasts known to be derived from the cryptic Midland Valley source and heralds the development of the more extensive Strathmore Basin. Further work on the provenance of the sedimentary rocks, including dating studies of heavy mineral suites, may identify more precisely the clastic sources. Geochemical studies may reveal the extent of the contribution from local volcanic, as opposed to other volcanic, sources. Sedimentological studies will refine our understanding of the relationship between the high-energy alluvial deposition and the penecontemporaneous volcanic eruptions. The development of the Crawton Basin, including syndepositional fault control on its subsidence, is a further topic of future research.

Conclusions

The Crawton Bay GCR site provides the best exposures of the Crawton Volcanic Formation and its type section. This is a significant marker horizon at the top of the Dunnottar–Crawton Group, and its relationship with the conglomerate formations above and below can be seen at the site. The intercalated lavas and conglomerates here have been studied since the late 19th century, because of the fine exposures, which demonstrate the relationship between volcanic and sedimentary rocks in a rifted marginal basin. The sedimentary rocks provide important evidence for the large size of the river systems, which were interrupted by calc-alkaline volcanic outpourings in Late Silurian to Early Devonian times. The Crawton Basin was the precursor to the larger Strathmore Basin that developed within the northern Midland Valley. However, unlike the Crawton Basin, the Strathmore Basin lacks clasts derived from cryptic sources within the Midland Valley.

NORTH ESK RIVER, ABERDEENSHIRE (NO 595 719–NO 604 703)

R.A. Smith

Introduction

The North Esk River section in the Gannochy Gorge (Figure 3.16) is in the Strathmore Group of the Lower Old Red Sandstone in the northern Midland Valley, about 500 m south-east of the Highland Boundary Fault. The Strathmore Group is the youngest Lower Devonian succession of the area and comprises a sequence of alluvial siltstones, sandstones and conglomerates. This GCR site provides the best sections in which to examine the variations in lithology and thickness of the strata, as well as the pebble content of the conglomerates. The change in clast content along the gorge reflects movements along the Highland Boundary Fault and the history of erosion of the Dalradian Supergroup, which formed land to the north during this part of the Devonian Period. This is also an important section for understanding the tectonic development of the Midland Valley and its relationship to the Grampian Highland Terrane at that time. The sedimentology and fluvial architecture of the succession can be studied in this impressive gorge, which is 20 m to 30 m deep.

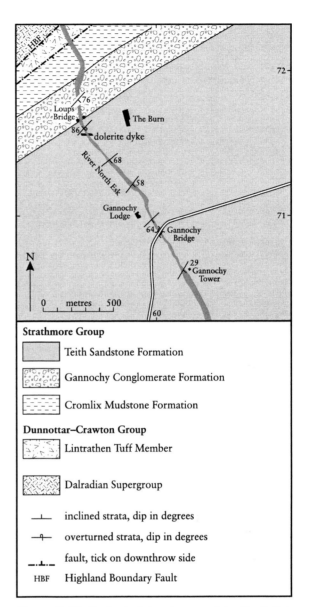

Figure 3.16 Geological sketch map of the North Esk River section.

Description

The North Esk River cuts through the Strathmore Group on the steep north-west limb of the Strathmore Syncline. To the north of the site, poorly exposed, reddish brown siltstones and poorly sorted fine- to medium-grained, argillaceous sandstones are correlated with the coarser-grained facies of the Cromlix Mudstone Formation, which is truncated against the Highland Boundary Fault. The Cromlix Mudstone Formation is the basal formation of the Strathmore Group and because it is relatively soft it is generally poorly exposed. To the south, the formation passes conformably up into the 250 m-thick Gannochy Conglomerate Formation, which is more resistant and forms the northern part of the gorge.

The sandstones and conglomerates upstream of Gannochy Tower (NO 6021 7065) were formerly correlated with the Garvock Group (Campbell, 1913; MacGregor, 1968), but later re-assigned to the Strathmore Group on the faulted north-west limb of the Strathmore Syncline (Armstrong and Paterson, 1970; MacGregor, 1996b). Armstrong and Paterson defined the Gannochy Conglomerate Formation at this site, extending from the base of the lowest conglomerate (NO 5947 7199) north of the disused Loups Bridge to the highest pebbly sandstone beds near Gannochy Tower (NO 6020 7064). They calculated the total thickness of the formation in the gorge to be 1400 m. Subsequent geological surveys (British Geological Survey, 1995; Carroll, 1995c) confirmed the correlation with the Strathmore Group, and refined the lithostratigraphical classification. Red, argillaceous beds of the Cromlix (formerly Edzell) Mudstone Formation are overlain by the Gannochy Conglomerate Formation, which is in turn overlain by the Teith (formerly Edzell) Sandstone Formation (Browne *et al.*, 2002). Carroll (1995c) re-defined the Gannochy Conglomerate Formation in the North Esk River, placing the top of the formation at the top of the clast-supported orthoconglomerate beds immediately downstream of Loups Bridge (NO 5948 7164).

The clast-supported conglomerates of the Gannochy Conglomerate Formation (Figure 3.17) appear to occur in lenses within pebbly sandstones, which dip almost vertically in this section. Locally, 50 m north of an E–W-trending Permo–Carboniferous dolerite dyke (NO 5951 7156), the beds are overturned, dipping 86° to the north-west. The conglomerates

Figure 3.17 Coarse clast-supported orthoconglomerate of the Gannochy Conglomerate Formation at Loups Bridge, North Esk River. (Photo: BGS No. D5347, reproduced with the permission of the Director, British Geological Survey, © NERC.)

are weakly stratified and poorly sorted, and contain rounded to subrounded pebbles in a medium- to coarse-grained sandstone matrix. Cross-bedded pebbly sandstone lenses occur within the conglomerate lenses and some of the sandstone bodies have basal, pebbly lags. Conglomerates continue south of Loups Bridge (NO 5948 7164), and are conformably overlain by fine- to medium-grained, pebbly sandstones of the Teith Sandstone Formation. The conglomerates have a decreasing proportion of quartzite and quartz pebbles in their upper part (MacGregor, 1968, 1996b).

The Teith Sandstone Formation forms the bulk of the gorge section and is relatively coarse-grained, containing several thick conglomerates within red-brown sandstones and mudstones. The sandstones are commonly trough cross-bedded, pebbly to medium-grained and include large detrital micas. Red mudstone beds up to 2.5 m thick and dipping about 60° to the south-east are well exposed about 230 m north of Gannochy Bridge. The conglomerate interbeds are polygenetic and although quartzite is the most common pebble type, vein quartz, porphyry, andesite, granite, gneiss, felsite and sedimentary types are also present (MacGregor, 1968, 1996b). The overlying part of the Teith Sandstone Formation south of Gannochy Tower, and about 270 m downstream from Gannochy Bridge, consists of mainly fine- to medium-grained, red sandstones, with some coarse-grained to pebbly sandstones with mudstone intraclasts. The beds dip about 30° to the south-east. Some of the finer-grained sandstones are cross-bedded or ripple cross-laminated, and there are thick, planar-bedded, silty sandstones with very thin mudstone interbeds (Carroll, 1995c).

Interpretation

The sequence of sandstones and conglomerates north of Gannochy Tower was originally placed in the Garvock Group (Campbell, 1913), but Armstrong and Paterson (1970) re-assigned it to the Strathmore Group on structural grounds, considering its position on the north-west limb of the Strathmore Syncline. The Strathmore Group exposed in Gannochy Gorge is coarser grained than that on the south-east side of the Strathmore Syncline, probably due to its proximity to the source of the detritus to the north.

The depositional environment of the Cromlix Mudstone Formation has been inferred from exposures at Mid Mains of Balfour (NO 624 740), a few kilometres to the north-east, to be an alluvial floodplain that periodically dried out, as indicated by local desiccation cracks. Trace fossils include *Skolithos*, *Arenicolites* and *Beaconichnus* (Carroll, 1995c), indicating a habitat suitable for burrow-feeders, possibly close to a lake shoreline. The Gannochy Conglomerate Formation marks the progradation of a high-energy alluvial-fan system, with reduced fluvial energies being represented by the overlying Teith Sandstone Formation. Only the lower part of the Teith Sandstone Formation is seen in this section, but it appears to fine upwards overall. Its base is probably coarser grained here, as it is gradational from the Gannochy Conglomerate Formation, and the facies on the north-west limb of the Strathmore Syncline is typically laterally variable and coarser grained than the more distal facies to the south-east. In the younger parts of the formation exposed elsewhere, the plant remains *Arthostigma* sp. and *Psilophyton* sp. are locally common, indicating plant colonization of the less active floodplains. Spores from plant beds in the Strathmore Group indicate an Early to Mid-Emsian age (Richardson *et al.*, 1984).

Haughton and Bluck (1988) recognized two alluvial styles in the Strathmore Group. Fine-grained alluvium was deposited in axial systems by very large rivers that probably drained areas of Scandian uplift. First-cycle conglomerates were deposited in fans close to the Highland Boundary Fault and derived from the north-west. The textural immaturity of the conglomerates suggests flash-flood deposition. The Strathfinella Hill Conglomerate Member, which crops out to the north-east, contains clasts of metamorphic rock of garnet- and higher grades. Haughton and Bluck (1988) related these grades to the Barrovian zones in the adjacent Dalradian rocks of the Grampian Highland Terrane to the north-west, inferring little displacement on the Highland Boundary Fault since Emsian times. Comparison of Lower Old Red Sandstone sequences immediately to the north and south of the Highland Boundary Fault (Bluck, 2000) suggests that the Grampian Highland Terrane pushed southwards, either during or after Lower Old Red Sandstone deposition, effectively removing the northern part of the Strathmore Basin and generating the steep northern limb of the Strathmore Syncline.

Peacock (1961) suggested that significant uplift of the source area caused the influx of conglomerates of the Gannochy Conglomerate Formation and noted a decrease in its maturity upwards (MacGregor, 1968, 1996b). The upper, less mature conglomerates contain gneisses (about 25%), as well as 'porphyry', 'granite' and sedimentary pebbles, presumed to have been derived from the highlands to the north. Armstrong and Paterson (1970) considered that the Gannochy Conglomerate Formation continued north-eastwards to link with the Strathfinella Hill Conglomerate Member, but detailed mapping by Carroll (1995c) indicates that they are the deposits of separate alluvial fans. The Lintrathen Tuff Member (Dunnottar–Crawton Group), which is exposed in a faulted wedge within the Highland Boundary Fault Zone farther north in the North Esk River section, also occurs to the north-west of the fault zone. Its presence there has been interpreted (Trench and Haughton, 1990) as confirmation that only a limited amount of lateral movement on the Highland Boundary Fault has occurred since the tuff erupted at 415.5 ± 5.8 Ma (Thirlwall, 1988).

Conclusions

The North Esk River GCR site is the best-exposed section through the Strathmore Group on the north-western side of the Strathmore Syncline and the type section for the Gannochy Conglomerate Formation. It provides good exposures of the conglomerate formation, as well as the underlying and overlying finer-grained formations. The pebbles in the conglomerate are mainly from the Grampian Highland Terrane to the north-west, in contrast to the older Lower Old Red Sandstone rocks sourced from more distant areas of Scandian uplift. The depositional model envisaged for these sedimentary rocks comprises an interplay of a large, axially draining alluvial system flowing to the south-west and the lateral build up of alluvial-fans draining southeast across the Highland Boundary Fault Zone during Emsian times. Because of its proximity to the clastic source, the facies of the Strathmore Group in this section is coarser grained than that on the south-east limb of the Strathmore Syncline. Detailed sedimentological and provenance studies at the site, compared to studies at Strathfinella, would test this simple model and provide important data on fault control on sedimentation and the timing of exhumation of the Grampian Highland Terrane.

MILTON NESS, ABERDEENSHIRE (NO 766 650–NO 771 648)

Potential GCR site

W.J. Barclay

Introduction

The coast section at Milton Ness near Montrose (Figure 3.18) provides the best example of mature Old Red Sandstone carbonate soil (calcrete) development in Scotland. The calcrete occurs in the Upper Devonian–Lower Carboniferous (Upper Old Red Sandstone) Kinnesswood Formation (of the Inverclyde Group), in a down-faulted block of predominantly fluvial sedimentary rocks. The near-horizontal rocks crop out in a wave-cut cliff up to 10 m high and about 500 m long, providing excellent lateral exposure. This has allowed a detailed examination of the morphology and development of a mature calcrete at the top of the section by Balin (2000). Palaeokarst cavities, reworked hardpan calcrete and rhizoliths were recognized by Balin for the first time in the Old Red Sandstone at this locality. In addition to the features commonly seen in Quaternary calcretes, the Milton Ness horizon preserves some features that are unusual in calcretes of any age. The

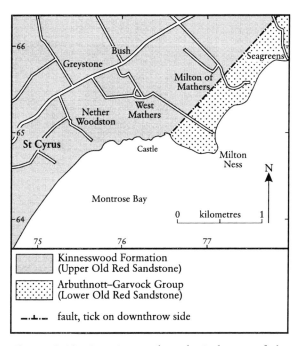

Figure 3.18 Location and geological map of the Milton Ness area. After Trewin (1987c).

Milton Ness

rhizoliths are some of the finest examples seen in Palaeozoic rocks. Mature calcrete development is largely restricted to the Kinnesswood Formation in Scotland, with minor amounts in the Lower Old Red Sandstone of the Orcadian Basin and almost none in the Middle Old Red Sandstone.

Description

The geology of Milton Ness (NO 766 650–NO 771 648) was described in a field guide by Trewin (1987c). The Upper Old Red Sandstone is down-faulted against Lower Old Red Sandstone lavas and conglomerates by a NE-trending fault at Rock Hall (Figure 3.18). The cliffs from here south-eastwards to Milton Ness expose an interesting succession of mudstones, sandstones, conglomerates and calcretes (Figure 3.19). Trewin (1987c) recognized three fining-upward cycles in the 10 m section. Intraformational conglomerate lenses form the base of the cycles, resting on the underlying scoured surface, locally filling the bases of channels and containing calcrete clasts. Cross-bedded and parallel-bedded sandstones overlying the conglomerates contain sporadic *Beaconites* burrows and calcified root tubules (rhizoliths). The tops of the cycles comprise red siltstones/mudstones with an abundance of carbonate nodules ('cornstone' of the older literature) interpreted as immature calcrete horizons. The nodules at the top of the uppermost cycle (Cycle 3, Figure 3.19) increase in abundance and size upwards and are capped by the laminated, brecciated and karstified calcrete described by Balin (2000).

Balin (2000) noted significant lateral variation in the amount and type of pedogenic alteration throughout the calcrete horizon, but locally,

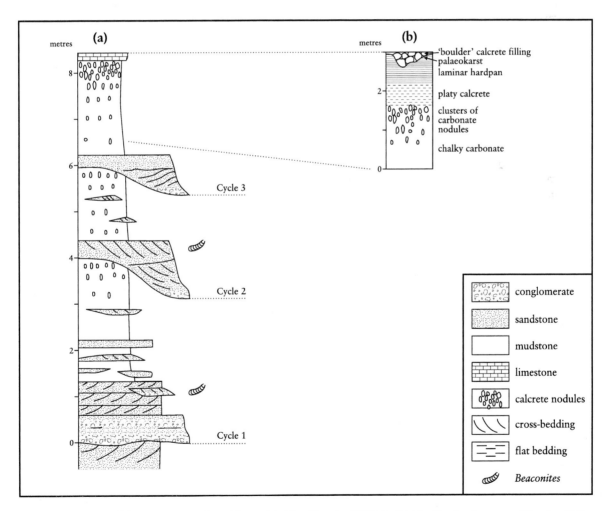

Figure 3.19 Log of the section at Milton Ness. (a) After Trewin (1987c); (b) idealized calcrete profile after Balin (2000).

development of the profile shows an idealized vertical zonation comprising: upper compact crust or hardpan (70 cm); platy, sheet-like carbonate (70 cm); clusters of carbonate nodules (120 cm); uniformly distributed chalky carbonate (55 cm); unaltered host mudstone. At the eastern end of the outcrop, karst cavities and reworked, intraformational calcrete-clast conglomerate represent additional stages, with a subsequent period of calcretization completing the development of the profile.

The hardpan layer is up to 1.5 m thick at the eastern end of the section, where it crops out as a resistant ledge extending for 150 m and forms the topmost part of a 4 m-thick calcrete profile. It is a brecciated limestone with irregular, sub-horizontal carbonate layers, horizontal, wavy to irregular chert layers and relicts of the original sandstone. Much of the limestone is an irregularly mottled dark and pale grey micrite, the latter grading to microspar, in a 'clotted' fabric typical of calcrete. Calcite spar fenestrae vein the micrite and sand and silt grains 'float' in the clotted fabric, surrounded by microspar fringes. The chert layers are of buff, yellow and pink, brecciated and laminated silica, locally with pisolites 0.1–2.5 mm in diameter and containing vugs filled with chalcedony, quartz and calcite.

Karstic dissolution and reworking of the hardpan horizon produced cavities and an intraformational calcrete clast bed respectively. At least ten karst cavities penetrate the hardpan at the easternmost end of Milton Ness, the largest measuring 1.2 m in depth and 1.2 m at its top. They formed in slight depressions in the hardpan, where water collected on the impermeable surface. The conglomerate forms a laterally persistent bed, draping the tops of the karstic cavities. It consists of clasts of relict host sandstone, as well as chert and carbonate clasts. Abrasion of the small chert pebbles indicates some transport and not in-situ modification of the underlying calcrete. Balin (2000) referred to the bed as a 'boulder calcrete', to denote clasts derived from the break-up of a hardpan calcrete during weathering, although later calcretization of the bed precludes determination of the original clast size.

Interpretation

The succession at Milton Ness is interpreted as a series of channel deposits built by laterally migrating river systems on an alluvial plain (Trewin, 1987c). Much of the Upper Old Red Sandstone in this part of the Midland Valley represents the distal part of an east-flowing axial drainage system. However, palaeocurrents at Milton Ness and nearby Boddin Point are south-directed, suggesting a tributary system in which the relatively small catchment area would have contributed to the low sedimentation rates necessary for calcrete formation (Balin, 2000).

The presence of calcrete in the finer-grained floodplain deposits points to a hot, semi-arid, seasonally wet climate. The laminated hardpan top to the calcrete at the top of the section represents a maturity (Stage IV of Machette, 1985) very rarely seen in Old Red Sandstone calcretes. Balin (2000) discussed the probable timescale of the horizon's formation, noting that Leeder (1975) proposed a minimum of 10 000 years for calcrete of similar maturity, but that a considerably longer period seems probable. Recent studies (e.g. Marriott and Wright, 2004) suggest that hundreds of thousands to millions of years may have been required for the formation of mature Old Red Sandstone calcretes. To this must be added the karstification, reworking and subsequent calcretization seen at Milton Ness. Given the semi-arid climate and presumed low discharge rates, Balin noted that carbonate dissolution rates would have been low, and that the karst cavities up to 1.2 m deep would have taken a minimum of several thousand years to form. The subsequent reworking of the karst surface and re-calcification of the resultant deposit extended the total period of stabilization and calcrete formation at this horizon, to perhaps hundreds of thousands of years.

Karst formation is generally ascribed to a wetter, more humid climate, in contrast to the semi-arid conditions required for calcrete formation. However, Balin (2000) interprets the Milton Ness example as the result of water collecting in hollows on the impervious hardpan surface under constant semi-arid conditions, possibly with tree-like roots initiating and contributing to the dissolution process.

The presence of rhizoliths in the sandy parts of the alluvial cycles suggests that the environment on the floodplain some distance from the rivers was less hospitable to plant growth because of a low water-table, or that the floodplain mudrocks were impermeable (Balin, 2000). The presence of horizontal, ramifying networks of rhizoliths rather than vertical tap

Aberlemno Quarry

roots points to plant colonization of sandy point bars. The large size of some of the tubules at Milton Ness (over 3 cm) points to relatively large, shrub- or tree-like plants, confirming that plant development was well advanced by Late Devonian times (Balin, 2000). The *Beaconites* trace-fossil burrows were probably made by arthropods that burrowed into the channel sands to escape the arid, dry season conditions (Trewin, 1987c).

Conclusions

The coast section at Milton Ness provides superb lateral exposure for about 500 m of a mature fossil carbonate soil (calcrete) in the Upper Old Red Sandstone. This is one of the best sections of Old Red Sandstone calcrete in Scotland, and one of very few in the Old Red Sandstone of Great Britain where soil development to a stage of laminated hardpan is recorded. The excellent preservation of the calcrete textures are comparable with some of the best Quaternary examples, and allow, by analogy with these, an interpretation of the contemporaneous climate. The karstic dissolution hollows are an unusual feature of calcretes of any age. The root traces left by the plants are among the finest examples seen anywhere in rocks of this age. These, and the carbonate soil fabrics make this a unique and important site worthy of protected status.

ABERLEMNO QUARRY, ANGUS (NO 526 551)

Potential ORS GCR site

M.A.E. Browne

Introduction

Aberlemno Quarry is one of a number of quarries (including Tillywhandland Quarry) worked on and around Turin Hill near Forfar, Angus for Arbroath Paving Stone during the eighteenth and nineteenth centuries (Mackie, 1980). Of these, Aberlemno Quarry and Clocksbriggs (or Wemyss) Quarry are well-known fossil plant localities, and Aberlemno is (as 'Turin Hill') already a confirmed GCR site for its Palaeozoic palaeobotany (Cleal and Thomas, 1995) and for its fossil osteostracan fish fauna (Dineley and Metcalf, 1999).

Description

Descriptions of the quarry were provided by Armstrong *et al*. (1978b), Cleal and Thomas (1995) and Dineley (1999c). It lies on the northeast flank of Turin Hill, forming an elongate entrenchment alongside a minor road (Figure 3.20). About 9.5 m of strata belonging to the Dundee Flagstone Formation and the overlying Scone Sandstone Formation (both of the Arbuthnott–Garvock Group) are exposed (Figures 3.21, 3.22). The succession exposed in a 300 m-long face is remarkably persistent laterally and comprises about 7.5 m of red-brown, medium- to coarse-grained, trough cross-bedded sandstones (Melgund Sandstone Member of the Scone Sandstone Formation) overlying about 2 m of fish-bearing, fine-grained sandstones and greyish green, fissile, laminated siltstones and mudstones (Dundee Flagstone Formation). The latter show much syn-sedimentary deformation and contain many ovoid, carbonate (non-ferroan micrite) nodules up to 30 cm in diameter.

Much of the fish remains collected from Turin Hill since the late 1800s are not identified as coming from specific quarries, but comparison of lithologies suggests that many came from this quarry. Heavily armoured cephalaspids dominate the fauna, in contrast to **Tillywhandland Quarry** (see GCR site report, this chapter) where free-swimming acanthodians dominate. Dineley (1999c) lists the cephalaspid species

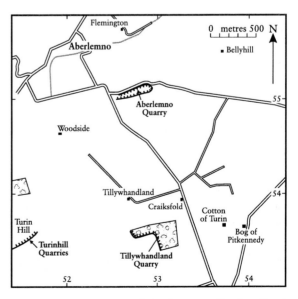

Figure 3.20 Locations of Aberlemno, Tillywhandland and Turin Hill quarries

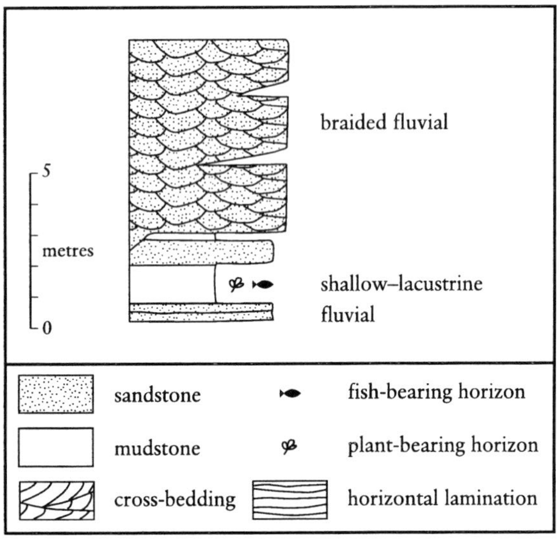

Figure 3.21 Section of Aberlemno Quarry. Based on Dineley (1999c) and Armstrong *et al.* (1978b).

that are most likely to have been collected from Aberlemno, but only *Cephalaspis pagei* is definitely attributed to the quarry. Acanthodian spines have been recorded, as well as the arthropod *Dictyocaris* and the eurypterids *Pterygotus* and *Erieopterus*.

The fossil plants in the basal laminated beds are particularly renowned and belong to the *Zosterophyllum* Zone of Banks (1980). Cleal and Thomas (1995) described them in detail. The commonest are impressions picked out by iron staining, but some coalified compressions and petrifactions also occur. *Prototaxites forfarensis*, *Parka decipiens*, *Pachytheca* sp., *Cooksonia caledonica* and *Zosterophyllum myretonianum* have been identified. Most significant are the well-preserved specimens of the alga or early land-plant *Parka*, the early vascular plant *Zosterophyllum* and the holotype of *Cooksonia caledonica*.

Figure 3.22 Aberlemno Quarry. Strike section in flaggy sandstones of the Dundee Flagstone Formation. (Photo: C.J. Cleal.)

Interpretation

The laminated beds at the base of the quarry were interpreted by Armstrong *et al.* (1978b) as the deposits of a shallow lake. The overlying sandstones of the Scone Sandstone Formation have a varied internal geometry consistent with deposition in a braided stream complex flowing to the west, and probably marking a switch from small local drainage systems to a larger regional one (Bluck, 2000). The switch was diachronous and transitional, the uppermost part of the Dundee Flagstone Formation interdigitating with the lowermost part of the Scone Sandstone Formation. This may have been due to blocking of local drainage by lava flows and the establishment of shallow lakes.

Armstrong and Paterson (1970) correlated the Aberlemno fish bed with No. iv of the eight (i to viii) fossil fish horizons they identified in the Arbuthnott–Garvock Group; five beds (i–v) lie in the Dundee Flagstone Formation. The flora recorded from the Arbuthnott–Garvock Group belong to the *Zosterophyllum* Zone (Banks, 1980) and suggest a Gedinnian (Lochkovian) or early Siegenian (Pragian) age (Edwards, 1980; Edwards and Fanning, 1985). Palynological and fish evidence supports an early Gedinnian age (Edwards, 1980; Edwards and Fanning, 1985). The fish and eurypterids indicate an age not older than the base of the Dittonian of the Anglo-Welsh Basin (Westoll, 1977). Richardson *et al.* (1984) correlated the Aberlemno horizon with the middle subzone of the *micrornatus-newportensis* Zone (= Lochkovian Stage) of the Anglo-Welsh Basin. A radiometric age of 407 ± 6 Ma was determined by Thirlwall (1983) for the lower part of the Arbuthnott–Garvock Group.

Conclusions

Aberlemno Quarry is of international importance for its fossil plant flora, having yielded one of the best *Zosterophyllum* Zone assemblages in the world. It is also important in being one of the two quarries on Turin Hill (the other being **Tillywhandland Quarry**) that remain open. A rich fossil fish fauna is dominated by the armoured cephalaspids, in contrast to the free-swimming acanthodians that dominate the fauna at Tillywhandland. The site's conservation value lies in providing an opportunity for further excavation and collection of its fauna and flora.

TILLYWHANDLAND QUARRY, ANGUS (NO 528 537)

Potential ORS GCR site

M.A.E. Browne

Introduction

Tillywhandland Quarry on Turin Hill near Forfar (Figure 3.23) is one of the best Early Devonian fossil fish sites in Scotland; it is an established GCR site for fossil fishes (Dineley and Metcalf, 1999). The site also provides a reference section for the Dundee Flagstone Formation, allows the detailed examination of the lithofacies and sedimentation of a typical Early Devonian fish bed, and is unique in the Midland Valley in exposing a clastic–carbonate–organic laminite similar to those of the Mid-Devonian Orcadian Basin. Trewin and Davidson (1996) provided a detailed description of the quarry, summarized by Dineley (1999c). The fauna includes a rich assemblage of primitive acanthodians, with some cephalaspids, collected from this and other quarries on Turin Hill, which worked the sandstones of the Dundee Flagstone Formation for over 200 years. Richardson and MacGregor (1986) ascribed the beds to the *micrornatus-newportensis* Zone of early Lockhovian age.

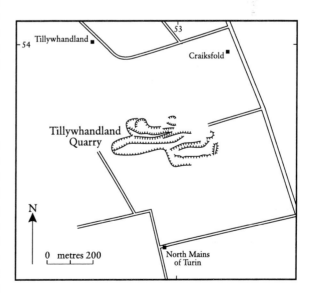

Figure 3.23 Location of Tillywhandland Quarry, Angus.

The Midland Valley of Scotland and adjacent areas

Description

Tillywhandland Quarry exposes beds of the Dundee Flagstone Formation (of the Arbuthnott–Garvock Group). Working probably ceased over 100 years ago and the quarry at the time of writing was overgrown. The Dundee Flagstone Formation consists mainly of medium- to coarse-grained, cross-bedded, fluvial sandstones. Numerous distinctive deltaic and lacustrine units up to 30 m thick of thinly bedded sandstones and intercalated siltstones and mudstones are particularly characteristic of the formation. They include a number of fish beds, including the one exposed in this quarry, that have also yielded arthropods and plant remains (Lang, 1927; Westoll, 1951).

The quarry exposes a fish-bearing laminite (Figure 3.24) and underlying sandstone, having worked the latter for building stone. Powrie (1861, 1864) first described the laminite (the Forfarshire Fish Bed), considering it to be the only such horizon and correlating it with similar occurrences elsewhere. However, later workers (Hickling, 1912; Armstrong and Paterson, 1970) envisaged several discrete beds. Armstrong *et al.* (1978b) provided a brief description of the quarry. Trewin and Davidson (1996) recorded a detailed section of the fish bed and adjacent strata (Figure 3.25), and the following account is a summary of their work. Dineley's (1999c) summary emphasizes the fish fauna and its palaeoecology, Wilson and Anderson (2004) describe the invertebrate fauna.

About 3 m of red to pale brown, medium- to coarse-grained sandstone are exposed beneath the fish bed. The lowest beds exposed are massive and trough cross-bedded in sets up to 0.5 m thick, with two pebble-lined erosion surfaces. The massive sandstones contain scattered pebbles and soft-sediment de-watering structures. The topmost 1 m of sandstone are parallel-laminated with primary current lineation. Sandstone dykes penetrate up to 0.5 m into the overlying laminite; they are up to 0.1 m wide at their base and folded by soft-sediment compaction. The fish-bearing laminite has a sharp contact with the underlying sandstone, is about 1.3 m thick and contains a 60 mm-thick pale green to buff, sticky bentonitic clay 1 m above its base. The laminites comprise a stacked sequence of repeated quadruplets of clastic siltstone, carbonate, organic matter, and green claystone averaging about 5 mm in thickness.

Figure 3.24 Part of Tillywhandland Quarry, showing lacustrine laminites. (Photo: M.A.E. Browne.)

Tillywhandland Quarry

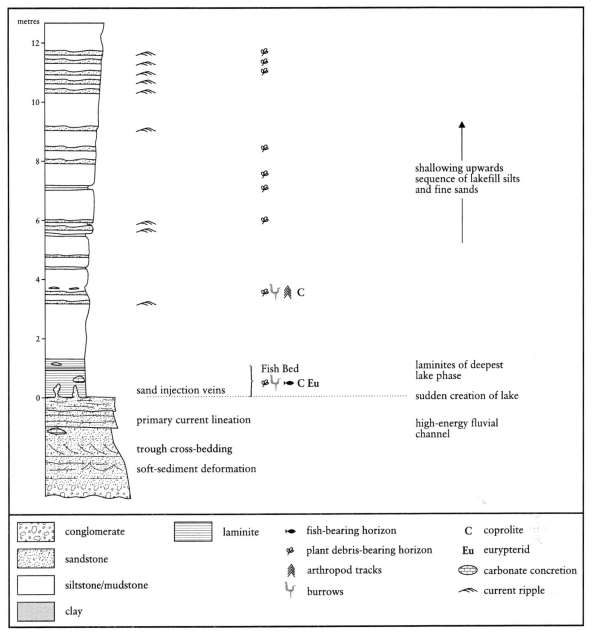

Figure 3.25 Section of Tillywhandland Quarry. Based on Trewin and Davidson (1996) and Dineley and Metcalf (1999).

Their colour varies from brown to green depending on the proportion of green claystone laminae present and the amount of carbonate and organic matter. Soft-sediment deformation produced folding and slide planes locally and there are some carbonate concretions. A very thin (0.01 mm) organic lamina commonly overlies the carbonate laminae and contains most of the fish fragments, with a few remains occurring also in the carbonate concretions. Coprolites appear to be concentrated in the organic-rich laminae and the finest laminae. The laminite grades upwards into green siltstones, which contain minor laminites with a few organic laminae. The sequence coarsens upwards generally to coarse siltstone–fine sandstone, with thin (up to 20 cm) sheets of fine-grained, muddy, current ripple-laminated sandstone that are commonly rich in plant debris.

Dineley (1999c) gave full details and classifications of the fish fauna recovered from the site. It comprises ?*Brachyacanthus scutiger*,

Cephalaspis pagei, C. powriei, C. sp., *Climatius reticulatus, Euthacanthus macnicoli, Euthacanthus* sp., *Ischnacanthus gracilis* (Eger), *Mesacanthus mitchelli, M.* sp. and ?*Uraniacanthus* sp.. Trewin and Davidson (1996) noted that acanthodians predominate, with *Mesacanthus* and *Ischnacanthus* the commonest species, the other acanthodians and the cephalaspids being limited to very few specimens. Preservation ranges from relatively complete specimens to disarticulated and fragmentary material. The invertebrates recovered are arthropods – the eurypterids *Parahughmilleria* sp. (Braddy, 2000) and *Pterygotus* and the millipede *Archidesmus macnicoli* (Wilson and Anderson, 2004). Trace fossils include small burrows and arthropod trackways (cf. *Diplichnites*).

Interpretation

The sandstones below the fish bed are interpreted as high-energy, fluvial channel deposits (Figure 3.22). The sharp contact with the fish bed represents a sudden end to fluvial deposition, the fish bed representing lacustrine deposition in a lake (or series of lakes) (named 'Lake Forfar' by Trewin and Davidson, 1996) subjected to a seasonally wet climate. Trewin and Davidson (1996) speculated that the sudden establishment of the lake may have been due to blocking of the drainage outflow to the south-west by the Ochil Lavas and progradation of alluvial fans from the north-west, probably fault-controlled, margin of the basin.

The laminites represent annual varved sediments, deposited from suspension in deep lake waters. They are similar to clastic–carbonate–organic triplets deposited in the hypolimnion of lakes in the Mid-Devonian Orcadian Basin (Donovan, 1980; Trewin, 1986), except that the green claystone laminae have not been seen in the Orcadian Basin and the dolomite laminae present there are absent in Lake Forfar. A period of about 2000 years is represented by the Tillywhandland laminite. The origin of the coarser laminae, whether by aeolian or distant fluvial input to the lake, remains unclear. The carbonate laminae were precipitated as a result of phytoplankton photosynthetic activity, the organic laminae were the product of seasonal decay of the phytoplankton, which settled on the lake bed to produce an organic sludge and the most anoxic bottom conditions.

The pale green to buff clay is an altered volcanic airfall tuff, and of potential use in correlation. Powrie (1864) correlated it, and thus also the Tillywhandland fish bed, with similar occurrences elsewhere, but Armstrong and Paterson (1970) recognized eight fish-bearing horizons. The coarsening-upward siltstones above the laminite record the infilling and shallowing of Lake Forfar, with the ripple-laminated sandstones indicating increased current activity.

Conclusions

Tillywhandland Quarry is one of the best Early Devonian fossil fish sites in Scotland, and has yielded a large number of specimens. It is also an important reference section for the Dundee Flagstone Formation and provides one of the few exposures of an early Devonian lake deposit, representing about 2000 years in the life of Lake Forfar. It has allowed detailed examination of the lacustrine strata, and will be a key section in the further study and correlation of these unique deposits.

WHITING NESS, ANGUS (NO 679 432–NO 659 409)

M.A.E. Browne

Introduction

First notified as an SSSI in 1961, this important GCR site (Figure 3.26) shows non-marine, Devonian sedimentary fluvial rocks. Extensive sea-cliffs expose late Devonian (Upper Old Red Sandstone) sandstones and conglomerates about 370 million years old resting with irregular unconformity on early Devonian (Lower Old Red Sandstone) sandstones about 410 million years old belonging to the Scone Sandstone Formation of the Arbuthnott–Garvock Group. The site shows the contrast between the far-travelled, early Devonian sandy alluvium and the locally derived pebbles in the late Devonian rocks. The basal breccias of the Upper Old Red Sandstone contain combinations of bedding types that show the control of topography of the underlying unconformable surface on braided stream morphology. The site is important in demonstrating that the Old Red Sandstone of the Midland Valley was formed

Whiting Ness

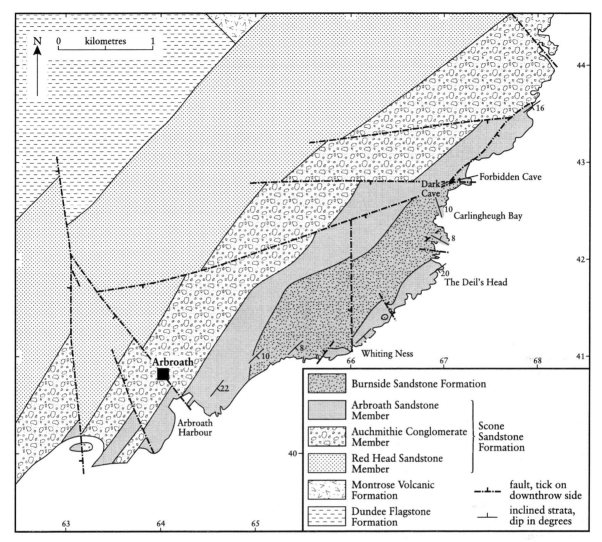

Figure 3.26 Geological sketch map of the bedrock geology of the area around Whiting Ness.

during two separate episodes of sedimentation separated by a break that represents the entire Mid-Devonian period. It is an important site in studies of the lithostratigraphy and palaeogeographic evolution of the Midland Valley. The constituent formations of the Arbuthnott–Garvock Group consist mainly of cross-bedded sandstones. These contain clasts of metamorphic and igneous rocks, as well as soil carbonate (calcrete) clasts of intrabasinal origin (cf. Balin, 2000; see **Milton Ness** GCR site report, this chapter), which are especially characteristic of the upper part of the group. There are interbedded conglomerates in the Arbroath area, and the topmost unit in Strathmore (the Scone Sandstone Formation) contains a persistent conglomeratic horizon with lenticular beds of nodular calcrete.

Description

The Arbroath Sandstone Member of the Scone Sandstone Formation consists of cross-bedded, fine- to medium-grained sandstones. These overlie the Auchmithie Conglomerate Member and are the youngest known strata of early Devonian age on the south-east limb of the Sidlaw Anticline. Named by Hickling (1908) and given formational status by Armstrong and Paterson (1970), the Arbroath Sandstone is now accorded member status (Browne et al., 2002). It is at least 365 m thick, and consists of bright purple-red sandstones on the coast near Arbroath, where it is overlain with marked unconformity by the Upper Old Red Sandstone. Near Carnoustie, however, 3 km to the south-west, the sandstones are generally green or purplish grey. The red

167

colour at Arbroath may be due to the proximity of the mid-Devonian unconformity, with deep weathering prior to the deposition of the Upper Old Red Sandstone. The Arbroath Sandstone Member dips to the south-east at about 20° on extensive wave-cut platforms north-east of Carnoustie and at Arbroath. Trough-cross-bedding is well displayed, with elongate troughs trending parallel to the strike of the rocks and showing predominantly SW-directed palaeo-currents.

The sandstones contain abundant pebbles and boulders (up to 0.3 m across) of nodular limestone at many levels. These clasts are considered to have originated as carbonate soil nodules (calcrete) in argillaceous overbank deposits that were subsequently almost completely destroyed as the result of river channel migration. The carbonate clasts commonly have the appearance of slightly abraded concretions and may not have been transported far before being incorporated in the sandy channel deposits. In some clasts, the carbonate appears to enclose mudstone that represents part of the original mud host. Intact mudstone beds are rare in the Arbroath Sandstone Member but one example with carbonate nodules may be observed at the foot of the cliffs (NO 662 412) 200 m east of Whiting Ness. It underwent partial penecontemporaneous erosion, producing a 'trail' of limestone clasts into an adjacent channel sandstone. The larger limestone clasts tend to occur at the base of the sandstone co-sets, probably representing lag deposits, the smaller ones tend to lie along the foresets. Many co-sets are relatively hard and calcareous in their upper parts, which protrude on the abraded wave-cut platforms. Their top surfaces locally display polygonal jointing that does not penetrate below the hard zone. These features, the formation of concretions in mudstones and the hardened upper calcareous zones in the sandstones are attributed to pedogenic processes in response to fluctuations in water-table levels.

East of Arbroath, in the general neighbourhood of Whiting Ness, a sequence of mainly red-brown and yellow conglomerates with subordinate sandstone beds and basal and marginal breccias rests with striking unconformity on the Arbroath Sandstone Member (Figure 3.27; Hickling, 1908). These beds are unfossiliferous, but the sandstones resemble those in the fossiliferous Late Devonian Stratheden Group in Stratheden and the Carse of Gowrie, and a Late Devonian age seems probable. The absence of calcrete ('cornstone'), suggesting a stratigraphical position below the lower Carboniferous Kinnesswood Formation of Fife, is consistent with this conclusion. The strata can be assigned to the Burnside Sandstone Formation, the lowest unit in the Fife succession. A general direction of transport towards the south and south-east is apparent (Ramos and Friend, 1982). Distinctive breccias composed of angular fragments of the Arbroath Sandstone Member, and clearly derived from the ancient bedrock slopes, accumulated in lenticular bodies. In places they rest on the unconformity, but elsewhere occur at a higher level and are intercalated with the overlying sandstones, or fill channels cut in them. In the area of the Steeple Rock (NO 6585 4095) SSW of Whiting Ness, blocks of the Arbroath Sandstone Member up to 2.5 m in length occur immediately above the unconformity.

Interpretation

The sedimentary rocks of the Arbroath Sandstone Member are typical of the Scone Sandstone Formation in the Strathmore Basin. Bluck (2000) provided a sedimentological interpretation for the formation based upon a road cutting at Crossgates–Burnside, south-west of Perth. At the west end of the 800 m-long cutting, lithic arenites above a basal mudstone-clast breccia are overlain by a single set of cross-bedded strata over 12 m thick. The complex was laid down in a single bar at least 12 m high in a river channel probably 15–20 m deep. Bluck concluded that the river was substantially deeper than those local (internal) streams of much steeper gradient that had deposited older conglomeratic formations in the Midland Valley up to that time. Its source was external to the Midland Valley, in the Scandian Orogen to the north-east.

The angular discordance between the Upper and Lower Old Red Sandstone strata at Whiting Ness is marked. The Lower Old Red Sandstone dips south-east at about 25°, whereas the Upper Old Red Sandstone dips approximately ESE at 10°. The latter was deposited against steep slopes forming part of the sub-late Devonian land surface, as can be seen in the cliffs (NO 6510 4100) on the east side of the Horse Shoe east of Whiting Ness. It is probable, however, that the palaeorelief of the unconformity seen in the irregular surface near Whiting Ness is small compared to that elsewhere in this area. The western limit of the Upper Old Red Sandstone

Whiting Ness

Figure 3.27 Conglomerates of the Burnside Sandstone Formation (Upper Old Red Sandstone) resting unconformably on the Arbroath Sandstone Member (Scone Sandstone Formation; Lower Old Red Sandstone) (NO 660 412). Note the steep angle of the unconformity. (Photo: BGS No. D2730, reproduced with the permission of the Director, British Geological Survey, © NERC.)

outcrop on the foreshore (NO 6510 4100) 900 m west of Whiting Ness is also an unconformable junction, and an ascending sequence broken by small faults can be followed on the intervening shore. Assuming that the stratification of the Upper Old Red Sandstone was originally horizontal, and on the basis of the prevailing dip (10° to the ESE) and the breadth of outcrop perpendicular to the strike of about 450 m in the ground between the two exposures of the unconformity, there is about 100 m of relief on the ancient land surface.

Most of the Upper Old Red Sandstone sediments appear to have accumulated in braided channels as lateral bars in the active part of an alluvial plain covered with sand and gravel. White, flat-bedded, fine-to medium-grained sandstones of sheet-flood type were laid down preferentially in topographically protected areas close to the steeper slopes on the surface of the unconformity near Whiting Ness (Ramos and Friend, 1982). Steep-sided gullies up to 1.5 m deep cutting into these beds are mostly filled with breccia derived from the adjacent steep slopes (Balin, 1993). Ramos and Friend (1982) deduced a south-westerly direction of transport for the breccias, with an axial drainage system flowing south-eastwards.

Conclusions

Whiting Ness and nearby cliff sections at Dickmont's Den and Forbidden Cave provide excellent exposures of a geological unconformity. This represents an ancient surface with a steeply dissected topography; the observed relief is of the order of 6 m, but is calculated to reach 100 m locally. The sections also show indurated braided river deposits of the Arbroath Sandstone Member that were uplifted, weathered and eroded in a semi-arid climate during mid-Devonian earth movements. Tropical weathering at that time probably caused the reddening of the strata. The overlying Burnside Sandstone Formation was mainly laid down by braided rivers flowing south and south-east, but contains breccias that were deposited by streams flowing south-west at the basin margin.

TAY BANK, PERTH AND KINROSS (NO 125 328)

M.A.E. Browne and W.J. Barclay

Introduction

This important site on the east bank of a meander loop of the River Tay south of Campsie (Figure 3.28) exposes Early Devonian (Lower Old Red Sandstone) sedimentary rocks and concretionary limestones of pedogenic origin (calcretes). It provides the type section of the Campsie Limestone Member and the best exposure of the Stanley Limestone, a group of calcretes in this member named from this locality. A concentration of mature calcrete soil profiles is developed at this stratigraphical level across the Midland Valley of Scotland, marking a period of river downcutting, floodplain stability and reduced sedimentation.

At the time of the original notification of the site in 1976, it was thought that comparison of this horizon with the similar Psammosteus Limestone in the Anglo-Welsh Basin might provide regional correlation and information on the climatic evolution of the British Isles during early Devonian times. The Stanley Limestone is, however, younger and may correlate with the Abdon limestones of the Anglo-Welsh Basin.

Description

The Lower Old Red Sandstone of the Strathmore region was described by Armstrong and Paterson (1970) and a compiled section of the rocks exposed in the river bed and both banks at the Tay Bank site was given by Armstrong *et al.* (1985). The lithostratigraphical classification of the succession was revised recently by Browne *et al.* (2002).

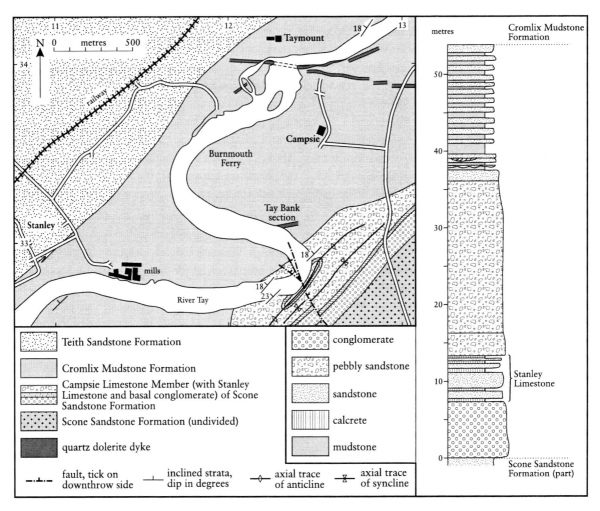

Figure 3.28 Geological sketch map of the area around the Tay Bank section and log of the Campsie Limestone Member exposed. After Armstrong *et al.* (1985).

Tay Bank

The rocks in the Tay Bank area belong to two formations – the Scone Sandstone Formation (of the Arbuthnott–Garvock Group) and the overlying Cromlix Mudstone Formation (of the Strathmore Group). The 2000 m-thick Scone Sandstone Formation consists largely of grey, yellow, brown, red, purplish and red-brown, medium- to coarse-grained, cross-bedded, arkosic to lithic sandstones characteristically containing intraformational mudstone and limestone clasts. Pebbles of metasedimentary and volcanic rocks are also present. Reddish brown or greenish grey silty mudstone and siltstone occur as impersistent interbeds, some more than 0.5 m thick. Pale grey, fine-grained limestone of pedogenic origin occurs as thin, lenticular beds, but more abundantly as reworked clasts, throughout the formation. Lava flows occur locally, as around Laurencekirk and west of Perth.

The Campsie Limestone Member forms the topmost part of the Scone Sandstone Formation. It is characterized by mature calcrete beds (the Stanley Limestone and equivalent Pittendriech Limestone) and conglomerates with intraformational and exotic clasts.

The Cromlix Mudstone Formation consists predominantly of bright red to brownish red, purplish brown and green, poorly sorted, soft, sandy and silty mudstones and siltstones. There are minor, poorly sorted, very fine-grained, argillaceous sandstones, and all of the mudstones and sandstones contain green reduction spots. The beds are typically massive and appear blocky as a result of pervasive fracturing, although weak planar bedding is common. Also present are thin beds of medium- to coarse-grained sandstone, and conglomerate sheets that pass into localized, very thick alluvial-fan bodies.

The Campsie Limestone Member, including the Stanley Limestone (Figure 3.29), is exposed on the banks of an incised meander in the River Tay just east of the village of Stanley (Figure 3.28). The strata lie on the north-west limb of an anticlinal inlier affected by minor NW-trending faults, the exposed beds dipping mainly about 20° to the north-west. The strata are accessible on the banks of the river, and in its bed when the water-level is low. The section compiled from the exposures by Armstrong *et al.* (1985) is shown graphically in Figure 3.28. The lowermost beds of the Campsie Limestone Member, including conglomerates, sandstones and calcretes (the Stanley Limestone), are best seen

Figure 3.29 The Stanley Limestone, a mature fossil soil carbonate (calcrete), exposed at the Tay Bank section. (Photo: M.A.E. Browne.)

in the river bed and in the east bank (NO 124 327). The clasts in the 7.5 m-thick conglomerate at the base of the Campsie Limestone Member include quartzite, felstone, greenstone, porphyrite, syenite and jasper (Armstrong *et al.*, 1985; J. Geikie, unpublished data). Geikie recognized a southerly direction of transport of the sediments based on imbrication of the clasts. The calcretes mainly occur in sandstone host rock, and are fine-grained, white, pale grey, pale greenish grey and pink, nodular limestones.

The uppermost beds of the Campsie Limestone Member, comprising intercalated sandstones and mudstones, dip below the basal beds of the Cromlix Mudstone Formation and are best seen on the west bank below Inchbervis Castle (NO 120 328). On the north side of the peninsula made by the meander loop, the basal argillaceous strata of the Cromlix Mudstone Formation are exposed along the river bank for about 500 m upstream of their junction with the Campsie Limestone Member. About 200 m from the junction, the mudstones are cut by a 0.5 m-wide, ENE-trending basaltic dyke, and thin, irregular basaltic intrusions up to 0.3 m thick occur a little farther east.

Interpretation

The sandstones of the Scone Sandstone Formation were deposited in a SW-flowing river system and are characterized by the presence of intraformational limestone clasts. The carbonate formed as concretionary soil (calcrete) nodules in floodplain silts and muds, reflecting a hot, seasonally wet, semi-arid climate appropriate to a location 10–20° south of the equator (Balin, 1993, 2000). Penecontemporaneous reworking of the floodplain deposits in migrating channels released the resistant carbonate nodules to become incorporated in the sandstones as clasts. However, the Campsie Limestone Member is rare in that the nodules and beds of calcrete in a horizon about 20 m thick largely escaped reworking, except for one or two conglomerate beds. Armstrong *et al.* (1978b) suggested that the calcretes formed during prolonged weathering when the previously well-developed axial river system was bypassed.

Further evidence of significant change in the basin's dispersal systems lies in the switch to the mudrock-dominated Cromlix Mudstone Formation. The formation does not appear to have been subjected to appreciable fluvial reworking (Armstrong *et al.*, 1978b). However, Phillips and Aitken (1998) suggested that the typical structureless silty mudstones in the formation are comparable with aeolian clay pellet deposits formed in landlocked basins in arid and semi-arid parts of south-east Australia (Dare-Edwards, 1984; Yang, 1997; Gibling *et al.*, 1998). The pellets form as pedogenic mud and silt aggregates in arid to semi-arid desert environments and are blown and deposited by winds.

Conclusions

This site is important in providing the best section of the Stanley Limestone and the type section of the Campsie Limestone Member, a regional marker horizon that defines the boundary between the Arbuthnott–Garvock and Strathmore groups. The Stanley Limestone is a concentration of carbonate soil (calcrete) profiles that developed at this stratigraphical level across the Midland Valley of Scotland. The soils represent shut-down of the basin and mark an important change in sedimentation patterns, the earlier river systems being replaced by predominantly arid to semi-arid floodplain environments. Little detailed work has been carried out on the sedimentology and petrography of the rocks at this site and further study is warranted.

GLEN VALE, FIFE (NO 171 068–NO 195 072)

Potential GCR site

M.A.E. Browne and W.J. Barclay

Introduction

The stream sections and crag exposures in the Glen Vale area, Fife and Kinross (Figure 3.30), are reommended for the GCR for the Upper Old Red Sandstone windblown (aeolian) facies sandstones that are exposed. The importance of these sandstones, which belong to the Upper Devonian Knox Pulpit Sandstone Formation (of the Stratheden Group), lies in their aeolian origin, in contrast to the fluvial origin of most of the Upper Old Red Sandstone succession in the Midland Valley. The exposures at John Knox's Pulpit (NO 1891 0582) and Dow Craig

Glen Vale

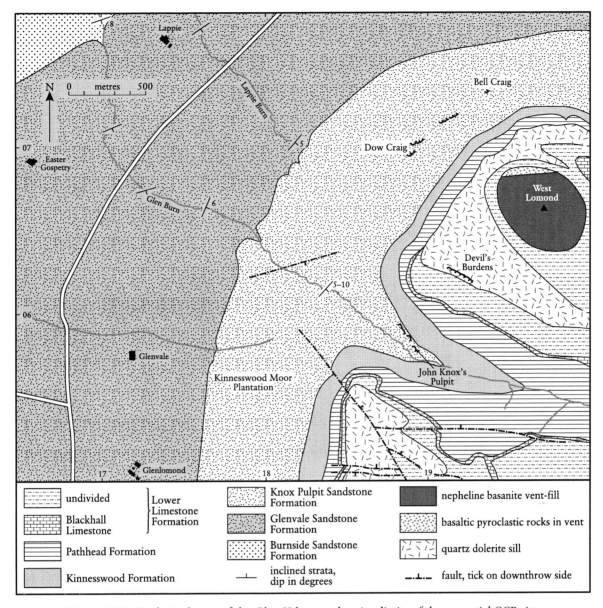

Figure 3.30 Geological map of the Glen Vale area showing limits of the potential GCR site.

(NO 1892 0702) (Figure 3.31) allow a three-dimensional study of many of the key sedimentological features of the formation. In the Glen Vale area, the Upper Old Red Sandstone succession comprises, in ascending order, the Burnside Sandstone, Glenvale Sandstone and Knox Pulpit Sandstone formations. The site thus allows the aeolian facies rocks to be placed in stratigraphical context, between the underlying Glenvale Sandstone Formation and the overlying lower Carboniferous Kinnesswood Formation (of the Inverclyde Group). Chisholm and Dean (1974) gave a comprehensive account of the Upper Old Red Sandstone succession in the area. Although they noted that an aeolian origin could not be discounted for the Knox Pulpit Sandstone Formation, these authors favoured a shallow marine origin. Subsequently, an aeolian origin was recognized (McAlpine, 1978; Mader and Yardley, 1985; Hall and Chisholm, 1987).

Description

The lowermost beds within the proposed site boundary belong to the Glenvale Sandstone Formation. This consists of white, yellow, brown, red and purple, fine- to coarse-grained, feldspathic sandstones. Clasts of red, green

The Midland Valley of Scotland and adjacent areas

Figure 3.31 Dow Craig in the Knox Pulpit Sandstone Formation. (Photo: J.I. Chisholm.)

and cream-coloured mudstone up to 0.15 m are common, but siliceous pebbles are rare to absent. Fish remains include fragments, as well as complete specimens, of *Holoptychius*. Large-scale trough cross-stratification, in sets up to 1.5 m thick is the commonest sedimentary structure, but low-angle cross-bedding and planar lamination are also found. Palaeocurrent flow was towards the east. Subsidiary beds of greenish grey and red silty claystone and siltstone are also present, some forming the upper parts of upward-fining cycles. The transitional junction with the underlying Burnside Sandstone Formation is exposed in the Glen Burn (NO 170 076) north of the site boundary, west of Lappie Farm. It is placed at the point above which the siliceous pebbles that are characteristic of the Burnside Sandstone Formation are absent. The formation includes the well-known Dura Den Fish Bed, which elsewhere has yielded abundant *Bothriolepis* (including *B. hydrophila*), *Glyptopomus* and *Holoptychius*. This fauna indicates a Famennian age (e.g. Westoll, 1977).

The Knox Pulpit Sandstone Formation consists of 130–180 m of soft, weakly cemented, white, buff and yellow, very fine- to coarse-grained, feldspathic sandstones. The transitional junction with the Glenvale Sandstone Formation is faulted and poorly exposed in Glen Burn (NO 181 063), where the lowest exposed strata include sporadic laminae of greenish grey, silty mudstone. The most characteristic feature of the formation is the marked grain-size lamination, with laminae 1 mm to 10 mm thick ('pin-stripe' lamination) of aeolian origin. Other distinctive features are the rarity of pebbles, a greenish grey, silty claystone near the base, ochreous, decomposed calcrete nodules near the top, and an absence of clastic mica flakes that are common in the other Upper Old Red Sandstone formations in the Midland Valley. Bedding forms include tabular planar cross-bedding and planar wedge-shaped cross-bedding, planar lamination and convoluted bedding. The cross-bedding sets are mainly about 1 m or less, with foresets dipping at 20° to 30°, although some sets in the lower part of the formation at Dow Craig are up to 2.5 m. The cross-bedding is predominantly directed to the west and north-west, but bi-modal, east- and west-directed palaeocurrents are also present. Herringbone pattern is seen locally, as at Dow Craig (NO 1892 0702), John Knox's Pulpit (NO 1891 0582) and Kilgour Craigs

(NO 226 078). Erosional, first-order bounding surfaces occur between some sets at Dow Craig. Convolutions affect the larger (up to 2.5 m) sets towards the base of the formation (Chisholm and Dean, 1974). Ripple lamination is rare, except near the top. Well-rounded millet seed grains are common in the coarser laminae. Trace fossils referred to *Skolithos* are most common towards the top of the formation, where they occur in ripple-bedded sandstones as simple, vertical tubes up to 2 cm across and 7–30 cm long. Concentrations of up to 1000 per m^2 are recorded, although they are generally much sparser.

The Kinnesswood Formation is the basal formation of the Inverclyde Group, which is characterized by the presence of sandstones with pedogenic carbonate ('cornstones') and by mudstones with thin beds of dolomite and limestone ('cementstone'). The base of the formation is placed at the appearance of carbonate-bearing strata. The junction with the Knox Pulpit Sandstone Formation is seen in the Glen Burn, at the lip of a small waterfall (NO 1908 0570), where there is a transition over a few metres. Miospores of Tournaisian (LN–PC biozones) age from near the base of the formation elsewhere (Smith, 1996) show that it straddles the Devonian–Carboniferous boundary, but that most of it is of early Carboniferous age. It consists predominantly of purple-red, yellow, white and grey-purple, fine- to coarse-grained sandstones that are mostly cross-bedded and arranged in upward-fining units. Fine-grained, planar bedded and poorly bedded sandstones, red mudstones and nodules and thin beds of calcrete also occur. The calcretes range from immature types, in which the host sandstones have a patchy carbonate matrix with ill-defined concretions, to mature types, in which well-defined nodules (glaebules) are elongate perpendicular to the bedding and overlain by laminar and pisolitic beds. Some of the laminar calcrete, in which the laminae are bedding-parallel, is brecciated and the carbonate is replaced by chert.

Interpretation

The Knox Pulpit Sandstone Formation was initially interpreted as marine in origin (Chisholm and Dean, 1974). However, compelling evidence for an aeolian origin was provided by Balin (1993), drawing on earlier aeolian interpretations by McAlpine (1978), Mader and Yardley (1985) and Hall and Chisholm (1987). The types of planar cross-bedding and heights of the sets are typical of modern aeolian dunes. Some are 2.5 m high in the lower part of the formation at Dow Craig (Figure 3.31). The grain-size lamination may be formed by grainfall (normal graded) and grainflow avalanche (reverse graded) deposition. Inter-set bounding surfaces, as seen for example at Dow Craig, are common in aeolian sands, with coarser sands occurring above them. They may be caused by shifts in wind direction, often accompanied by a slight increase in wind velocity, which introduces coarser sand (Balin, 1993). The planar lamination is produced by traction currents at high wind velocities.

The convoluted, soft-sediment deformation and the opposed (bi-modal) cross-bedding directions (including herringbone cross-bedding) in the sandstones suggested a tidal origin to Chisholm and Dean (1974). Balin (1993) noted that contorted bedding is present in all types of aeolian dune and suggested that the bi-modal cross-bedding was caused by alternating east- and west-directed winds, although bi-modality can also be created in seif (longitudinal) dunes by unidirectional winds as a result of changes in the slip faces of their linear crests. However, Balin (1993) argued that the bounding surfaces in the Knox Pulpit Sandstone Formation are too shallow-dipping to be related to such well-developed steep crests. She further noted that modern, large star dunes form from multi-modal wind directions, but that multiple slip faces are not usually preserved in the stratigraphical record because they are best developed in the easily eroded upper part of the dune. However, they are surrounded by much smaller crescentic and reversing dunes, which form where larger crescentic and linear dunes (the precursors to the star forms) advance into areas affected by variable wind directions. On this basis, Balin concluded that the bi-modal foreset was due to deposition in reversing dunes by opposing wind directions. Adjusting the palaeocurrent data for polar rotation (about 25° anti-clockwise), Balin (1993) found that the prevailing wind direction was to the WSW, with a secondary direction to the south-east. Neither direction is compatible with the presumed north-west trade wind direction, and Balin suggested that an exposed early Devonian volcanic terrane in the Ochil Hills exerted a topographic and orographic influence. Rapid heating of dark volcanic rocks in the daytime may have generated the main winds, with the weaker opposing winds produced by rapid air cooling over the hills at night.

Wavy lamination, small-scale convex-up lamination, climbing translatent ripple lamination (in co-sets up to 2 cm) and the *Skolithos* burrows were interpreted by Balin (1993) as features of interdune sedimentation. Thin wavy laminae are thought to have formed in wet interdune areas as a result of weak, variable wind conditions or falling water-table, leading to modification of the ripples. The convex-up laminae may be adhesion warts, formed where small mounds of sand adhere to a damp substrate. The climbing translatent ripple forms show slight upward coarsening of grain size and lack internal foreset lamination, which are apparently only associated with a dry substrate. These suggest aeolian, rather than waterlain, deposition. *Skolithos* is generally regarded as a marine trace fossil, and its presence was used by Chisholm and Dean (1974) to argue for a tidal origin for the Knox Pulpit Sandstone Formation. Balin (1993) favoured a wet interdune habitat.

The strata below (Glenvale Sandstone Formation) and above (Kinnesswood Formation) the Knox Pulpit Sandstone Formation are of fluvial origin. The cross-bedded sandstones of the Kinnesswood Formation were deposited in river channels and the fine-grained sandstones and mudstones were formed on the adjacent floodplains. Evidence of an arid to semi-arid palaeoclimate and inhospitable terrestrial environments includes desiccation cracks, pedogenic carbonates, the scarcity of fauna (other than fish) and an almost total absence of micro- and macro-floral remains. It is difficult to assess the aridity of the climate, since it is unclear as to how evolved the terrestrial vegetational cover was in the Midland Valley during Late Devonian times. However, large rhizoliths in pedogenic carbonates in the Kinnesswood Formation elsewhere in the Midland Valley (Balin, 2000; see **Milton Ness** GCR site report, this chapter) demonstrate the presence of substantial trees in Late Devonian to Early Carboniferous times. The Knox Pulpit Sandstone Formation shows evidence of hot desert environments, with higher water-table conditions in the interdune areas. The presence of small, ochreous carbonate nodules in the upper part of the formation may herald a less arid climate, the pedogenic carbonates in the Kinnesswood Formation showing that the climate during its deposition was semi-arid, with seasonal (?monsoonal) rainfall.

Conclusions

The natural stream sections in the Glen Burn and nearby crag exposures, including John Knox's Pulpit and Dow Craig, are proposed for GCR site status. The crags provide fine exposures of the Upper Devonian (Famennian) Knox Pulpit Sandstone Formation. This formation represents strata that were deposited by winds, in contrast to most of the Old Red Sandstone succession of the Midland Valley, which is of fluvial origin. The underlying (Glenvale Sandstone) and overlying (Kinnesswood) fluvial formations are well exposed in Glen Burn. The site is therefore important in providing evidence of the changing sedimentary environments of Late Devonian to Early Carboniferous times and in interpretation of the climate and palaeogeography at that time.

WOLF'S HOLE QUARRY, STIRLINGSHIRE (NS 790 981)

Potential ORS GCR site

M.A.E. Browne and W.J. Barclay

Introduction

The importance of Wolf's Hole Quarry, at Bridge of Allen in Stirlingshire (Figure 3.32), lies mainly in the fossil fish specimens recovered from a coarse-grained, pinkish brown quartz- and feldspar-rich arkose. It is already an established GCR site for its fossil fishes (Dineley and Metcalf, 1999) and only a brief summary is provided here. No fossils have been found recently, but the lithology of museum specimens matches the sandstones in the basal and middle parts of the section exposed today. The strata belong to the basal part (the Sheriffmuir Sandstone Member) of the Scone Sandstone Formation of the Arbuthnott–Garvock Group. A thin, porphyritic, andesitic lava flow at the top of the quarry is the highest occurrence of lava (of the Ochil Volcanic Formation) on the southern side of the Strathmore Basin.

No fossils have been found since the quarry closed in 1898, most of the fish specimens being found in the 1860s and 1870s when the sandstones were worked for building stone. The fish may have been preserved in one or more lenses, in a similar fashion to the preservation of some cephalaspids and pteraspids in the Welsh Borderland (Dineley,

Wolf's Hole Quarry

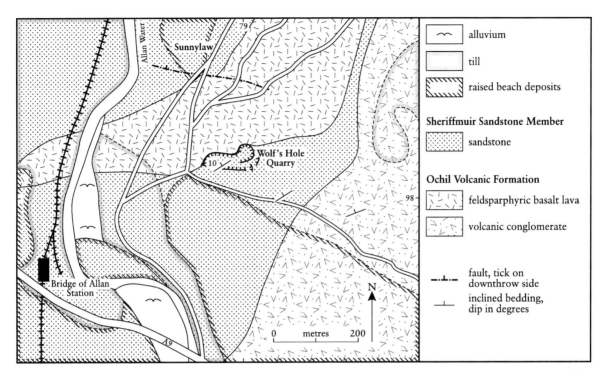

Figure 3.32 Geological map of the area around Wolf's Hole Quarry. After British Geological Survey 1:10 560 Sheet NS 79NE (1976).

1999c). The quarry is primarily selected as a GCR site for its small, but unique fish assemblage, including the type specimens of four agnathan species. Dineley (1999c) summarized the details of the fauna, which includes *Pteraspis mitchelli*, *Cephalaspis scotica*, *Securiaspis waterstoni* and *Securiaspis caledonica*.

Description

Wolf's Hole Quarry (Figure 3.33) exposes 25 m of cross-bedded sandstones with some siltstones, overlain by a flow of basaltic-andesite amygdaloidal lava (Francis *et al.*, 1970). The lowest 6 m of strata exposed in the quarry comprise several fining-upward sandstone–siltstone–mudstone cycles, each about 2–3 m thick and grading upwards from gritty, cross-bedded sandstones to siltstones and then green mudstones. The base of each unit rests on an erosion surface veneered by intraformational conglomerate. Sandstones become more dominant upwards, the cyclic beds passing into 10 m of cross-bedded sandstones. Within these, a persistent, thin (50–150 mm) grey mudstone can be traced across the width of the quarry face. At the top of the section 2 m of flaggy sandstones are truncated at the uneven base of the overlying

Figure 3.33 Wolf's Hole Quarry. A lava flow at the top of the section rests on thick-bedded sandstones. (Photo: M.A.E. Browne.)

lava. This is accessible at the north-east end of the quarry, where the lava base is amygdaloidal with infilled pipe vesicles. The lava flow is generally altered, rotten and feldspar-phyric.

Interpretation

Comparison can be made with the key section in the Scone Sandstone Formation at Crossgates–Burnside south-west of Perth described by Bluck (2000) (see **Whiting Ness** GCR site report, this chapter). None of the very large (> 12 m) bar forms seen at Crossgates–Burnside can be recognized at Wolf's Hole, perhaps because the section is smaller. However, it is likely that after a high-sinuosity fluvial regime represented by the fining-upward cycles at the base of the section, braided river channel systems became dominant, with minor overbank or floodplain mud/silt deposition. Mudstone drapes on cross-sets are only rarely preserved, and mudstone rip-up clasts in the sandstones represent the destruction and reworking of most of the fine sediment. The uneven basal surface of the lava may reflect the palaeotopography, the flow appearing to infill channels locally. The flow may have been emplaced from the 'south', from the volcanic terrane represented by the Ochil Volcanic Formation.

Dineley (1999c) discussed the fossil fishes fauna, its affinities and habitat. *Pteraspis mitchelli* is a very rare pteraspid occurrence in Scotland, although pteraspids are common in the Dittonian of the Welsh Borderland. The taxonomy of *Pteraspis mitchelli* remains in some doubt, Blieck (1981, 1984) noting affinities with *Protopteraspis* and similarities to *Pteraspis rostrata*. The affinity with *P. rostrata* suggests a connection with the Fintona Beds of County Fermanagh (Harper and Hartley, 1938), from which *P. rostrata* is recorded, the only pteraspid known from Ireland. *Securiaspis* is confined elsewhere to the Dittonian (Lochkovian–Pragian) of the Welsh Borderland and Spitsbergen, and this is the only occurrence in Scotland. The fauna as a whole suggests a correlation with the Dittonian of the Anglo-Welsh Basin. Palynological investigation of the grey and green beds at Wolf's Hole may offer a further correlation. The preservation of the fish in fluvial facies is also more akin to the Anglo-Welsh Basin occurrences, unlike the predominantly lacustrine facies in which most of the Lower and Middle Old Red Sandstone fossil fish of Scotland are found.

Conclusions

The conservation value of Wolf's Hole Quarry lies mainly in its important fish fauna, including the type specimens of four species. The occurrence of *Pteraspis* is unusual for Scotland and may indicate a link with northern Ireland and the Anglo-Welsh Basin at that time. The occurrence of the fish in fluvial facies is also unusual in Scotland, most of the fossils occurring in lake deposits. The site has potential for further examination for remains of fishes, as well as for plant microfossil study. It provides a good section of the Sheriffmuir Sandstone Member of the Scone Sandstone Formation, the strata being interpreted as the products of meandering and braided river environments. The lava flow exposed is the highest such unit in this part of the Midland Valley.

AUCHENSAIL QUARRY, WEST DUNBARTONSHIRE (NS 342 799)

Potential ORS GCR site

M.A.E. Browne and W.J. Barclay

Introduction

Auchensail Quarry (Figure 3.34) near Cardross, West Dunbartonshire exposes sandstones and mudstones/siltstones of the Teith Sandstone Formation (Strathmore Group) of the Lower Old Red Sandstone. It has yielded some of the best-preserved plant fossils of Emsian (Early Devonian) age in Britain (Morton, 1976; Scott *et al.*, 1976). The sandstones contain an abundance of plant material, including exceptionally well-preserved examples of *Sawdonia* and the youngest examples of *Prototaxites* known in Britain, some of which reveal considerable anatomical detail (Rayner, 1983, 1984, 1995). The site is already an established GCR site for its Palaeozoic palaeobotany (Cleal and Thomas, 1995) and a brief summary is presented here. In plant material of similar age from Ballanucater Farm GCR site, the outer protective skin (cuticle) of the plants is preserved (Rayner, 1995).

The geology of the quarry was described by Scott *et al.* (1976) and summarized by Rayner

Auchensail Quarry

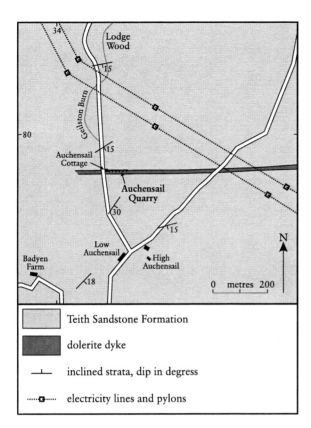

◀Figure 3.34 Location and geology of Auchensail Quarry. After British Geological Survey 1:10 560 manuscript map NS 37NW (1984).

(1995). The quarry exposes strata of the Teith Sandstone Formation of the Strathmore Group (Armstrong and Paterson, 1970). The strata are typical of the 1000 m-thick Teith Sandstone Formation, and illustrate the sedimentological character of this fluvial unit (Paterson et al., 1990). Spores recovered from the formation belong to the *annulatus-sextantii* Biozone (Emsian) (Richardson et al., 1984).

Description

The quarry exposes a thin, east-trending, basaltic dyke intruding about 20 m of interbedded sandstones and argillaceous beds (Figures 3.35, 3.36). The sandstones are green, fine- to medium-grained and well indurated, and arranged in upward-fining units up to 2 m thick, some with trough cross-bedding and lateral accretion surfaces. They have conglomeratic bases with

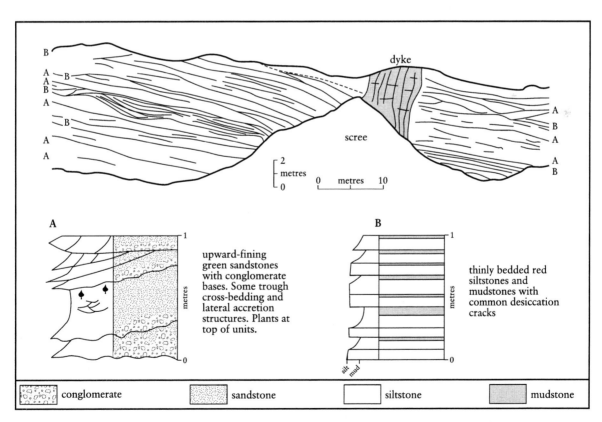

Figure 3.35 Sketch of Auchensail Quarry and its facies associations. After Scott et al. (1976), reproduced by Cleal and Thomas (1995).

The Midland Valley of Scotland and adjacent areas

Figure 3.36 Auchensail Quarry: sandstones, siltstones and mudstones of the Teith Sandstone Formation. Note the igneous dyke just to the left of centre of the quarry. (Photo: C.J. Cleal.)

cobbles and pebbles of intraformational red mudstone. Large rafts of spiny plant axes at the top of units are aligned parallel to the palaeocurrent direction. Primary current lineation trending N60°E is recorded at a nearby exposure. Plant fossils are abundant in the sandstones. Mats of coalified plant compressions occur mainly within the upper part of the sandstone units, whereas discrete pyrite petrifactions occur throughout. The floral assemblage comprises *Drepanophycus spinaeformis*, *Dawsonites* sp., *Prototaxites* sp., *Sawdonia ornata*, ?*Sporogonites* sp. and *Zosterophyllum* sp.. It is typical of the Emsian Strathmore Group and similar to those in grey, coarse-grained sandstones and blue-green mudstones at Ballanucater Farm, Callander. A discussion on the plant fossil assemblages is given by Rayner (1995).

The argillaceous beds are of red siltstone and mudstone up to 2 m thick, comprising stacked fining-upward units 0.2–0.3 m thick. Most of the units are tabular, but some lenticular ones fill abandoned palaeochannels. No plants are preserved in this facies, but a single specimen of the trace fossil *Beaconites* Vyalov is recorded in a loose block and fish remains are recorded nearby. Desiccation-cracked bedding surfaces are common.

Interpretation

The lithofacies in Auchensail Quarry are interpreted as fluvial deposits (Scott *et al.*, 1976). The sandstones were the point-bar deposits of medium-size, meandering rivers; the mudrocks were floodplain deposits. The plants occur only in the sandstones, suggesting that they lived within the margins of the river channels, perhaps on the point bars. No plants occur in growth position. However, post-depositional oxidation may account for their absence in the red floodplain beds. *Beaconites* has been interpreted as the locomotory or temporary resting burrow of amphibians or reptiles. The burrows may be similar to those in the beds of wet-dry, seasonal streams in modern deserts, such as the dry season burrows of the lungfish (Scott *et al.*, 1976).

Conclusions

Auchensail Quarry is accorded GCR site status because it has yielded an exceptionally well-preserved suite of Early Devonian (Emsian) land plants. The quarry also provides a good section of the interbedded sandstones and mudstones/siltstones of the Teith Sandstone Formation, the youngest Early Devonian formation in the Midland Valley of Scotland.

SICCAR POINT TO HAWK'S HEUGH, SCOTTISH BORDERS (NT 813 710–NT 790 714)

M.A.E. Browne and W.J. Barclay

Introduction

Accorded SSSI status in 1961, the Siccar Point unconformity (known as 'Hutton's unconformity') is an internationally famous place of geological pilgrimage. It is of great historical importance in the development of the science of geology. Although James Hutton had previously observed unconformities in Arran (see **North Newton Shore** GCR site report, this chapter) and Jedburgh (see Figure 4.2, Chapter 4), the unconformity at Siccar Point is the most spectacular. It was here in 1788 that Hutton, accompanied by James Hall and John Playfair, was the first person to appreciate the significance of unconformities in the geological record (Hutton, 1795). He used the locality to demonstrate the cycle of deposition, folding, erosion and further deposition that the unconformity represents. He understood the implication of unconformities in the evidence that they provided for the enormity of geological time and the antiquity of the earth, in contrast to the biblical teaching of the creation of the Earth (Repcheck, 2003). All three of 'Hutton's unconformities' have been proposed as Global Geosites on the basis of their historical importance in the development of geology (Cleal *et al.*, 2001). A casting of the Siccar Point unconformity is housed in the American Museum of Natural History, New York.

Field guides to Siccar Point were provided by Craig (1960, 1986) and Greig and Davies (1978). Greig (1988) and Balin (1993) gave detailed accounts, from which the following description is largely derived. The geology of Siccar Point is placed into a wider context of the geology of the area between Redheugh and the Hawk's Heugh SSSI to the west (Figure 3.37), including

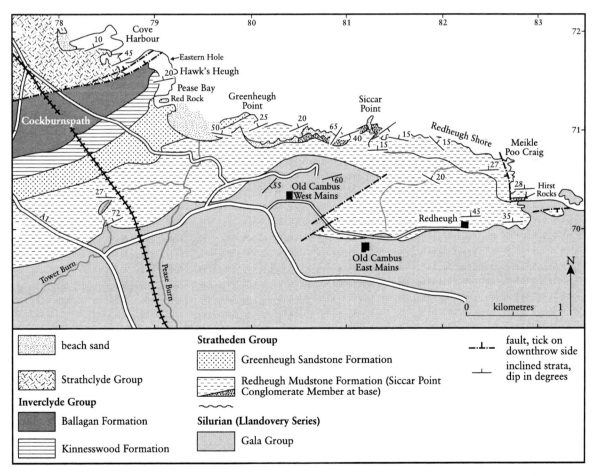

Figure 3.37 Geological map of the area around Siccar Point, from Cove Harbour to Hirst Rocks. After British Geological Survey 1:50 000 Sheet 34 (Scotland), Eyemouth (1982).

the superb cliff sections of Upper Old Red Sandstone in Pease Bay (Craig, 1960; Greig and Davies, 1978; Fyfe, 1985; Clarkson, 1986; Greig, 1988; Salter, 1992; Balin, 1993). The 4 km coast section from Siccar Point westwards to Hawk's Heugh and Cove Harbour provides a unique, magnificently exposed transect through the succession of Upper Devonian to Lower Carboniferous rocks and merits protected status in its entirety. The uppermost part of the transect lies within the Cove GCR site and is described in the GCR volume on Lower Carboniferous stratigraphy (Cossey et al., 2004)

Description

Siccar Point

Siccar Point (Figure 3.38) is a coastal promontory of gently dipping Upper Devonian (Upper Old Red Sandstone) beds resting discordantly on folded, vertical, Lower Palaeozoic strata. The latter are turbiditic, dark grey, fine-grained wacke sandstones and interbedded finely laminated, fissile mudstones of the Gala Group, of early Silurian (Llandovery Series) age. The sandstone beds are up to 0.4 m thick and the mudstones are up to 0.15 m. The beds are locally contorted by tight folds and cut by closely spaced, parallel fractures produced by intense compressive stresses during the Caledonian Orogeny. In the south-east of the outcrop, the beds dip to the north-west at almost 90°, but a synclinal axis results in dips to the south-east of about 60° in the north-west of the site. They are truncated by the sharp, angular unconformity, which is overlain by Upper Devonian strata comprising reddish brown conglomerates and sandstones of the Redheugh Mudstone Formation of the Stratheden Group (Figure 3.39). These beds were deposited in the Oldhamstocks Basin (Lagios, 1983), which was possibly a sub-basin of the Scottish Border Basin. The beds are undeformed, with only a gentle inclination produced during the later, here less intense Variscan Orogeny. Siccar Point is the only well-exposed example of this unconformity in the Southern Uplands and clearly demonstrates the contrasting effects of the Caledonian and Variscan orogenies in this part of Scotland.

The Stratheden Group comprises the Redheugh Mudstone Formation and the overlying Greenheugh Sandstone Formation (Browne et al., 2002). At Siccar Point, the Redheugh Mudstone Formation comprises a basal conglomerate (the Siccar Point Conglomerate Member) overlain by argillaceous sandstones fining upwards generally into red, sandy mudstones with mainly thin intercalated sandstones. The basal conglomerate is locally up to 6 m thick at Siccar Point, but is absent higher up on the south-eastern side of Siccar Cove. It is present at Hirst Rocks (NT 830 705) to the east and at one locality in Tower Burn south of Pease Bay to the west, but elsewhere the lowest Old Red Sandstone rocks are red-brown sandstones (Greig, 1988).

The unconformity surface at Siccar Point is very irregular because of differential rates of pre-Late Devonian weathering and erosion of individual beds in the Silurian succession. On the north side of the point, the overlying conglomerate forms an area of wave-cut slope about 100 m long and up to 20 m wide. The conglomerate dips seawards (north-west)

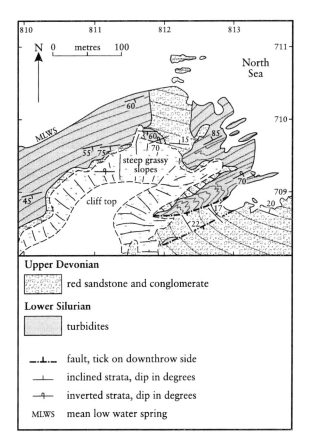

Figure 3.38 Geological map of Siccar Point: Hutton's Unconformity. After Greig (1988).

Siccar Point to Hawk's Heugh

Figure 3.39 Hutton's Unconformity at Siccar Point. (Photo: BGS No. D1471, reproduced with the permission of the Director, British Geological Survey, © NERC.)

between 15° and 20°. The surface on which the conglomerate rests is in detail very uneven, some of the beds of wacke sandstone standing up more sharply and prominently than others. The overlapping of the beds of conglomerate against the wacke sandstone surface is well displayed on centimetre- to metre-scale, with at least 3–10 m of palaeorelief on the unconformity surface. Relief of 10 m is observable towards the western margin of the main outcrop at Siccar Point. Similar relief is also seen in the faulted, 60 m-high cliff face forming the eastern limit of the exposure, where the unconformity in part of this excellent 125 m-long view has a scalloped surface (Balin, 1993).

The conglomerates were deposited preferentially in hollows on the original land surface (Greig, 1988). Beds of crumbly red mudstone and siltstone with ribs of sandstone rest on the unconformity above the small inlier of Silurian in Tower Burn (NT 758 702). In Pease Burn, red sandstones dipping at 35° to the north rest unconformably on Silurian rocks.

West of Siccar Point, the cliffs of Silurian rocks are capped by conglomerates that are up to 3 m thick in depressions in the palaeosurface. The unconformity descends to the beach south-east of Kirk Rigging, striking ENE on the shore, where there is little basal conglomerate. To the south-east of Siccar Point, a set of NE-trending faults downthrows the conglomerate to the south-east to sea level in the bay. Near high-water mark in the corner of the bay is a small Silurian inlier overlain by 5 m of conglomerate and 6 m of red, pebbly sandstones. Further faulting leads to exposure of higher sandstones on the shore in front of the landslip-scarred Marly Brae (NT 8130 7075), where the unconformity can be seen. Further exposures of the unconformity, but with little palaeorelief, occur at the east end of Redheugh Shore on the beach south of Hirst Rocks (NT 8265 7030) and in a small outlier in the cliffs above.

The conglomerates are poorly sorted and framework-supported with a matrix of red, medium- to coarse-grained sandstone. The angular, generally tabular clasts are of grey, wacke sandstone of pebble- to boulder-grade up to 0.56 m, with a few vein quartz pebbles up to 0.07 m (Balin, 1993). Greig (1988) noted strong imbrication indicating transport by south- to SE-flowing palaeocurrents. Individual beds are sheet-like and range from about 0.9 m to 2 m in thickness, with apparently planar tops and bases, except where adjacent to basement channel margins. Weak normal and reverse grading are present locally. The sandstones that overlie the conglomerates, and locally lie within them, are horizontally laminated and trough cross-bedded. The laminated beds are up to 1.6 m thick and comprise centimetre-scale laminae with scattered cobbles. The trough cross-bedded sandstones comprise sets up to 0.5 m high and over 1 m wide, commonly stacked into multi-storey units. Palaeocurrents are mainly to the south-west, as also inferred from most of the pebble imbrication in the conglomerate. There are also a few tabular sheets of cross-bedded sandstone about 0.15 m thick with low-angle foresets.

The Siccar Point Conglomerate Member passes up into the main, mudrock-dominated part of the Redheugh Mudstone Formation. The formation is about 200 m thick and comprises red-brown, sandy mudstones interbedded with mainly thin, red-brown, pale yellowish grey, green, purple and cream sandstones that become more numerous and massive upwards. One bed contains calcareous concretions up to 1 m in diameter (Greig, 1988). The proportion of mudstone to sandstone is of the order 3:1 at Meikle Poo Craig (NT 822 708) 1.5 km east of Siccar Point. Fossil fish remains indicating a Famennian age include *Bothriolepis hicklingi* found in loose blocks at Redheugh, Siccar Point and Greenheugh Point, and fragments of *Grossilepis brandi*, *Bothriolepis* sp. and *Holoptychius* sp. from Hazeldean Burn (NT 8138 7019) (Miles, 1968).

Pease Bay

Pease Bay, 2 km west of Siccar Point (Figure 3.37) is a sandy bay dramatically rimmed by cliffs of Upper Old Red Sandstone strata that dip 25°–35° to the north (Clarkson, 1986). At its eastern end, Greenheugh Point (NT 800 711), the transition between the Redheugh Mudstone Formation and overlying Greenheugh Sandstone Formation is exposed. The latter comprises mainly medium- to coarse-grained sandstones and thin, red, mudstone and siltstone interbeds. About 70 m of the formation is present at Greenheugh Point, with a further 25 m at Red Rock (NT 791 711) on the west side of Pease Bay. Salter (1992) and Balin (1993) gave detailed sedimentological accounts of the beds. The sandstones are mainly red-brown, but with some pale yellow and pale green beds. Fragments of *Bothriolepis* have been found in the lower sandstones and scales of *Holoptychius nobilissimus* have been found in abundance (Dineley, 1999d). A large fallen block crowded with intact specimens of *Bothriolepis* was recorded by Clarkson (1986), the fish probably washed out of a river channel during flooding and stranded in a drying pool.

Red, medium- to coarse-grained, cross-bedded sandstones at Red Rock have silty mudstone interbeds containing yellowish green ribs and coarse sandy layers. In the corner of the bay 150 m to the north, the highest mudstone is exposed on the shore and is succeeded by red sandstones forming the high cliff at the Deil's Hole. These dip 17° to the north and comprise alternating parallel-bedded and cross-bedded sandstone bodies with well-rounded grains. The basal beds of the sandstone are trough cross-bedded, the troughs trending south. Above, both fluvial and aeolian facies (Figure 3.40) have been identified (Salter, 1992).

Hawk's Heugh (NT 790 714)

About 240 m north of Red Rock (NT 7910 7134) three E–W-striking faults, seen particularly well in the cliff, throw down to the north, bringing in the Kinnesswood Formation (the Cockburnspath Formation of Balin, 1993). This formation is exposed in an extensive wave-cut platform and high cliff at Hawk's Heugh and is characterized by the presence of calcrete. Accorded protected status because of the occurrence of *Remigolepis*, the only British example of this fish (Dineley, 1999d), the site provides superb exposures of this calcrete-rich topmost part of the Upper Old Red Sandstone. Strongly cross-bedded sandstones at the base of the section contain an abundance of veins and layers of bright red and purple-brown, iron-stained carbonate nodules, the latter commonly aligned along the horizontal

Figure 3.40 Interbedded fluvial and aeolian sandstones at Red Rock, Pease Bay. The aeolian sandstones show sandflow and wind ripple lamination, the fluvial sandstones are sheet-flood deposits (Photo: B.P.J. Williams).

bedding planes. Above, grey beds appear locally in the red beds and there is an increasing upwards abundance of carbonate in the form of red-brown, iron-stained irregular lenses and layers of nodules that form prominent ridges trending seawards across the wave-cut platform. Vertical, cylindrical concretions also occur. The sandstones are predominantly red, particularly in the lower part of the formation; many of the higher beds are greyish white with red flecks of ferroan dolomite. The sandstones occur in trough cross-bedded, tabular cross-bedded and massive bodies. Intraformational calcrete-clast conglomerates are interbedded in the succession and thin, red-white mottled, locally micaceous, argillaceous interbeds occur sporadically (Balin, 1993).

The Hawk's Heugh SSSI is described in the companion GCR volume on fossil fishes (Dineley and Metcalf, 1999). *Remigolepis* was found in a loose block of intraformational conglomerate similar to a bed in the cliff above about 6 m below the top of the Kinnesswood Formation. *Holoptychius* and *Bothriolepis* are also recorded from the bed.

The top of the Kinnesswood Formation is in Eastern Hole (NT 7900 7157), the bay to the north of Hawk's Heugh. It was placed at the top of a prominent deep-red, 0.9 m-thick calcrete with cream chert veins and lenses by Salter (1992), and 1–1.5 m above it by Greig (1988) and Balin (1993), at the top of a bed of trough cross-bedded sandstone of Old Red Sandstone facies that overlies the calcrete. Above this, yellow and dark green, plant-bearing mudstones and ferroan dolostones (cementstones) of the Ballagan Formation make their first appearance. Described as a sandstone by Greig (1988), the calcrete is a silty mudstone with coarse, ferroan, dolomite crystals (Andrews *et al.*, 1991). A cementstone with layers of conglomerate containing yellow, angular dolomite clasts and fish fragments (the Eastern Hole Conglomerate) lies about 13 m above.

Interpretation

James Hutton was probably the first natural philosopher to recognize and understand the significance of the relationships between rocks separated by an angular unconformity. He understood the long time-period represented by the unconformity and also, to a degree, the events that led to the formation of the observed features. The first of the unconformities he studied was at **North Newton Shore** on Arran in 1786 (see GCR site report, this chapter), followed by the one at Inchbonny, Jedburgh (formerly known as 'Aller's Mill', now a RIGS site; Figure 4.2, Chapter 4). Hutton and John Playfair were obviously struck by the significance of the exposure at Siccar Point – 'the mind seemed to grow giddy by looking so far into the abyss of time' (Playfair, 1805).

The planar-bedded conglomerates and interbedded sandstones above the unconformity at Siccar Point were interpreted by Balin (1993) as sheet-flood deposits, formed mainly by ephemeral floods, which created pulses of sediment-charged water that spread out from the mouth of a channel. Localized thickening of the conglomerate beds results from the infilling and draping of the uneven basement topography at the unconformity surface. Balin (1993) considered that the coarse grain-size of all the Late Devonian lithofacies at Siccar Point was consistent with deposition by high-velocity, high-gradient flows. Framework-supported gravels are laid down by high-energy water flow that prevents, or partially prevents, deposition of sand from suspension. This fact, together with the clast imbrication and the framework-supported structure of the conglomerates, suggests deposition from traction currents or as flash-flood sheets. Reduced-flow regime allowed sand to settle in the spaces in the gravels and deposition of the plane-bedded sandstone lithofacies (Balin, 1993).

The generally impermeable bedrock contributed to the high velocity of the flood discharges. Also important was the bedrock topography, the steep-sided gully at the western side of Siccar Point, for example, being responsible for the local SE-directed imbrication (Balin, 1993). Pipe-like burrows, about 1 cm wide and 10 cm long, in the planar-bedded sandstones at Hirst Rocks suggest periods of lower flow-energy and non-deposition. The conglomerates form part of an upward-fining succession, suggesting that uplift of the source area ceased. With complete draping of the palaeotopography and burial of remnant highs, floodplain mud- and silt-dominated deposition became established and the remainder of the Redheugh Mudstone Formation was laid down widely in the Scottish Border Basin.

The Greenheugh Sandstone Formation has been interpreted as entirely fluvial by Balin (1993) and as mixed fluvial and aeolian by Salter (1992). The basal part of the formation at Greenheugh Point was interpreted by Salter (1992) as the deposits of shallow, braided streams and overbank sheet-flooding across a broad, flat alluvial-plain. Balin (1993) noted that the absence of vegetation resulted in easily eroded channel banks, facilitating the migration of stream channels. Silts were deposited in ephemeral lakes on the floodplain at the terminations of some channels. Minor aeolian reworking of exposed fluvial bar forms also occurred. The sandstone-dominated succession at Red Rock was interpreted by Salter as the product of deposition in ephemeral, shallow channels that became the sites of aeolian deflation during periods of increased aridity. Salter also suggested that stabilized aeolian dune-fields formed during times of maximum aridity, resulting in aeolian bedforms up to 2 m high at Red Rock (Figure 3.40). Balin (1993) favoured an entirely fluvial origin, but noted that the textures of some grains (Fyfe, 1985) suggest that they may have been involved in a phase of aeolian transport. The presence of pin-stripe lamination and the alternation of laminae with marked contrast in grain size is regarded as characteristic of aeolian deposition.

Calcrete palaeosol development in the upper part of the Upper Old Red Sandstone (the Kinnesswood Formation) is seen throughout the Midland Valley of Scotland and Southern Uplands. The carbonate concretions at Hawk's Heugh are interpreted as pedogenic calcrete, their formation along foresets, bedding planes and desiccation cracks being aided by increased groundwater permeability. The vertical, cylindrical concretions are interpreted by Balin (1993) as rhizocretions formed around plant roots. The 0.9 m-thick calcrete close below the top of the Kinnesswood Formation contains horizontal, cream-coloured chert lenses, indicating a mature stage of palaeosol development (Stage 4 of Leeder, 1975; Stage VI of Machette, 1985), and a period of formation of up to 1.5 million years (Salter, 1992). Low sedimentation rates and a semi-arid climate are requisite for calcrete formation, although the climate may have been less arid than during deposition of the Greenheugh Sandstone Formation. The prolonged period of tectonic stability during formation of the mature calcrete near the top of the Kinnesswood Formation preceded a major change in palaeogeography when coastal floodplain deposition of the Ballagan Formation was introduced. Eustatic sea-level rise at the start of the Carboniferous Period, a wetter climate and tectonic factors may all have contributed to the change.

The fish from Pease Bay (*Bothriolepis* and *Holoptychius*) point to a Famennian age for the Greenheugh Sandstone Formation (Dineley, 1999d). *Remigolepis* from the Kinnesswood Formation at Hawk's Heugh occurs elsewhere in the world in strata ranging from Frasnian to

Early Carboniferous in age. The position of the Devonian–Carboniferous boundary at the Hawk's Heugh GCR site is not clear. The base of the Eastern Hole Conglomerate and the basal cementstone beds have variously and arbitrarily been taken as the boundary in the past, but late Tournaisian (CM Zone) spores have been recovered from the lowermost cementstones (Andrews *et al.*, 1991). The Devonian–Carboniferous boundary may therefore lie at a level lower than its previous, arbitrary position, within or near the base of the Kinnesswood Formation (Browne *et al.*, 1999).

Conclusions

Siccar Point is a world-renowned site forever associated with James Hutton, in particular, but also with Sir James Hall and Professor John Playfair. It is arguably the most important SSSI in Scotland and a place of international geological pilgrimage. In addition to its main, historical, importance, it is also important in providing excellent exposures of the Silurian and Late Devonian sedimentary rocks that lie below and above Hutton's unconformity respectively. The unconformity is beautifully displayed in three dimensions at Siccar Point, providing one of the best exposures of an angular unconformity in Scotland.

The coastline from Siccar Point westwards through Pease Bay to the Hawk's Heugh fossil fishes GCR site and the Cove Lower Carboniferous Stratigraphy GCR site presents a magnificently exposed, complete transect of the entire Late Devonian (Famennian) succession and the transition into the overlying strata of Early Carboniferous (Courceyan) age. Both river and wind-borne sediments are beautifully displayed in the cliffs and foreshore and the beds have yielded important fish remains, including the only occurrence of *Remigolepis* at Hawk's Heugh and complete specimens of *Bothriolepis* from Pease Bay. The whole section is frequently visited by students and professional geologists. It is an important teaching resource in terms of the sedimentary and contemporaneous pedogenic structures displayed, and the depositional environments they represent. There is also the potential for future fossil fish discoveries, and the section between the two existing GCR sites at Siccar Point and Cove is eminently worthy of protected status.

LARGS COAST, NORTH AYRSHIRE (NS 191 633–NS 192 619)

A.A. Monaghan

Introduction

The section of Late Devonian rocks on the west coast of Scotland north of Largs (Figure 3.41) is an important GCR site for understanding the internal structures of Upper Old Red Sandstone braided river deposits. Outcrops of intercalated pebbly sandstones and finer-grained sandstones with thin silty mudstone interbeds are well exposed along the shore. These rocks demonstrate the lateral and vertical facies variations and the scale of bar forms in a braided river channel, and the relationship of the bar forms to the river floodplain deposits.

The sedimentology and tectonic setting of the Largs section and others along the coast of the Firth of Clyde have been studied by Bluck (1967, 1978, 1980a,b, 1986, 1992, 2000). This account is based on Bluck's detailed analysis of the geometry of the sandbodies at the site. He is one of a group of workers who have made major advances in the understanding of Old Red Sandstone fluvial facies, sedimentary processes and environments by comparing the rocks with modern analogues.

The Upper Old Red Sandstone rocks north of Largs belong to the Kelly Burn Sandstone Formation of the Stratheden Group (British Geological Survey, 1990; Paterson *et al.*, 1990; Browne *et al.*, 2002). The base of the Kelly Burn Sandstone Formation is transitional with the Skelmorlie Conglomerate Formation. Its top is a sharp, possibly disconformable junction with the Lower Carboniferous Kinnesswood Formation. The Late Devonian age of the Stratheden Group is constrained by a rare Famennian fish fauna (Browne *et al.*, 2002), the Tournaisian age of the overlying Kinnesswood Formation (Browne *et al.*, 1999) and the Early Devonian age of the underlying Lower Old Red Sandstone.

The Midland Valley of Scotland is interpreted to have been delimited in Late Devonian times by the Highland Boundary Fault in the north and the Southern Upland Fault to the south, with a central, western high (Bluck, 1978; Browne *et al.*, 1985). In the Firth of Clyde area, extensional half-grabens formed by sinistral strike-slip movement on the Highland Boundary Fault are thought to have controlled sediment dispersal

The Midland Valley of Scotland and adjacent areas

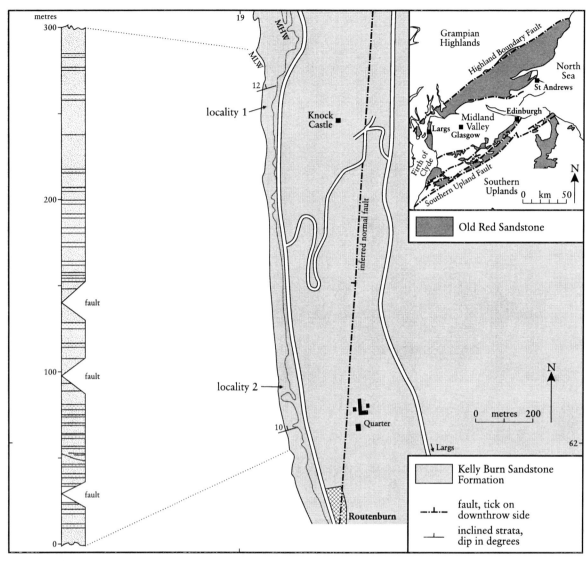

Figure 3.41 Geological sketch map of the Largs coast section and summary graphic log. The inset shows the location of Largs, Ayrshire. After Bluck (1980b).

and accumulation (Bluck, 1978, 1992). Sediments were laid down in these basins by river systems that flowed to the east and northeast on the north side of the Midland Valley (Bluck, 1978). The Largs site exposes finer-grained sedimentary facies that are distinct from the proximal, coarser conglomerates of the underlying Skelmorlie Conglomerate Formation (Bluck 1967, 1992; Browne *et al.*, 2002).

Description

The Largs coast section, 3 km WNW of the town of Largs, comprises approximately 300 m of cross-stratified sandstones, pebbly sandstones and conglomerates interbedded with planar bedded, fine-grained sandstones and thin silty mudstones (Figure 3.41). Together, these facies form upward-fining cycles 2–10 m thick (Bluck, 1980b, 1992). The sedimentary rocks are cut by several faults and numerous Tertiary felsic alkaline and dolerite dykes (British Geological Survey, 1990).

A typical cycle begins with coarse-grained, pebbly, cross-stratified sandstone that rests on an erosion surface cut in the underlying fine-grained sandstone. The coarse-grained sandstone is generally overlain by finer-grained sandstones with complex sedimentary structures, lateral facies variations and numerous internal erosion surfaces. Two examples of such cycles are described in detail.

Locality 1. West of Knock Castle (NS 1913 6303)

This outcrop (Figure 3.42) commences with an upward-coarsening unit 2.8 m thick resting on a basal erosion surface and comprising 0.5–1 m-thick beds of cross-stratified, pebbly sandstone. The foresets of the cross-stratification have a comparatively low spread of dip directions. Above this, a 4 m-thick, upward-coarsening unit consists of four facies that interfinger down-dip and have transitional boundaries. The facies are:

1. An eastward-thinning sandstone wedge with soft-sediment folding and re-folding structures.
2. A cross-stratified sandstone that thins to the east and has converging dips.
3. A sheet of upward-coarsening, coarse-grained, pebbly sandstone that caps the outcrop.
4. A coarse-grained sandstone of uncertain affinity.

Cross-bedding indicates a north-westerly palaeoflow and the whole outcrop becomes finer-grained in that direction.

Locality 2. WNW of Quarter (NS 1917 6215)

Two groups of facies can be observed at this locality. The lower group comprises alternations of planar sheets of sandstones and thin, subordinate, silty mudstones with desiccation cracks. The sandstones exhibit cross-stratification, low-angle cross-stratification and parallel lamination, and locally rest on erosion surfaces. These beds dip in the same direction as the coarser overlying group and have similar north-easterly palaeocurrents.

An erosion surface separates the two groups. The basal part of the upper group comprises pebbly sandstones with large-scale cross-stratification that dips radially outwards. Smaller-scale cross-stratification migrates along and down the larger foresets. Above this, a mudstone bed grades laterally to the east and north into clay-rich sandstones and westwards into a wedge of pebbly sandstones. In turn, these partly interfinger with, and are overlain by, pebbly sandstones that form a tabular cross-stratified unit with much internal re-activation and ripple cross-lamination.

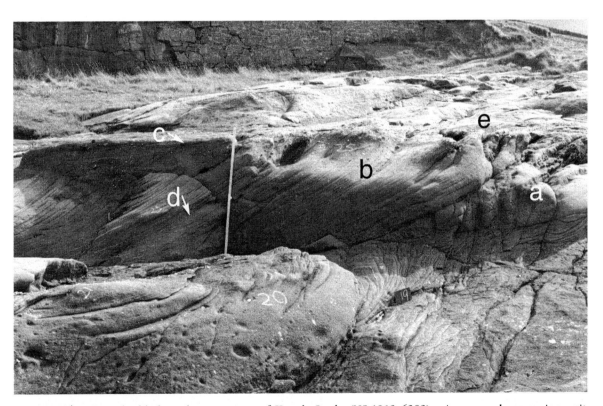

Figure 3.42 Cross-bedded sandstones, west of Knock Castle (NS 1913 6303). An upward-coarsening unit interfingering between coarse sediment (e) and fine sediment (d) at (a); gradational contact between coarse and fine sediment at (b); sharp, erosive contact at (c), and counter-current ripples at (d); scale rule 1 m. (Photo: B.J. Bluck.)

Interpretation

A key feature in the recognition of bar form in braided river systems is the upward coarsening of the sands from a sharp base, caused by the downstream migration of the coarse-grained head of the bar over the finer-grained tail (Bluck, 1992; Figure 3.43). Also, braided pebbly alluvium has palaeocurrents that vary laterally and vertically and a facies distribution that indicates the influence of the bar form on river flow (Figure 3.43; Bluck, 1980b; Miall, 1996). The Knock Castle section and the upper part of the Quarter section can be interpreted as different types of braided river-bar deposits. At the Knock Castle outcrop, the consistent dips and volume of coarse-grained, cross-stratified sandstones in the lower part are attributed to bedform deposition within a river channel (Bluck, 1980b). The upper part is interpreted as a mid-channel bar form, where the bar head (represented by the coarse-grained, pebbly sandstone sheet) migrated north-west over the bar tail (represented by the deformed sandstone; see Figure 3.43). The converging palaeocurrents in the cross-stratified sandstones towards the bar tail are observed on modern bar forms that have channels to each side (Bluck, 1980b, 1992). The variability of facies and palaeocurrents implies that the bar split the flow of the river and was subject to variations in water-level. The bar sequence thickens downstream and indicates a minimum water depth of about 3 m. The Knock Castle outcrop is therefore an overall fining-upward cycle interpreted as the growth of a mid-channel bar over an earlier mid-channel bedform (Bluck, 1980b).

At the outcrop WNW of Quarter, the structure and palaeocurrents of the lower pebbly sandstones indicate a linguoid bar within the river (Bluck, 1980b; Figure 3.44, Stage 1). The facies arrangements and geometry in the upper part of the section suggest that the linguoid bar later acted as a lateral bar at least 1 m high, separated from the bank by an abandoned channel in which muds accumulated (Bluck, 1980b; Figure 3.44, Stage 2). During flood events, the lateral bar acted as a 'chute bar' when the river flowed across it. The direction of bar migration, as indicated by the sandstone sheets, was normal to the general flow of the river (Figure 3.44). In addition to the bar facies, the lower part of the Quarter section, comprising finer-grained,

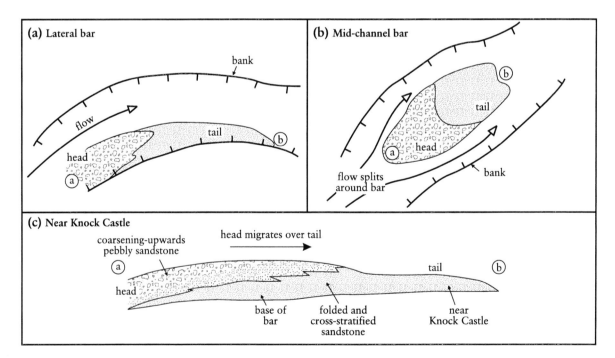

Figure 3.43 (a) Flow around lateral bar; and (b) mid-channel bar; (c) cross-section through a bar. After Bluck (1992).

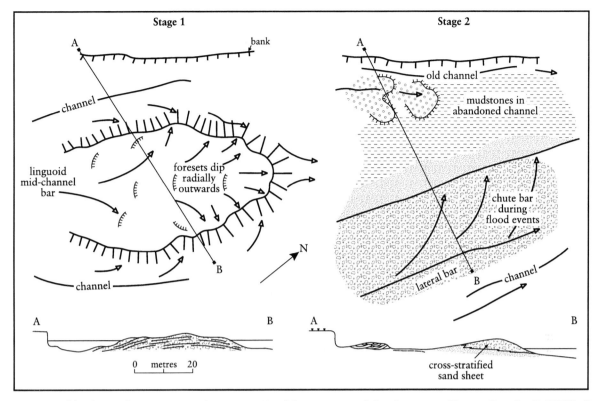

Figure 3.44 Plan and cross-sectional views (A–B) of the interpreted development of bars at Locality 2, WNW of Quarter. Stage 1 records the growth of a linguoid mid-channel bar, Stage 2 records the development into a lateral and chute bar. After Bluck (1980b).

planar sheet-like sandstones and thin mudstones, is interpreted as the deposits of a floodplain that was periodically subaerially exposed (Bluck, 1980b). The sandstones were ripple-topped, lobate sand sheets deposited by decelerating flows. Palaeocurrents suggest that the sheet flooding occurred on a broad floodplain, with flows in the same direction as the channel and no levees to the channel. Preservation of fine lamination and the lack of iron reduction suggest that that the floodplain had a meagre plant cover or was unvegetated (Bluck, 1980b).

The sedimentary rocks of the Largs section occur in fining-upward cycles with sharp, basal erosion surfaces, and are interpreted as the deposits of a braided river system (Bluck, 1967). The cyclicity is thought to result from the migration of the active river channel over the floodplain. The overall dip of cross-stratification implies flow to the north-east, but the accretion directions of bars and bedforms are commonly to the north-west. The Largs site is therefore interpreted to have been at the north-west margin of a 'cone' of sediment prograding to the ENE (Bluck, 1980b).

Conclusions

The well-exposed Largs Coast GCR site is a key locality for observing and understanding the typical three-dimensional structures and processes of sediment deposition in an ancient braided stream system. Detailed sedimentological study at the site has allowed the reconstruction of a braided river system subject to changing water-levels and channel switching. Of the two sections discussed in detail, one illustrates the change from a channel bedform to a migrating mid-channel bar, and the other shows the transition from a linguoid mid-channel bar to a lateral bar at the side of the river channel (Figures 3.43, 3.44). Floodplain deposits are identified between the barforms. The thickness of the bars suggests a minimum water depth of 1–3 m at this locality.

NORTH NEWTON SHORE, ISLE OF ARRAN (NR 933 518)

S.L.B. Arkley

Introduction

The North Newton Shore GCR site (Figure 3.45) in the north-west of Arran is a locality of international and historical importance, exposing what is widely known as 'Hutton's Unconformity'. The discovery of this site by James Hutton in the summer of 1787, and its recognition as an unconformity, resulted in a major advance in the concept of geological time and helped form the foundations of modern geology. This site is one of several in Scotland made famous by Hutton, which have subsequently been the focus of attention by many generations of geologists. Hutton later discovered similar unconformities in the River Jed, near Jedburgh (Figure 4.2, Chapter 4) and at Siccar Point on the Berwickshire coast (see Siccar Point to Hawk's Heugh GCR site report, this chapter). At all three, steeply dipping low- to medium-grade Caledonian metamorphic rocks are overlain by shallow dipping Late Devonian–Carboniferous sedimentary rocks. The North Newton Shore site displays an angular unconformity between steeply inclined metasedimentary rocks of the Precambrian to early Cambrian Dalradian Supergroup and younger sedimentary rocks of the latest Devonian–earliest Carboniferous Kinnesswood Formation of the Inverclyde Group. Subsequent studies have revised the precise position of the unconformity from that originally described by Hutton, and identified calcretization at the unconformable surface.

Description

North Newton Shore lies approximately 1.5 km north of the village of Lochranza, forming part of the prominent post-glacial emerged beach. The unconformity is exposed along the foreshore for

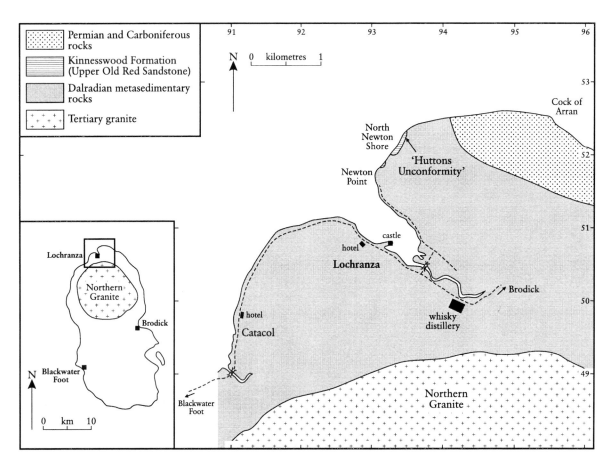

Figure 3.45 Geological map of the area of North Newton Shore and Hutton's Unconformity, Isle of Arran. Inset shows location of map.

North Newton Shore

a distance of 325 m, about 0.5 km north-east of Newton Point, with a series of scattered smaller outcrops just north of Newton Point itself (Figures 3.45, 3.46).

The first published description of the site appeared in Hutton's 'The Theory of the Earth' (Hutton, 1795). He described the unconformity thus: 'the schistus and the sandstone strata both rise inclined at an angle of about 45 degrees; but these primary and secondary strata were inclined in almost opposite directions; and thus they met together like the two sides of a lamda (λ), or the rigging of a house, being a little in disorder at the angle of their junction.'

The site was described and illustrated in more detail by Geikie (in Hutton, 1899) and Tyrrell (1928), the former based on Hutton's original unpublished manuscripts. The metamorphic rocks underlying the unconformity were described by Gunn (1903) as 'Highland Schists' and their micaceous nature was commented on by Jameson (1800). They are greenish grey, schistose metasedimentary rocks that are steeply inclined (40°–60°) to the south-east and contain common quartz veins. The lithology is dominantly a greywacke grit, with bands of finer-grained, more micaceous, chloritic slaty material. The metamorphic rocks are overlain by red and yellowish sandstones containing lenticular calcareous beds, breccias and distinctive beds of white, pedogenic concretionary carbonate (calcrete). The sedimentary rocks stretch for approximately 325 m along the shore and continue about halfway up the width of the emerged beach. At the south end of the exposure, they dip about 30° to the NNW and at the north end 25° to the WNW, the strike curving across the outcrop. The intervening plane of unconformity appears planar and sub-parallel to the overlying sedimentary rocks.

The metamorphic rocks beneath the unconformity form part of the North Sannox Grits of the Southern Highland Group, the uppermost stratigraphical unit of the Dalradian Supergroup. These rocks are tentatively assigned to the Lower Cambrian Series (British Geological Survey, 1987). The overlying sedimentary rocks have been attributed to the Kinnesswood Formation of the Inverclyde Group of latest Devonian–earliest Carboniferous age, based on the predominant upward-fining cycles of sandstones and red-brown, silty mudstones with nodules and thin beds of calcrete (Hall *et al.*, 1998; Browne *et al.*, 1999).

Interpretation

North Newton Shore is one of the most important localities in the development of modern geology. Hutton recognized that the juxtaposition of two

Figure 3.46 Hutton's Unconformity. Steeply dipping Dalradian metasedimentary rock in the foreground (lower half of image), overlain by more gently inclined thicker-bedded Upper Old Red Sandstone sedimentary rocks dipping towards the sea. (Photo: S.L.B. Arkley.)

lithological units displaying very different geological characteristics could not be explained by normal continuous rock-forming processes. This led him to propose that a gap in deposition had occurred, during which erosion took place, prior to deposition of the overlying rock mass. This interpretation and the concept of 'unconformability' eventually became widely accepted by the geological community, and unconformities are recognized as fundamental to the study of geology and geological time.

Several workers examined the site during the twentieth century, resulting in a re-assessment of the precise location of the plane of the unconformity. Anderson (1944, 1954) established that the lower 1.2 m of the calcareous sedimentary rocks were in fact 'calcitised' grits belonging to the underlying metamorphic rocks, thereby re-positioning the plane of unconformity at the base of the overlying 0.6 m-thick calcrete (cornstone) that underlies the calcareous, red, pebbly sandstones. This was subsequently confirmed and illustrated by Tomkeieff (1953) (Figure 3.47).

More recently, the importance of the 'calcitised' horizon has been recognized and it has been re-interpreted as having formed by carbonate pedogenesis during Late Devonian to Early Carboniferous times. Intensive fragmentation of the metamorphic rocks at and below the unconformity surface resulted from displacive and replacive carbonate formation, which is strongly developed along the schistosity planes in a bed up to 50 cm thick (Bucheit *et al.*, 1997; Donovan *et al.*, 1998). This zone of altered and calcretized metamorphic rock is parallel to the overlying unconformable surface. Interpreted by earlier workers as a 'sedimentary' cornstone deposit containing clasts of gritty material, it is a calcified regolith. Donovan *et al.* (1998) identified two higher calcretized horizons. The lower overlies the calcified regolith and is a 25 cm-thick calcretized schist pebble breccia containing angular clasts up to 8 cm in length, supported within a carbonate matrix dominated by dolomite spar. The uppermost pedogenic layer is a calcretized quartz pebble conglomerate containing abundant rounded clasts up to 3 cm in diameter, and a few schist clasts.

The geological features exposed at the site provide clear evidence for a time interval between the formation of the underlying deformed and metamorphosed Dalradian rocks and the deposition of the overlying latest Devonian–earliest Carboniferous sedimentary rocks. In addition, the unconformity exhibits features that indicate the geological environment and climate immediately prior to deposition of the overlying sediments. The interpretation of the carbonate-rich horizons as calcretized regolith and pedogenic horizons invokes a long period of subaerial weathering in a semi-arid environment with seasonal rainfall prior to the deposition of the mineralogically mature alluvial deposits of the Kinnesswood Formation.

Conclusions

The North Newton Shore GCR site is recognized internationally as an important site in the history of geology, discovered by James Hutton in the 18th century. The site provides a superb example of an angular unconformity where the age relationships between two distinct superimposed rock masses can be demonstrated

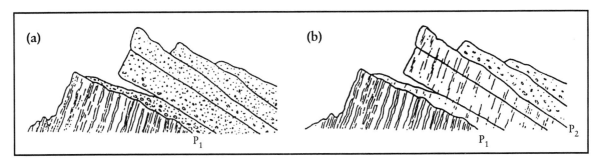

Figure 3.47 Sketches of Hutton's Unconformity at North Newton Shore from Tomkeieff (1953) showing his re-interpretation of the precise contact between the two rock formations. Height of illustrated exposure is about 1 m. (a) – position of plane of unconformity as placed by Hutton (P_1); (b) – new position as suggested by Tomkeieff (P_2).

clearly. The steeply inclined Caledonian metasedimentary rocks of the Dalradian Supergroup are directly overlain by much younger sedimentary rocks belonging to the latest Devonian–earliest Carboniferous Kinnesswood Formation. The more recent recognition of a calcrete horizon at the unconformity has resulted in the re-positioning of the unconformable surface and increased the understanding of the pedogenic processes that occurred at the surface during a prolonged period of non-deposition. 'Hutton's Unconformity' has long been used as a teaching site to demonstrate a typical angular unconformity, first recognized over two centuries ago. The site has provoked some controversy and has been the subject of evolving interpretation of the detail of the unconformable junction ever since.

Chapter 4

Southern Scotland and the Lake District

A.A. McMillan

Introduction

INTRODUCTION

Old Red Sandstone rocks crop out in the Scottish Borders and eastern Dumfriesshire (Mykura, 1991; Trewin and Thirlwall, 2002). Situated south of the Southern Upland and Lammermuir faults, these isolated outcrops lie in the Southern Uplands Terrane (Figure 4.1; Browne et al., 2002). Together with small outcrops of conglomerates fringing the Vale of Eden, these strata provide important evidence for the interpretation of the Old Red Sandstone palaeogeography of Britain. The oldest strata, which are of Late Silurian to Early Devonian age, comprise predominantly fluvial conglomerates and sandstones assigned to the Reston Group. Volcanic rocks in the Reston Group include basaltic and andesitic lavas and tuffs of the Eyemouth Volcanic Formation between St Abbs Head and Eyemouth in Berwickshire (Greig, 1988; Stephenson, 1999) and the Cheviot Volcanic Formation in Roxburghshire (Browne et al., 2002). It is likely, but unproven, that the strata of the Reston Group were deposited in basins separated from those of the Midland Valley of Scotland (Browne et al., 2002). One of the most extensive of the conglomerates, the Great Conglomerate Formation of the Reston Group (Davies et al., 1986), crops out in the Dunbar, Lauder and Haddington areas. It comprises boulder conglomerates with wacke sandstone clasts and interbedded thin sandstones. Although no diagnostic fossils have been recorded, the strata are considered to be of Early Devonian age on the basis of correlation with pre-volcanic sedimentary rocks at Bell Hill, St Abbs (Greig, 1988; cf. Rock and Rundle, 1986).

No rocks of Mid-Devonian age are known in the Southern Uplands Terrane. Following a period of uplift and denudation at the end of the Caledonian Orogeny, alluvial basins developed in the Scottish Borders during Late Devonian times. Their fill is largely assigned to the Stratheden Group (Paterson and Hall, 1986; Browne et al., 2002). The Stratheden Group rests with angular unconformity on Old Red Sandstone of Silurian to Early Devonian age or on Early Palaeozoic turbidites of the Southern Uplands. The unconformity at Jedburgh (the Jedburgh Unconformity; Figure 4.2) is a Regionally Important Geological and Geomorphological Site (RIGS).

The largest of the Late Devonian basins is the Scottish Border Basin (Leeder, 1973, 1974, 1976), which extends from Berwickshire south-westwards to Jedburgh and from there as a narrow, discontinuous belt to Kirkbean on the Solway coast. In the Eyemouth and Jedburgh districts, the Stratheden Group comprises a clastic red-bed succession of mainly fluvial pebbly sandstones, siltstones and some conglomerates. The group's thickest development (up to 200 m) is in the Langholm, Jedburgh and Cheviot districts. The conglomerates were mainly sourced from the Galloway highlands (Leeder, 1973, 1976). The sandstones provided a good source of local building material and many of the Borders abbeys and houses of the Tweeddale district are constructed of them (MacGregor and Eckford, 1946).

Nodules and beds of dolomite and chert are common in the Upper Old Red Sandstone and are particularly well-developed in Liddesdale (see **Palmers Hill Rail Cutting** GCR site report, this chapter). They are interpreted as calcretes and silcretes, indicative of contemporaneous pedogenesis. The calcrete-bearing strata are referred to the Kinnesswood Formation of the basal Carboniferous Inverclyde Group, as at Pease Bay (see **Siccar Point to Hawk's Heugh** GCR site report, Chapter 3) and at **Milton Ness** near Arbroath in the Midland Valley (see GCR site report, Chapter 3). Up to 30 m of these distinctive strata are present at Kirkbean, in Annandale and Liddesdale and at Burnmouth and Cockburnspath (Smith, 1967, 1968; Leeder, 1973, 1974; Paterson et al., 1976; Browne et al., 2002). At Kirkbean, and in Annandale and Liddesdale, the Kinnesswood Formation is overlain by weathered vesicular olivine basalt lavas of the Birrenswark Volcanic Formation (Stephenson et al., 2003). In Berwickshire, it passes up conformably into the Ballagan Formation (Cementstone Group of Smith, 1967, 1968).

In north-west England, possible equivalents of the Reston Group on the southern margin of the Northumberland and Solway basins (Figure 4.1) include the Polygenetic Conglomerate of the Penrith and Cross Fell areas and the Mell Fell Conglomerate Formation (see **Pooley Bridge** GCR site report, this chapter) on the north-eastern flanks of the Lake District mountains (House et al., 1977). The age of these conglomerates is uncertain, field relationships indicating only a post-Silurian–pre-marine Carboniferous age (Bluck et al., 1992).

There is little evidence for strata equivalent to either the Stratheden Group or the overlying

Southern Scotland and the Lake District

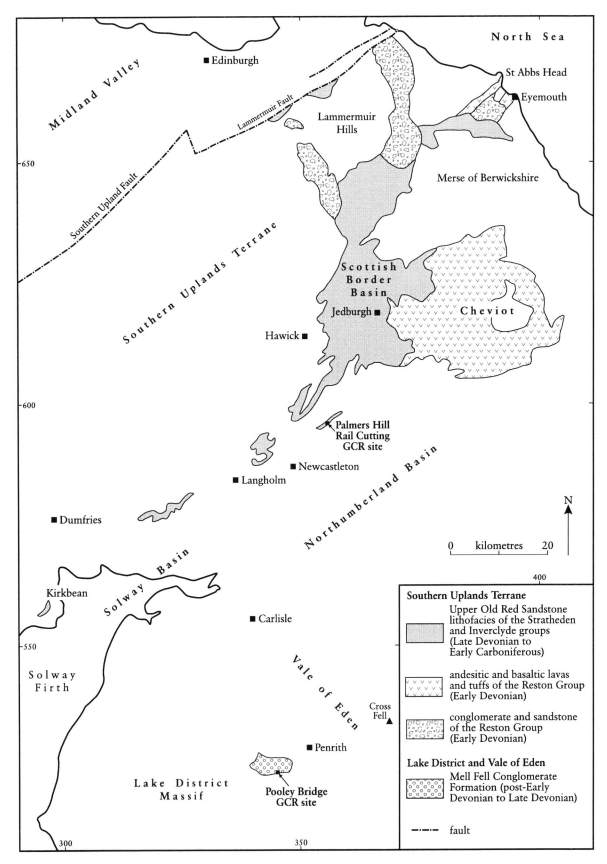

Figure 4.1 Distribution of Old Red Sandstone strata of the Southern Uplands and the Lake District.

Introduction

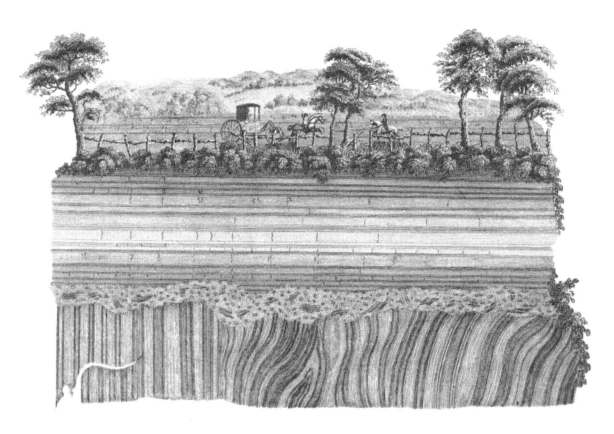

Figure 4.2 Engraving after a drawing of the unconformity at Jedburgh (NT 652 198) by John Clerk of Eldin (1787), used for Plate III of the *Theory of the Earth*, Volume 1, by James Hutton (1795). Vertical Silurian greywackes and shales are unconformably overlain by Upper Old Red Sandstone basal breccia and overlying sandstones. From Craig *et al.* (1978), reproduced by permission of Sir R.M. Clerk Bt.

Kinnesswood Formation on the margins of the Lake District Massif. Interestingly, however, Capewell (1955), in his regional study of pre-marine Carboniferous 'Basement Series' sedimentary rocks on the east side of the English Lake District, refers to characteristics in the alluvial 'Red Sandstones' of the Birk Beck valley that may be similar to those of the Ballagan Formation 'cementstone' facies of the Midland Valley of Scotland. The Birk Beck valley succession comprises Lower Conglomerates (possible correlatives of the Pinksey Gill Beds of Tournaisian age (Westoll, 1977), Red Sandstone and Upper Conglomerates, conformably overlain by Lower Carboniferous marine strata. Thus, there is local evidence in the Lake District for the development of Old Red Sandstone lithofacies of Late Devonian to Early Carboniferous age.

The transition from continental Old Red Sandstone to marine deposition may have taken place at slightly different times in the Early Carboniferous in the basins of Southern Scotland. The debate over the age of the Old Red Sandstone was fuelled by the discovery in the 19th century of fossil fishes (Jameson, 1805; Milne, 1843; Nicol, 1847; Powrie, 1870). The discovery of *Holoptychius noblissimus* Agassiz at Wauchope Burn, SSW of Jedburgh (Milne, 1843) and fish fragments believed to have come from Tudhope Quarry, Jedburgh (Nicol, 1847) resulted in the beds being assigned to the 'old red sandstone'. Powrie (1870) assigned strata with similar fish fragments at Denholm Hill, Hawick to a 'Passage Group' between the Upper Old Red Sandstone and the Carboniferous strata. Strata bearing *Holoptychius noblissimus* Agassiz at Dinley Burn, Langholm were referred to the Upper Old Red Sandstone by Peach and Horne (1903). Lumsden *et al.* (1967) noted that neither the presence of cornstone (calcrete) or *Holoptychius* was diagnostic of precise age, and that the strata now assigned to the Kinnesswood Formation could be Late Devonian or Early Carboniferous in age.

PALMERS HILL RAIL CUTTING, SCOTTISH BORDERS (NT 549 965)

Introduction

Scattered outcrops of Upper Old Red Sandstone extend from Kirkbean, south-west of Dumfries (Leeder, 1976; Leeder and Bridges, 1978) north-eastwards through Annandale to Liddesdale (Lumsden *et al.*, 1967; Leeder, 1973, 1976). In Liddesdale, the strata range in thickness from less than 15 m to over 180 m in a narrow and sinuous outcrop that extends for 20 km north-eastwards from Langholm to Robert's Linn and around the outlier of Carboniferous rocks on Arkelton Hill. To the north-east, the outcrop of Upper Old Red Sandstone fluvial sandstones broadens and extends from Jedburgh to the Merse of Berwick-shire and to Cockburnspath (see **Siccar Point to Hawk's Heugh** GCR site report, Chapter 3).

The Palmers Hill Rail Cutting in Liddesdale (Figure 4.3) exposes about 23 m of Upper Old Red Sandstone fluvial sandstones. The uppermost 7–8 m of strata contain pedogenic carbonate (calcrete) nodules and two regionally significant, mature calcrete horizons (Lumsden *et al.*, 1967; Leeder, 1976). The calcrete-bearing strata are referred to the Kinnesswood Formation of the Lower Carboniferous Inverclyde Group and the underlying strata to the Upper Devonian Stratheden Group (Paterson and Hall, 1986; Browne *et al.*, 2002). In the Palmers Hill Rail Cutting, the strata rest with angular unconformity on wacke sandstones of the Lower Palaeozoic (Wenlock) Riccarton Group (Lumsden *et al.*, 1967). They are overlain by weathered vesicular olivine basalt lavas of the Tournaisian Birrenswark Volcanic Formation.

Description

The Palmers Hill Rail Cutting lies on the north-eastern flank of Arnton Fell in Liddesdale and is one of several sections including those in the Dinley Burn, Laidlehope Burn, Riccarton Burn, Dawston Burn, Caddroun Burn and Robert's Linn (Figure 4.4) in which calcrete is recorded in the upper part of the Upper Old Red Sandstone succession. About 120 m of thinly bedded, massive and cross-bedded, brownish red, medium- to coarse-grained sandstones are exposed in Riccarton Burn, south-west of Palmers Hill Rail Cutting.

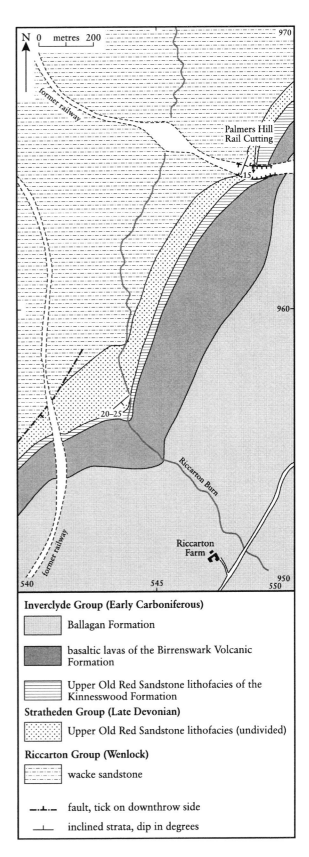

Figure 4.3 Geological map of the area around Palmers Hill Rail Cutting.

Palmers Hill Rail Cutting

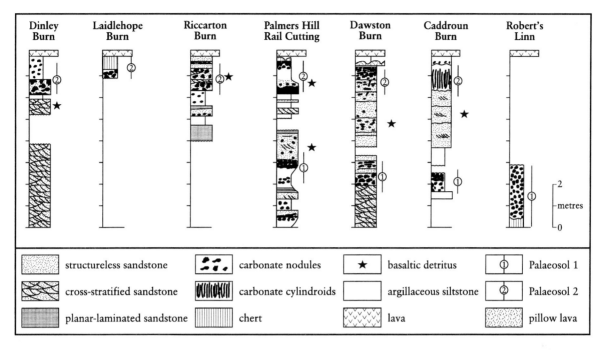

Figure 4.4 Graphic logs of the Kinnesswood Formation of Liddesdale showing the distribution of calcrete. After Leeder (1976).

The cutting exposes about 23 m of strata, which dip about 15° ESE. The lowermost 15 m are here assigned to the Stratheden Group. They comprise coarse-grained, massive, reddish brown sandstones containing well-rounded quartz pebbles and rounded clasts of deep reddish purple siltstone and silty mudstone. The lowest 1.8 m of sandstone have yielded unidentified fish scales (Lumsden *et al.*, 1967). The overlying Kinnesswood Formation is 7–8 m thick, contains nodular and massive calcrete (Leeder, 1976; the 'sandy cornstones' of Lumsden *et al.*, 1967) and is overlain by weathered lavas of the Birrenswark Volcanic Formation (Figure 4.4; Leeder, 1976). The strata comprise cross-bedded and planar-laminated sandstones interbedded with silty mudstone. The lower 4 m contain numerous carbonate nodules and a dolomitic calcrete horizon (Palaeosol 1 of Leeder, 1976) developed in a sub-arkosic sandstone. The sandstones between this horizon and a higher dolomitic calcrete (Palaeosol 2 of Leeder) are lithic-rich and contain chloritized basaltic detritus. In thin section, the calcrete is invariably seen to be a ferroan dolomite sparite exhibiting replacive and displacive fabrics of carbonate after quartz sand and silt (cf. Steel, 1974; Balin, 2000). Basaltic detritus is present above the higher calcrete, in the uppermost metre of the section, and the top few centimetres immediately below the basaltic lavas contain stringers and nodules of chert, interpreted by Leeder (1976) as silcrete.

Interpretation

The red beds of the Upper Old Red Sandstone have been variably assigned a Late Devonian and Early Carboniferous age. In the Langholm district, the presence of *Holoptychius noblissimus* Agassiz in red beds with 'cornstones' led earlier workers (e.g. Peach and Horne, 1903) to assign the strata to the Late Devonian (Lumsden *et al.*, 1967), but these criteria are not diagnostic of age. Indirect evidence of a minimum age of the strata is provided by radiometric age determinations on the overlying basalts. An olivine basalt flow of the Birrenswark Volcanic Formation at Watch Hill near Langholm and an aphyric basalt from a plug at Mellerstain Hill near Kelso have been dated by the K–Ar whole-rock method as 361 ± 12 Ma and 361 ± 7 Ma respectively, close to the Devonian–Tournaisian boundary (De Souza, 1982).

The calcrete horizons and volcanic detritus in the sandstones at these other localities provide the basis for the palaeogeographical reconstruction of a shallow, alluvial, inland basin, the Scottish Border Basin. During Early Carboniferous times, this became the northern margin of the larger successor basins – the Northumberland and Solway basins. The Upper Old Red Sandstone of the Scottish Border Basin is interpreted as a fluvial succession of sandstones and pebbly sandstones deposited in braided streams (Leeder, 1973; 1976). Palaeocurrent data are consistent with dominantly south-westerly derivation and compositional characteristics indicate an igneous and sedimentary sourceland in Galloway. However, in the eastern part of the basin, derivation from the south-east and north-east has been postulated by Smith (1967) and Paterson *et al.* (1976). A semi-arid continental climate and periodic emergence is indicated by the presence of wind-rounded sand grains, pedogenic calcrete and silcrete, and numerous desiccation-cracked surfaces.

The presence of calcretes in the Kinnesswood Formation of the Scottish Border Basin, similar to those in south Ayrshire (Burgess, 1961), indicates periodic uplift, emergence and dissection of the alluvial plains. The uplift probably took place in response to partial melting in the upper mantle, which eventually resulted in basaltic volcanism (Leeder, 1974, 1976). The first regionally recognizable calcrete (Palaeosol 1) was partially buried by lava flows during subsequent local volcanic activity, and volcanic detritus was reworked into the overlying fluvial sandstones. Further uplift resulted in the development of a second calcrete (Palaeosol 2), to be followed by the more extensive fissure eruption of basalt lavas of the Birrenswark Volcanic Formation, the outcrop of which extends discontinuously from the Kirkbean district to Kelso.

In thin section, the replacive and displacive fabrics of the dolomitic carbonate are typical of many modern and ancient pedogenic carbonates, and an analogy may be made with modern accumulations of carbonate in soil profiles of semi-arid regions (e.g. Steel, 1974; Balin, 2000). Peach and Horne (1903) noted that the cornstone (calcrete) is commonly accompanied by a lenticular red chert bed in the Riccarton area. The chert, which replaced dolomite, is interpreted as a pedogenic silicification phase (Leeder, 1976). Detailed petrological research on the origin of the dolomite at this site has yet to be undertaken.

Conclusions

The Palmers Hill Rail Cutting GCR site exposes strata that were river sediments which filled a regionally extensive, shallow inland basin, the Scottish Border Basin. The section is important because it exposes excellent examples of two fossil carbonate soil (calcrete) horizons, and provides evidence for the onset of volcanic activity. The soils developed in response to the changing tectonic setting and cessation of sedimentation of the basin. Their association with volcanic detritus provides compelling evidence for uplift of the land prior to and during the eruption of basaltic lavas of the Birrenswark Volcanic Formation.

The site is also important in providing evidence to allow reconstruction of the palaeogeography, sedimentary environments and climate of the Scottish Borders in Late Devonian to Early Carboniferous times about 362 million years ago.

POOLEY BRIDGE, CUMBRIA (NY 465 243)

Introduction

The Pooley Bridge GCR site comprises a shore section and road cutting at the northern end of Ullswater (Figure 4.5). It exposes the conglomerates of the lower part of the Mell Fell Conglomerate Formation, named after Great Mell Fell on the eastern side of the Lake District (Dakyns *et al.*, 1897; Green, 1918). The conglomerates were mentioned by early investigators, including Playfair, Otley, Sedgwick and Nicolson, all of whom referred them to the Old Red Sandstone. They form part of a 48 km-long outcrop of clastic strata on the eastern side of the Lake District from near Penrith to Ravonstonedale (Ward, 1876; Harker and Marr, 1891; Dakyns *et al.*, 1897; Oldham, 1900; Green, 1918). Originally referred to as 'Basement Beds' by Dakyns *et al.* (1897), the strata separate the Carboniferous Limestone from the underlying Lower Palaeozoic rocks. Recent studies of the Mell Fell Conglomerate Formation include those by Capewell (1954, 1955), Wadge (1978) and Kimber and Johnson (1984). The outlier has been re-mapped recently by the British Geological Survey as part of the

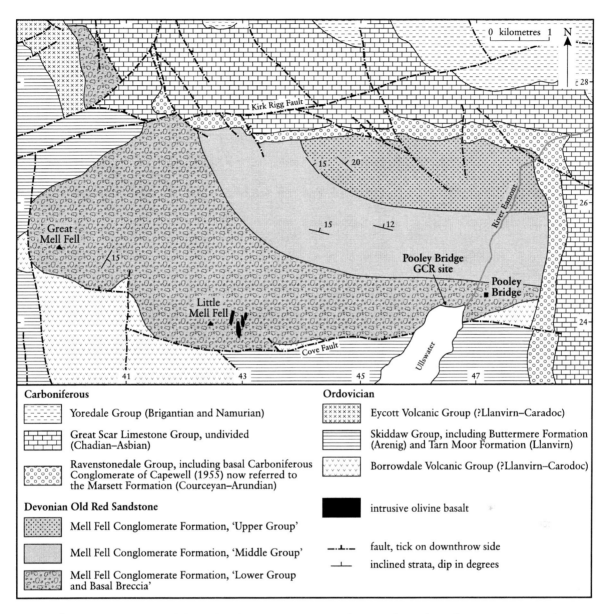

Figure 4.5 Geological map of the Mell Fells–Ullswater area. After British Geological Survey 1:50 000 Sheet 30 (England and Wales), Appleby (2003).

revision of the Appleby 1:50 000 geological sheet 30 (England and Wales) (British Geological Survey, 2003; McCormac, 2001; Millward *et al.*, 2003). Hillier and Williams (in press) provide a summary and Soper and Woodcock (2003) discuss the regional tectonic setting of the Mell Fell Trough.

The age of the Mell Fell Conglomerate Formation is uncertain, but it is thought to be Late Devonian to Early Carboniferous in age. Soper and Woodcock (2003) assign it a post-Acadian (Givetian–Frasnian) age on tectonic considerations. The Pooley Bridge site is significant in interpreting the regional geology, being the best exposed example of Old Red Sandstone lithofacies in north-west England. The conglomerates accumulated in a fault-bounded trough, the Mell Fell Trough, which covers an area of about 20 km² north of Ullswater. The strata rest unconformably on rocks of the Ordovician Borrowdale Volcanic Group. They are overlain unconformably by conglomerates of the Marsett Formation, the basal part of the Ravenstonedale Group of Early

Carboniferous age (British Geological Survey, 2003). The Mell Fell Conglomerate is correlated with identical cobble conglomerates in Heltondale Beck (NY 506 207), the Shap Wells Conglomerate Formation of the Shap Wells Trough (McCormac, 2001) and the 'Polygenetic Conglomerate' (Marr, 1899; Shotton, 1935). The last is confined to three outcrops in the Cross Fell Inlier (Burgess and Wadge, 1974) and one near Greystoke, west of Penrith (Arthurton and Wadge, 1981). Estimates of the thickness of the Mell Fell Conglomerate range from about 275 m (Dakyns et al., 1897) to under 1000 m (McCormac, 2001) and 1500 m (Capewell, 1955).

Description

The Pooley Bridge GCR site consists of part of the foreshore of Ullswater and a cutting on the north side of the B5320 to about 500 m west of Pooley Bridge. The road cutting was overgrown at the time of writing, but a nearly continuous section occurs on the lake shore. The exposed beds belong to the Mell Fell Conglomerate Formation. On the basis of its field relationships with surrounding rocks, it is of post-Early Devonian age, and older than Early Carboniferous marine deposits. The Shap Wells Conglomerate and the Mell Fell Conglomerate formations are the only representatives in the Lake District of the Old Red Sandstone lithofacies, deposited in a continental desert setting subject to seasonal, tropical rain storms.

The Mell Fell Conglomerate Formation outlier was described in detail by Capewell (1955), who divided it into three units: (i) 'Lower Group and Basal Breccia' dominated by wacke sandstone clasts, (ii) 'Middle Group' characterized in the eastern part of the outcrop by minor acid intrusive and intermediate volcanic rocks and in the west by volcanic rocks and wacke sandstones and (iii) 'Upper Group' comprising volcanic rocks and wacke sandstones. The Pooley Bridge site provides an excellent section of the 'Lower Group'. Here, the strata consist of a very poorly sorted, dark red-brown, boulder- to cobble conglomerate (Figure 4.6) with irregular finer beds (Capewell, 1955). Rudimentary sub-horizontal to gently north-dipping bedding is seen in alignment of subangular to subrounded clasts. These range from about 0.1 m to 0.3 m in diameter, with some reaching 1.0 m. Oldham (1900) commented on the poor sorting and absence of well-rounded pebbles. The clasts consist of about 95% wacke sandstone (Harker and Marr, 1891; Capewell, 1955) and sandstone. The presence of the Lower Ludlovian graptolite *Monograptus colonus* in the clasts shows that they were derived from the Windermere Supergroup. Minor constituents of the conglomerate are lavas and pyroclastic rocks of Borrowdale Volcanic Group affinity, fine-grained, apparently unfossiliferous limestone and vein-quartz. A calcite-hematite vein cuts a wacke cobble. The matrix of smaller pebbles and coarse sand is purple-red with a calcite or iron oxide cement. There is evidence that the conglomerates infill channels, and thin interbeds of fine-gravel are cross-bedded.

Interpretation

The sedimentary characteristics of the Mell Fell Conglomerate Formation suggest that it is of fluvial origin. The deposits were laid down in generally arid conditions, but with occasional violent storms generating very high levels of run-off. Capewell (1955) envisaged the formation as a bajada deposit (cf. Walker, 1967), laid down in coalescing, low-angle fans. These may have accumulated at the base of a fault scarp. Wadge (1978) noted that the dips in the formation converge to the north and east, supporting its interpretation as the product of several coalescing debris fans. Kimber and Johnson (1984) commented that the alternations of coarse, clast-supported conglomerates with thin, well-sorted pebbly sandstones in the west and north of the Mell Fell Trough was consistent with fluvial deposition in braided channels in a mid-fan environment. A north-eastwards change, from proximal conglomerates to distal, finer-pebble conglomerates and pebbly sandstones, and palaeocurrent directions indicate north-east-flowing drainage.

The clasts in Capewell's (1955) 'Lower Group' of the Mell Fell Conglomerate Formation at Pooley Bridge consist largely of sedimentary rocks of the Windermere Supergroup. They are predominantly siltstones, mudstones, and wacke sandstones of the Bannisdale Formation and Coniston Group. The nearest present outcrops of these rocks lie about 22 km to the south

Figure 4.6 Conglomerate of the Mell Fell Conglomerate Formation on the shore of Ullswater near Pooley Bridge. (Photo: D. Stephenson.)

around Lake Windermere. The presence of a high percentage (95%) of texturally immature wacke sandstone clasts, coupled with the paucity of locally exposed rocks such as those of the Borrowdale Volcanic Group, has led to much debate on the source of the conglomerate (see discussion in Capewell, 1955). The sedimentological characteristics of the conglomerate indicate local provenance and deposition from rapid fluvial run-off from high hinterland, but the lithologies point to presently distant sources. Wadge (1978) considered that the lack of abrasion of the clasts suggested a source of Silurian and late Ordovician rocks much closer to Ullswater. McCormac (2001) noted that, although the conglomerate is texturally immature, the degree of rounding of the large clasts is unusual for such proximal deposits, and they may have been reworked from a desert regolith.

The age of the Mell Fell Conglomerate Formation has also been debated. The conglomerate contains cleaved Silurian (Ludlow) pebbles, and thus its deposition is considered to be post-Early Devonian. The likely presence of an unconformity between the Mell Fell Conglomerate and a younger group of conglomerates (the Basement Beds of the Penrith district (Arthurton and Wadge (1981) and the Marsett Formation (formerly Basal Beds) of the Ravonstonedale Group (McCormac, 2001; British Geological Survey, 2003) points to a period of uplift and erosion (Capewell, 1955). Thus Capewell (1955) favoured a pre-Late Devonian age for the Mell Fell Conglomerate, citing the evidence elsewhere in Britain for a conformable but diachronous boundary between the uppermost Old Red Sandstone and lowest Carboniferous beds. The Marsett Formation of Early Carboniferous age crops out to the north and east of Pooley Bridge. Although contact relationships are largely concealed, it is inferred that the junction with the Mell Fell Conglomerate is an unconformity, from a comparison of the regional dips of the two formations. The Marsett Formation is succeeded by strata which have yielded faunas of Chadian age (Dean, 2001). Thus the deposition of the Mell Fell Conglomerate Formation is constrained between the Early Devonian (post-cleavage) and the Chadian. Evidence further constraining its age to the Courceyan or older is provided by the intrusive basalts of Little Mell Fell (Figure 4.5) (Capewell, 1955; Millward, 2003). These rocks, intruded into the Mell Fell Conglomerate Formation, are chemically similar to the Cockermouth Lavas (MacDonald and Walker, 1985) and are considered to be the easternmost manifestation of this volcanism (Millward, 2003). The age of the Cockermouth Lavas is tightly constrained to the Courceyan, based on the presence of CM Zone spore assemblages in overlying sedimentary rocks in Gill Beck GCR site, 4.5 km north-east of Cockermouth.

Conclusions

The Pooley Bridge GCR site exposes the Mell Fell Conglomerate Formation and provides the best section of strata of Old Red Sandstone lithofacies in north-west England. It is correlated with the Shap Wells Conglomerate Formation and the Polygenetic Conglomerate of the Penrith district and Cross Fell Inlier and provides important palaeogeographical and sedimentological evidence for the uplift and erosion of the Lake District Massif. The strata comprise poorly bedded, boulder- to cobble-grade conglomerates with some lenticular sandstone beds. The precise timing of deposition of the Mell Fell Conglomerate is uncertain but a post-Early Devonian to Late Devonian age is most likely.

Chapter 5

The Anglo-Welsh Basin

Introduction

INTRODUCTION

W.J. Barclay

The Anglo-Welsh Basin formed on the southern margin of the newly amalgamated Laurussian (Old Red Sandstone) continent and the northern margins of the Rheic Ocean (Figure 1.3, Chapter 1). The basin lay in a distal ('external') setting relative to the main Caledonian Orogen in Late Silurian and Early Devonian times (Allen, 1979). Anglesey may have been the site of a small isolated internal basin within the orogen at the start of Old Red Sandstone deposition, but similar facies in the higher part of the succession point to contiguity with the main basin (Allen, 1965a). The basin lay in sub-tropical latitudes, palaeomagnetic data suggesting a latitude of $17 \pm 5°S$ for the Lower Old Red Sandstone (Channel *et al.*, 1992). The abundance of calcrete palaeosols in the basin-fill points to a warm, semi-arid climate with seasonal rainfall.

Local variations on the regional subsidence pattern were exerted by synsedimentary extensional and transtensional faults. The former were particularly active during deposition of the Late Silurian–Early Devonian Milford Haven Group in south-west Pembrokeshire. The latter may have caused the intermittent emergence of a landmass in the Bristol Channel area in Early and Mid-Devonian times. Previous models of the basin's evolution invoked post-Caledonian flexural subsidence (e.g. King, 1994; Friend *et al.*, 2000; Woodcock, 2000a), but orogen-wide, sinistral pre-Acadian transtension and tectonic subsidence have been proposed recently (Dewey and Strachan, 2003; Soper and Woodcock, 2003).

The Old Red Sandstone is present in a narrow outcrop on Anglesey and in small outliers in Clun Forest and Long Mountain in central Wales. In south and south-central Wales, it is widely distributed in the Carmarthen Fans, Black Mountains and Brecon Beacons, around the northern and eastern rims of the South Wales Coalfield, and in the Welsh Borderland. It also crops out on the limbs of several Variscan folds in south-west Pembrokeshire, where it is magnificently exposed in sea cliffs. The main basin extended south-eastwards into south-east England where the Old Red Sandstone is concealed by younger strata. It may also have extended northwards to the southern Lake District, where a cover of at least 3.5 km of Old Red Sandstone sediments is estimated to have been removed during Acadian inversion (Soper and Woodcock, 2003). The outliers in Anglesey, Long Mountain and Clun Forest are remnants of this former cover.

Old Red Sandstone red-bed sedimentation generally began in Late Silurian Přídolí times, when the Lower Palaeozoic marine Welsh Basin was finally filled, then inverted during the Acadian Orogeny. However, there are earlier occurrences of Old Red Sandstone facies at Marloes Sands, south-west Pembrokeshire, where it is of late Wenlock to early Ludlow age, and in the Llandovery area, where the Trichrûg Formation is of mid-Ludlow (late Gorstian) age. The Old Red Sandstone extends throughout the Devonian and into the Early Carboniferous, with the Mid-Devonian largely unrepresented in the rock record, except perhaps for the southerly derived Ridgeway Conglomerate Formation of south-west Pembrokeshire and the northerly derived Hangman Sandstone Formation of north Devon. For the most part, the southerly advancing Acadian deformation front resulted in uplift and erosion of the Anglo-Welsh Basin in Mid-Devonian times, now represented by a regional unconformity at the base of the Late Devonian Upper Old Red Sandstone.

Přídolí and Lochkovian–Pragian strata are widespread in south-central Wales and the Welsh Borderland, higher (Emsian and Famennian) strata occur mainly as narrow outcrops around the South Wales and Forest of Dean coalfields and as outliers, as for example in the Clee Hills of Shropshire. Good summaries of the Old Red Sandstone succession of the Anglo-Welsh Basin are given by Allen (1974a, 1977) and Hillier and Williams (in press). Bluck *et al.* (1994), Woodcock (2000a) and Soper and Woodcock (2003) presented recent regional overviews. A maximum of about 4.3 km of strata is present north of the Ritec Fault in Pembrokeshire, with up to about 2 km elsewhere. Two major Old Red Sandstone megasequences are recognized, traditionally referred to the Lower Old Red Sandstone and the Upper Old Red Sandstone. The Lower Old Red Sandstone succession is mainly of Late Silurian (Ludlow–Přídolí) to Early Devonian (Emsian) age, the Upper Old Red Sandstone of Late Devonian (late Frasnian–Famennian) to Early Carboniferous age.

The Anglo-Welsh Basin

Figure 5.1 shows a correlation of the Old Red Sandstone successions and their lithostratigraphical subdivisions. Figure 5.2 shows the location of the GCR sites described in this volume. The basal formation of the Old Red Sandstone (Downton Castle Sandstone/ Tilestones/Long Quarry Sandstone) is of early Přídolí Series, Silurian age, and the GCR sites in these formations are described by Lane (2000a) in the companion volume on Silurian stratigraphy (Aldridge *et al.*, 2000). A large part of the Raglan Mudstone Formation and its equivalents are also of Přídolí age. Some of the GCR sites in this formation are also described by Lane (2000a) (Table 1.4, Chapter 1) and only brief descriptions are given in this volume.

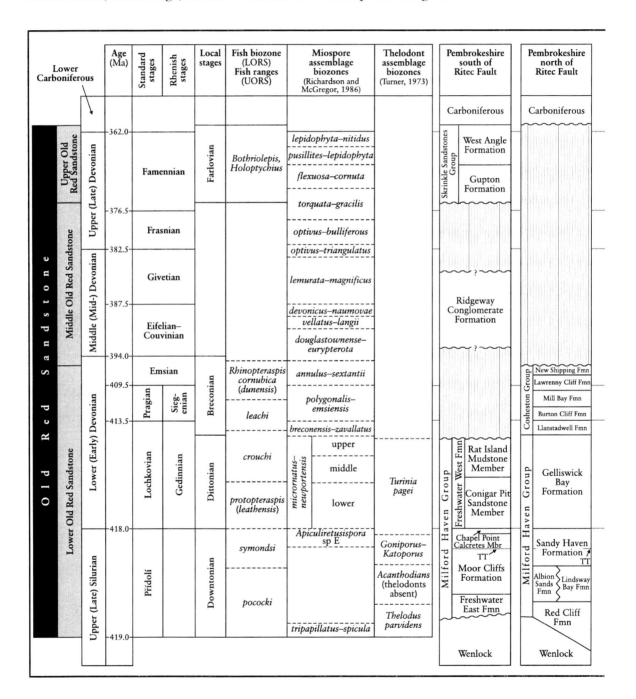

Figure 5.1 Correlation of the successions and biostratigraphical classifications of the Old Red Sandstone in the Anglo-Welsh Basin. Ages in millions of years ago (Ma) are from Williams *et al.* (2000). Broken lines denote imprecisely located boundaries.

Introduction

The transition from marine, nearshore shelf deposition to continental red-bed Old Red Sandstone deposition was generally regarded as taking place in latest Silurian–Early Devonian time. However, the age of the lowermost Old Red Sandstone beds in the Marloes Peninsula of south-west Pembrokeshire has generated controversy. Sanzen-Baker (1972) suggested that they were Ludlow in age. Allen *et al.* (1976) and Allen and Williams (1978) proposed that an unconformity with no angular discordance separated the underlying Ludlow marine strata from red beds of Přídolí age. Hillier (2000) recorded a transition from Wenlock shallow marine strata (the Gray Sandstone Group) into the Old Red Sandstone Red Cliff Formation, and

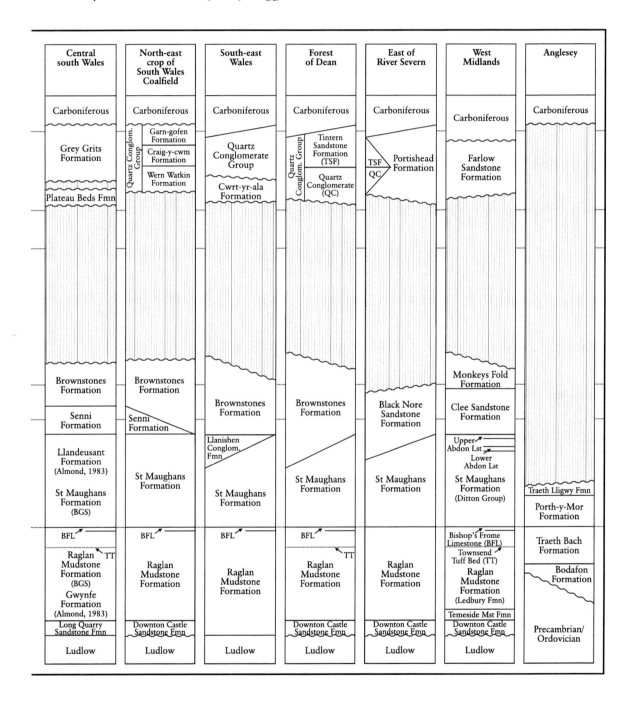

The Anglo-Welsh Basin

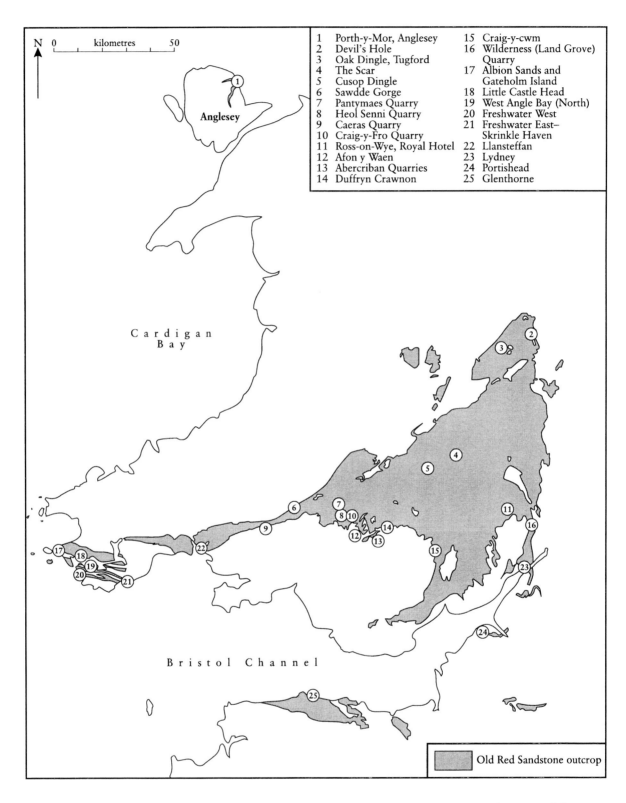

Figure 5.2 Map of the Old Red Sandstone outcrops in the Anglo-Welsh Basin showing location of the GCR sites described in this chapter. After British Geological Survey 1:625 000 Solid Geology Map UK South Sheet, 4th edn (2001).

Introduction

the recent discovery of Ludfordian spores from the basal beds of the Red Cliff Formation (Hillier and Williams, 2004) confirms a transitional boundary.

The Přídolí–Lochkovian (Silurian–Devonian) boundary remains imprecisely located (see 'Introduction', Chapter 1), but is tentatively placed on the basis of spore assemblages within the uppermost part of the Downton Group (in the Raglan Mudstone Formation and the Moor Cliffs and Sandy Haven formations, its equivalents in Pembrokeshire), at about the level of the Psammosteus Limestone (Bishop's Frome Limestone/Chapel Point Calcretes Member). This horizon marks a major facies change in the Old Red Sandstone, when marginal marine, coastal mudflats and marine-influenced alluvial floodplains were replaced by medial to proximal alluvial environments. The mudstone/siltstone-dominated sequences of the Přídolí Series are succeeded by one in which sandstones occupy about 30% of the succession, this proportion increasing towards the top of the Lochkovian Stage. Also, the composition of the sandstones changes, with a decrease in mica and a change in heavy-mineral assemblages. A more distant northern metamorphic terrane during the Přídolí was replaced by a more proximal source of earlier Palaeozoic sedimentary and igneous rocks of what is now north Wales. Although there is evidence of local marine influence in the Welsh Borderland, the Lochkovian succession is otherwise entirely terrestrial, representing alluvial riverine and floodplain deposition, the transition from Silurian marine to Devonian continental environments being essentially completed.

The Lower Old Red Sandstone is subdivided into three loosely defined local chronostratigraphical stages, locally applied to this area because of the general lack of faunas and floras directly correlatable with the standard European marine stages (Figure 5.1). The stages are the Downtonian (now incorporated into the Silurian Přídolí Series; the name is retained in a lithostratigraphical sense as the Downton Group), the Dittonian (converted to lithostratigraphical usage by some authors as the Ditton Group (e.g. Dineley, 1999f), and of Lochkovian to Pragian age); and the Breconian (Croft, 1953) of late Pragian to Emsian age. The Upper Old Red Sandstone is ascribed to the Farlovian local stage and is of Frasnian–Famennian to early Tournaisian (Carboniferous) age.

The Lower Old Red Sandstone broadly comprises an upward-coarsening offlap succession, consisting, in upward succession, of marginal marine, coastal floodplain, alluvial channel and floodplain and alluvial-fan environments of increasing energy and proximity to the advancing Acadian front (Figure 5.3). Continuing southwards progradation of the facies belts and migration of the fluvial fall-line at the Acadian deformation front resulted in erosion and non-deposition throughout Mid-Devonian time in most of the basin, with the alluvial systems depositing sediment in north Devon. Interrupting this overall southward progradation is the regionally developed Psammosteus Limestone, representing a period of widespread (over 20 000 km^2) and prolonged carbonate soil (calcrete) formation. It represents basin shut-down and sediment starvation, and resulted in a change in basin-fill architecture and fish faunas. A regional airfall tuff, the Townsend Tuff Bed, lies about 100 m below the Psammosteus Limestone (Allen and Williams, 1981a) and provides a further, synchronous correlative tool (Figure 5.1).

A series of extensional faults in south-west Pembrokeshire defined structural blocks and basins. The Benton and Ritec faults acted as intermittent sources of basin-margin alluvial-fans and sites of condensed deposition, with deposition of thick successions in their hanging-walls. The Lower Devonian Cosheston Group is confined to the Winsle area (Thomas, 1978; Marshall, 2000a,b), which is bounded to the north by the Benton Fault and to the south by the Ritec Fault. The ?Middle Devonian Ridgeway Conglomerate Formation is confined to the Tenby–Angle Basin to the south of the Ritec Fault. Correlation of the successions is achieved by the presence of the Townsend Tuff Bed and Psammosteus Limestone (Chapel Point Calcretes Member), but the latter is absent or poorly developed to the north of the Ritec Fault and west of the Llandyfaelog–Pontesford Lineament. Exotic clasts in the Brownstones Formation of the upper Swansea Valley may be related to transpressive movement on the Swansea Valley Fault (Tunbridge, 1980a). The pebbly Caeras

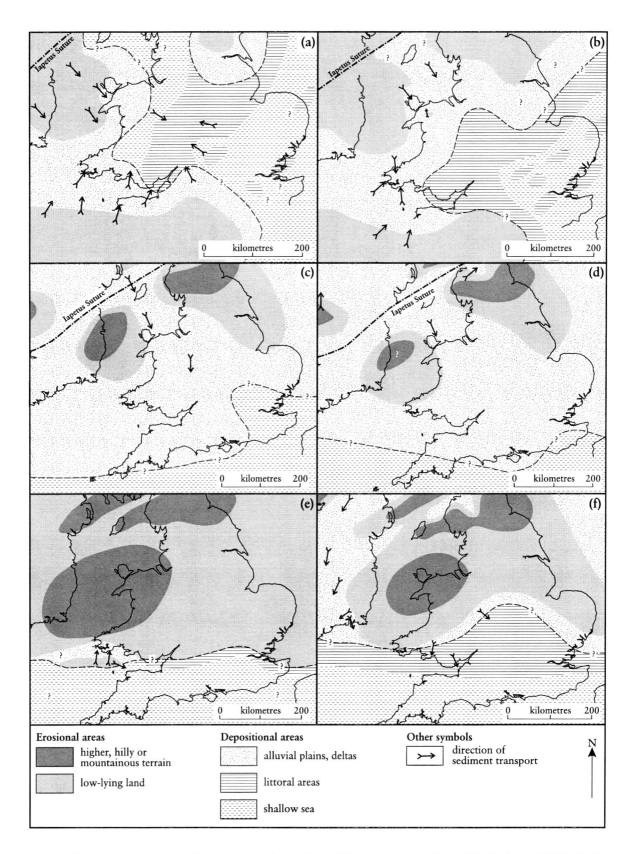

Figure 5.3 Palaeogeographical evolution of the Anglo-Welsh Basin. (a) Earliest Přídolí; (b) mid-Přídolí; (c) Lochkovian; (d) late Pragian–early Emsian; (e) Givetian; (f) Frasnian–early Famennian. (a) and (b) after Bassett *et al.* (1992); (c)–(f) after Bluck *et al.* (1992).

Introduction

Beds of the Cennen Valley (**Caeras Quarry** GCR site) may have a similar origin in relation to the Carreg Cennen Disturbance.

Most of the early deposits of the Lower Old Red Sandstone succession were derived from the Caledonian uplands north of the Highland Boundary Fault (e.g. Allen and Crowley, 1983). However, the exact nature of the dispersal systems remains controversial. Orogen-parallel drainage from the region of Scandian uplift to the north-east of the UK, through the Midland Valley of Scotland and southwards to the Anglo-Welsh Basin has been proposed (e.g. Simon and Bluck, 1982; Bluck et al., 1992; Sherlock et al., 2002), but Scandian drainage may have terminated in the Midland Valley or Northern Ireland (Friend et al., 2000), with much of the later (post-Psammosteus Limestone) Lower Old Red Sandstone fill of the Anglo-Welsh Basin derived from closer Lower Palaeozoic sources in southern Scotland, southern Ireland and north Wales (Allen, 1974b; Allen and Crowley, 1983; Haughton and Farrow, 1989; Soper and Woodcock, 2003).

The Přídolí Tilestones Formation of south-central Wales (Almond, 1983) was derived from the south. Palaeocurrents in the Ludlow–Přídolí Red Cliff and Lindsway Bay formations of Pembrokeshire also indicate derivation from the south (Hillier, 2000; Hillier and Williams, 2004), although white micas in the former are said to have a Laurentian (Highland) origin (Sherlock et al., 2002). Parts of the Lindsway Bay and the coeval Albion Sands formations were derived from the west, and probably deposited in a fault-controlled axial valley. The southerly source of sediment is thought to have been the remnants of the Early Palaeozoic landmass of Pretannia (Cope and Bassett, 1987; Bluck et al., 1992), which formed the south-east shoulder of the Welsh Basin.

The ?Mid-Devonian Ridgeway Conglomerate Formation of Pembrokeshire, limited to the south (hanging-wall side) of the Ritec Fault, and the Pragian Llanishen Conglomerate Formation of the Cardiff district (Allen, 1975; Waters and Lawrence, 1987) were derived from the south. The Ridgeway Conglomerate Formation forms part of the **West Angle Bay (North)** and **Freshwater East–Skrinkle Haven** GCR sites, as well as part of the **Freshwater West** site. The source of these rocks is believed to have been an emergent area in what is now the Bristol Channel, known as the 'Bristol Channel Landmass' (Mechie and Brooks, 1984; Tunbridge, 1986; Cope and Bassett, 1987; Marshall, 2000a,b). South of the Bristol Channel, the Hangman Sandstone Formation, seen at the **Glenthorne** GCR site, is the only unit of northerly derivation and proven Mid-Devonian age.

The lowermost Přídolí strata, traditionally referred to the Lower Old Red Sandstone, comprise the Downton Castle Sandstone Formation, and the Tilestones (Long Quarry Sandstone) Formation (its lateral equivalent in south-central Wales, seen at the **Sawdde Gorge** GCR site). These are shallow marine, barrier and shore-face deposits, with red-bed Old Red Sandstone deposits overlying them. Locally, a green, lacustrine/back-barrier facies (the Temeside Mudstone Formation) intervenes; the Pont-ar-llechau Member in the Sawdde Gorge is of similar facies, although red (Almond, 1983; Almond et al., 1993). Higher Přídolí strata (Sandy Haven, Moor Cliffs and Raglan Mudstone formations) are predominantly red mudstone/siltstone-dominated successions with common nodular calcretes. Representing marginal marine, coastal mudflat and marine-influenced alluvial-floodplain environments, these strata are represented at the **West Angle Bay (North)**, **Freshwater West**, **Freshwater East–Skrinkle Haven**, **Sawdde Gorge**, **The Scar** and **Cusop Dingle** sites. The **Little Castle Head**, and **Albion Sands and Gateholm Island** GCR sites are described in detail in the Silurian stratigraphy GCR volume (Aldridge et al., 2000) and briefly described in the present volume. The mature calcretes of the Psammosteus Limestone at the top of the Přídolí succession (the Chapel Point Calcretes Member in Pembrokeshire (Williams et al., 1982) and the Bishop's Frome Limestone (Brandon, 1989) in south-east Wales, the Welsh Borderland and south-central Wales) are seen at the **Devil's Hole**, **Freshwater West**, **Freshwater East–Skrinkle Haven**, **Sawdde Gorge**, **Llansteffan**, **Cusop Dingle** and **Lydney** GCR sites.

The Lochkovian–Pragian (Dittonian) rocks (upper part of the Milford Haven Group, Gelliswick Bay, Freshwater West and St Maughans formations; Ditton Group; Figure 5.1) are characterized by the cyclic arrangement of their component facies (mudstone/siltstones, sandstones and intraformational conglomerates) in upward-fining, commonly calcretized alluvial cycles (Allen, 1964a). The thickest development (1000–1500 m) is north of the Ritec Fault in south-west Pembrokeshire. Overall, the succession coarsens upwards, with the sandstones becoming thicker and more dominant. Fish fragments are common in the conglomerates, plant microfossils and megafossils are present locally (e.g. Wellman *et al.*, 2000). There are a few instances of complete fish specimens being discovered, such as in the St Maughans Formation at the Cwm Mill GCR site near Abergavenny (SO 311 156) and the Wayne Herbert Quarry site (SO 335 320) in Herefordshire farther north (Dineley, 1999f). These fish occur in finer lithologies and were probably entombed in overbank deposits during flooding. The swimming and resting traces of fish (*Undichna*) are recorded at Tredomen Quarry near Talgarth (Morrissey *et al.*, 2004). Arthropod traces include the ubiquitous burrow trace *Beaconites barretti* (Morrissey and Braddy, 2004) and crawling traces (*Diplichnites*; Smith *et al.*, 2003; Morrissey *et al.*, 2004).

The intraformational conglomerates, along with some extraformational varieties, are interpreted as channel-lag deposits. The sandstones exhibit a range of architecture (Williams and Hillier, 2004), and although predominantly comprising the in-channel sandbodies of high-sinuosity streams, laterally accreted bodies and sheet-flood deposits also occur. The fine-grained lithologies are thought to represent deposition in floodplain environments that were subjected to frequent desiccation and carbonate soil formation; at least some of the mudrocks may have originated as wind-blown dust, perhaps reworked into lacustrine sediment in floodplain ponds (Marriott and Wright, 2004). Some mudrocks are interbedded with thin point-bar type gravelly lenses of calcrete clasts in low-angle cross-bedded sets and have pelleted fabrics, indicating deposition of pedogenic, pelleted mud aggregates from bedload in small sinuous channels (Ékes, 1993; Marriot and Wright, 1996, 2004). In addition to the immature calcretes, more mature, massive to rubbly calcretes occur sporadically, including the Coldra, Ruperra and Pontypool limestones of south-east Wales (Squirrell and Downing, 1969), the Ffynnon limestones of the Black Mountains and the Abdon limestones of the Clee Hills (Ball and Dineley, 1961; Allen, 1961, 1974c). Some tuffs are present in the Gelliswick Bay Formation in south-west Pembrokeshire and in the St Maughans Formation at Tredomen Quarry (Morrissey *et al.*, 2004).

Sites representing the Dittonian rocks include **Oak Dingle**, **Freshwater West** (potential GCR site), **Freshwater East–Skrinkle Haven**, **Llansteffan**, **Cusop Dingle** (potential GCR site) and **Lydney**. In addition, the disused quarry at Pantymaes near Sennybridge exposes a fine section through a thick channelized sandstone complex and overlying floodplain mudrocks (Owen and Hawley, 2000), as well as a suite of arthropod crawling traces (Smith *et al.*, 2003), and is recommended for GCR status.

Above the St Maughans Formation near Cardiff and Newport in south-east Wales is the Llanishen Conglomerate Formation. Its pebbles include Llandovery quartzites and ?Silurian volcanic rocks, in a suite unlike the northerly derived pebbles of most of the Old Red Sandstone succession, and a nearby southerly source is thus favoured (Allen, 1975; Waters and Lawrence, 1987).

In Anglesey, about 500 m of Old Red Sandstone rocks crop out from the coast between Dulas Bay and Lligwy Bay (see **Porth-y-Mor** GCR site report, this chapter) inland to near Llangefni. The succession records the burial of an already deformed and dissected Anglesey platform by Caledonian molasse shed from the mountain belt to the north. Asymmetric folding and an associated cleavage are attributed to the late Emsian Acadian deformation. The correlation and age of the succession, probably deposited at least initially in a palaeovalley isolated from the main part of the Anglo-Welsh Basin, and with no fossils yet found, are uncertain. Basal conglomerates and pebbly sandstones (the Bodafon Formation) are interpreted as the localized deposits of a series of coalescing alluvial-fans banked against a north-west-facing valley side. The Traeth Bach Formation comprises mainly red-brown siltstones with abundant calcrete nodules which are correlated by Allen (1965a) with the Přídolí Raglan Mudstone Formation of south Wales, although he interprets them as playa-lake deposits, unlike the floodplain alluvial facies of the Raglan Mudstone Formation. The Porth-y-Mor Formation is correlated with the Dittonian succession of south Wales. Thick calcretes at the top of the Traeth Bach Formation invite comparison with the Psammosteus Limestone and support this correlation. The succession of fining-up alluvial cycles and sedimentary structures, including epsilon cross-stratification (first identified by Allen (1963a) in ancient deposits here), is interpreted as the deposits of south-eastward-flowing, meandering

Introduction

streams and their floodplains. The Traeth Lligwy Formation consists of bioturbated sandstones and siltstones and conglomerates, interpreted by Allen (1965a) as the deposits of permanent lakes, a facies not seen in south Wales, except for some minor examples in the Senni Formation at Ferryside (Almond, 1983) and Laughorne (B.P.J. Williams, pers. comm.).

Strata of probable Emsian age are referred to the Breconian local stage. Its type area is the Brecon Beacons, where it comprises the Senni and Brownstones formations (Croft, 1953). The Senni Formation, and the equivalent Cosheston Group in south-west Wales and Clee Sandstone Formation in Shropshire, are characterized by their olive-green colour. The Senni Formation is known particularly for its early vascular plant remains, including *Gosslingia breconensis*, *Hostinella heardii*, *Krithodeophyton croftii*, *Sennicaulis hippocrepiformis*, *Tarella trowenii*, *Uskiella spargens* and *Zosterophyllum llanoveranum* (Cleal and Thomas, 1995). The **Heol Senni Quarry** GCR site provides a characteristic section of the formation, which is interpreted as the deposits of low-sinuosity, seasonally flowing, sandy, braided streams in a mid-fan setting (Owen, 1995). The Cosheston Group (Thomas, 1978; Wellman *et al.*, 1998) is much thicker (1500–1800 m), having been deposited in a zone of active rifting in the hanging-wall of the Benton Fault. The alluvial deposits of braided to meandering, generally southerly flowing stream systems contain, in their lower part, soft-sediment deformation structures that probably reflect seismic activity associated with the rifting of the basin. High water-table conditions are invoked for the preservation of the plant remains and the predominantly green colour of these formations, due to the presence of chloritized micas. A possible climatic cause of the high water-table has yet to be investigated. Only four vertebrate localities are known – the Breconian index fossil fish *Rhinopteraspis dunensis* (= *cornubica*) at Primrose Hill Quarry, Crickhowell (SO 207 200). *R. dunensis* occurs in the Mid-Siegenian of mainland Europe (Dineley, 1999f); *Althaspis senniensis* at Heol Senni Quarry (SN 9145 2210; see GCR site report, this chapter); *Pteraspis dixoni* and *Cephalaspis* sp. at Pengau (Pen-y-gau) Farm (SN 3732 0850) near Ferryside, Pembrokeshire; and *Protopteraspis gosseleti* at Allt ddu near Brecon (SO 027 242) (Habgood, 2000; Edwards and Richardson, 2004).

The Brownstones Formation (the Black Nore Sandstone Formation east of the Severn) is the highest Lower Old Red Sandstone formation in the Anglo-Welsh Basin. It is characterized by red-brown, fine- to coarse-grained, fluvial sandstones, with red-brown, locally green mudstone and siltstone interbeds. Maximum development is in the Forest of Dean, where 1200 m are present; the **Ross-on-Wye, Royal Hotel** GCR site is a typical section. In the type area of the Brecon Beacons and Black Mountains, the sandstones generally form extensive sheets, as seen in the magnificent north-facing scarps of the Brecon Beacons, the succession tending to become more sandstone-dominated in its upper parts (Tunbridge, 1981a). The succession is interpreted as the deposits of a prograding fan system formed in a semi-arid, seasonally wet climate. Channelized sandstones are the product of mid-fan, wet season, flashy deposition, with more distal floodplain environments represented by sheet-flood sandstones and mudstones and siltstones. Gravelly, pebbly and conglomeratic sandstones are locally common, particularly in the upper part of the formation in the Forest of Dean and parts of the north and east crops of the South Wales Coalfield. Pebbly beds at Llyn-y-Fan Fawr (SN 834 216), 4 km west of the Swansea Valley are sourced from the east and perhaps attributable to uplift on the Swansea Valley Fault (Tunbridge, 1980a); similarly, the pebbly Caeras Beds in the Cennen Valley between Llandybie and Kidwelly may have been deposited as a result of uplift on the Carreg Cennen–Llandyfaelog Fault (**Caeras Quarry** GCR site). Pebble suites at Ross-on-Wye and in the Clee Hills comprise igneous, metamorphic and sedimentary rocks of Ordovician to Devonian age thought to have been derived from north Wales (Allen, 1975). The Brownstones Formation of Wales has yet to yield animal body fossils, but the Dittonian index fossil fish *Althaspis leachi* is recorded from the basal Brownstones Formation at the **Wilderness Quarry** GCR site in the Forest of Dean (Allen *et al.*, 1968). East of the Severn Estuary, the Black Nore Sandstone Formation, seen at the **Portishead** GCR site, provides the only occurrence of sandstones of aeolian origin at this level in the basin (Dodd, 1986).

No Middle Devonian strata are yet proved in south Wales, except perhaps for the southerly sourced Ridgeway Conglomerate Formation of south-west Pembrokeshire (see above). Mid- and Early Devonian ages have been suggested, but, bounded by unconformities and lacking in diagnostic fossils (only some crossopterygian fish fragments and plant fragments have been found), the age remains uncertain. It consists of petromict conglomerates, sandstones, and siltstones with calcretes. The clasts in the conglomerates are mainly of quartzite, lithic greywacke, siltstone and vein quartz, with a marked influx of phyllite clasts in the higher beds. The formation is interpreted as the deposits of a braided stream/alluvial-fan complex sourced from a nearby, southerly Lower Palaeozoic–?Precambrian source. A northward-prograding alluvial-fan occupied the area of **Freshwater West**, with more distal braided stream environments at **West Angle Bay (North)** (Williams *et al.*, 1982). Some large-scale, cross-bedded siltstones have been interpreted as bedload-transported pedogenic mud aggregates (Ékes, 1993).

Late Devonian (late Frasnian–Famennian) Upper Old Red Sandstone rocks, referred to the Farlovian local stage, overlie the unconformity that truncates the Breconian and older successions. The **West Angle Bay (North)**, **Freshwater West** (potential), **Freshwater East–Skrinkle Haven** and **Portishead** GCR sites provide sections of the unconformity. The strata are predominantly fluvial, deposited on a southerly facing palaeoslope, but lacustrine, possible aeolian and marginal marine facies occur locally in the Plateau Beds Formation (**Afon y Waen** and **Duffryn Crawnon** potential GCR sites) and aeolian sands are present in the Portishead Formation at the **Portishead** GCR site. The succession is thin in comparison with the Lower Old Red Sandstone, with up to a maximum of about 350 m in south-west Dyfed.

In the Pembroke peninsula, the Skrinkle Sandstones Group (see **Freshwater West**, **West Angle Bay (North)** and **Freshwater East–Skrinkle Haven** GCR site reports, this chapter) accumulated as a synrift succession of alluvial-fan, plain and lacustrine deposits in the hangingwall of the Ritec Fault. Its distribution is confined to the Tenby–Angle Basin, thickening southwards from 100 m near the Ritec Fault to 330 m at Freshwater West (Marshall, 2000a,b). The beds lie unconformably on the Ridgeway Conglomerate Formation, overstepping it eastwards to rest on the Milford Haven Group, and there is a transition into grey Carboniferous beds at the top of the succession. The group is subdivided into the Gupton Formation and overlying West Angle Formation. The Gupton Formation is interpreted as comprising two axial basin-fill, fluvial, coarsening-upward sequences of mature, clean sandstones. The West Angle Formation is characterized by red sandstones, conglomerates rich in igneous, sandstone and phyllite clasts, and calcretes. The pebbles in the conglomerates were probably derived from the Precambrian Pebidian volcanic complex and associated Ordovician volcanic rocks of the northern south-west Pembrokeshire coast, either directly, or by recycling of the clasts of the Cosheston Group. The upper part of the West Angle Formation (the Red-Grey Member) comprises fining-upward fluvial sequences with mudstone and calcrete tops similar to the meandering channel deposits below, but shows an upward increase in grey beds and grey-green sandstones with plant remains (**West Angle Bay (North)** GCR site). The grey beds contain initially non-marine fossils, but increasing marine influence heralds the main transgression at the start of the Carboniferous Period.

In the Carmarthen Fans, Black Mountains and Brecon Beacons, the Upper Old Red Sandstone is represented by the Plateau Beds Formation, best known for its sporadic vertebrate and marine brachiopod faunas. Fish-bearing conglomerates include the Afon y Waen Fish Bed (see **Afon y Waen** GCR site report, this chapter). The succession represents a broadly transgressive sequence from fluvial (and possibly aeolian) deposition to marginal marine environments (Lovell, 1978a,b; **Duffryn Crawnon** potential GCR site). Fossils recovered include the brachiopods *Cyrtospirifer verneuili*, *Lingula* spp., *Ptychomaletoechia omaliusi*, the bivalves *Leptodesma* cf. *lichas*, *Pterinopecten* sp. and *Sanguinolites* sp. and fragments of the fish *Bothriolepis*, *Coccosteus*, *Holoptychius*, *Pseudosauripterus anglicus*, cf. *Rhinodipterus* and *Sauripterus* (Taylor and Thomas, 1975).

The Plateau Beds Formation is overstepped by the Grey Grits Formation (seen at the **Abercriban Quarries** GCR site), which consists mainly of quartzitic sandstones. Thin, green mudstone interbeds and quartz pebble layers occur sporadically. Fauna is restricted to fish fragments and the bivalve *Sanguinolites*, along with some burrows, including forms resembling *Skolithos* and *Arenicolites*. Cross-bedding indicates a predominant southerly or south-easterly flow, but north-east currents are also recorded, as well as some herring-bone cross-bedding. Braided stream deposition is suggested, although a shallow marine origin for at least part of the formation cannot be discounted (Lovell, 1978a,b).

The Upper Old Red Sandstone is represented by the Quartz Conglomerate Group on the north-east, east and south-east crops of the South Wales Coalfield (see **Craig-y-cwm** potential GCR site report, this chapter). In the Cardiff area, a 114 m-thick succession of thinly bedded quartzitic sandstones, siltstones and mudstones with subordinate thick, commonly pebbly sandstones and calcretes (the Cwrt-yr-ala Formation; Waters and Lawrence, 1987) underlies the Quartz Conglomerate Group. Quartzitic sandstones at the base of the group on the north-east of the coalfield (the Wern Watkin Formation) correlate with the Grey Grits Formation. They are overlain by quartz pebble conglomerates (the Craig-y-cwm Formation) that contain sporadic fish fragments, including *Bothriolepis*. The topmost part of the formation comprises micaceous, feldspathic and garnet-rich sandstones and interbedded red mudstones (the Garn-gofen Formation). Vertebrate remains from these beds include a fragment of *Osteolepis macrolepidotus*. The bivalve *Archanodon jukesi* was collected from the Quartz Conglomerate Group in Monmouthshire, and *Holoptychius nobilissimus* is recorded at Tongwynlais. The group crops out around the Forest of Dean Coalfield in Monmouthshire and Gloucestershire, where it comprises basal quartz conglomerates overlain by the Tintern Sandstone Formation. East of the Severn, the sandstone and conglomerate facies become indistinguishable as the succession thickens southwards and the beds are named the Portishead Formation. In the Clee Hills, the Upper Old Red Sandstone is represented by the Farlow Sandstone Formation, which consists of lower yellow, and upper grey, sandstones.

The uppermost beds of the Quartz Conglomerate Group in south-east Wales locally show a transition into the lower Tournaisian Tongwynlais Formation (Avon Group) of the Carboniferous Limestone, with interbedded marine limestones and red-brown sandstones. Spores from 18 m below the top of the group at Tongwynlais are now regarded as earliest Famennian in age. Elsewhere, there is generally a depositional break between the uppermost beds of the group and the Carboniferous Limestone, but interdigitation of shallow marine beds occurs in Pembrokeshire, as seen at the **West Angle Bay (North)** and **Freshwater East–Skrinkle Haven** GCR sites.

PORTH-Y-MOR, ANGLESEY (SH 490 885–SH 493 876)

J.R. Davies

Introduction

The foreshore sections to the north and south of Porth-y-Mor on the north-east coast of Anglesey (Figure 5.4) afford some of the best and most accessible exposures through Old Red Sandstone fluvial fining-upward cycles in the Anglo-Welsh Basin. The sections occupy an important place in the history of research into fluvial sedimentology, for it was here that Allen (1965a) first recognized ancient examples of epsilon cross-bedding and showed it to be formed by lateral accretion on the point bars of meandering river channels. Also, the sections form part of the only outcrop of Old Red Sandstone strata preserved in north Wales. The GCR site is the type locality of the Porth-y-Mor Formation and, at its southern end near Traeth Lligwy, exposes the overlying Traeth Lligwy Formation. Coastal cliffs at Traeth Bach 800 m north-west of Porth-y-Mor, and 200 m north-west of the GCR site boundary, provide good exposure of the underlying unit of the Old Red Sandstone (the Traeth Bach Formation) and also merit protected status. This folded and cleaved Anglesey succession lies about 110 km north-west of the nearest Devonian outcrop in the Welsh Borderland, providing an important constraint on Devonian palaeogeographical reconstruction, and on the timing and nature of late Caledonian orogenic events.

Description

The Anglesey succession of Devonian red beds was first described in detail by Greenly (1919), who, although recognizing its gross similarities with the Lower Old Red Sandstone of south Wales and the Welsh Borderland, viewed it as the deposits of a separate northern basin. Allen (1965a) subdivided the succession into four formations (which he referred to as 'Beds'). These are, in ascending order, the Bodafon Formation, Traeth Bach Formation, Porth-y-Mor Formation and Traeth Lligwy Formation. Apart from the localized, conglomeratic Bodafon Formation, the formations are exposed in the coastal cliffs between Traeth Dulas (SH 487 888) and Traeth Lligwy (SH 487 877). The Porth-y-

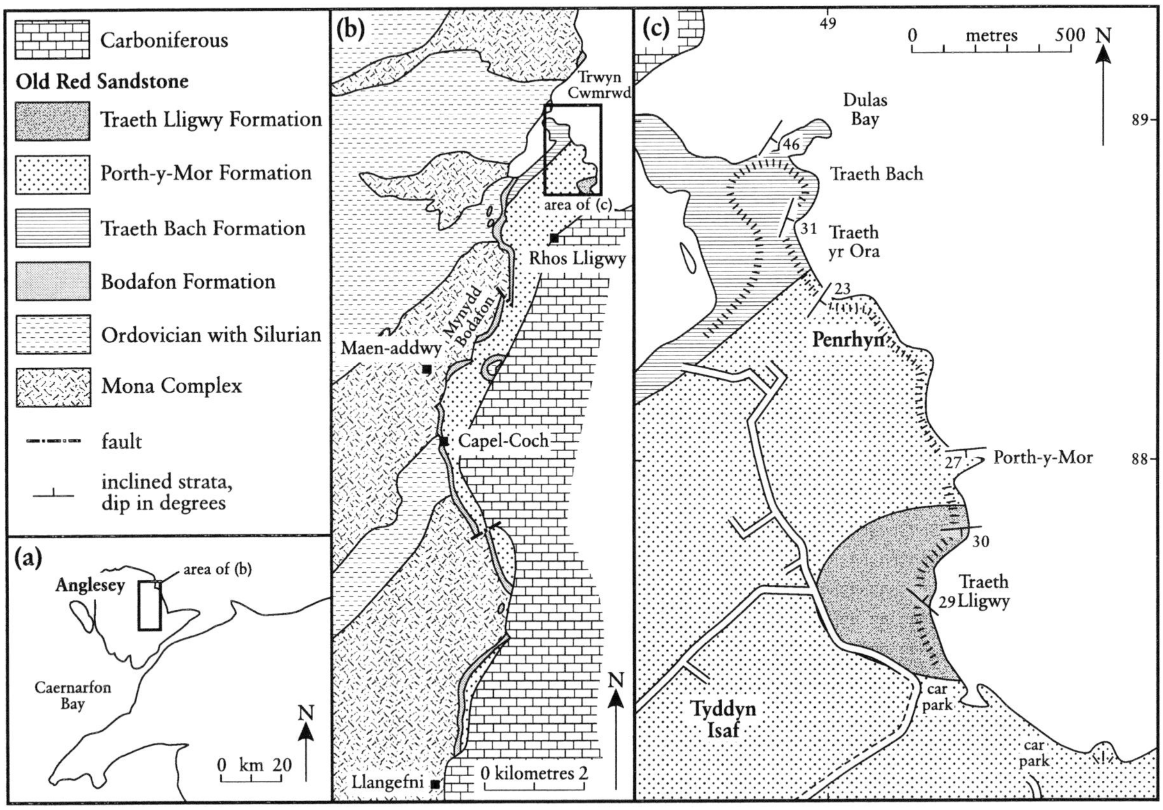

Figure 5.4 Geology of the Porth-y-Mor GCR site, Anglesey. Map (a) shows location of Map (b). Map (b) shows the main outcrop of Old Red Sandstone in Anglesey and location of Map (c). Map (c) shows the geology of the Porth-y-Mor site. Maps (b) and (c) after Allen (1965a).

Mor site lies within this tract, exposing the eponymous Porth-y-Mor Formation and the overlying Traeth Lligwy Formation. Allen gave details of the petrology and sedimentary structures of the rocks, documenting heavy-mineral assemblages, palaeocurrent indicators and the distribution of trace fossils. He also provided palaeoenvironmental and palaeogeographical interpretations, invoking alluvial-fan, fluvial in-channel, alluvial-floodplain and lacustrine depositional environments (see also Allen, 1970).

The Bodafon Formation ranges from 3 m to 45 m in thickness and comprises conglomerates and pebbly sandstones with minor red siltstone lenses. It is exposed on the coast at Trwyn Cwmrwd (SH 492 902), 2 km NNW of the GCR site, where up to 7.6 m of pebbly sandstones and thin conglomerates rest unconformably on vertically foliated gneisses and dip gently seawards. The conglomerates are extraformational ('exotic'), the clasts comprising mainly quartzite, schist, gneiss and vein quartz derived from the local basement and ranging from cobble to pebble grade, with a few boulders.

The Traeth Bach Formation is at least 130 m thick, the lowest 100 m being exposed in the low cliffs at Traeth Bach (SH 487 888–SH 489 887). The formation is dominated by red calcrete-rich siltstones, with minor extraformational conglomerates, pebbly and thin sandstones (Figure 5.5). The calcrete is mainly in the form of nodules, but massive mature dolomite and limestone calcretes occur at several levels, the thickest (5.2 m) lying about 95 m from the base of the section. The pebbles in the conglomerates are mainly of local Precambrian and Ordovician rocks.

The Porth-y-Mor Formation is estimated to be about 347 m thick and is the thickest unit within the local Devonian succession. The section commences on the south side of Traeth yr Ora and dips consistently to the south at about 27°. The contact with the underlying Traeth Bach Formation is concealed beneath the beach to the north. The exposed succession, about 225 m thick, comprises a series of fining-upward cycles. Allen (1965a) recognized 43 cycles, each constructed of a lower conglomerate and sandstone and an upper muddy siltstone. Calcareous nodules (calcrete glaebules) are common throughout the siltstone units and locally form the dominant element, coalescing to form continuous beds of impure limestone and dolomite. In the following description of the section, the cycle numbers are those of Allen (1965a), as shown on his log of the section (Figure 5.5). A 1.4 m-thick calcrete lies at the base of the exposed succession, underlying Cycle 1. There are substantial exposure gaps between cycles 10 and 11 at Porth-y-Mor (SH 4922 8830–SH 4933 8806) and in the cove to the south (SH 4935 8803) between cycles 12 and 13. Apart from the pale grey limestones and brown-weathering dolomites, the lithologies in the Porth-y-Mor Formation are typically red or purple. The siltstones exhibit a northerly dipping cleavage throughout.

Conglomerates are normally present at the base of each cycle, where they rest on an erosion surface cut in the underlying siltstone, ranging from a few centimetres to over 1 m in thickness. Exotic and intraformational types are represented, but the former only occurs at the base of Cycle 1, where a massive 3.4 m-thick unit comprises rounded to angular clasts, ranging up to cobble grade in size, of quartzite, schist, gneiss, vein quartz and cleaved mudstone. These clasts are readily matched with rock types in adjacent Precambrian and Lower Palaeozoic outcrops. In contrast, the intraformational conglomerates, which occur throughout the remaining part of the section, are composed of rounded pebbles and granules of red and green siltstone, fine-grained, red sandstone and limestone, as well as reworked calcrete nodules. They record the reworking of Old Red Sandstone sediments from elsewhere within the basin of deposition.

In every cycle, the basal conglomerate is overlain by, and commonly grades upwards into, a much thicker sandstone unit. Coarse- to fine-grained sandstones are commonly interbedded in the lower parts of these units, but fine- and very fine-grained sandstones with sporadic siltstone partings generally dominate the upper levels. This fining upwards is reflected in the range of tractional sedimentary structures in the sandstones. The coarser lower sandstones and basal conglomerates commonly exhibit large-scale tabular or trough cross-bedding in sets up to 0.45 m thick. In contrast, the overlying finer-grained sandstones are characterized by co-sets of small-scale cross-lamination separated by planar-laminated intervals. This succession of sedimentary structures is seen particulary well in cycles 1 and 2, which contain the thickest sandstone bodies in the succession.

In ten cycles, the basal sandstone and conglomerate unit exhibits a distinctive set of shallow-dipping bedding surfaces. Each surface is tangential with the top of the sandstone, and, in an opposing sense, with the scoured base of the unit, giving a sigmoidal form (Figure 5.6). This pattern of bedding surfaces is epsilon cross-bedding (Allen, 1963a), first recognized in an ancient sedimentary sequence here (Allen, 1965a).

The surfaces cross-cut the vertical lithological changes within the conglomerate and sandstone units, but the upward-fining character, and the upward succession of tractional structures seen in the cycles overall is visible within each cross-bed of the epsilon sets. Critically, palaeocurrent directions indicated by the internal tractional features show that current flow was consistently parallel to the strike of the epsilon cross-bedding surfaces. The latter dip predominantly towards the east, but epsilon cross-sets with either westerly or southerly dipping surfaces are also present.

The upper argillaceous parts of the Porth-y-Mor cycles are typically devoid of tractional sedimentary structures, apart from thin beds of cross-laminated sandstone, which are common at the base and locally present at higher levels. Calcrete nodules, commonly cylindrical in shape and aligned perpendicular to bedding, are scattered throughout. They regularly form part of vertical profiles in which their size and frequency increases upwards until the host siltstone is absent from capping beds of massive limestone or dolomite (Allen, 1965a, 1974d). In thin section, the limestones display a complex range of textures and fabrics (Allen, 1965a). Brecciation is common, with darker masses of inclusion-rich, microcrystalline calcite cross-cut by intersecting veins of clear, coarsely crystalline calcite. Tubular root-related structures (rhizoliths) are also present.

Palaeocurrent indicators in the Porth-y-Mor Formation show that depositing currents largely flowed from the north-east. This is consistent with the orientation of associated linear channel and scour features, and of primary current

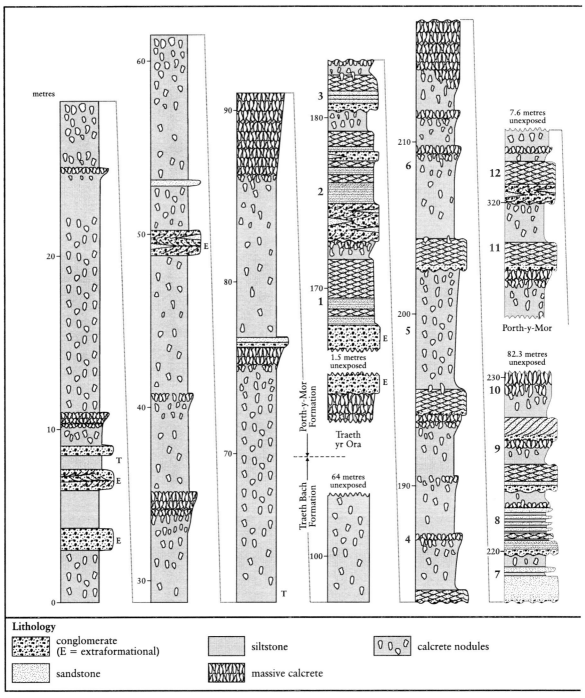

Figure 5.5 Graphic log of section of coastal exposures between Traeth Dulas and Traeth Lligwy measured by Allen (1965a). The Traeth Bach Formation is from 0 m to 129.54 m (425 feet), the Porth-y-Mor Formation is from 129.54 m to 477.01 m (1565 feet) and the Traeth Lligwy Formation is from 477.01 m to 498.96 m (1637 feet). Thicknesses above the base of the section are shown in metres. The numbers in bold are Allen's cycle numbers in the Porth-y-Mor Formation. After Allen (1965a).

lineation present in the plane-bedded sandstones. Trace fossils include a few sporadic burrows throughout, and some arthropod tracks high in the formation (SH 4940 8779), but no body fossils have yet been discovered. The conformable top of the formation is placed at the base of the lowest of the distinctively bioturbated sandstones that characterize the succeeding Traeth Lligwy Formation, in which burrow traces are abundant.

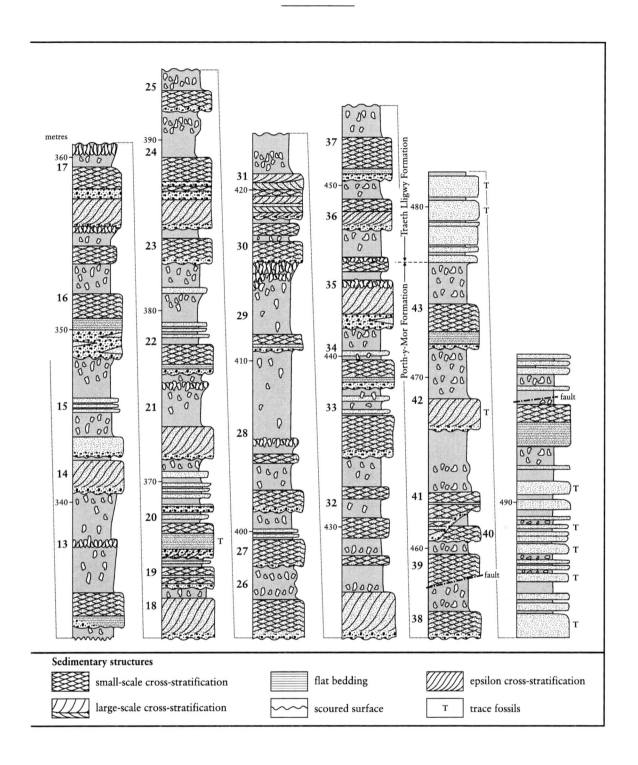

The Traeth Lligwy Formation forms the cliffs on the north side of Traeth Lligwy (SH 494 878–SH 493 875) at the southern end of the GCR site. The basal 24 m of the formation are exposed, preserved in the axial area of an E–W-trending syncline. They are of sandstones and siltstones, in thinner beds than the Porth-y-Mor Formation below. Red, very fine-grained sandstones ranging from 0.15 m to 2.1 m thick, but averaging 0.53 m, are interbedded with red, sandy, siltstones from

The Anglo-Welsh Basin

Figure 5.6 Epsilon cross-bedding in Cycle 42 (see Figure 5.5) of the Porth-y-Mor Formation (SH 4940 8780). Hammer for scale (circled). (Photo: J.R. Davies.)

0.07 m to 1.2 m thick. Calcrete is confined to a few horizons of nodules and conglomerates are absent, but organic burrowing structures are abundant in the sandstones. The thickest sandstone is flat-bedded in its lower part and small-scale cross-bedded in its upper part.

Interpretation

Allen's (1965a) synthesis of the sedimentary processes and depositional environment of the Anglesey Old Red Sandstone succession was an important early study of fluvial sedimentology. The succession accumulated in a depositional basin flanked to the south-west by an upland area of Precambrian and Lower Palaeozoic rocks. These older rocks were the source of alluvial-fan gravels (the Bodafon Formation), which accumulated along the basin margin. Away from this margin, these interdigitate with the overlying Traeth Bach and Porth-y-Mor formations. In the GCR site calcareous muddy siltstones of the Traeth Bach Formation were interpreted by Allen as the deposits of ephemeral, non-saline playa lakes. Subsequently, a broad meandering, SE-trending river belt was established, within which the Porth-y-Mor Formation was deposited.

The overlying Traeth Lligwy Formation is interpreted as the deposits of more permanent flood basin lakes, the reduced amount of calcrete and abundance of burrowing structures suggesting a higher water-table and less prolonged periods of subaerial exposure of the sediments.

The conglomerate and sandstone units of the Porth-y-Mor Formation represent river channel deposits. The pattern of upward-fining above a basal erosion surface, and the associated succession of sedimentary structures, are consistent with an upward reduction in the velocity and bed shear rate of the depositing currents. These conditions are encountered on the point bars of a meandering stream where the fastest and most-powerful currents are associated with the deepest part of the channel, close to the outer bends of the meanders. Here, the river erodes into its earlier floodplain sediments and deposits its coarsest bedload. In contrast, the currents flowing across the upper parts of point bars formed on the inside bends of the meanders are much slower, and deposit finer-grained sediment. Lateral migration of the river channel and lateral accretion of its point bars generate the upward-fining channel sequence.

The channel conglomerates and sandstones of the Porth-y-Mor Formation are noteworthy in being the first rocks in the stratigraphical record in which epsilon cross-bedding and its significance were recognized. Allen (1965a) demonstrated that the sigmoidal surfaces, dipping perpendicular to the flow direction shown by associated tractional structures, represent the surfaces of meander point-bars, and therefore, that each epsilon cross-set records the process of incremental, sustained lateral accretion. At a time when the understanding of ancient sedimentary, fluvial fining-upwards units was in its infancy, the discovery and interpretation of these features showed that they were the products of laterally migrating, meandering channels of mixed bedload rivers (Allen, 1965a, 1970). On the basis that the thickness of the epsilon cross-sets should correspond to the bankfull depth of the channel in which they were deposited, Allen (1965a) estimated that the rivers that supplied the sediment of the Porth-y-Mor Formation were likely to be at least 60 km long and 20 m wide, and perhaps up to 600 km long and 90 m wide.

The upper siltstones of the Porth-y-Mor cycles record the accumulation of fine-grained sediment on extensive river floodplains. The calcareous nodules and limestone and dolomite beds display many of the features of modern calcretes and dolocretes. They record the diagenetic growth of carbonate within soil profiles during periods of prolonged subaerial exposure in a semi-arid, seasonally wet climate (Allen, 1974d). Allen's (1965a) description of the Porth-y-Mor examples was the first and most detailed account of Old Red Sandstone calcrete in Wales, following its recognition by Burgess (1961) in Ayrshire and Pick (1964a) at **Portishead** (see GCR site report, this chapter). The Traeth Lligwy Formation is interpreted as the deposits of perennial lakes, a facies not seen elsewhere in the Anglo-Welsh Basin.

Early palaeogeographical models envisaged the Anglesey Old Red Sandstone succession as deposited in a narrow gulf opening to the northeast, connected to the Midland Valley of Scotland, and structurally isolated from the sequences of south Wales and the Welsh Borderland (Greenly, 1919; Wills, 1952). Allen (1965a) noted the similarity in lithofacies and cyclicity of the Porth-y-Mor Formation to the Dittonian and Breconian rocks of Pembrokeshire and the Welsh Borderland. He also noted that although the heavy-mineral assemblages are similar to those of Pembrokeshire, they are different to those of the Welsh Borderland. He viewed the Anglesey succession as lying at the margin of a broad depositional tract connected to the main basin to the south, supplied with sediment sourced some distance to the north-west by south-easterly flowing rivers. This remains the favoured palaeogeographical interpretation (Allen, 1974a; Allen and Crowley, 1983; Bluck *et al.*, 1992), although the age of the succession remains unproved in the absence of fossil or other direct dating evidence. Structural and stratigraphical constraints provide a timespan only between the Ludlow and Dinantian, inclusive (Allen, 1965a; Allen and Williams, 1979a). A Mid-Devonian age has been suggested (Hurst *et al.*, 1978), but the presence of the folding and cleavage makes this unlikely (Allen and Williams, 1979a; Hurst *et al.*, 1979). Allen (1965a) attributed the deformation to the widely recognized Mid-Devonian (late Caledonian–Acadian) tectonic event and correlated the Porth-y-Mor Formation with the Dittonian (Lochkovian) succession of south Wales on the characteristic cyclicity of the facies (Allen, 1974b, 1977). The thick calcretes on both sides of Traeth yr Ora, including the one at the base of the exposures on the south side, are reminiscent of the thick, regionally developed Psammosteus Limestone in south Wales and the Welsh Borderland, which would support Allen's correlation. However, no fish remains have yet been found in the Porth-y-Mor Formation, in contrast to their common presence in the Dittonian succession in south Wales and the Welsh Borderland, and the age of the Anglesey succession remains to be proved.

Conclusions

The Porth-y-Mor site is one of the best, accessible sections of Old Red Sandstone alluvial cyclic deposits in the UK. It occupies a unique place in the history of sedimentological research, the excellent exposures including examples of epsilon cross-bedding, first recognized in an ancient sedimentary sequence here. The well-developed profiles of soil carbonate (calcrete) were amongst the first to be described in detail in the Anglo-Welsh Basin. The importance of the site also lies in the presence of the unique lake deposits of the Traeth Lligwy Formation.

The site, set within the cleaved and folded Anglesey Old Red Sandstone succession, provides evidence to suggest that this northern sequence shared the same sediment source as parts of the Lower Old Red Sandstone of south Wales. The two areas may have been part of an extensive, continuous depositional tract prior to Mid-Devonian (Acadian) deformation and uplift. However, the Porth-y-Mor rocks have yet to yield fossils to indicate their precise age, and the site offers opportunities for further palynological and microfossil research.

DEVIL'S HOLE, SHROPSHIRE (SO 672 929)

Potential ORS GCR site

W.J. Barclay

Introduction

The Devil's Hole site in Shropshire (Figure 5.7) is an established GCR site for its fossil fishes (Dineley, 1999f). A brief summary is presented here, along with a geological map of the site. The opportunity is also taken to correct the vertical section (Figure 5.8) given in Dineley's account (fig. 4.8).

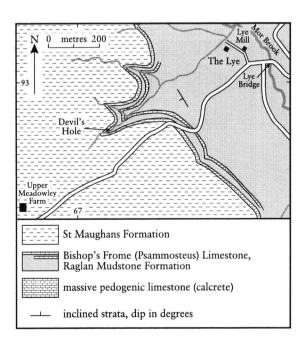

Figure 5.7 Geological map of Devil's Hole GCR site. Based on British Geological Survey 1:10 560 manuscript map Shropshire 58NW (1960) and 1:10 000 manuscript map SO 69SE (2002).

The site provides a stream section across the 'Downtonian'–'Dittonian' boundary, with fish remains recovered from above and below the Bishop's Frome (Psammosteus) Limestone. Excavations by the GCR Unit of the Nature Conservancy Council between 1980 and 1982 enabled a detailed sedimentological and palaeontological analysis by M.A. Rowlands and P. Tarrant (Tarrant, 1991). The section is now poorly exposed and difficult to access.

The section was originally collected by the [British] Geological Survey (Whitehead and Pocock, 1947) and later by Ball and Dineley (1961). Reference to the geology of the site was made by Ball and Dineley (1961), Allen and Tarlo (1963), Banks (1980), Richardson *et al.* (1981), Allen (1985), Blieck (1985), Halstead (1985) and Jenkins (1998). The fish faunas were also referred to by Wills (1948, 1950), White (1950), Denison (1956), Robertson (1957), Turner (1973), Blieck (1981, 1984, 1985), Tarrant (1981) and Vergoossen (2000).

Description

A mature 2.5 m-thick calcrete cropping out in a lichen-coated waterfall (SO 6710 9284) is correlated with the Bishop's Frome (Psammosteus) Limestone and marks the top of the Raglan Mudstone Formation (Jenkins, 1998). The succession below the limestone comprises mainly red mudstones/siltstones with sporadic thin sandstones and several thin calcretes. The St Maughans Formation (Ditton Group of Dineley, 1999f) lies above the Bishop's Frome Limestone and comprises a succession of fining-upward cyclic sandstone–siltstone–mudstone units. The basal coarse members of the cycles include intraformational conglomerate lenses, in which most of the disarticulated fish remains are found. The commonest fish below the limestone is *Traquaraspis (Phialaspis) symondsi*. Above the limestone, pteraspids dominate, including several species of *Protopteraspis*.

Interpretation

The Raglan Mudstone Formation is interpreted as the deposits of a coastal alluvial-floodplain subject to frequent desiccation and soil carbonate formation and crossed by minor distributary channels. The Bishop's Frome Limestone represents a prolonged period of basin-wide non-deposition and soil carbonate formation. The St

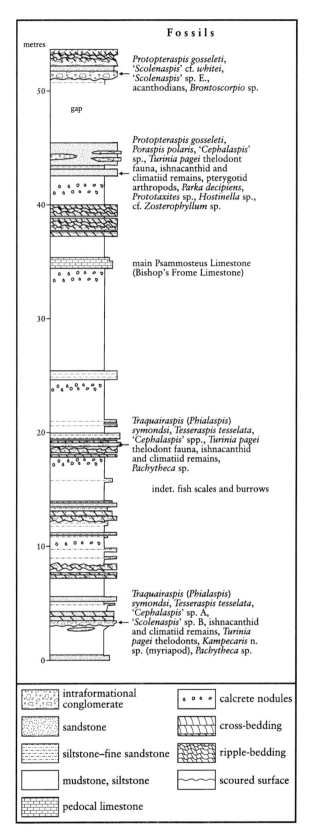

Figure 5.8 Vertical section of the strata at Devil's Hole. Based on Dineley (1999f, fig. 4.8), after M.A. Rowlands (MS).

Maughans Formation represents a medial alluvial environment, with the sandstones being mainly channelized, high-sinuosity stream deposits and the argillaceous lithologies being floodplain deposits.

Conclusions

Historically, this has been an important site for the large amount of fossil fish material it has yielded. The assemblages span the 'Downtonian'–'Dittonian' boundary. Although poorly exposed and difficult to access, the site presents opportunity for further excavation and may help in pinpointing the Silurian–Devonian boundary within the Old Red Sandstone.

OAK DINGLE, TUGFORD, SHROPSHIRE (SO 566 871)

Potential ORS GCR site

W.J. Barclay

Introduction

Oak Dingle near Tugford, Shropshire is already an established GCR site for its fossil fishes (Dineley, 1999f). A brief summary is presented here. The stream section (Figure 5.9) exposes a near-strike section in strata in the lower part of the St Maughans Formation (Ditton Group of Dineley, 1999f). Its main conservation interest is in the rich fossil fish fauna recovered from intraformational conglomerates. Another important feature not discussed by Dineley is the occurrence of modiolopsid bivalves. White (1935) first reported fish from this locality, since when it has been described by Ball and Dineley (1961), Allen (1964a), Greig *et al.* (1968) and Collinson (1978). Turner (1973) collected a thelodont fauna, and other microvertebrate remains from the site are listed by Vergoossen (2000). Greig *et al.* (1968) described modiolopsid bivalves from the site, first reported by Eyles (1953).

Description

Figure 5.10 shows a graphic log of the strata compiled by Allen (1964a). Greig *et al.* (1968) provided details of the section. The strata lie in the lower part of the St Maughans Formation, close above a group of calcretes in about 35 m of

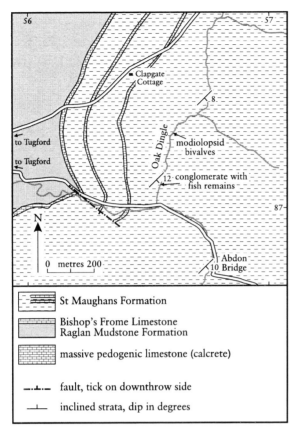

Figure 5.9 Geological map of Oak Dingle, Tugford. After British Geological Survey 1:10 560 manuscript map Shropshire 65NW (1960)

strata named the 'Psammosteus Limestones' by Ball and Dineley (1961) and Greig *et al.* (1968). In this area, there appears to be no single laterally persistent, thick limestone to mark the junction between the underlying Raglan Mudstone Formation and overlying St Maughans Formation. However, a thick (4.3 m) calcrete, the lowest of three, is present to the west of Oak Dingle and can be traced to a 1.3 m calcrete in the stream (SO 563 870) between Oak Dingle and Tugford. Above, it is separated by 0.6 m of mudstone from 0.6 m of mudstone- and calcrete-rich intraformational conglomerate, which is unusual in also containing round quartz pebbles. Greig *et al.* (1968) placed the junction (between their Ledbury Group and Ditton Series) at the base of this bed, which accords with the base of the St Maughans Formation in current lithostratigraphical classification.

The section mainly comprises sandstones showing evidence of multiple channel incision and filling in their basal part (Figure 5.10). Fish fragments, including *Pteraspis rostrata*, occur in a grey intraformational conglomerate (SO 5656 8709) and in the base of a 2.1 m-thick sandstone upstream (SO 5661 8730); at the latter, *Modiolopsis complanata* and *Modiolopsis* sp. occur in a grey siltstone about 0.6 m below the sandstone (Eyles, 1953). Dineley (1999f) summarizes the fish fauna as comprising the pteraspids *Pteraspis rostrata* var. *trimpleyensis* (this is the best site known for this species), *Wiegeltaspis* n. sp. (the earliest record in Britain); the cephalaspids *Cephalaspis* n. spp. and *Stensiöpelta* sp.; and indeterminate acanthodian spines. A *Turinia pagei* thelodont fauna is also recorded (Turner, 1973).

Interpretation

Figure 5.10 shows the sedimentological interpretation of the rocks at the site by Allen (1964a). The sandstones are interpreted as the deposits of a meandering, high-sinuosity river. Most of the sandstones are in-channel deposits, including some laterally accreted point-bars. The intraformational conglomerates were channel-bottom lag deposits. The finer-grained alluvial-floodplain muds and silts were largely destroyed by the meandering river system, but are preserved at the base and top of the section. Invertebrate burrows and calcrete nodules in the topmost bed point to levee, floodplain and backswamp environments. The presence of modiolopsid bivalves indicates sporadic marine influence, which had largely ceased by this time in the Anglo-Welsh Basin, but is known elsewhere at the same stratigraphical level (Barclay *et al.*, 1994).

Conclusions

The Oak Dingle site allows three-dimensional examination of the sandbodies of an early Devonian fluvial, meandering river complex and its overbank environments. The fish fragments recovered from the site comprise an important early Devonian assemblage and include new species and the earliest examples of some taxa. The site is also one of only a handful in the Anglo-Welsh Basin to have yielded modiolopsid bivalves from a level above the Psammosteus Limestone, providing rare evidence of marine influence in early Devonian times.

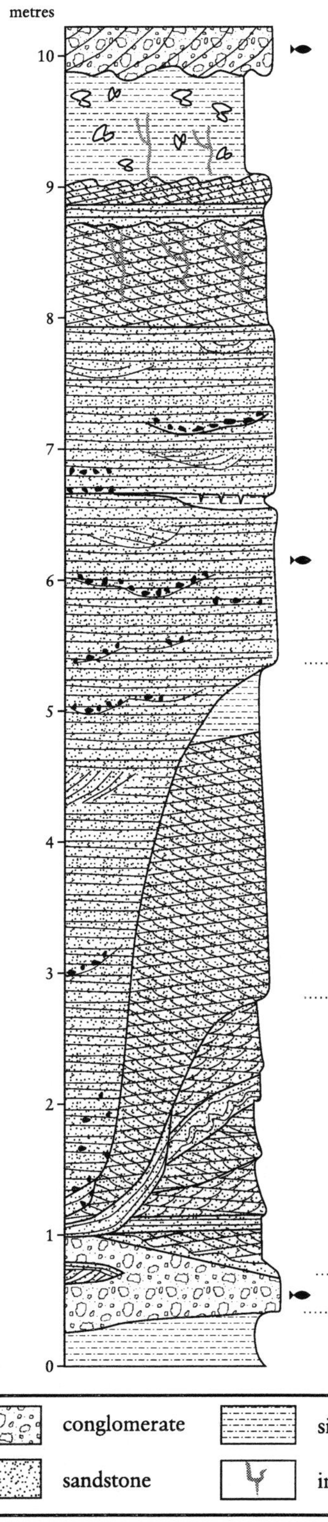

Oak Dingle

Main facts	Interpretation
Red coarse siltstones with invertebrate burrows and abundant calcium carbonate concretions, above red, very fine, ripple- or flat-bedded sandstones with invertebrate burrows.	Vertical accretion deposit from overbank floods. Levee overlain by backswamp deposits. Fluctuating groundwater table during times of exposure.
Fills and covers channel. Red, flat-bedded, fine sandstone with parting lineation, scour-and-fill and local scoured surfaces. Scattered siltstone clasts. Local cross-stratification. Lenticular sun-cracked siltstone.	Channel-fill and lateral accretion deposit. Sand transported as bed-load and reworked over shifting channel floor of flat-topped banks. Exposure of higher banks.
Cut on siltstone or very fine sandstone. In form of channel. Relief ~4.8 m.	Erosion at floor of wandering river channel. Extent of wandering possibly controlled by earlier channel plug.
Red, flat- or ripple-bedded, very fine sandstone passing up into red, coarse siltstone with carbonate concretions. Scattered siltstone clasts at base.	Channel-fill deposit. Overbank floods plug cut-off channel almost to top. Fluctuating groundwater table and periodic exposure.
Cut on very fine sandstones. In form of channel. Relief ~2.2 m.	Attempt to re-open plugged channel.
Repeated graded units overlying scoured surface with small-scale channels. Mostly ripple-bedded, very fine sandstone. Some siltstone.	Channel-fill deposit. Repeated intrusions of suspended sediment down a sloping surface, probably at times of higher stage.
Wedging intraformational conglomerate with sandstone lenticles.	Lag deposit formed at deepest parts of floor of wandering river channel.
Cut on siltstone. Low relief.	Erosion at floor of wandering river channel.

Legend: conglomerate, sandstone, siltstone, invertebrate burrows, cross-lamination, cross-bedding, parallel lamination, fossil fish fragments

Figure 5.10 Vertical section of the strata in Oak Dingle, Tugford, showing the sedimentary facies and their interpretation by Allen (1964a). After Dineley (1999f, fig. 4.11).

THE SCAR, HEREFORDSHIRE (SO 354 444)

W.J. Barclay

Introduction

The Scar GCR site is a river cliff on the north bank of a large meander loop of the River Wye 2 km north-west of the village of Monnington-on-Wye, Herefordshire (Figure 5.11). The cliff provides one of the largest vertical and lateral inland sections of the Přídolí Series Raglan Mudstone Formation in the Anglo-Welsh Basin.

Description

Clarke (1952) gave a brief description of the section and later (Clarke, 1955) described the geology of the area immediately to the west of the site on the Merbach Ridge. Brandon and Hains (1979) and Brandon (1989) gave details of the lithologies of the Raglan Mudstone Formation in the Hereford area to the east. Other good sections are present downstream near Breinton (SO 4518 3994), where about 11 m are exposed, and at Redbank Cliff, Holme Lacy (SO 5560 3614–SO 5552 3604), where about 32 m are exposed (Brandon and Hains, 1979). The Scar GCR site lies about 3 km to the east of a NW-trending, NE-facing escarpment that is capped by basal sandstones of the St Maughans Formation. The section is estimated to lie approximately 140 m below the top of the Raglan Mudstone Formation, which is marked by the Bishop's Frome Limestone.

Clarke (1952) noted that it was possible to see the cliff (named 'Brobury Scar') from the Deepwell (a source of tufa) on the opposite bank of the Wye. He noted somewhat lyrically that 'this is one of the biggest as well as one of the most beautiful exposures of the Old Red Sandstone in Herefordshire. There are some hundred feet of solid rock exposed and this consists of

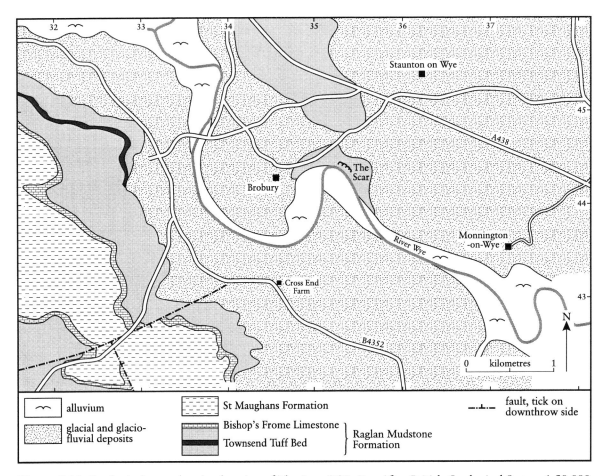

Figure 5.11 Geological map showing location of The Scar GCR site. After British Geological Survey 1:50 000 Sheet 197 (England and Wales), Hay-on-Wye (in press).

purple shales at the river level with two persistent green bands followed by a massive sandstone. There is then a great thickness of marl with a thinner sandstone band near the top. The shales show sun cracking and also worm casts.'

The river cliff is of Raglan Mudstone Formation. It is largely inaccessible and dangerous, Clarke presumably making most of his observations from the Deepwell. The base of the eastern sector is accessible with difficulty by descending from the cliff-top path east of the cliff and working back westwards through thick vegetation to the base of the cliff. The cliff exposes a 30 m-thick succession dominated by red and purple mudstones and siltstones, all of which are pedogenically altered to some degree. The beds dip about 5°–6° to the north-east. Thin lenses and tabular sheets of sandstone up to a maximum thickness of about 1 m occur sporadically.

Pale green, leached mottling and layers occur throughout the succession. The siltstones/mudstones are dull red, purplish and red-brown and micaceous. Lamination is evident at the bases of the units, but upwards they become massive, blocky and pedogenically disrupted. Incipient calcretization is ubiquitous in the topmost parts of the argillaceous layers, with prismatic vertisol fabrics and calcrete glaebule formation. All of the mudstones are distinctly calcareous. Locally, the calcrete nodules attain walnut-size. The most pedogenically altered horizons are clearly distinguished by their purple colour and pale green, prismatic vertisol mottling.

The sandstones are largely confined to the basal part of the section (Figure 5.12). The highest sandstone seen lies about 10 m above the base of the section. The sandstones are red-brown and purple, fine- to medium-grained and highly micaceous. They have gradational tops and sharp bases resting on erosion surfaces on the underlying mudstones. Some intraformational mudstone clasts are present in the bases of the sandstones. Cross-bedding seen is mostly planar, with minor synsedimentary deformation of the foresets seen at one point. One set of foresets indicates currents from the NNE. One bedding plane of a fallen block is littered with burrow traces.

Interpretation

The succession of strata of the Přídolí Raglan Mudstone Formation exposed at Monnington (or Brobury) Scar is interpreted as the deposits of an alluvial floodplain, formed under semi-arid tropical conditions and subject to seasonal

Figure 5.12 Sandstone overlying mudstone at the base of The Scar. (Photo: W.J. Barclay.)

wetting (e.g. Allen, 1974d; 1986; Brandon, 1989). Purple horizons are interpreted as cyclic pedogenic soil horizons, with incipient vertisol and calcrete formation. The deposits show a cyclicity typical of alluvial-floodplain deposits. The sandstones were either the channel-fills of shallow braided streams, perhaps subject to sporadic crevassing, or sheet-flood deposits. Detailed examination will refine the sedimentological interpretation. The environment was not entirely hostile to animal life, burrow traces being preserved locally.

Conclusions

The Scar GCR site provides one of the best inland exposures of Přídolí red-bed strata in the Anglo-Welsh Basin. About 30 m of beds mainly comprise a stacked succession of cyclic mudstone/siltstone-dominated units, which show evidence of soil formation in a semi-arid, seasonally wet climate.

CUSOP DINGLE, HEREFORDSHIRE–POWYS (SO 233 421–SO 257 384)

Potential GCR site

D.J. Hawley

Introduction

Cusop Dingle (Figure 5.13) is a prominent deep valley that cuts into the foot-slopes of the Black Mountains, immediately south-east of Hay-on-Wye, Powys. For much of its length it forms the border between England and Wales. The stream sections exposed by the Dulas Brook and its tributaries, the Crigiau Stream and Esgryn Brook, provide approximately 215 m of vertical succession, making up one of the two most continuously exposed inland sections of late Přídolí–early Devonian Old Red Sandstone succession in the Welsh Borderland–Black

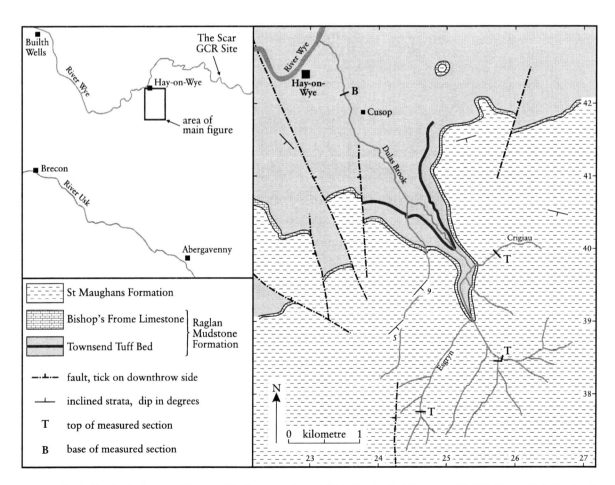

Figure 5.13 Geological map of Cusop Dingle. Based on British Geological Survey 1:50 000 Sheet 214 (England and Wales), Talgarth (2004).

Cusop Dingle

Mountains area. The **Sawdde Gorge** section (see GCR site report, this chapter) provides the other (Almond, 1983). The section exposes the upper part of the Raglan Mudstone Formation (late Downtonian–Přídolí) and the lower part of the St Maughans Formation (Dittonian) (Figure 5.14), the top of the former marked by a thick pedogenic limestone (the main 'Psammosteus' (Bishop's Frome) Limestone). The lower part of the succession is dominated by red siltstones, commonly showing pedogenic characteristics, and punctuated by grey, green and red sandstone bodies. It includes the volcanogenic Townsend Tuff Bed (Allen and Williams, 1981a). The upper part of the succession is dominated by green and grey fluvial sandstones with subordinate green and red siltstones. The lithologies illustrate a facies change in the Lower Old Red Sandstone from distal (marginal marine) mudflats to fluvial-dominated alluvial plains. Descriptions of geological localities in Cusop Dingle and the adjacent areas have featured in literature and other records, but the following account presents the first comprehensive description of the succession.

Murchison (1834) provided the first geological sections and notes on the area. King (1934) summarized the geology and stratigraphy of the area, allocating the succession to lithostratigraphical zones. M'Caw (1936) gave a brief description of the geology of the Black Mountains area based on King's work, together with a rudimentary geological map. From the 1930s to the 1950s, Cusop Dingle was visited by palaeontologists, including H.A. Toombs, W.N. Croft and R.H. Denison from the British Museum (Natural History), in search of early vertebrate remains. Brief lithological descriptions of the localities collected were recorded in their field notebooks. White (1946) provided outline descriptions of two localities in Cusop Dingle, together with lists of the fauna found. Turner (1973) noted a thelodont assemblage 9 m below the 'Psammosteus' Limestone in Cusop Dingle. The geology and stratigraphical succession of Merbach Ridge and Merbach Brook, immediately east of the Black Mountains and Cusop Dingle, was described by Clarke (1955). Clarke described and named two lithologically distinct marker units within the Downtonian Red Marls, the Newton Marlstone and the Middlewood Sandstone. He also mentioned a bed of 'volcanic ash' about 95 m below the main 'Psammosteus' Limestone, later correlated by Allen and Williams (1981a) with the Townsend Tuff Bed.

The geology of the Black Mountains area to the south of Cusop Dingle is not well documented, but Turner *et al.* (1995) described the microvertebrate fauna from a section exposed at Pwll-y-Wrach, near Talgarth, first recorded by Hawley (1991). Barclay and Wilby (2003) provide a brief description of the geology of the Talgarth 1:50 000 geological map.

Description

The lower part of the succession (Figure 5.13), below the Bishop's Frome Limestone, exposes about 160 m of strata dominated by packages of siltstones (65–75%), with the remaining 25–35% consisting of two types of sandstone bodies. The siltstones are red-brown, micaceous and generally poorly sorted. They contain some primary sedimentary structures and many secondary features, including palaeosols. The most abundant primary structures are parallel bedding planes, parallel lamination, low-angle ripple cross-lamination, wrinkle marks on bedding planes, desiccation cracks and a range of trace fossils, including surface trackways, shallow horizontal burrows, vertical lobate burrows and vertical *Skolithos*-type burrows. A high proportion of the siltstones contain pedogenic (vertisol) structures, including concentrations of pale subspherical carbonate (calcrete) nodules, most common in the upper part of some siltstone units, where they are locally concentrated and form horizons that are more resistant to erosion. Below horizons of carbonate nodules, or where concentrations are low or absent, many siltstones display large bowl-shaped, curved, slickensided fracture planes (pseudo-anticlines), blue-grey drab haloes, prismatic ped structures and wedge-shaped ped forms. Desiccation cracks occur in the tops of some beds with weak carbonate nodule development. The vertical profile of pedogenic features is truncated and repeated in many siltstone units.

The sandstones (Figure 5.15) are mainly green or grey, although purple varieties also occur and the finer-grained sandstones are commonly red. Many sandstones in the lowest part of the succession contain high proportions of mica, giving them a grey or silver colour, but increasing amounts of quartz in the higher sandstones result in them being dominantly

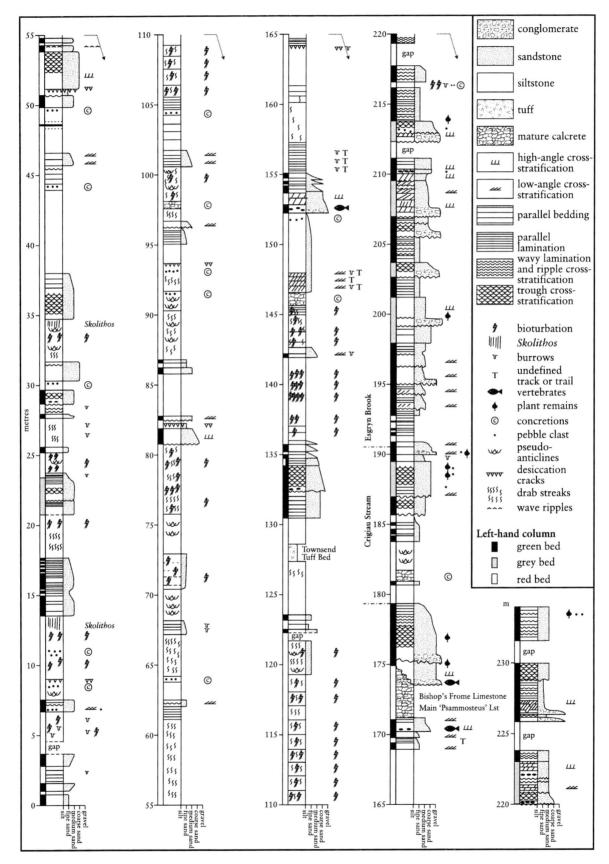

Figure 5.14 Graphic log of the section in Cusop Dingle.

Cusop Dingle

Figure 5.15 Distributary channel sandstone body in the Raglan Mudstone Formation at Cusop Mill fall (SO 239 413). Fining-upward tabular and trough-cross-stratified beds overlie a scoured surface cut in parallel-bedded fine-grained sandstones that conformably overlie siltstones. The coarser sandstones are more resistant and cap a waterfall. (Photo: D.J. Hawley.)

green or grey. Two distinct sandstone bodies can be recognized. Coarse- to fine-grained, fining-upward sandbodies are 3 m to 5 m thick. Their basal parts consist typically of fine- to medium-grained sandstone sets with parallel bedding, low-angle cross-stratification or ripple cross-lamination, capped by a thin, bioturbated siltstone with desiccation cracks. The succeeding main sandstone unit overlies a sharp, mild erosion surface with strings of siltstone clasts or laterally impersistent lenses of poorly sorted conglomerate containing clasts of red and green siltstone, calcrete pebbles and poorly preserved fish fragments. The sandstone is coarse grained and well-sorted, and comprises tabular and/or trough cross-stratified units. The trough cross-sets are thinly laminated where there is a high mica content. At the top of the sandbodies are fine-grained, tabular, parallel-laminated or ripple-laminated, fining-upward sandstone sets, capped by thin red siltstones or truncated by a scour surface. Palaeocurrent directions are variable, but predominantly to the south-east.

Medium- to fine-grained, well-sorted, red and green interbedded sandstone bodies overlie siltstone units. They range between 0.4 m and 0.8 m in thickness, and typically have a very low-relief, undulating base or one comprising interdigitating lenses of siltstone and sandstone. The beds commonly comprise low-angle cross-stratified beds or decimetre-thick tabular sheets with alternately coloured lamination. Some of the beds are bioturbated and contain shallow cylindrical burrows and trails on the upper surfaces.

The 'Psammosteus' Limestones Group (Squirrell and Downing, 1969) includes three mature calcretes and is the correlative of the Chapel Point Calcretes Member of Pembrokeshire, where there are up to eight profiles in 50 m of mudrock. The main 'Psammosteus' Limestone of Ball and Dineley (1961) and the Bishop's Frome Limestone of Brandon (1989) is 1.5 m to 4 m thick (its base is not exposed in the Esgryn Brook), with less well-developed calcretes 11.5 m above and about 20 m below.

A fine-grained, green, splintery tuff about 1.5 m thick with a distinctive jointing/fracture pattern (SO 250 400) occurs 40 m stratigraphically below the base of the main 'Psammosteus' Limestone, at 190 m above OD, and is overlain by two thinner beds of hard, pale purple rock, both with some green mottling and separated by a thin, red, coarse-grained sandstone. The lowest bed is very fine-grained, whereas the upper one is coarser grained and contains glassy fragments. These beds are correlated with the Townsend Tuff Bed (Allen and Williams, 1981a), which crops out at similar stratigraphical levels in the Digedi Brook and River Enig to the west, and in Scotland Dingle and Merbach Brook to the east.

The upper part of the succession (above the main 'Psammosteus' Limestone) exposes about 55 m of strata correlated with the St Maughans Formation. These are dominated by green and grey, coarse- to fine-grained sandstones and conglomerates comprising 65% of the succession, with the remainder being siltstones. In addition to the stream section, exposures also occur in small outcrops and old quarries scattered across the upper slopes of the valley.

The sandstones are multi-storey bodies up to 9 m thick made up of discrete channel-fill and fining-upward units of variable thickness. The base of each unit lies on an erosion surface of varying relief, overlain by lenses and thin beds of poorly sorted conglomerate with planar or low-angle cross-stratification. A scatter or string of disc-shaped siltstone clasts is common where conglomerate is absent. The overlying sandstones are generally well-sorted and preserve an upward progression of sedimentary bed types reflected in the decrease in grain size, from planar and trough cross-stratification through low-angle cross-beds to horizontally bedded and laminated beds with parting lineation, passing up into ripple cross-lamination showing common re-activation surfaces. This sequence is rarely found in full and is commonly truncated by an erosion surface and overlain by a succession of units displaying one or more of the bed types.

Poorly preserved plant fragments commonly occur in the coarser-grained sandstones, scattered along the bases of troughs. Concentrations of more comminuted plant debris are found at the tops of laminated beds. Similarly, fish fragments occur in basal conglomerates and coarse-grained sandstones, with better-preserved specimens occurring in the laminated beds. Burrows occur in the troughs of ripples. Palaeocurrents are variable, but have a predominantly southerly drainage direction.

The sandbodies grade up into fine-grained, micaceous beds up to 4.5 m thick. These are dominated by green, very fine-grained sandstones and green siltstones with subordinate thinner beds of intercalated green and red siltstones. In thicker developments, these pass up into red siltstone beds. Strong parallel lamination is dominant, with coarser-grained layers displaying low-relief undulations or low-amplitude ripple-lamination, sporadically disturbed by bioturbation. Some thicker red siltstones contain weakly developed pedogenic features.

Interpretation

The lithologies and facies displayed in the lower part of the succession in Cusop Dingle represent the distal component of an extensive fluvial system, deposited in a semi-arid climate. The environment is interpreted as a broad, low-relief aggradational alluvial-floodplain, largely composed of silt and crossed by shallow, through-flowing, sandy river channel systems. Mica-rich sediments and distinctive suites of heavy minerals suggest that the source of sediments was the metamorphic terrane of north-west Britain (Allen, 1974c).

The two types of sandstone bodies correspond to different types of fluvial drainage across the area. The thicker sandbodies represent channels of the main distributary systems. These were high-sinuosity, laterally migrating rivers that deposited broad bar forms and low-relief dunes and were subject to highly variable discharge rates, with channels drying completely at some times. Frequent flooding caused water to spill on to low-gradient interfluve areas, resulting in shallow outbreak channels and poorly channelized or unconfined ephemeral flows from which the thinner, sheet-like sandstone bodies were deposited (e.g. Williams and Hillier, 2004). The vertical distribution of the sandstone bodies and the development of pedogenic horizons suggest regular avulsion of the main channels, over distances of several kilometres (Love and Williams, 2000).

The packages of siltstones record episodes of subaqueous deposition interspersed with significant periods of subaerial exposure and pedogenesis. Thick sequences of well-laminated siltstones indicate frequently repeated and rapid inundation of the floodplain by shallow sheets of water followed by very short intervals of subaerial exposure. In some instances the sediment remained damp enough for colonization by burrowing invertebrates, although the bioturbation density is generally low, suggesting that residence time was limited by relatively rapid sedimentation rates. Marriott and Wright (2004) suggest that burrowed laminated facies in stratigraphically equivalent beds in Pembrokeshire may indicate more prolonged periods of subaqueous deposition in semi-permanent lakes formed in depressions in the floodplain. Long interruptions to these aggradational periods are recorded by desiccation cracks and other pedogenic features, which show that sedimentation rates slowed or stopped and the floodplain frequently dried out. Variation in the maturity of the pedogenic features and calcrete palaeosols indicates non-depositional periods of the order of between 500 and 30 000 years (Retallack, 1990). The main 'Psammosteus' Limestone represents the longest hiatus in sedimentation, for which Allen (1985) proposed a maximum period of 30 000 years. He postulated that the 'Psammosteus' Limestone facies records the commencement of final uplift of the former Welsh Basin and Irish Sea Ridge, producing a change in sediment provenance and depositional style. During this time, the Anglo-Welsh Basin was sediment-starved and effectively shut down. The Townsend Tuff Bed is evidence of a Plinian-type eruption that deposited volcanic ash across the floodplain. It is widespread throughout Pembrokeshire, central south Wales and the Welsh Borderland, providing a valuable stratigraphical marker horizon across the Anglo-Welsh Basin (Allen and Williams, 1981a). Its source is not known, although Allen and Williams (1981a) suggested it might have been either to the west or east along the strike line of the developing Rheic Ocean, with dispersal by winds from one of these directions.

The succession above the main 'Psammosteus' Limestone shows the change in depositional style consequent on the uplift to the north-west (Allen, 1985). This resulted in the establishment of predominantly fluvial conditions on a more proximal floodplain characterized in part by laterally accreting, sinuous, meandering rivers draining southwards (cf. Williams and Hillier, 2004). The overall climate may have become wetter, the region experiencing sub-tropical monsoonal conditions.

The sandstone bodies display the fining-upward features of point-bar deposits, arranged in (incomplete) multi-storey packages with common erosion surfaces and repetition of bedsets, which would be expected from lateral, downstream meander migration during aggradation (Bridge and Diemer, 1983). The prevalence of parallel-laminated beds suggests currents of relatively large stream power. Cross-cutting re-activation surfaces within ripple cross-laminated beds point to frequent fluctuations in water-levels. Individual bedsets were probably deposited during individual, major, flood events, but the rivers were perennially charged. The regularity of siltstone clasts in basal conglomerates and cross-beds and the low proportion of siltstone to sandstone indicate regular reworking of the floodplain. The increased amounts of plant material suggests more favourable conditions for plant growth, with stands of vegetation growing on the floodplain marginal to the channels. The dominance of the green and grey colour of the lithologies suggests that the area experienced relatively high water-tables during this period.

Conclusions

The stream sections in the Dulas Brook and its tributaries, the Crigiau Stream and Esgryn Brook, expose approximately 215 m of vertical succession, making up the most continuous and representative section of the late Přídolí–early Devonian Old Red Sandstone succession in the Welsh Borderland and Black Mountains area. The sections provide evidence of the changing conditions in the Lower Old Red Sandstone Anglo-Welsh Basin, illustrating a shift in environment from low-lying mudflats created by ephemeral distributary channels at the distal end of a river system, through a hiatus in sedimentation to alluvial plains dominated by perennially charged meandering rivers. The section includes exposure of two important stratigraphical marker beds in the Lower Old Red Sandstone, the volcanic Townsend Tuff Bed and the main 'Psammosteus' Limestone, which allow correlation and comparison with other Lower Old Red Sandstone outcrops in the Anglo-Welsh Basin.

The Anglo-Welsh Basin

SAWDDE GORGE, POWYS
(SN 729 245–SN 733 237)

P.R. Wilby

Introduction

The section described here lies in the Afon Sawdde gorge in Powys, south-central Wales, and is an extension to the 2 km-long section of strata of Llandovery to Ludlow age described in the companion GCR volume on Silurian stratigraphy (Aldridge *et al.*, 2000). It commences at Pont-ar-llechau bridge (SN 7285 2448) and extends for approximately 1 km upstream (Figure 5.16). Almost complete exposure in the bed and banks of the Afon Sawdde provides one of the most continuous sections in Wales and the Welsh Borderland through the upper part of the Ludlow Series, all of the Přídolí Series and the basal part of the lower Devonian (Lochkovian) succession. Crucially, it includes a complete section through the early Přídolí Tilestones Formation (the Long Quarry (Sandstone) Formation of Potter and Price, 1965), a probable lateral equivalent of the Downton Castle Sandstone Formation of the Welsh Borderland. The southern boundary of the original GCR site extended 150 m upstream of Pont-ar-llechau bridge, but it is recommended that it should be extended southwards to beyond Turkey Cottage (SN 7331 2365) to include the basal part of the St Maughans (Llanddeusant) Formation. It is a key section in the study of the nature and the

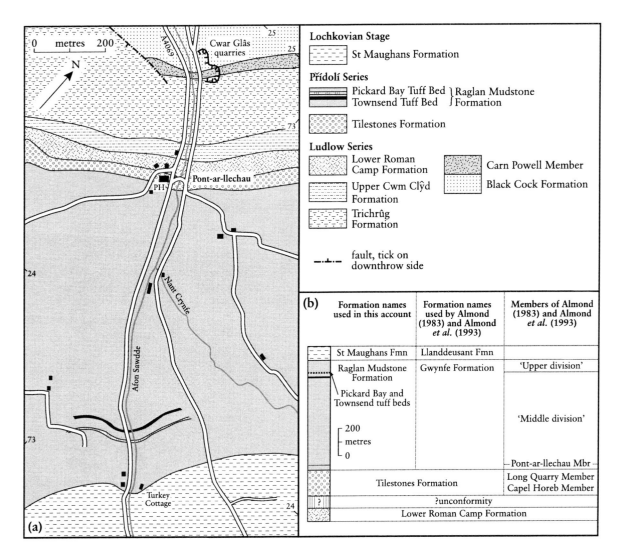

Figure 5.16 Geology of the upper part of the Sawdde Gorge: (a) – geological sketch map (north-west part after Derek J. Siveter, 2000); (b) – lithostratigraphical classification of the succession.

timing of the late Silurian marine to terrestrial transition. In addition, it is important for the interpretation of the depositional environments of the rocks exposed. It also provides one of the few inland exposures of the stratigraphically important Townsend Tuff Bed (Allen and Williams, 1981a), as well as the younger Pickard Bay Tuff Bed and a number of other lesser tuffs.

Since Murchison's (1839) visit, the section has been included in several regional studies (Strahan et al., 1907; King, 1934; Potter and Price, 1965; Squirrell and White, 1978; Bassett et al., 1982). The sedimentology and palaeontology of the Ludlow sequence are well known (e.g. Stamp, 1923; Potter and Price, 1965; Richardson and Lister, 1969; Burgess and Richardson, 1995; David Siveter, 2000) and have been summarized in field guides by Simpson (1971), Bassett (1982a), Siveter et al. (1989) and Almond et al. (1993). The site played a key role in the previous definition of the Silurian–Devonian boundary and of the base of the Old Red Sandstone in central Wales (see Potter and Price, 1965, fig. 2 for a summary of the most important views). The section formed a major component of Almond's (1983) PhD study of depositional environments (Almond et al., 1993; Figure 5.17). Almond (1983) retained the name 'Tilestones' for the Long Quarry (Sandstone) Formation, named the mid- to late Přídolí rocks the 'Gwynfe Formation', and the early Devonian rocks the 'Llanddeusant Formation'. The name 'Tilestones Formation' is used by the British Geological Survey (BGS) in this area and in this account. The other names used by the BGS (Raglan Marl Group (now Raglan Mudstone Formation) and St Maughans Formation respectively; Squirrell and White, 1978) have found more general acceptance (Siveter et al., 1989; David Siveter, 2000; Lane, 2000b) and are used in this account. Figure 5.16 shows these and the names erected by Almond. Most recently, Jenkins (1998) examined the uppermost part of the section in an attempt to establish the extent of marine influence on Old Red Sandstone deposition around the Silurian–Devonian boundary in the Anglo-Welsh Basin.

Description

The strata in the gorge dip steeply (up to 70°) towards the SSE. The base of the Old Red Sandstone (and of the Tilestones Formation) is exposed in the river bed 4 m downstream (north) of Pont-ar-llechau bridge, and in a cutting behind the former Three Horse Shoes Inn (SN 7283 2447; Figure 5.18). It is marked by a sharp change in colour and lithology, but with no angular discordance in bedding. Blue-grey, calcareous mudstones and siltstones (Lower Roman Camp Formation) are overlain by yellowish grey and green, micaceous sandstones and siltstones (Tilestones Formation). A regional unconformity has been postulated at the base of the latter (e.g. Potter and Price, 1965). About 14 km to the north-east of the gorge, the Tilestones Formation is said to rest unconformably on the late Ludfordian Upper Roman Camp Formation (e.g. Lane, 2000b), with the latest Ludfordian (Whitcliffe Formation) absent. In the Sawdde gorge, the Tilestones Formation rests on strata assigned by Potter and Price (1965) to the Lower Roman Camp Formation (e.g. David Siveter, 2000), both the Upper Roman Camp Formation and the Whitcliffe Formation apparently being cut out.

The upper part of the Lower Roman Camp Formation consists of hard, blue-grey, calcareous mudstones with abundant hummocky siltstone and sandstone interbeds 5–20 cm thick. It contains a restricted, but abundant, marine fauna, much of which is concentrated in shelly lags within the coarser units (Potter and Price, 1965; Siveter et al., 1989). The biostratigraphically important ostracod *Neobeyrichia lauensis* appears in the lower part of the formation, suggesting correlation with the Upper Leintwardine Formation (early Ludfordian) of the Ludlow Anticline (Potter and Price, 1965; David Siveter, 2000).

The Tilestones Formation is 18.5 m thick and consists predominantly of yellowish or green-grey, mica-rich, flaggy siltstones and fine- to medium-grained sandstones. A 0.5 m-thick mudstone occurs near the base. Bedding is conspicuously thicker (0.15–0.5 m) than that in the Lower Roman Camp Formation. A limited faunal assemblage includes brachiopods (e.g. *Lingula minima*, *Protochonetes ludloviensis* and *Microsphaeridiorhynchus nucula*) (Potter and Price, 1965; Almond, 1983), bivalves (e.g. *Modiolopsis complanata*), gastropods (e.g. *Turbochelius helicites*, *Loxonema gregaria*), crinoid ossicles, nautiloids (*Orthoceras* sp.), carbonaceous plant fragments and an assemblage of kloedinine beyrichiacean ostracods (Siveter et al., 1989; David Siveter, 2000). Most

The Anglo-Welsh Basin

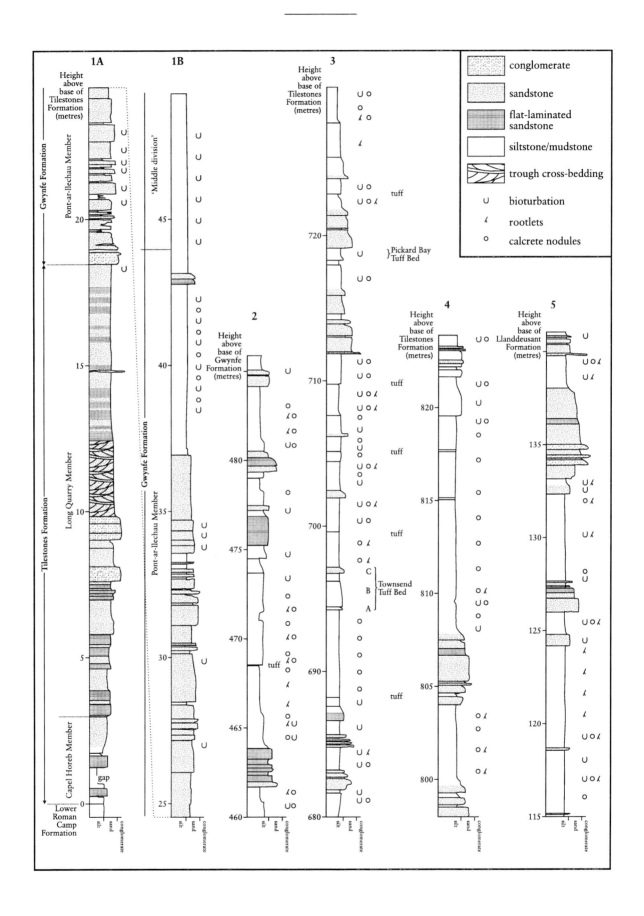

Sawdde Gorge

◀**Figure 5.17** Graphic logs of parts of the Old Red Sandstone succession in the Afon Sawdde. 1A,1B – continuous section of the Tilestones Formation and basal part (Pont-ar-llechau Member) of the Gwynfe (Raglan Mudstone) Formation (SN 7282 2452–SN 7286 2449); 2 – part of the 'Middle Division' of the Raglan Mudstone Formation (SN 7308 2400); 3 – part of the 'Middle Division' containing the Townsend Tuff Bed and Pickard Bay Tuff Bed (SN 7325 2385); 4 – section typical of the 'Upper Division' of the Raglan Mudstone Formation (SN 7330 2381–SN 7332 2372); 5 – section typical of the St Maughans Formation (SN 7340 2364). After Almond *et al.* (1993). Note that the logs have different scales.

of the sandstones are planar laminated or low-angle cross-laminated, some have bases resting on erosion surfaces. Many contain concentrations of shell fragments, green intraformational mudstone clasts (up to 3 cm) and faecal pellets; rare acid igneous clasts also occur. Locally, particularly near the top of the formation, communities of bivalves are preserved in life position and, in some of the thinner beds, burrowing (*Skolithos* sp.) destroyed all of the original sedimentary structures.

To the south-west of the Afon Sawdde, the Tilestones Formation is overstepped by the 'Green Beds' (the Capel Berach Beds of Potter and Price, 1965), the local basal unit of the overlying Raglan Mudstone Formation (Squirrell and White, 1978). In the Afon Sawdde, approximately 7 m upstream (south) of Pont-ar-llechau bridge, the formation is conformably overlain by the Raglan Mudstone Formation (Figure 5.17). The basal 26 m (the Pont-ar-llechau Member of Almond, 1983) are distinguished by very fine- to medium-grained, highly quartzitic sandstones with thin green siltstones and mudstones. The lowest 5 m of these comprise an upward-coarsening sequence of stacked, grey and green-grey, micaceous sandstones, which preserve horizontal to low-angle cross-lamination where not homogenized by bioturbation. *Skolithos* burrows are preserved locally and a bedding plane near the top exposes large *Fucusopsis*-type burrows. Some of the sandstones are composite bodies. Many of them rest on erosion surfaces, their bases containing scattered quartz granules, intraformational mudstone clasts

Figure 5.18 Section behind the former Three Horse Shoes Inn (SN 7283 2447) exposing the junction of the basal beds (Capel Horeb Member of Almond *et al.*, 1993) of the Tilestones Formation and the underlying topmost beds of the Lower Roman Camp Formation. The hammer marks the junction. (Photo: R.A. Waters.)

and/or shell fragments; one shows excellent soft-sediment deformation structures. Interbedded with the sandstones are dull green and yellowish green lenticular siltstones and mudstones.

Above are 15 m of red, micaceous, ripple cross-laminated and wavy bedded, very fine-grained sandstones and homogeneous siltstones and mudstones, some arranged in fining-upward packages. Bioturbation is widespread and the beds have yielded a poor, low-diversity, nearshore, shallow marine fauna, including the brachiopod *Lingula cornea*, the gastropod *Cymbularia carina*, the ostracod *Leperditia* sp., and a range of forms also present in the Tilestones Formation (e.g. *Modiolopsis complanata*, *Turbochelius helicites*, *Lingula minima*) and older strata (*Loxonema conicum*) (Potter and Price, 1965). Almond (1983) recovered a cornua of *Cyathaspis banksi*, an ostracoderm known from the Downton Castle Sandstone Formation (Bassett *et al.*, 1982).

The topmost 6 m of Almond's Pont-ar-llechau Member consists of dull red, massive mudstones and very fine-grained, ripple cross-laminated sandstones interbedded with wavy, flaser and lenticular bedded siltstones. Some of the sandstones are gritty and pebbly and all of the lithologies are commonly intensely bioturbated, with modiolopsids and lingulids seen in life position.

The succeeding 750 m of strata (the 'Middle division' and 'Upper division' of Almond's Gwynfe Formation) are typical of much of the Raglan Mudstone Formation in south and central Wales and the Welsh Borderland. They consist predominantly of massive, bright red, silty mudstones/siltstones with moderately well-developed calcrete profiles, and minor purple, grey and pink, fine- to medium-grained, lithic sandstones. The sandstones are typically 1 m to 3 m thick, locally pebbly, and have sharp bases resting on erosion surfaces. Internally, they generally show a fining-upward trend, with planar cross-bedding at the base passing up into horizontal lamination and ripple cross-lamination with lenticles of siltstone and mudstone at the top. Desiccation cracks and Skolithoform burrows are common at their tops, and some sandstones pass up into thin units of ripple cross-laminated silty mudstone.

Except for trace fossils, particularly *Beaconites antarcticus*, the 'Middle division' and 'Upper division' of the Raglan Mudstone Formation are poorly fossiliferous and have yielded only a few specimens of *Modiolopsis* sp., *Pachytheca* sp. and disseminated plant debris. Airfall tuffs are common (Almond, 1983; Almond *et al.*, 1993), the thickest of nine recorded being the Townsend Tuff Bed (2.9 m). As at **Little Castle Head** (see GCR site report, this chapter; Lane, 2000c), and at most localities in south-central Wales (Allen and Williams, 1981a), the Townsend Tuff Bed comprises three closely spaced airfall tuffs (A, B and C), which are sheared here. Fall A (0.25 m) is a muddy dust tuff with its upper surface characteristically strewn with faecal pellets. Fall B comprises fine-grained, cream, purple and yellow, crystal- and crystal-lithic tuff passing up into a siliceous dust tuff. It is truncated by an erosion surface overlain by Fall C, which consists of coarse- to medium-grained crystal-lithic tuff grading up into dull red and green mottled dust tuff. A tuff 23 m above is correlated by Almond (1983) with the Pickard Bay Tuff Bed of south Pembrokeshire. It is 0.75 m thick and consists of two superimposed tuffs, the lower fine-grained and crystal-lithic, the upper a muddy dust tuff. Both the Townsend and Pickard Bay Tuff beds outcrop in weathered recesses on the west bank of the river (*c*. SN 7325 2385) and the outcrop of the Townsend Tuff Bed can be traced in the adjacent field in a waterlogged gully.

Although still dominated by mudstones, the overlying 100 m of strata (the 'Upper division' of Almond, 1983) show a general coarsening-upward trend, through an increase in the thickness and the frequency of the sandstones, until the base of the St Maughans Formation is reached about 20 m south of Turkey Cottage. Almond (1983) and Jenkins (1998) identified two types of cyclic facies sequences in these beds. The first (5–8 m thick) comprises sandstones above a basal erosion surface, fining upwards through trough- and low-angle cross-bedded, fine- to medium-grained sandstones into micaceous, planar-laminated siltstones (up to 1.5 m thick) and then blocky siltstone with calcrete nodules. The second (4.4–8.7 m thick) is characterized by a coarsening- and then fining-upward trend. Typically, it consists of ripple cross-laminated siltstone (0.3–1 m thick) overlain by low-angle and trough cross-bedded sandstones (2.5–3 m thick) that fine upwards into thin (up to 1 m), planar-laminated and ripple cross-laminated siltstones. These are in turn overlain by a micaceous, blocky siltstone unit (up to 4.6 m thick), commonly containing calcrete nodules.

There are numerous moderately mature calcrete horizons in the uppermost beds of the Raglan Mudstone Formation, although the thick limestone development of the Psammosteus (Bishop's Frome) Limestone as seen elsewhere is absent. A 10.2 m-thick horizon of siltstone with calcrete nodules (Stage I–II calcrete of Machette, 1985) 10–20 m south of Turkey Cottage (SN 733 238) marks the top of the formation (Jenkins, 1998).

The basal 36 m of the St Maughans (Llanddeusant) Formation consists of interbedded sandstones and siltstones with minor conglomerates arranged in three fining-upward sequences (Jenkins, 1998). The sandstones include some laterally accreted units. The conglomerates rest on erosion surfaces and are intraformational, except for four thin (2–7 cm) extraformational beds near the base of the formation, which have small (up to 5 mm) quartz pebbles in addition to intrabasinal mudstone and siltstone clasts (Jenkins, 1998).

Interpretation

The period between the late Ludlow and late Přídolí was one of dramatic palaeogeographical and environmental change within the Welsh Basin (Woodcock, 2000b). The differentiation between basin and shelf that had existed throughout Ordovician and Silurian times (Holland and Lawson, 1963; Cherns, 1988) broke down as Avalonia finally docked with Laurentia, initiating basin inversion (Allen, 1985; Woodcock and Gibbons, 1988). The Sawdde Gorge is situated on the south-east flank of the Tywi Anticline, a major Caledonian lineament which formed part of a fault system that defined the southern margin of the basinal area (Woodcock and Gibbons, 1988). To the south of the fault system, the strata of the Ludlow Series formed mainly in shallow shelf and nearshore marine environments and show an overall shallowing trend that culminated in a regional hiatus at the base of the Přídolí Series (Bassett, 1982a; David Siveter, 2000).

This unconformity (Straw, 1930; Walmsley, 1962; Squirrell and White, 1978; Bassett et al., 1982) is placed at the base of the Tilestones Formation in the Sawdde Gorge (Potter and Price, 1965; David Siveter, 2000), although there is no angular discordance and no basal lag deposit, as reported elsewhere at this level.

The Tilestones Formation is interpreted as the product of shallow marine deposition, probably sourced in part from volcanic rocks to the south-west (Allen, 1985; cf. Almond, 1983). Sedimentation occurred in a range of nearshore to marginal marine and lagoonal settings against the NW–SE-trending shore of the 'Pretannia' landmass to the south (Potter and Price, 1965; Squirrell and White, 1978; Cope and Bassett, 1987). Almond (1983) interpreted the basal 3 m (the Capel Horeb Member) as the deposits of a brackish-water lagoon or protected shallow marine embayment, the sandstones representing washover storm events.

The upward increase in grain size and frequency of higher-energy bedforms were interpreted by Almond (1983) to be caused by increased wave action, or by encroachment of a tidal-flat complex into the area. Above is a transgressive sequence of lower to middle shoreface sediments (the Long Quarry Member of Almond, 1983) that were deposited on a high-energy coastline periodically affected by storm waves. The shelly sandstones were possibly the deposits of tidal channels.

Almond (1983) interpreted much of the Raglan Mudstone Formation as the deposits of the lower reaches of a muddy, low-lying coastal plain (cf. Allen, 1985). However, the basal 5 m (of the Pont-ar-llechau Member) are interpreted by her as the infilling of a shallow lagoon by washover sheet sands. The succeeding 15 m-thick, heterolithic sequence is interpreted as the deposits of small tidal channels. The uppermost 6 m of the Pont-ar-llechau Member are interpreted as a stacked, generally regressive sequence of subtidal to supratidal deposits, the latter containing evidence of pedogenesis in the form of small calcrete nodules.

The sandstones in the succeeding mudrock-dominated, calcrete-prone 'Middle division' and 'Upper division' are interpreted by Almond (1983) as the deposits of low- to moderately sinuous river channels, perhaps subjected to tidal influence. The calcretes point to carbonate soil formation on the alluvial floodplain. Jenkins (1998) distinguished fining- and coarsening-upward sequences in the uppermost 100 m of the Raglan Mudstone Formation, emphasizing the importance of rapid run-off, variable discharge and the flashy nature of sedimentation in the Old Red Sandstone. The fining-upward sequences are interpreted as the bars and fills of wide, shallow channels prone to overtopping, the overlying sediments probably being deposited during one discharge event. Siltstone-based, coarsening-upward sequences may have been the products of flood events, with initial deposition of silt from dense hyper-concentrated sediment- and soil-rich flood waters followed by sand deposition from streams derived from the upper catchment.

The absence of a thick, mature calcrete at the top of the Raglan Mudstone Formation equivalent to the Psammosteus Limestone was ascribed by Almond *et al.* (1993) to more continuous subsidence and sedimentation in this area, perhaps controlled by movement on the nearby Carreg Cennen Disturbance (Cope, 1979).

The St Maughans Formation represents a range of fluvial channel, overbank and floodplain facies similar to those at the top of the Raglan Mudstone Formation, but with channelized sandstones becoming thicker and more common at the expense of both finer-grained overbank and sheet-flood sandstones and argillaceous calcretized floodplain mudrocks. A thin, silt-stone-based, coarsening-upward unit at the base of the formation may have been the deposit of a two-peak, single discharge event similar to those in the Raglan Mudstone Formation.

Conclusions

The Sawdde Gorge site provides a unique transect through the Silurian and Early Devonian rocks of south-central Wales. Together with the sections through earlier Silurian rocks described in the GCR volume on Silurian stratigraphy (Aldridge *et al.*, 2000), the site and recommended extension present one of the most complete successions through the Silurian strata of the Welsh Basin and succeeding Old Red Sandstone of the Anglo-Welsh Basin. The part described here is a key section in the understanding of the transition from late Silurian offshore marine to littoral and terrestrial environments. The nature of the junction between the rocks representing the offshore and littoral environments warrants further examination. The site is one of only a handful of inland exposures of the strati-graphically important Townsend Tuff Bed. The recommended extension to the GCR site provides a complete section through the late Silurian (Přídolí Series) Raglan Mudstone Formation into the early Devonian St Maughans Formation, and provides an opportunity for further investigation of the Silurian–Devonian boundary.

PANTYMAES QUARRY, POWYS (SN 914 265)

Potential GCR site

W.J. Barclay

Introduction

Pantymaes Quarry, 3 km south of Sennybridge, Powys (Figure 5.19) is an important site in providing a superb inland exposure of a fluvial sandstone complex within the St Maughans Formation of the Lower Old Red Sandstone. The level of exposure has allowed a detailed analysis of the facies and sedimentology of the sandstone unit (Owen and Hawley, 2000). The quarry is also internationally known for its early arthropod trackways (Smith *et al.*, 2003).

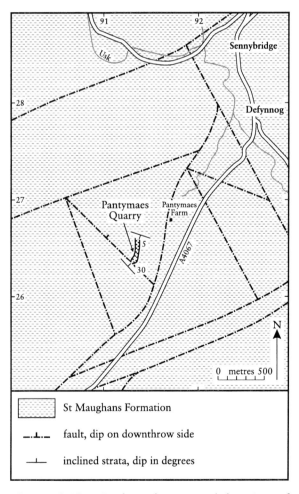

Figure 5.19 Geological map and location of Pantymaes Quarry. After British Geological Survey 1:50 000 Sheet 213 (England and Wales), Brecon (in press).

Pantymaes Quarry

Description

The long-abandoned Pantymaes Quarry (Figure 5.20) exposes a main face 25 m high and 275 m long in strata belonging to the Lower Old Red Sandstone. It has been described in detail and interpreted by Owen and Hawley (2000). These authors referred the section to the upper (Dittonian) Red Marl Group; more recent mapping by the British Geological Survey (Barclay *et al.*, in press) assigns it to the St Maughans Formation, of Lochkovian age. A description of arthropod myriapod trackways at the quarry has been given by Smith *et al.* (2003), who also provide sedimentological logs of eight vertical transects along the quarry face. Hassan (1982) carried out a palynological study.

The strata dip gently southwards in most of the quarry, but are horizontal along a synclinal axis in the southern part, with north-easterly dips to its south in the hanging-wall of a small NW-trending fault at the southern end of the quarry (cf. Owen and Hawley, 2000). The succession comprises a lower sandstone unit (the Sandstone Facies Association of Owen and Hawley, 2000) up to about 15 m thick (its base is not exposed) and an overlying red mudstone about 15 m thick (the Mudstone Facies Association of Owen and Hawley) (Figure 5.21). An erosion surface separates these units. Thin sandstone beds overlie the mudstone unit at the top of the quarry face.

The following account summarizes that of Owen and Hawley (2000). The sandstone unit consists predominantly of grey-green, micaceous, fine- to coarse-grained sandstone (Facies S2 of Owen and Hawley). Bedding is mostly planar, horizontal to gently inclined parallel lamination, with some trough cross-bedding. Erosive bounding surfaces of individual sandstone bodies define a complex of nested channels. Some bodies fine upwards, with intraformational calcrete clast conglomerate lenses (Facies S1 of Owen and Hawley) at the

Figure 5.20 Main face of Pantymaes Quarry. Lower sandstones are overlain by mudstones with thin sandstone beds at their top. The lower sandstones are 15 m thick. (Photo: D.J. Hawley.)

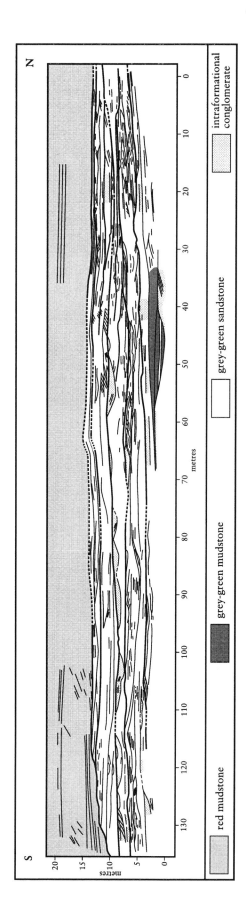

◀Figure 5.21 Main face at Pantymaes Quarry showing the main sedimentary bounding surfaces in the Sandstone Facies Association. After Owen and Hawley (2000).

base and cross-lamination at the top. Plant debris, including *Parka decipiens*, occurs in flaggy, micaceous sandstones and at the bases of fining-upward units. Arthropod tracks are recorded from the finer sandstones. Grey siltstone (Facies S3 of Owen and Hawley) occurs as clasts in intraformational conglomerate and as interbeds up to 1.5 m thick.

The sandstone body is traversed by a series of erosion surfaces, the most extensive of which can be traced for over 100 m. These are major (first-order) bounding surfaces that separate discrete channelized complexes, within which are less extensive second-order erosion surfaces. Five channel complexes (A to D) are recognized and described in detail by Owen and Hawley (2000).

The 15 m-thick red mudstone unit overlying the erosion surface that truncates the lower sandstone unit comprises, in detail, three facies. Facies M1 is a finely laminated mudstone with repeated horizons of bioturbation in the form of vertical burrows and horizons of calcrete nodules. Desiccation cracks are also present and seen best in fallen blocks. Facies M2 is predominant and comprises red, purple and blue, massive, friable mudstone with scattered calcrete nodules and a prismatic to blocky ped structure. There are some shallow, curved, slickensided slip surfaces and bioturbation by *Skolithos*-type vertical burrows. Facies M3 is a succession of upward-fining, red, micaceous siltstone–mudstone units, each about 0.55–0.7 m thick and consisting of a basal centimetre-thick sandstone, weakly cross-laminated, burrowed siltstone, and an upper 0.25–0.35 m-thick bed of upward-fining, intensely bioturbated siltstone–mudstone with thin blue-grey vein-like structures and desiccation cracks. *Skolithos*-type vertical burrows occur in the cross-laminated siltstones, as well as some ovate, horizontal burrows packed with pellets. Lobate sub-horizontal burrows are present at the top of some of the siltstones and arthropod trackways are present on parting surfaces. The topmost units have vertical burrows with a range of diameters and lengths of between

0.25 m and 0.35 m. Arthropod trackways are seen in fallen blocks of facies similar to these units.

The majority of arthropod trackways described in the quarry are seen in an isolated west-facing outcrop in the floor of the quarry. The outcrop is 5 m long and 3 m high and surrounds a pool in the centre of the quarry (Smith *et al.*, 2003). It consists of grey planar-laminated and ripple-laminated, micaceous, fine-grained sandstones and siltstones. The trackways (Figure 5.22) occur in 0.5 m of siltstones, which are overlain by cross-bedded, medium- to coarse-grained sandstones. They are assigned to two types (A and B) of *Diplichnites gouldi*, Type A probably being a kampecarid myriapod and Type B a coarthropleurid myriapod.

Interpretation

The intraformational conglomerate (Facies S1) lenses and beds are interpreted by Owen and Hawley (2000) as the deposits of high-energy channel-fills and bars. The sandstones (Facies S2) are interpreted as the fluvial deposits of braided river channelized flows with variable discharge, the dominance of planar lamination over cross-bedding suggesting a high-energy, perhaps flash-flow regime. Owen and Hawley interpret the grey siltstones (Facies S3) as wet, alluvial-floodplain deposits, high water-table allowing preservation of plant debris. It also allowed preservation of the trackways of the arthropods that inhabited and crawled over the floodplain.

The red mudstone with calcretized horizons (Facies M1) is interpreted as an overbank deposit, its occurrence above the erosion surface that truncates the underlying Sandstone Facies Association suggesting accumulation in an abandoned channel. The limestone nodules are Stage II calcrete soils (Machette, 1985), indicating prolonged carbonate soil formation and a semi-arid, seasonally wet climate. The red, purple and blue calcretized mudstone (Facies M2) represents flood basin overbank deposition of muds, the sediments being subject

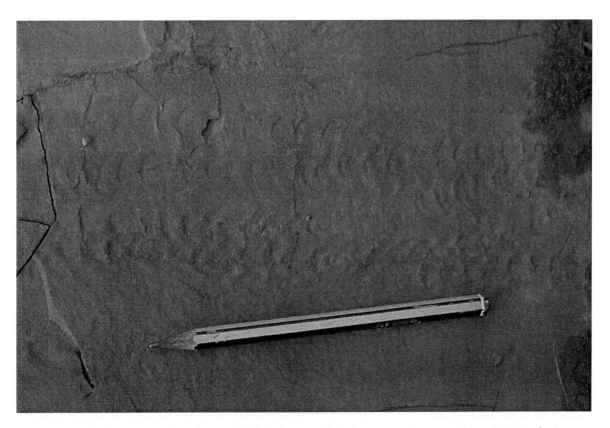

Figure 5.22 Arthropod trackways *Diplichnites gouldi* in Pantymaes Quarry. (Photo: D.J. Hawley.)

to repeated wetting and drying and carbonate soil formation. The upward-fining red siltstone units (Facies M3) are interpreted by Owen and Hawley as flood basin overbank deposits. Marriott and Wright (2004) suggest rapid deposition from crevassing events for similar heterolithic, fining-upward sequences in the Moor Cliffs Formation of Pembrokeshire. These authors stress the importance of deposition of pedogenic mud aggregate bedload-transported sediment in the accumulation of mudrock sequences in the Old Red Sandstone, and a similar origin for some of the mudrocks at Pantymaes is possible. They also suggest the presence of semi-permanent depressions on the floodplain in which ephemeral lakes formed after floods and which trapped dust from frequent dust storms. It is possible that some of the laminated mudstones (Facies M3) may have had a similar origin.

The quarry is important in exposing a sequence of facies that is markedly different from the fining-upward, autochthonous sandstone–mudstone alluvial cycles that characterize the St Maughans Formation elsewhere in south-east Wales and the Welsh Borderland (e.g. Allen, 1963b, 1970) and equivalent successions in Pembrokeshire (Conigar Pit Sandstone Member (Williams et al., 1982) and Gelliswick Bay Formation (Allen and Williams, 1978)). Owen and Hawley (2000) suggest an allocthonous, external, tectonic trigger for the change from the Sandstone Facies Association to the Mudstone Facies Association and the intervening erosion surface.

The channel complexes of the Sandstone Facies Association show major differences in fluvial style, suggesting to Owen and Hawley (2000) that it is not a multi-storey channel sandbody representing the punctuated infilling of a single channel, but that it represents distinct phases of fluvial development. Northward-stepping of successive erosion surfaces in several channel complexes further suggests tectonic control, each complex representing a response of the fluvial system to a pulse of tectonic movement. Owen and Hawley (2000) suggest that the sporadic movements on the Carreg Cennen Disturbance about 2 km to the north-west may have disrupted the general southerly drainage, producing an increase in stream gradient across the fault zone, braiding of the drainage to the south of the fault, and input of coarser intraformational bedload. Dextral strike-slip movement along the fault may have caused the migration of successive channels towards the north. Recent mapping by the British Geological Survey (in press) suggests the presence of several faults of north-east trend between Pantymaes and the Carreg Cennen Disturbance, and these may also have exerted control on sedimentation.

Conclusions

Pantymaes Quarry is a site of national and international importance, and worthy of consideration for protected status. Its conservation value lies in it being almost unique in the Anglo-Welsh Basin in providing a large, accessible inland section of the St Maughans Formation where lateral facies variations in a sandstone complex can be demonstrated. It also shows a style of facies architecture quite different from that seen elsewhere at this level in the basin, perhaps indicating pulsed movement on the nearby Carreg Cennen Fault and local tectonic control on drainage development. The well-preserved, internationally known arthropod traces also merit protection, providing valuable information on our understanding of the earliest known land animals that colonized the alluvial plains of the Early Devonian.

HEOL SENNI QUARRY, POWYS (SN 915 221)

W.J. Barclay

Introduction

Heol Senni Quarry lies on the north-east face of Fan Bwlch Chwyth (Figure 5.23), Powys, 1.8 km south-west of the village of Heol Senni. The quarry provides a fine section of the Lower Devonian Senni Formation, exposing about 40 m of a succession dominated by grey-green sandstones (Figure 5.24). In addition to the site's lithological and sedimentological value, it is the type locality of *Althaspis senniensis*, which is known only from this site, providing unique dating evidence of the Senni Formation. A wide range of fossil plant remains and miospores adds to the site's importance.

Heol Senni Quarry

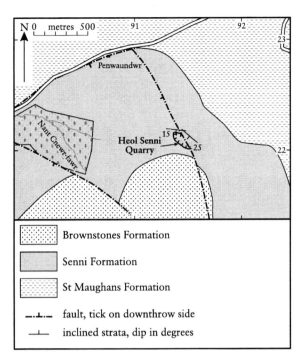

Figure 5.23 Geological sketch map and location of Heol Senni Quarry. After British Geological Survey 1:50 000 Sheet 213 (England and Wales), Brecon (in press).

Description

The quarry was first described by Edwards *et al.* (1978) for a field guide produced for a symposium on the Devonian System. These authors summarized the lithologies, palaeobotany and palynology. During the symposium field visit to the quarry, fossil fish remains (*Althaspis senniensis*) were discovered (Loeffler and Thomas, 1980). Almond *et al.* (1993) gave a brief description of the quarry. The site's importance as a fossil fish locality is detailed in the GCR fossil fishes volume (Dineley and Metcalf, 1999).

The succession comprises grey-green sandstones, with minor siltstones, mudstones and intraformational conglomerates. The beds dip generally about 5° to the SSW. A NW-trending fault, which throws down to the west, cuts the south-east corner of the quarry, producing a steeper (*c.* 25°) south-west dip in its footwall, with dips of 5° in a minor hanging-wall anticline. There is no obvious cyclicity to the arrangement of the facies. The sandstones are fine- to medium-grained, lithic arenites and are generally

Figure 5.24 View looking into the north-west corner of Heol Senni Quarry. (Photo: W.J. Barclay.)

massive, but have low-angle cross-bedding and parallel lamination. The latter is defined by grain size, and mica and carbonaceous concentrations. Some large-scale planar and trough cross-bedding is also evident. The finer-grained sandstones are ripple cross-laminated and linguoid ripple marks are present on some bedding surfaces. The sandbodies overlie erosion surfaces, with lenses of intraformational conglomerates at their base. The conglomerates, most of which are calcitic, contain grey-green siltstone intraclasts up to 450 mm in length and rolled calcrete nodules. No exotic pebbles are recorded. Some sandstone beds are rich in mica and show primary current lineation. Shredded plant debris is also common. Soft-sediment deformation structures are prominent and include a laterally persistent ball-and-pillow unit and large-scale slump folds. The thin mudstone interbeds are mostly grey-green and blue-grey, but red-brown siltstones, some containing calcrete nodules, are present towards the top of the exposed succession. Some of the mudstones have sandstone-filled desiccation cracks.

The quarry is one of only four localities to have yielded fossil fish remains in the Senni Formation discovered to date. The species (*Althaspis senniensis*) is unique to this site (Loeffler and Thomas, 1980; Dineley, 1999f). The other three localities are Primrose Hill Quarry, Crickhowell (White, 1938; Barclay, 1989), which yielded *Rhinopteraspis dunensis* (= *cornubica*), Ferryside, Carmarthenshire, which yielded *Pteraspis dixoni* and *Cephalaspis* sp., and Allt ddu, Brecon (SO 027 242), which yielded *Protopteraspis gosseleti* (Habgood, 2000; Edwards and Richardson, 2004). Dineley (1999f) gave a detailed account of the faunal significance of the Heol Senni site and it is not repeated here.

Plant fragments are common (Edwards *et al.*, 1978). Vascular plant axes several centimetres long, cf. *Sawdonia* and cf. *Cooksonia* are recorded. Edwards *et al.* (1978) recorded well-preserved miospore assemblages from blue-grey siltstone beds. One assemblage from a fissile bed near the top of the succession contains large numbers of distally sculptured *Emphanisporites* specimens and numerous examples of *Apiculiretusispora*. Thomas (1978) recognized over fifty taxa of dispersed miospores. The important constituents of the assemblages are listed by Edwards *et al.* (1978).

Interpretation

Loeffler and Thomas (1980) provided the most detailed interpretation of the sedimentary environments of the Senni Formation (cf. Thomas, 1978). Allen (1974a, 1979) provided a general overview. Loeffler and Thomas (1980) interpreted the deposits as those of a comparatively high-discharge, mixed bedload, sand-dominated, braided stream complex in a medial alluvial setting. Marginal ponds on the river floodplains allowed colonization by land plants. Although most of the succession comprises in-channel sandbodies, overbank siltstones and mudstones are preserved as thin drapes and interbeds. Relatively high sedimentation rates and water-table levels during accumulation are indicated by the combination of the vertical succession of litho-facies units, the prevalence of grey-green beds, the suites of stratification types and other sedimentary structures, and the preservation of the vascular plant remains and miospores.

The in-channel facies recognized include channel lags, channel-fill deposits, units formed by the migration of linguoid, transverse, lateral and rhomboid mid-channel bars and low-amplitude sand waves, bar delta wedges and bar top sequences. Thick, multi-storey sand-bodies containing interbeds and/or lenses of intraformational conglomerate that overlie basal erosion surfaces imply repeated channel super-position, filling and excavation. Many channel units are draped by thin mudstones deposited from suspension during waning flood stages. Local, rapid, channel shifting and complete or partial channel abandonment was common. Ponding of flood waters occurred in channel cut-offs and (between stages) in parts of the secondary and tertiary distributaries of the active channel network. This local ponding, combined with decomposition of plant material (with the subsequent lowering of redox potentials), and the presence of reworked calcrete glaebules and calcitized plant fragments, giving high pH and bicarbonate ion concentrations, resulted in the calcification of many channel-lag intraformational conglomerates.

The extra-channel and floodplain sediments comprise proximal and distal crevasse-splay deposits, some possible levee sediments, and thin, fine-grained, fluvio-lacustrine units laid down in temporary floodplain lakes.

Sandstones interbedded with these mudstone-dominated fluvio-lacustrine units are interpreted as incursions of crevasse-splay sands deposited during river-avulsion episodes. The Heol Senni pteraspid may well have been carried from its habitat in the main channel system and deposited within a floodplain lake or channel-fill mud unit during one such flood event. Strongly reducing post-depositional conditions are indicated for most of the fluvio-lacustrine sediments by their typically blue-grey to grey-green colour, and the preservation within them of pyrite nodules, macroplant cuticles and miospores.

Vascular plants flourished along the shores of temporary lakes, and colonized abandoned channel-fills and near-channel overbank deposits. High sedimentation rates and water-table levels, and post-burial reducing conditions resulted in a high preservation potential for the plants.

In the Brecon Beacons the Senni Formation–Brownstones Formation boundary has been drawn at the junction between the predominantly grey-green sequence and the overlying red-brown succession, even though there are apparently no major sedimentological differences between the lithofacies immediately below and above the colour change. The facies also interfinger locally (Barclay, 1989) and Owen and Hawley (2000) note a similar interfingering of green and red facies in the underlying St Maughans Formation. However, the Brownstones higher in the succession were laid down by sandy braided streams of a more shallow and ephemeral nature than those which deposited the sediments of the Senni Formation (Tunbridge, 1981a).

The Senni Formation is the lithostratigraphical and chronostratigraphical equivalent of the lower and middle portions of the 560–600 m-thick Mill Bay Formation of the Cosheston Group of south-west Wales, which Thomas (1978; Loeffler and Thomas, 1980) and Wellman et al. (1998) interpreted as the deposits of a braided–meandering river system. Owen (1995) proposed a similar interpretation for the Senni Formation of the Llansteffan peninsula, suggesting low-sinuosity channels with seasonal flow in a seasonally wet, semi-arid climate.

Loeffler and Thomas (1980) and Dineley (1999f) provided the most comprehensive stratigraphical reviews. The palynomorphs indicate a late Lochkovian to Pragian age and the pteraspid *Althaspis senniensis* is a late Dittonian to Breconian form. The Primrose Hill pteraspid *Rhinopteraspis dunensis* occurs in the type area of the Siegenian (Pragian) stage. This, combined with the Heol Senni pteraspid and the palynomorphs, indicates that the Lochkovian–Pragian boundary lies within the Senni Formation. *Rhinopteraspis dunensis* is taken as the index fossil of the poorly defined, local Old Red Sandstone Breconian stage.

Conclusions

Heol Senni Quarry is of great importance for the sedimentological and stratigraphical information it provides. It has the thickest exposed inland vertical succession of Senni Formation in south Wales. The architecture of the stacked sand-dominated units and the relationship and types of the lithofacies present allow the anatomy of the contemporaneous floodplain and river channel environments to be modelled. Rapid sedimentation and a high water-table subsequent to burial of plant fossils provided high preservation potential and the site has yielded early vascular plant remains, as well as a rich miospore assemblage. The latter is particularly important in stratigraphical correlation in view of the lack of body fossils, except for the discovery at the quarry of the only specimen (the holotype) of the ostracoderm fish *Althaspis senniensis*. This is one of only four localities in the Senni Formation to yield fossil vertebrates, and together with its miospore assemblage, makes the site of great importance, fully justifying its protected status.

CAERAS QUARRY, CARMARTHENSHIRE (SN 607 167)

W.J. Barclay

Introduction

Caeras Quarry, 2 km north-west of the village of Llandybie (Figure 5.25), exposes conglomeratic sandstones at the top of the Brownstones Formation of the Lower Old Red Sandstone. Quartz pebbles predominate in the beds, but some larger pebbles are locally derived and provide evidence of tectonic movements

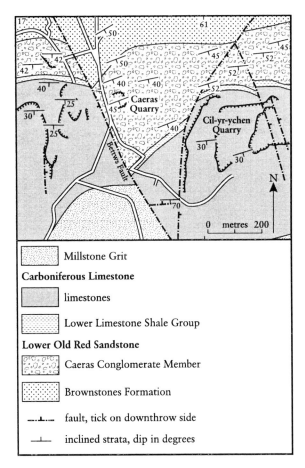

Figure 5.25 Geological map of the area of Caeras Quarry. After British Geological Survey 1:10 560 Sheet SN 61NW (1973).

causing local basement uplift and erosion. This contrasts with most of the Brownstones, which had a source area far to the north. This pebbly facies is only locally developed on the north crop of the South Wales Coalfield at this level, and this quarry provides the best exposure in the area. Similar pebbly sandstones and conglomerates are present to the east around Llyn-y-Fan Fawr (SN 834 216) in the Carmarthen Fans (Tunbridge, 1980a).

Description

Caeras Quarry provides the best exposures of conglomeratic sandstones that occur at the top of the Brownstones Formation in the west of the Cennen Valley, between Llandybie and Kidwelly, south-east Carmarthenshire (Strahan *et al.*, 1907; Squirrell and White, 1978). The conglomerates were formerly termed the 'Pebbly Beds' (Institute of Geological Sciences, 1977; Squirrell and White, 1978), but are here named the 'Caeras Conglomerate Member' to distinguish them from pebbly beds that occur at different levels elsewhere in the Brownstones. The quarry illustrates the variable nature of the succession, although overgrown at the time of writing, with the highest beds in the south wall providing the best exposures.

The Brownstones Formation in the area has been described by Squirrell and White (1978). It comprises a succession of interbedded red-brown sandstones and siltstones/mudstones, with rare, intraformational, calcrete-clast conglomerates. A few of the siltstone/mudstone beds contain calcrete nodules. Conglomerates are common in the west of the Cennen Valley area, and at Caeras Quarry they are interbedded with siltstone/mudstone beds containing abundant calcrete nodules. The overall proportion of sandstones and conglomerates in the succession is generally 50–65%, although ranging from 40% to 75%.

The Brownstones Formation forms the highest beds of the Early Devonian Lower Old Red Sandstone. Although unfossiliferous, it is assigned a late Pragian to Emsian age. It also belongs to the local Breconian stage, and is unconformably overlain by the Lower Carboniferous Lower Limestone Shale Group at Caeras. In the region of Caeras Quarry, the Brownstones are thickest in the east, reaching 610 m in the Sawdde Fechan and Clydach valleys, but attenuating markedly westwards to about 210 m south of Carreg Cennen Castle. To the west of there, they thicken again to about 370 m in the vicinity of Caeras Quarry. The Caeras Conglomerate thickens markedly from its feather edge at Blaengweche Farm about 3.5 km east of Caeras Quarry to its maximum development of 170 m at Caeras. It then thins from Caeras westwards to 90 m at Carmel (SN 5865 1660) and to about 50 m west of the Afon Dulais.

Squirrell and White (1978) recorded the section at Caeras Quarry. Figure 5.26 shows the section graphically, modified in the light of a recent site visit, where only the topmost beds (Figure 5.27) remain well exposed. The beds dip about 45° SSE. Harrison (in Squirrell and White, 1978, their Appendix 1) gave details of some of the lithologies present.

About 13 m from the top of the section, within 5.2 m of sandstones noted by Squirrell

Caeras Quarry

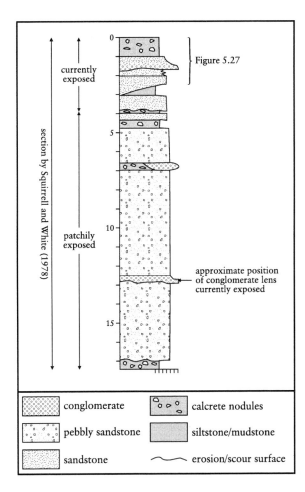

◀Figure 5.26 Graphic section of part of Caeras Quarry. After Squirrell and White (1978).

Cantrill (in Strahan *et al.*, 1907, p. 63) recorded a conglomerate ('conglomeratic cornstone') in this quarry 8.2 m below the base of the Lower Limestone Shale Group. It contained a few pebbles of 'earthy limestone' containing what are probably *Camarotoechia nucula*, *Chonetes striatella*? and *Orthis* (*Dalmanella*) *elegantula*. Cantrill noted that the pebbles are from the Ludlow Series. Squirrell and White (1978) noted that the quarry is now partially filled and that no conglomerate is visible.

Interpretation

The Brownstones of south Wales and the Welsh Borderland are broadly interpreted as the proximal fluvial deposits of a piedmont plain on the southern margin of the evolving Caledonian mountain chain. They represent the culmination of a Lower Devonian coarsening-upward off-lap succession (Allen, 1979; Allen and Williams, 1979b). Continuing uplift and southerly migration of the facies belts led to non-deposition and erosion in the Anglo-Welsh basin in Mid-Devonian times. Allen (1974a) provided a general interpretation of the sedimentation of the Brownstones. Allen and Crowley (1983) demonstrated that the source of the sediments was the outcrop of Lower Palaeozoic rocks in the Irish Sea–north Wales area. Tunbridge (1980a, 1981a) interpreted the sheet sandstones of the Brecon Beacons area as semi-arid sheet-flood deposits, deposited on extensive alluvial-fans. The mudstones and siltstones are interpreted as floodplain deposits, either deposited from flash-floods, or from slow-moving or still water bodies, although an aeolian origin for some is also possible (Tunbridge, 1981a). The carbonate nodules in the finer lithologies are interpreted as pedogenic calcrete formed on the alluvial floodplains and are indicative of a tropical, seasonally wet, semi-arid climate (e.g. Marriott and Wright, 2004).

and White, a grey-green, brown-weathered, quartzitic sandstone has a conglomeratic, pebbly base. A lens of grey mudstone with enclosed quartz pebbles occurs near the base of the sandstone, but it is not clear from the limited exposure whether it is a large reworked clast or the remains of an in-situ bed. Palynological analysis of the mudstone produced no dateable spores (M. Stephenson, pers. comm.).

The highest strata of the Caeras Conglomerate Member were formerly exposed at the nearby Cil-yr-ychen Quarry (SN 6165 1686), where basal olive-green sandstones of the Carboniferous Lower Limestone Shale Group rest on:

	Thickness (m)
Sandstone, red, massive, fine- to coarse-grained, commonly conglomeratic (quartz pebbles)	17.0
Siltstone, red, with many calcrete nodules up to 5 cm long	0.6
Sandstone, red, massive, medium- to coarse-grained, with quartz pebbles	4.0

Local pebbly facies similar to the Caeras Conglomerate Member occur elsewhere in the higher parts of the Brownstones Formation in south Wales and the Welsh Borderland. Allen (1974b) described the clasts in the Ross-on-Wye area. The pebble suite there comprises a range

The Anglo-Welsh Basin

Figure 5.27 Topmost beds exposed in Caeras Quarry; red mudstone/siltstone with calcrete nodules rests on sandstone. The hammer is 0.3 m in length. (Photo: W.J. Barclay.)

of igneous, sedimentary and igneous rocks ranging from Ordovician to early Devonian in age, with perhaps some Precambrian, and concluded that they were derived from north Wales. A similar pebble suite occurs in the Woodbank Series of Shropshire (Allen, 1974b), which is the stratigraphical equivalent of the Brownstones. The pebbles in the conglomeratic facies in the topmost 100 m of the Brownstones around Llyn-y-Fan Fawr (SN 834 216) about 24 km north-east of Caeras Quarry (Tunbridge, 1980a) are angular, and hence of local derivation. They include acid volcanic rocks, lithic arenites and vein quartz derived from the east, suggesting relatively short-lived uplift of older Lower Palaeozoic rocks. Tunbridge concluded that this is evidence of uplift caused by strike-slip movement along the Swansea Valley Fault 4 km to the south. Cope (1981) suggested that a Precambrian source lay in the vicinity, and may have supplied the volcanic and vein quartz pebbles, with Cambrian outcrops supplying the lithic arenites. The Caeras Conglomerate occurs south-east of, and in close proximity to, the Carreg Cennen–Llandyfaelog Fault, which, along with the Swansea Valley and Neath disturbances, makes up a suite of major NE-trending basement fractures. It is possible that the conglomerate provides evidence of late Early Devonian movement on the Carreg Cennen–Llandyfaelog Fault (Tunbridge, 1986), although the source of the pebbles remains problematic. The quartz pebbles may have been derived locally, as in the case of the Llyn-y-Fan Fawr deposits, or from north Wales, as deduced by Allen (1974b) for those in the Ross-on-Wye area, but the Ludlow pebbles at Cil-yr-ychen Quarry may have been derived from a closer, unknown source. The conglomerates such as those at Caeras and Llyn-y-Fan Fawr occur high in the Brownstones Formation and may herald the onset of the Acadian Orogeny.

Some doubt remains concerning the age and correlation of the Caeras Conglomerate Member. Allen (1974a) noted that the pebbly beds vary widely in position relative to the top of the Senni Beds, which underlie the Brownstones, but everywhere underlie the Carboniferous sequence. He suggested that they might therefore belong to the Upper Old Red Sandstone and to be of late Devonian to early Carboniferous age. Squirrell and White (1978), however, retain the beds as part of the Brownstones Formation.

Conclusions

Caeras Quarry provides the best exposure of locally developed pebbly beds (the Caeras Conglomerate Member) lying at the top of the Brownstones Formation. There is some debate as to whether the beds belong to the topmost Brownstones or the overlying Upper Old Red Sandstone, and the site merits palynological examination to resolve this, as well as a detailed sedimentological study. The section illustrates the variable nature of the beds, comprising interbedded sandstones, pebbly and conglomeratic sandstones and siltstones/mudstones with calcrete nodules. Some of the pebbles may have been of local derivation, providing evidence of late Early Devonian movement on the major NE-trending lineaments and erosion of Precambrian and older Lower Palaeozoic basement outcrops.

CRAIG-Y-FRO QUARRY, POWYS (SN 972 207)

Potential ORS GCR site

W.J. Barclay

Introduction

Craig-y-Fro Quarry is a classic Old Red Sandstone fossil plant site, the occurrence of well-preserved early Devonian land plants from here being the first recorded in southern Britain (Heard, 1926, 1927). It is already an established GCR site for its Palaeozoic palaeobotany (Cleal and Thomas, 1995) and only a brief summary is presented here. The quarry has yielded one of the best-preserved Devonian floral assemblages in Britain and is the type locality for several new species discovered by Professor Dianne Edwards. It is situated by the A470, 10 km south-west of Brecon, Powys, and 1 km north-west of the Storey Arms Centre (Figure 5.28). Variously known as 'Brecon Beacons Quarry' and 'Storey Arms Quarry', Cleal and Thomas (1995) gave it the geographically more precise name 'Craig-y-Fro Quarry'. It lies near the top of the Senni Formation of the Lower Old Red Sandstone, about 60 m below the junction with the overlying Brownstones Formation. A brief geological description was given by Robertson (1932). Edwards and Richardson (1978) provided a detailed description of the section (Figure 5.29) and its plant and miospore assemblages. Hassan (1982) also reported on the miospore assemblages from the site.

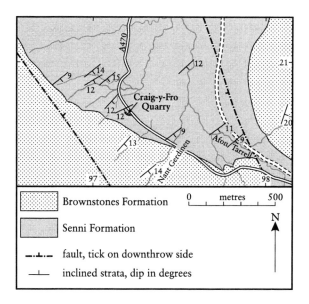

Figure 5.28 Geological map of the area around Craig-y-Fro Quarry. After British Geological Survey Scale Sheet SN 92SE (1973).

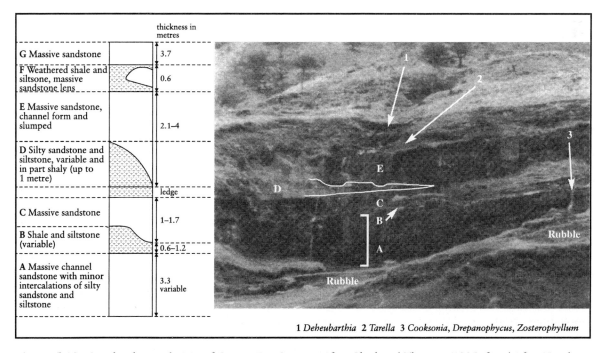

Figure 5.29 Graphic log and view of Craig-y-Fro Quarry. After Cleal and Thomas (1995, fig. 4.16). Numbers (1), (2), and (3) are three of the main plant beds. (Photo: D. Edwards.)

Description

The quarry exposes up to about 14 m of beds that dip 12° to the south-east. They consist mainly of massive, lenticular, channelized green sandstones, with siltstone interbeds that are truncated by erosion surfaces at the base of the overlying sandstones. Figure 5.29 (from Cleal and Thomas, 1995) is based on the section recorded by Edwards and Richardson (1978). Four plant-bearing horizons were recorded by these authors. The lowest lies about 7 m above road level and has yielded the type specimen of *Gosslingia breconensis*. A silty sandstone 2 m above contains *Cooksonia*, *Drepanophycus* and *Zosterophyllum*. An exposure of a weathered siltstone above the quarry at its southern end has yielded a new possible zosterophyll, *Tarella trowenii*. The overlying sandstone has yielded another new zosterophyll, *Deheubarthia splendens*. The macroplant taxa obtained from the quarry are detailed by Cleal and Thomas (1995) and summarized by Wellman *et al.* (1998).

Interpretation

Cleal and Thomas (1995) discuss the significance and importance of the macroplant assemblages. Miospore assemblages belong to the *polygonalis–emsiensis* biozone, providing a Pragian age (e.g. Wellman *et al.*, 1998). The strata are interpreted as the deposits of a high-energy, braided fluvial system, with most of the fine-grained floodplain sediments being reworked during subsequent fluvial incision and channel migration (Thomas, 1978; Owen, 1995). The green colour of the rocks and the preservation of the plants points to relatively high water-table levels and perhaps a more humid, wetter climate. Red-brown mudstones with calcrete nodules and intra-formational conglomerates with calcrete clasts above the section point to re-establishment of more arid conditions.

Conclusions

Craig-y-Fro Quarry is a site of national importance, its conservation value lying in the well-preserved, Early Devonian, fossil-plant remains it has yielded. It also provides an easily accessible section in which to study the sedimentological features of the Senni Formation.

ABERCRIBAN QUARRIES, POWYS (SO 064 123; SO 064 127)

W.J. Barclay

Introduction

These quarries on the slope east of the Taf Fechan Reservoir, 5 km north of Merthyr Tydfil (Figure 5.30), provide sections typical of the Grey Grits Formation of the Upper Old Red Sandstone. The formation ranges from Late Devonian to early Carboniferous in age, and marks the regional change from continental,

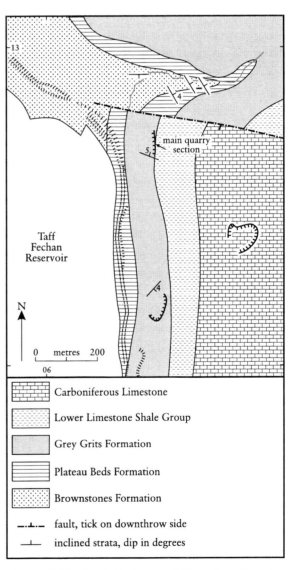

Figure 5.30 Geological map of Abercriban Quarries and vicinity. After British Geological Survey 1:10 560 Sheet SO 01SE (1980).

mainly fluvial, red-bed deposition of Late Devonian times to the shallow marine carbonate shelf environment of the early Carboniferous. The more northerly quarry (SO 0635 1272) shows a complete section of the formation, along with its contacts with the underlying Plateau Beds Formation and overlying Lower Limestone Shale Group. The more southerly quarry (SO 0635 1225) exposes about 8 m of the formation. The quarries are important in providing exposures revealing the nature of the basal and upper boundaries of the formation; in providing sedimentological evidence for the depositional environments of the Grey Grits Formation; and for allowing comparison of the Upper Old Red Sandstone of the north crop of the South Wales Coalfield with the thicker successions of Pembrokeshire. A third quarry (SO 066 124), also known as 'Abercriban Quarry', is higher in the succession and entirely in the overlying Carboniferous Limestone (the type locality of the Abercriban Oolite; Barclay et al., 1988).

Description

The more northerly quarry (SO 0635 1272) (Figure 5.31) is the best-known Grey Grits Formation locality and is referred to variously as 'Abercriban Grit Quarry' (Robertson, 1932), 'Abercriban Quarry' (Hall et al., 1973; Taylor and Thomas, 1975; Barclay et al., 1988) and 'Abercriban grit quarry' (Lovell, 1978a,b). It exposes the junction of the Grey Grits Formation with the overlying Lower Limestone Shale Group and with the underlying Plateau Beds Formation (Taylor and Thomas, 1975; Lovell, 1978b; Barclay et al., 1988), thus providing a complete (11 m) section of the formation. Following descriptions by Robertson (1932) and Hall et al. (1973) and a brief mention by Taylor and Thomas (1975), Lovell (1978a,b) provided the most detailed description and sedimentological analysis. Barclay et al. (1988) and Almond et al. (1993) gave summary descriptions.

The quarry has a north–south face about 120 m long and about 12 m high. The beds are almost horizontal.

The Plateau Beds Formation forms the lowest 1.5 m exposed in the quarry and comprise thinly bedded, greyish green and red-brown siltstones with small, rounded quartz pebbles and granules. Lingulids have been recovered from the red-brown siltstones (Hall et al., 1973).

The main lithologies of the Grey Grits Formation here are pale greenish grey to white, quartzitic, lithologically uniform, fine- to medium-grained sandstones (mainly quartz arenites), with lesser amounts of siltstone and mudstone forming thin interbeds. Conglomerates are mainly absent, with only a few quartz pebble lenses at some levels. Some intraformational mudstone and siltstone clasts line cross-bedding foresets locally and also occur at the bases of sandbodies. There are some parallel-laminated bodies, but the sandstones are mainly cross-bedded, with trough and planar types. The beds are mainly tabular, with sharp bases, but only a little scouring of the underlying beds. Cross-bedding is usually of medium scale, but some co-sets are up to 1 m thick. The palaeocurrent grand vector-mean for the formation is to the south-east, but both SSE and NNE palaeo-currents have been recorded in beds at different, discrete levels. The best example of this is seen at the northern end of the quarry where a 1.15 m-thick bed (3.85 m above the base of the formation) shows ENE-directed cross-bedding and is overlain by a bed with SE-directed cross-beds. The intervening erosion surface can be traced through most of the quarry (Lovell, 1978a,b).

No body fossils have been recovered from the formation here, although Taylor and Thomas (1975) recorded a calcareous brachiopod fragment near the base of the formation nearby. Trace fossils are rare, except for some simple, *Skolithos*-like, vertical burrows at some levels, and a few slightly sinuous simple burrows (epichnial casts) on one surface near the base of the formation. Elsewhere, a thin basal intraformational conglomerate has yielded fish fragments and the bivalve *Sanguinolites* sp. (Hall et al., 1973; Taylor and Thomas, 1974; Lovell, 1978a,b).

The basal bed of the Castell Coch Limestone Formation, which is the basal formation of the Lower Limestone Shale Group, overlies an erosion surface that truncates the Grey Grits Formation. A green, 0.9 m-thick siltstone bed at the north end of the quarry is absent to the south. The Lower Limestone Shale Group begins with a thin (0.2 m) lag conglomerate containing quartz pebbles, some altered acid-lava pebbles (including jasper), intraformational clasts and some phosphatized clasts. Fish teeth and spiriferid brachiopods have been recovered (Lovell, 1978a).

Interpretation

There were formerly differing interpretations of the relationship of the Grey Grits Formation and the underlying Plateau Beds Formation. Robertson (1932) thought that the two formations interfinger, but Allen (1965b) suggested that there is evidence of uplift and erosion between the two formations. Taylor and Thomas (1975) confirmed Allen's interpretation, noting that the Grey Grits form a wedge resting unconformably on various levels of the Plateau Beds throughout their outcrop, depending on the amount of pre-Grey Grits erosion. The nature of the boundary between the Grey Grits Formation and the overlying Lower Limestone Shale Group has also been the subject of different interpretation. Allen (1965b) and Hall *et al.* (1973) suggested a transitional contact. Taylor and Thomas (1975) noted a lack of discordance and no apparent break between the formations, and lent support to the suggestion of Cantrill (in Strahan *et al.*, 1904) that the Grey Grits represent commencement of Carboniferous sedimentation in the region. However, the widespread lag conglomerate at the base of the Lower Limestone Shale Group indicates at least a disconformity, with erosion having preceded deposition of the conglomerate (Burchette, 1981; Lovell, 1978a,b; Barclay *et al.*, 1988). The truncation of the topmost siltstone bed of the Grey Grits at the main Abercriban quarry (see above) confirms this interpretation.

The depositional environment of the Grey Grits has been interpreted variously as shallow marine to fluvial. Allen (1965b) proposed a dominantly fluvial origin. Regionally, Taylor and Thomas (1975) noted the presence of conglomerate layers at the bases of sandbodies resting on erosion surfaces, and small channels cut in calcareous sandstones and infilled with discoidal pebbles. They suggested depositional environments that varied from fluvial to marginal marine. Lovell (1978a,b) supported Allen's (1965b) fluvial interpretation, proposing deposition in shallow, sandy braided streams. It is possible, however, that the fluvial channels were tidally influenced (B.P.J. Williams, pers. comm.).

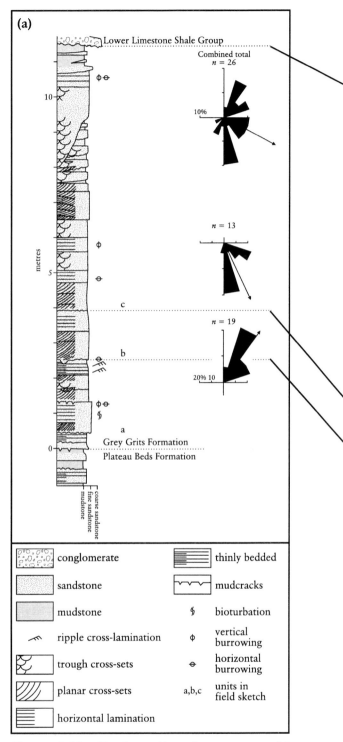

Figure 5.31 Northern Abercriban quarry. (a) composite log from the north and south ends of the quarry showing palaeocurrent directions. After Lovell (1978a,b). Continued on page 261.

Figure 5.31 – *continued*. Northern Abercriban quarry. (b) – view of part of the northern Abercriban quarry showing the basal beds of the Lower Limestone Shale Group and the upper part of Grey Grits Formation; (c) – sketch section of lower part of quarry (after Lovell, 1978a,b). (Photo: BGS No. A11993, reproduced with the permission of the Director, British Geological Survey, © NERC.)

The exact age of the Grey Grits remains unknown, but a Late Devonian to Early Carboniferous age is likely (Lovell, 1978a,b). The underlying Plateau Beds are thought to be late Frasnian to early Famennian in age and the base of the Lower Limestone Shale Group here appears to be of Early Carboniferous (Tournaisian (Tn) 2a) age (Lovell, 1978a,b).

Conclusions

The larger of the two quarries in the Grey Grits Formation (SO 0635 1272) provides a complete section through the formation, exposing its upper and lower boundaries. The quarry is thus important in providing evidence of the nature of these boundaries. It is also important in the detailed sedimentological evidence it provides for a fluvial, shallow braided stream origin for the Grey Grits Formation, thereby putting the formation into the regional context of Late Devonian to Early Carboniferous sedimentation when compared with, for example, the broadly coeval Skrinkle Sandstones Group of Pembrokeshire.

AFON Y WAEN, POWYS (SN 976 147)

Potential ORS GCR site

W.J. Barclay

Introduction

The section of strata exposed in the bed of the Afon y Waen in Powys, south-central Wales (Figure 5.32) is characteristic of the Upper Devonian (Frasnian–Famennian) Plateau Beds Formation of the Upper Old Red Sandstone. The site is an established GCR site for its fossil fishes (Dineley and Metcalf, 1999) on account of the Late Devonian fish remains recovered from a conglomerate – the eponymous Afon y Waen Fish Bed (Hall *et al.*, 1973). The importance of the site is enhanced in being one of the best sections of the Plateau Beds Formation, which elsewhere contains a shallow marine fauna that is unique in the predominantly continental Late Devonian Old Red Sandstone facies. The formation is also important in containing an aeolian sandstone facies, not known elsewhere in the Anglo-Welsh Basin, and seen, for example, in a key section at **Duffryn Crawnon** (SO 095 150) (see GCR site report, this chapter; Lovell, 1978a,b).

Description

Hall *et al.* (1973) provided a detailed description of the section, following the discovery of fish remains by Taylor (1972). In the Afon y Waen valley to the north of the site, the Plateau Beds are repeated by a NE-trending fault. The basal part of the Plateau Beds is exposed north of the fault (SN 972 172–SN 973 174), where a variable succession comprises 0.9–1.8 m of purplish red, flaggy sandstones and red mudstones, truncated by an erosion surface and overlain by 0.6–2.7 m of quartz pebble conglomerate (Figure 5.33). The section south of the fault is shown in Figure

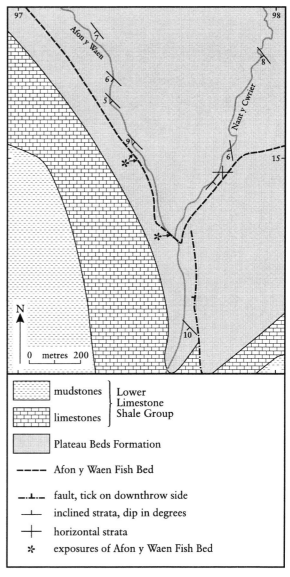

Figure 5.32 Geological map of Afon y Waen potential GCR site. After British Geological Survey 1:10 560 manuscript maps SN 91NE and SN 91SE (both 1973).

Afon y Waen

Figure 5.33 Cwar Llwyd, Afon y Waen. The basal bed of the Plateau Beds Formation comprises 3 m of massive quartz pebble conglomerate above 1.5 m of thinly bedded sandstones and siltstones (SO 9720 1730). (Photo: BGS No. A12010, reproduced with the permission of the Director, British Geological Survey, © NERC.)

5.34. About 6.1 m of thinly bedded, purplish red, argillaceous sandstones and siltstones with a few mudstone interbeds become more massive upwards, a thin grey sandstone lying 4.5 m from their top. The topmost bed of this group is a fine-grained, reddish purple, argillaceous sandstone with buff sandstone-filled burrows. It is overlain by the Afon y Waen Fish Bed, which is exposed at two localities in cliffs on the west bank of the stream (SN 9747 1500; SN 9761 1476). At the first, a 1.52 m-thick, purplish grey, trough cross-bedded, quartzitic sandstone has lenses of quartz conglomerate at its base containing fish fragments. These also occur in the bases of the cross-bedded sets. At the second locality (Hall *et al.*, 1973; Taylor and Thomas, 1974), the section is:

	Thickness (m)
Conglomerate with mudstone clasts, abundant fish fragments and a few quartz pebbles in a purple, fine-grained sandstone matrix	0.1
Sandstone, fine-grained, purple to grey, with a layer of pebbles and fish fragments at the base, which is channelled down into the underlying unit	0.6–1.0
Sandstone, fine-grained, purple to grey, trough cross-bedded	0.7
Conglomerate, friable, with quartz pebbles, red-brown mudstone clasts and abundant fish fragments	0.15

The Anglo-Welsh Basin

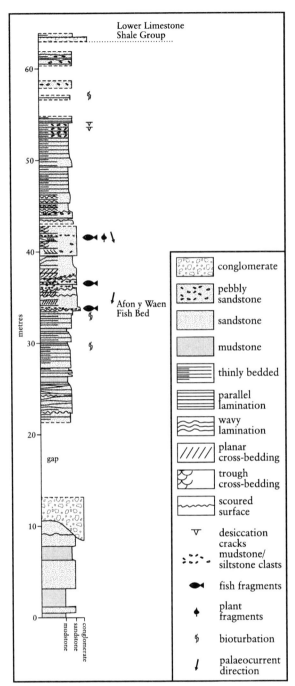

Figure 5.34 Vertical section of the strata in the Afon y Waen. Basal 13 m after Hall *et al.* (1973), upper part after Lovell (1978a).

Holoptychius sp. and *Bothriolepis* sp. have been found at these localities (Taylor, 1972; Hall *et al.*, 1973). No marine fossils have been found in the section, but lingulids occur above the level of the Afon y Waen Fish Bed in the nearby Cefn Esgair South Borehole (SN 9845 1353) and in Nant Mawr (SN 9519 1572) 2.5 km to the north-west. The fish bed was taken by Hall *et al.* (1973), Taylor and Thomas (1974, 1975) as a marker bed, subdividing the Plateau Beds into Lower and Upper units, and these units are shown on the Institute of Geological Sciences (1979) map of the area.

Above the fish bed, 4.57 m of fine- to medium-grained quartzitic sandstone are overlain by thinly bedded, deep purple, fine-grained, argillaceous sandstones and red-brown mudstones. Blue-grey, fine-grained, ripple cross-laminated sandstones close above were correlated with the Grey Grits Formation by Hall *et al.* (1973), but included in the Plateau Beds by Lovell (1978a).

Brachiopods recovered elsewhere from the Plateau Beds include *Lingula* sp., *Cyrtospirifer verneuili*, cf. *Ptychomaletoechia omaliusi*, *Leptodesma* cf. *lichas* and ?*Pterinopecten* sp.. In addition to *Bothriolepis* sp. and *Holoptychius* sp., fish remains include cf. *Pseudosauripterus anglicus* and cf. *Sauripterus* sp.. Trace fossils include cf. *Planolites* and *Rusophycus*. All of the brachiopods were recovered from the topmost part of the formation.

Interpretation

In a detailed sedimentological analysis of the Plateau Beds, Lovell (1978a) recognized three subdivisions (A, B and C). Division A is present only to the west of Afon y Waen, where it comprises a laterally variable succession of trough cross-bedded, pebbly sandstone and conglomerate with red mudstone interbeds. It was tentatively interpreted by Lovell (1978a) as the deposits of a southerly flowing braided stream system. Division B consists mainly of thin- and medium-bedded, red-brown sandstones, with local planar cross-bedded sandstones indicating southerly derivation. Lovell (1978a,b) suggested an aeolian origin for the latter, with the former being fluvial, possibly wadi sediments. The Afon y Waen section lies entirely within Division C. This is a heterolithic unit of interbedded, channelized, pebbly and conglomeratic sandstones, fine-grained sheet sandstones and mudstones. Cross-bedding indicates SE-directed palaeocurrents. A marginal marine environment is suggested (Allen, 1965b; Taylor and Thomas, 1975; Lovell, 1978a,b), with evidence of supratidal, tidal-flat and possibly subtidal environments (Lovell, 1978b). All of the marine fauna recovered from the Plateau Beds comes from this uppermost division. Lovell (1978a) noted that the Afon y Waen Fish Bed is a lenticular horizon of limited lateral extent. Similar channel-bottom lag deposits elsewhere, of which there are several, probably occur at slightly different stratigraphical levels, and the bed is not a continuous marker bed as envisaged by Hall *et al.* (1973) and Taylor and Thomas (1974, 1975). Similar facies occur above and below the Afon y Waen Fish Bed. The brachiopods and fish remains recovered from the formation suggest a Frasnian to Famennian age for the Plateau Beds.

Conclusions

The Afon y Waen section exposes beds characteristic of the upper part of the Plateau Beds Formation of the Upper Old Red Sandstone. The section is the type locality of the Afon y Waen Fish Bed, a lenticular conglomerate that has yielded fragments of the Late Devonian fish *Bothriolepis* and *Holoptychius*. The strata are of continental, Old Red Sandstone red-bed facies, but at nearby localities include marginal marine deposits that presage the marine transgression which led to the establishment of marine environments in Early Carboniferous time. Future research at the site will include the search for marine fossils known to occur at this level elsewhere.

DUFFRYN CRAWNON, POWYS (SO 095 150)

Potential GCR site

W.J. Barclay

Introduction

Accessible cliff exposures at the head of the Duffryn Crawnon Valley, Powys (Figure 5.35) between the headwaters of Nant ddu and the Afon Crawnon provide a complete transect through the Late Devonian Plateau Beds Formation. These strata are unique in the Anglo-Welsh Basin in containing shallow marine deposits and possible aeolian sandstones, both facies being present here. Following the discovery of fish remains during mapping by the [British] Geological Survey (Hall *et al.*, 1973; Taylor and Thomas, 1974, 1975), a detailed sedimentological analysis was carried out by Lovell (1978a,b). The site's importance is enhanced in it being the type locality for the Plateau Beds (Lovell, 1978a).

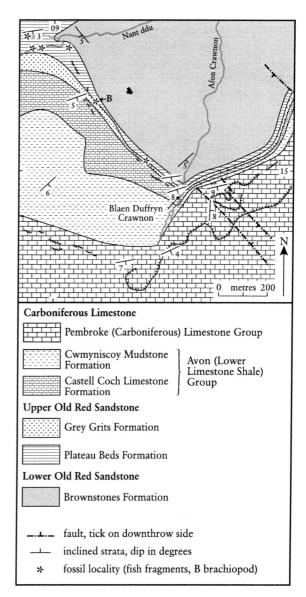

Figure 5.35 Geological map of Duffryn Crawnon potential GCR site. Based on British Geological Survey 1:10 560 manuscript map SO 01NE (1973).

Description

The following account is based largely on the work of Lovell (1978a,b). Earlier descriptions were given by Hall *et al.* (1973) and Taylor and Thomas (1974, 1975). The base of the formation (SO 0908 1547) is placed at the base of a 0.75 m-thick red mudstone containing thin layers of quartz granules and sand. From there, the entire formation can be seen by following the public footpath that contours round the bottom of the steepest part of the cliffs, or by climbing a grassy slope about 80 m south of

The Anglo-Welsh Basin

where the base is exposed. The underlying Brownstones Formation comprises interbedded red-brown sandstones and mudstones.

Figure 5.36 shows the graphic section compiled by Lovell (1978a,b). The Plateau Beds Formation is 36 m thick, its upper boundary (SO 0930 1518) marked by an erosion surface that truncates a 2.5 m-thick red mudstone unit and is overlain by 0.2 m of grey intraformational fish-bearing conglomerate at the base of the Grey Grits Formation. The formation broadly comprises two units. The lower one is sandstone-dominated and consists mainly of red-brown, sparsely micaceous, fine- to medium-grained sandstone (division 'B' of Lovell), the upper (division 'c') is more heterolithic and comprises interbedded finer-grained sandstones and mudstones.

A thin, channelized small-pebble extraformational conglomerate overlies the basal granular mudstone bed. It has yielded fragments of *Holoptychius* and cf. *Bothriolepis* (Hall *et al.*, 1973). It is succeeded by 18 m of red-brown and purple-grey, tabular beds of planar large- and medium-scale cross-bedded sandstones, with most of the cross-bedding directed to the northwest. These sandstones are pebble-free, with no mudstone interbeds or clasts, and are either non-micaceous or very slightly micaceous. The overlying heterogeneous beds (division 'c') are laterally variable, but can be subdivided into a

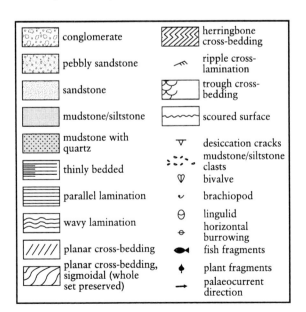

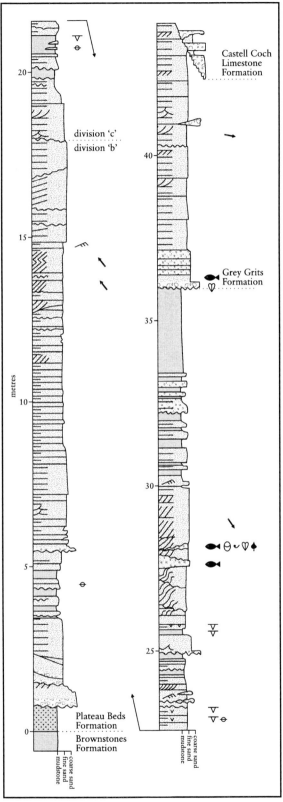

Figure 5.36 Vertical section of the strata at Duffryn Crawnon. After Lovell (1978a,b).

Duffryn Crawnon

lowermost part of 9 m consisting mainly of small, fining-upward, fine-grained sandstone–mudstone cycles 0.75 m to 1.75 m thick. The unit bases are slightly erosional or sharp, and parallel lamination is present in a number of beds. Some channelling is seen locally (Figure 5.37). A few beds have rippled surfaces, and desiccation cracks occur at some levels. Bioturbation is present in some beds in the form of horizontal hypichnial casts. Sediment dispersal in both these beds and the overlying 9 m was towards the south. The upper 9 m commence with a siderite clast and fish-bearing sandstone that passes up into cross-bedded sandstones or is overlain (and laterally cut out) by a channel-fill, cross-bedded sandstone. There are several superimposed and cross-cutting channelized sandbodies, and the bases of two channels contain brachiopod-bearing mudstone clasts. This channel-dominated sequence is overlain by a mudstone-dominated sequence, which, in turn, passes into the uppermost sandstones. Fossils recorded from division 'c' are lingulids (including Lingulid sp. B of Butler, in Taylor and Thomas, 1975), *Ptychomaletoechia omaliusi*, and fragments of fish and plant. Cf. *Bothriolepis* and *Holoptychius* sp. are recorded from an intraformational conglomerate about 12 m below the top of division 'c' (Hall *et al.*, 1973), in a horizon initially correlated by Hall *et al.* (1973) as the Afon-y-Waen Fish Bed (see **Afon y Waen** GCR site report, this chapter), but the correlation was later abandoned (Taylor and Thomas, 1975).

Figure 5.37 South face of Blaen Duffryn Crawnon (SO 0936 1507). Sandstones and siltstones fill a channel in the upper part of the Plateau Beds Formation. Brachiopods are present in the channel-fill. (Photo: BGS No. A12015, reproduced with the permission of the Director, British Geological Survey, © NERC.)

The Grey Grits Formation is up to 10 m thick and consists predominantly of grey-green, cross-bedded quartzitic sandstones showing SE-directed palaeocurrents. A siderite- and green mudstone-clast conglomerate marks the base of the formation and has yielded fish fragments and the bivalve *Sanguinolites*. Small channel-fill and scour features characterize the immediately overlying beds and are lined with discoidal quartz pebbles.

Interpretation

The Brownstones Formation is interpreted to be of sheet-flood and braided river origin (Tunbridge, 1981a). It is truncated by a regional unconformity marking basin inversion and erosion during the culmination of the Acadian Orogeny. Although no angular discordance is discernible at Duffryn Crawnon, the unconformity marks a major hiatus in which the Mid-Devonian succession is unrepresented.

The granule-rich mudstone at the base of the Plateau Beds is recognized widely (Hall *et al.*, 1973; Lovell, 1978a,b) and is interpreted by Lovell (1978a) as an alluvial mudflow deposit. Hall *et al.* (1973) subdivided the Plateau Beds into lower and upper units at the level of the Afon-y-Waen Fish Bed, but Lovell (1978a) concluded that fish-bearing conglomerates such as this are lenticular and not laterally persistent, and probably occur at different levels. The main differentiation of the succession is lithological, with a lower sandstone-dominated unit (division 'b') and an upper heterolithic unit (division 'c').

Some cross-bedded sandstones of division 'b' show NW-directed palaeocurrents, opposite to the regional drainage direction. Lovell (1978a,b) considered that these may represent wind-blown dune sets, either interbedded with unconfined waterlain (?wadi) sands or forming part of a coastal dune sequence. Lovell (1978a) noted, however, that if an aeolian environment is represented, dune formation appears to have been relatively limited. If Lovell's aeolian interpretation is correct, this is the first recorded occurrence of wind-blown sand deposition in the Anglo-Welsh Basin (but see also **Portishead** GCR site report, this chapter). The more heterogeneous deposits of division 'c' were interpreted as marginal marine sediments, with evidence of supratidal, tidal-flat and possibly subtidal deposition (Allen, 1965b; Lovell, 1978a,b). The fish fragments and brachiopods recovered from here and elsewhere constrain the age of the Plateau Beds to the late Frasnian–Famennian. The Grey Grits were interpreted by Allen (1965b) and Lovell (1978a) as fluvial deposits, although Taylor and Thomas (1975) suggested marginal marine deposition. At least some of the fluvial channels may have been affected by the influx of shallow marine waters (B.P.J. Williams, pers. comm.).

Conclusions

Duffryn Crawnon is the type locality of the Late Devonian Plateau Beds Formation and provides an excellent, complete transect through the entire formation. These strata are unique in the Anglo-Welsh Basin in containing shallow marine deposits and aeolian sandstones, and the site's importance lies in the presence of both of these facies. The site also exposes the lower and upper boundaries of the Plateau Beds, thereby providing evidence on the inter-relationships of the formations present. Fossil fish fragments from the site have provided information on the age of the Plateau Beds, and the sporadic presence of shallow marine fossils adds further importance.

CRAIG-Y-CWM, TORFAEN (SO 283 089)

Potential GCR site

W.J. Barclay

Introduction

Craig-y-cwm lies on the east-facing escarpment of the South Wales Coalfield 3.5 km south-east of Blaenavon (Figure 5.38). It provides the most complete section of the Upper Devonian (Upper Old Red Sandstone) Quartz Conglomerate Group of the eastern part of the South Wales Coalfield. It is also the type locality for the Craig-y-cwm and Garn-gofen formations of the Quartz Conglomerate Group (Barclay, 1989; Lovell, 1978a). Although these formations are not completely exposed, the Wern Watkin Formation, the basal formation of the group, is exposed in its entirety. A similar succession is exposed to the north at Craig-yr-Hafod (SO 2780 0994).

Craig-y-cwm

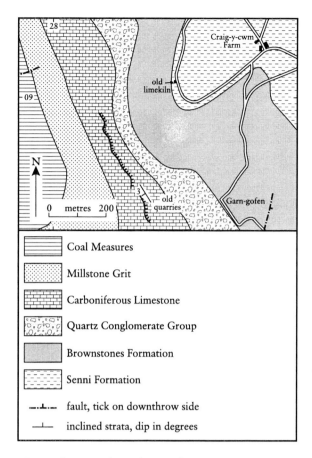

Figure 5.38 Geological map of Craig-y-cwm potential GCR site. After British Geological Survey 1:10 560 manuscript map SO 20NE (1984).

Description

The Quartz Conglomerate Group is 42 m thick at Craig-y-cwm (Figure 5.39). Strahan and Gibson (1900) first gave details of the succession at the site (repeated by Robertson, 1927). Barclay (1975) recognized three units within the group, later formalized as formations by Lovell (1978a), who gave a detailed description of the site and sedimentological interpretations of its constituent formations (summarized by Barclay, 1989). The following account is based largely on Lovell (1978a).

The base of the Wern Watkin Formation rests with no obvious discordance on the Lower Old Red Sandstone Brownstones Formation. The junction is placed where very fine- to fine-grained, ripple cross-laminated, thinly bedded, non-micaceous, non-calcareous, pale greenish sandstones of the Wern Watkin Formation rest on the pinkish brown, micaceous, fine- to medium-grained, friable sandstones of the Brownstones Formation. The latter is leached and weathered to pale green sand immediately below the junction. The Wern Watkin Formation is 19 m thick and comprises mainly pale green-grey to yellow-white, non-micaceous, fine-grained, flat-bedded, quartzitic sandstones (Lovell, 1978a). The sandstones are mainly in tabular sheets ranging from 0.2 m to 0.4 m in thickness, although one bed is 1.2 m thick. Mudstones are uncommon, but two thin beds are present, and intraformational green siltstone

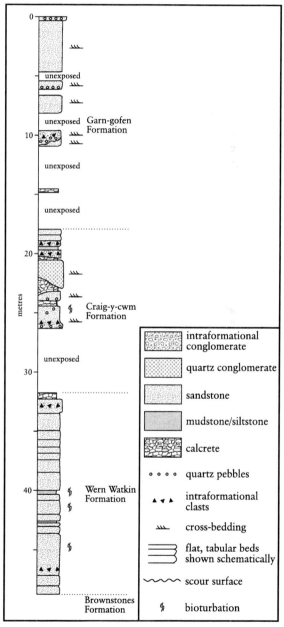

Figure 5.39 Graphic log of strata at Craig-y-cwm. After Barclay (1989).

clasts occur at some levels. The sandstone sheets show equal amounts of parallel lamination and cross-bedding. Both planar and trough cross-bedding occur. The 1.2 m-thick sandstone can be traced for over 300 m and its bedding characteristics change from a single, planar cross-bed set to a series of superimposed trough cross-beds. Measurements of the cross-bedding give a vector mean towards the north-west, which contrasts with southerly current directions measured at other localities. The top of the formation is marked by a massive calcrete, with the succeeding 5.3 m of beds (probably mudstones at the base of the Craig-y-cwm Formation) unexposed.

The Craig-y-cwm Formation is characterized by the presence of well-rounded, extraformational pebbles that range from 20 mm to 50 mm in long diameter. Quartz pebbles predominate (Figure 5.40) and give the Quartz Conglomerate Group its name. Lovell (1978a) noted also quartzose sandstones, and rare lava and jasper pebbles. Intraformational red and green mudstone, siltstone and calcrete clasts also occur. The conglomerate beds are lenticular and interbedded with pebbly and non-pebbly sandstones. Some intraformational conglomerate lenses containing mudstone, siltstone and calcrete clasts are also present. The quartz conglomerates are framework-supported, with a matrix of greenish grey to white, medium- to coarse-grained sandstone. Red-brown and green mudstone interbeds, some showing incipient calcrete development, are locally present. The formation is about 8 m thick at Craig-y-cwm, with the basal and topmost parts unexposed. The formation shows lateral variability, with the lithologies occurring as lenticular units, but some fining-upward cyclicity is noted, as well as a general fining upwards of the formation. Some scouring and erosion is present beneath the bases of some of the units. The vector mean of palaeocurrents measured by Lovell (1978a) is 148°.

The Garn-gofen Formation consists mainly of greenish grey to white, cross-bedded, strongly micaceous, calcareous sandstones. It is the least exposed part of the section, with unexposed

Figure 5.40 Cross-bedded pebbly sandstones above calcrete, Craig-y-cwm Formation, Craig-y-cwm. Hammer for scale. (Photo: BGS No. 13449, reproduced with the permission of the Director, British Geological Survey, © NERC.)

parts probably comprising mudstone or soft sandstone. Lovell (1978a) noted one thick bed of mudstone and one intraformational conglomerate. The sandstones show medium- to large-scale trough and planar cross-bedding. A thin (10 cm) sandstone 3 m below the top of the formation has convolute bedding that can be traced laterally for over 100 m. Palaeocurrents measured by Lovell (1978a) show a bi-modal distribution to the west and north-west, with a vector mean of 254°. The top of the formation is sharply overlain by the basal lag conglomerate of the Avon (Lower Limestone Shale) Group.

Interpretation

The Wern Watkin Formation is interpreted as the deposits of a generally southerly flowing sandy, braided, shallow fluvial system. The widespread mature calcrete at its top indicates a prolonged period of non-deposition and carbonate soil formation. The Craig-y-cwm Formation is interpreted as the deposits of mixed gravel–sand bedload braided streams. Overbank or floodplain fine sediment was largely reworked, but mudrock beds occur locally, as do mature calcretes, indicating periodic stability. The Garn-gofen Formation is interpreted as the deposits of meandering streams flowing from the north-east, with the abundance of clastic mica, garnet and feldspar indicating a metamorphic rock source.

Regionally, the Quartz Conglomerate Group crops out on the eastern rim of the South Wales Coalfield, extending from about Daren Cilau on the north crop eastwards to the Blorenge, and from there southwards to the Newport area. It also crops out on Pen-Cerrig-Calch and the Sugar Loaf to the north of the coalfield. To the west of Daren Cilau, the basal formation (Wern Watkin Formation) is correlated with the Grey Grits Formation (see **Duffryn Crawnon** GCR site report, this chapter), the upper two formations being absent. In the Forest of Dean, the Upper Old Red Sandstone is represented by basal quartz conglomerates (the Quartz Conglomerate, probably equivalent to the Craig-y-cwm Formation) and the overlying Tintern Sandstone Formation (probably equivalent to the Garn-gofen Formation). No fish remains have yet been recovered at the site, but fragments such as *Bothriolepis* sp. recovered elsewhere indicate a Late Devonian to Early Carboniferous age (Barclay, 1989).

Conclusions

Craig-y-cwm provides an excellent opportunity to examine easily the facies and formations that comprise the Upper Old Red Sandstone Quartz Conglomerate Group of the South Wales Coalfield. It is the type locality for the Craig-y-cwm and Garn-gofen formations, and a reference section for the Wern Watkin Formation. The detailed sedimentological analysis carried out at this site, among others, has led to an increased understanding of the fluvial depositional environments of the Late Devonian.

ROSS-ON-WYE, ROYAL HOTEL, HEREFORDSHIRE (SO 597 241–SO 595 239)

W.J. Barclay

Introduction

This site along Wilton Road, Ross-on-Wye (Figure 5.41) contains easily accessible, roadside cliff exposures of the Brownstones Formation, which forms the highest part of the Lower Old Red Sandstone magnafacies of south Wales and the Welsh Borderland. There are more extensive, but less accessible sections in nearby road cut-

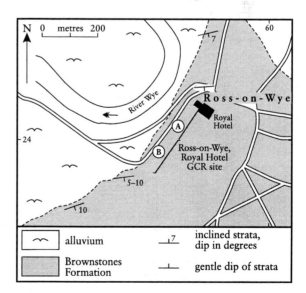

Figure 5.41 Location of Ross-on-Wye, Royal Hotel GCR site. After British Geological Survey Ross-on-Wye Special 1:10 000 Sheet (1980).

tings on the A40 at Glewstone (SO 567 223) and the A449 (SO 598 252); this site comprises an almost continuously exposed, 300 m-long section in which the architecture of the sandbodies can be easily seen. The Brownstones here are pebbly, cross-bedded sandstones, which formed in an alluvial, low-sinuosity, braided river system. The section has facilitated a reconstruction of bar morphologies, river channel sizes and palaeocurrent directions, and demonstrates the vertical stacking of the facies. The architecture of the stacked sandbodies, representing in-channel bars and dunes, is typical of the eastern outcrops of the Brownstones Formation and contrasts with those farther west in south-central Wales, in which unconfined sheet sandstones are more typical. The sedimentological study of this site and the nearby road cuttings by J.R.L. Allen (1983a), and that of the beds between the Townsend and Pickard Bay Tuff beds in Pembrokeshire by Allen and Williams (1982) are the most detailed analyses of fluvial architecture carried out in the Anglo-Welsh Basin.

Description

Allen (1971, 1974b, 1978a, 1980, 1983a,b) provided detailed descriptions and illustrations of the section (Figure 5.42a,b). Smith (1980) provided a summary. The following description is based largely on Allen's work. The cliff section exposes about 25 m of beds over a distance of about 300 m. They lie in the upper part of the Brownstones Formation and dip gently to the SSE. The beds comprise mainly red-brown, fine- to coarse-grained, locally gravelly, cross-bedded, lenticular sandbodies, grouped into multi-storey complexes in which a hierarchical structure is defined by erosion surfaces. At the north-east end of the site, near the entrance to the Royal Hotel, a near-dip section shows cross-bedded, coarse-grained sandstones and pebbly, conglomeratic sandstones arranged in sets about 0.3–0.5 m thick, with foresets directed consistently to the south-west. There are also some parallel-laminated, fine- to medium-grained sandstones, many being the topsets of the cross-bedding foresets. Towards the south-west of the site, a

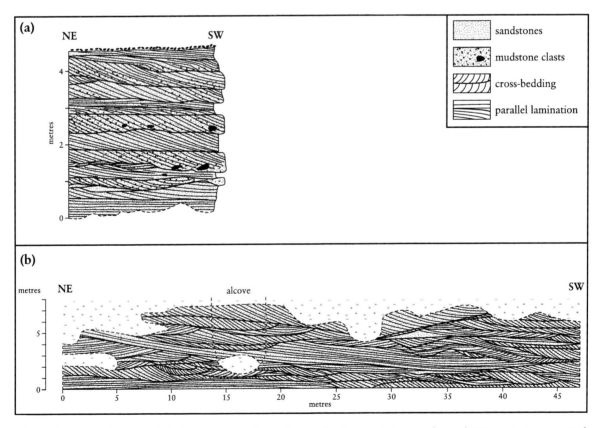

Figure 5.42 Graphic logs of the Brownstones Formation at the Ross-on-Wye Royal Hotel GCR site. (a) – vertical section at Point A on Figure 5.41; (b) – elevation showing sedimentary structures at Point B on Figure 5.41. After Allen (1971, 1978a).

near-strike section shows similar cross-bedded, pebbly sandstones and parallel-laminated sandstone bodies that have less consistent palaeocurrent directions (Figure 5.43). Parallel-laminated units extend laterally for up to 45 m. Most of the sandbodies rest on erosion surfaces and the pebbles include exotic types, as well as intraformational calcrete, mudstone and siltstone clasts, some of which are calcified, and a few of which are up to cobble or boulder size.

The exotic pebbles are mainly vein quartz, quartzite, cataclasite, jasper, acid lava, fine-grained green to red and yellow sandstones (?Lower Old Red Sandstone), a few with ?Silurian brachiopods, fine- to coarse-grained ?Silurian greywackes, pink sandstones with a Llandovery fauna and dark grey to black, fine-grained, acid lavas and tuffs with a mid-Ordovician shelly fauna, chert and 'oolite'. A detailed analysis of the pebbles is given by Allen (1974b).

Interpretation

Allen (1971) compared the rocks to those of modern sand-bed streams. He later (Allen, 1974b, 1983a,b) gave a detailed sedimentological interpretation of the section, as well as interpreting dune morphologies in exposures nearby as having formed by differentiation of a mixed bedload by gravel overpassing of hump-back bars (Allen, 1983b). The cross-bedded units originated as channel dunes or bars that migrated downstream with accompanying scouring. The absence of argillaceous floodplain deposits, other than as reworked clasts in the conglomeratic and pebbly layers, suggests low-sinuosity, high-energy streams with steep gradients (Allen, 1974b, 1983a). The prevalence of siltstone clasts, but absence of siltstone interbeds points to deposition of silt drapes on floodplains and in abandoned channels, but constant channel switching led to their destruction. This, plus the low variance in cross-bedding directions, points to braided streams, with flashy behaviour suggested by the scoured surfaces at the bases of the sandbodies and the laterally impersistent fining-upward units. In addition, Allen (1983a) noted some epsilon cross-bedding in most of the complexes, indicating the presence of laterally accreted bars as well as the predominant, larger, downstream-migrating forms.

Allen (1971, 1974b) noted that except for the acid-lava pebbles, the pebble suite resembles that of the stratigraphically equivalent Monkeys Fold Formation of Brown Clee Hill in Shropshire and concluded that the pebbles were derived from Ordovician, Silurian, early Lower Old Red

Figure 5.43 Cross-bedded sandstones of the Brownstones Formation at the Ross-on-Wye, Royal Hotel GCR site. (Photo: W.J. Barclay.)

Sandstone and perhaps Precambrian outcrops in north Wales and Anglesey.

The markedly lenticular, channelized sandstones present in this section, typical of the Brownstones Formation of the Welsh Borderland and south-east Wales, contrast with the more heterogeneous Brownstone successions farther west in the Brecon Beacons, where, in addition to the braided channel facies seen in the Ross-on-Wye section, interbedded sheet-like sandstone bodies and floodplain siltstones were interpreted by Tunbridge (1981a) as representing deposition in a more distal setting on an extensive alluvial plain.

Conclusions

The site comprises easily accessible roadside cliff exposures of the Brownstones Formation. The section has been the subject of a detailed sedimentological analysis and interpreted as the deposits of braided, steep, flashy rivers. Its study has allowed reconstruction of bar morphologies, river channel sizes and palaeocurrent directions. The far-travelled pebbles in the sandstones and conglomerates were derived from a northerly Welsh source, consistent with the general southerly flow of the streams that carried them.

WILDERNESS (LAND GROVE) QUARRY, GLOUCESTERSHIRE (SO 672 185)

W.J. Barclay

Introduction

This site near Mitcheldean, Gloucestershire is a working building stone quarry in the Early Devonian Brownstones Formation of the Lower Old Red Sandstone (Figure 5.44). The site is important in displaying a fine section of alluvial facies that represent deposition in an environment that was intermediate between the more distal environments of the underlying Dittonian St Maughans Formation and the more proximal, sand-dominated environments of the higher parts of the Brownstones. The vertical arrangement of the facies in the quarry were included in Allen's (1964a) detailed study of cycles in the Dittonian rocks of the Welsh Borderland, and thus instrumental in the

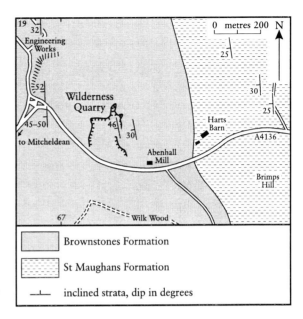

Figure 5.44 Geological map of the area around Wilderness Quarry. After British Geological Survey 1:10 560 manuscript map Gloucs. 23SE (1956).

modelling of Early Devonian Old Red Sandstone alluvial environments. The site is of national and international importance in being the only locality in the Brownstones Formation to have yielded remains of the late Dittonian index fish *Althaspis leachi*.

Description

The quarry exposes beds dipping steeply (*c.* 30°) westwards (Figure 5.45). They lie about 150 m above the base of the Brownstones Formation of the Forest of Dean (Welch and Trotter, 1961). Allen (1962, 1964a,b, 1971) and Allen *et al.* (1968) gave detailed descriptions of this quarry. Currently, about 23 m of beds are exposed, the section summarized as:

	Thicknesses (m)
Thinly bedded, red-brown sandstones and mudstones	*c.* 10 m
Red-brown sandstones in large channels, some sandstone beds up to 1 m thick; thin red-brown mudstone interbeds	*c.* 6 m
Mudstone, red-brown; prominent persistent bed	*c.* 1 m
Sandstone, red-brown, in beds of about 1 m average thickness	*c.* 4 m
Mudstone, red-brown, with pale green layers and thin sandstones	*c.* 2 m
Sandstone; top forms prominent bedding plane at eastern limit of workings	

Wilderness (Land Grove) Quarry

Figure 5.45 Wilderness Quarry. Main face showing major channelized sandstone bodies and intervening thin mudstones: (Photo: R.T. Mogridge.)

The following account of the succession (Figure 5.46) is based largely on the descriptions of Allen (in Allen *et al.*, 1968; Allen, 1971). The sandstones are predominantly red-brown, but pale green reduced zones occur at the bases of beds and in local random patches. Some red/green colour layering is present locally. Fine- to medium-grained, red-brown sandstones predominate, with some very fine-grained sandstones. Minor lithologies include coarse-grained, red-brown siltstones, intraformational conglomerate and red-brown siltstones/mudstones with calcrete nodules. The sandstones are mainly parallel-laminated, but cross-bedding occurs at some levels. The finer-grained sandstones are commonly cross-laminated and have rippled bedding surfaces. The siltstones are poorly bedded, although some are cross-laminated and most are strongly bioturbated. The intraformational conglomerates consist of rounded to angular clasts of siltstone and mudstone, very fine-grained sandstone and calcrete set in a sandy matrix. Intraformational mudstone/siltstone clasts are common, mainly occurring as concentrations at the bases of channel-fill sandstones, but they also occur sporadically throughout the sandstones.

The beds are arranged in fining-upward cycles. Each has a basal scoured surface overlain by sandstones, and locally conglomerates, in turn overlain by siltstones interbedded with thin sandstones. Fine-grained sandstone-filled desiccation cracks are present locally towards the tops of the siltstones. The scoured surfaces locally cut steeply down into the siltstones and thin sandstones of the underlying cycle. Broad, shallow channels are common in the siltstones, locally extending down to the underlying thick sandstones.

Wilderness Quarry is noted also for yielding one of the highest ostracoderm faunas known from the Lower Old Red Sandstone of south Wales and the Welsh Borderland (Allen *et al.*, 1968). The assorted, disarticulated remains of the ostracoderm *Althaspis leachi* (White) and, very rarely, the articulated remains of the thelodont *Turinia pagei* (Powrie) were collected from a distinctive bed of purplish to greenish white sandstone within siltstones 29 m above the base of the section recorded in 1968. This bed occurs high up on the western face of the quarry, but was formerly exposed by the entrance to the quarry 2 m above the quarry floor. In addition, the quarry yields several

The Anglo-Welsh Basin

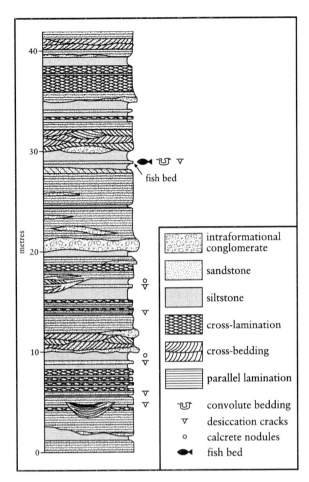

Figure 5.46 Generalized section of the Brownstones Formation at Wilderness Quarry. After Allen (1971).

different types of trace fossil, which are particularly abundant in the siltstones. The traces include irregularly, randomly orientated, branching tubes with longitudinally crinkled walls and less common smooth, sinuous tubes on bedding planes. *Planolites* occurs in the fish-bearing sandstone.

The fish-bearing sandstone ranges between 14 cm and 25 cm thick and overlies a red siltstone that fills and extends beyond a shallow depression in the top of the underlying sandstone. At its thickest, the bed is a tough, green, fine-grained, parallel-laminated sandstone with scattered siltstone clasts at its base. Over most of its outcrop in the quarry, the bed is a distinctive, fine- to very fine-grained, pinkish to mauve-brown sandstone that is parallel laminated at its base and cross-laminated in its finer-grained upper part. Convolute bedding is widespread.

Interpretation

The succession is interpreted as one that accumulated on an alluvial floodplain (e.g. Allen, 1964a, 1971; Allen *et al.*, 1968). The scoured surfaces at the bases of the cycles were cut by stream channels migrating laterally across the floodplain. The channel-fill sandstones were deposited in high-energy streams, with planar sandstone beds being more common than dune forms. The fine-grained, argillaceous lithologies are interpreted as overbank, extra-channel floodplain deposits, as indicated by their desiccation cracks and pedogenic carbonate (calcrete) nodules. The fish bed, occurring within fine-grade, floodplain facies, probably originated as a flash-flood deposit, either as a crevasse-splay sheet-flood or levee.

Fossil fish remains collected by Sibly in 1915 were first identified as *Pteraspis* cf. *dunensis* (Trotter, 1942) and later as the Breconian index fossil *Rhinopteraspis dunensis* (Welch and Trotter, 1961). These, as well as a large collection of pteraspid fragments obtained in 1965 were assigned to *Althaspis leachi* (Allen *et al.*, 1968). The presence of this late Dittonian index fossil in beds previously assumed to be Breconian in age emphasizes the need for more accurate definition of the local Anglo-Welsh stages, but may also demonstrate the diachronism of the Brownstones Formation.

Conclusions

Wilderness Quarry provides the best exposure in the Welsh Borderland at which to examine a facies of the Brownstones Formation that is transitional between those of the underlying St Maughans Formation and the overlying, more typical, Brownstones. Although sandstones dominate, mudstone interbeds are common and divide the succession into typical cycles comprising alluvial meandering channel deposits and overbank floodplain facies.

The site is internationally important in yielding the only fossil vertebrate remains from the Brownstones Formation. The occurrence of abundant disarticulated fragments of the late Dittonian index fossil *Althaspis leachi* in a floodplain deposit is one of a few such occurrences in the Anglo-Welsh Basin, and provides an insight into the contemporaneous habitats and sedimentary environments of the Anglo-Welsh Basin during Early Devonian times.

Lydney

LYDNEY, GLOUCESTERSHIRE (SO 652 015–SO 655 023)

P.R. Wilby

Introduction

This site is situated on the north-west bank of the River Severn in Gloucestershire and extends northwards from Lydney Harbour (SO 652 015) to a point mid-way between Fairtide Rock (SO 655 021) and Cliff Farm (SO 656 025) (Figure 5.47). It consists of laterally extensive cliff and foreshore sections in the uppermost part of the Raglan Mudstone Formation and the lowermost few metres of the St Maughans Formation. The Raglan Mudstone Formation is mostly of Přídolí (Silurian) age and the St Maughans Formation is Early Devonian (Lochkovian–Dittonian stage), with the Silurian–Devonian boundary thought to lie within the uppermost part of the Raglan Mudstone Formation. The locality has played a key role in establishing the extent of both fluvial and marine influences in the deposition of the Lower Old Red Sandstone, and includes a regionally important fish fauna, for which it is independently selected as a GCR site (Dineley and Metcalf, 1999). It is also one of only a few sites in the Anglo-Welsh Basin that has yielded modiolopsid bivalves at this stratigraphical level. In addition, the site contains several mature fossil soil carbonate (calcrete) profiles, including the regionally important Psammosteus Limestone (the Bishop's Frome Limestone of the Welsh Borderland and the Chapel Point Calcretes Member of Pembrokeshire).

Description

The section was initially documented by Welch and Trotter (1961), who described its geological context and identified the Psammosteus Limestone. Subsequently, Allen (1964a) examined the locality in greater detail and recognized the palaeoenvironmental significance of a complex cyclothem, which contributed to the establishment of a fluvial, meandering stream model for the deposition of large parts of the Lower Old Red Sandstone. The limestones at the site were among those used to develop a model for soil carbonate (calcrete) formation in the Old Red Sandstone of the Anglo-Welsh Basin (Allen, 1974d).

Allen (1964a, 1971) placed the cyclothem he described, and thus a large part of the sequence at Lydney, in the St Maughans Formation. Later, however (Allen, 1974d, 1978b), he recognized that the cyclothem actually lies beneath the Psammosteus Limestone, which forms the topmost bed of the underlying Raglan Mudstone Formation. Confusion over the stratigraphical position of the cyclothem, and that of the associated fish faunas, has since been compounded by Dineley (1999e, fig. 3.19). He produced an erroneous composite section by combining two different versions (Allen, 1964a, fig. 5 and Allen, 1978b, fig. 34) of the same cyclothem, one of which he placed above the Psammosteus Limestone and the other beneath it.

The sequence is gently folded and cut by several NW-trending faults. Between two prominent faults (SO 652 016; SO 655 023), the first about 200 m north of the harbour wall, the topmost mudstones of the Raglan Mudstone Formation are downthrown to the level of the foreshore and can be examined in laterally continuous exposures over several hundred metres (Figure 5.47). The Psammosteus Limestone and a few metres of the overlying St Maughans Formation are well exposed above about halfway up the cliff face just to the north of Fairtide Rock.

The section in the Raglan Mudstone Formation is typical of the unit elsewhere in Wales and the Welsh Borderlands. It consists predominantly of thick, red, micaceous, calcretized mudstones and siltstones, interbedded with thin, very fine-grained sandstones. It is, however, notable at this site in containing bivalve molluscs, one of only a few occurrences known at this level (Barclay *et al.*, 1994). The bivalves (*Modiolopsis complanata* Sowerby var. *trimpleyensis* Reed) are preserved in abundance and in life position, most with their valves tightly closed, in two thin layers within a 5 m-thick interval of red, well-bedded, coarse siltstones on the foreshore not far beneath the Psammosteus Limestone (Allen, 1973a). Scattered fish debris and abundant sub-vertical burrows are also present. The bivalves were examined by Allen and a British Geological Survey party in 1980, but were not located during a visit to the site in 2000 and are now probably concealed beneath estuarine mud. Among the fish remains recovered from intraformational conglomerate layers, Dineley (1999e) lists the heterostracan *Tesseraspis tessellata* Wills and the unusual acanthodian *Sabrinacanthus arcuatus* Miles.

The Anglo-Welsh Basin

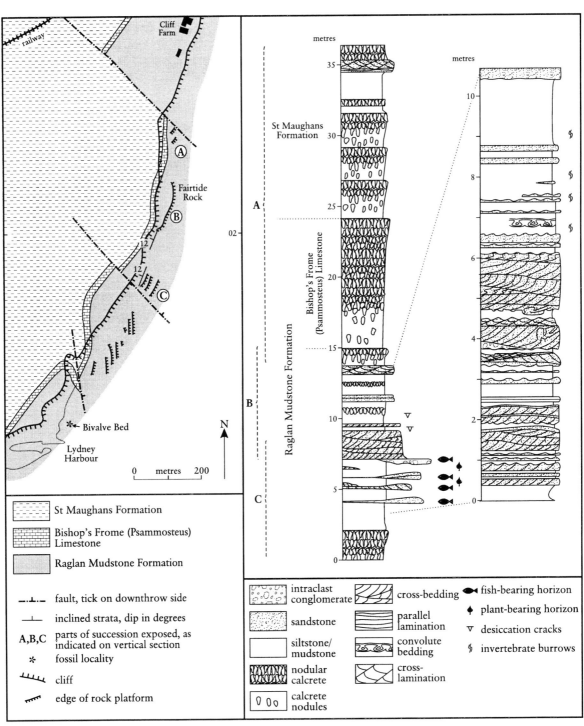

Figure 5.47 Geological sketch map of Lydney GCR site (inset) and composite vertical section of the strata exposed. Map after British Geological Survey 1:10 560 manuscript map SO 60SE (1973). Section based on Allen (1978b) and Dineley (1999e, fig. 3.19).

The cyclothem immediately below the Psammosteus Limestone consists of three facies arranged in a fining-upward sequence, and is best exposed on and around Fairtide Rock. The lowermost facies rests on a scoured surface of red silty mudstones with abundant calcrete nodules. It consists of finely interbedded pale green siltstones, some of which yield plant fragments, and white to green, fine- to coarse-grained lenticular sandstones. The sandstones have sharp bases, locally resting on erosion surfaces. They contain mudstone rip-up clasts, are cross-stratified, and have sharp, rippled or smooth tops. They are generally 5–25 cm thick and locally contain fish debris, *Pachytheca* and other carbonized plant fragments. These beds are sharply overlain by red, green and purple, micaceous, cross-bedded, fine- to medium-grained sandstones of the second facies. Most of the sets in this facies are planar, but some are contorted owing to extensive de-watering. This facies everywhere lies on a scoured surface, and locally, a 20 cm-thick intraformational conglomerate. The topmost facies comprises red, coarse-grained, micaceous siltstones interbedded with thin (less than 16 cm), red, mauve and white, ripple-bedded sandstones. The sandstones range from clayey and fine-grained to clean and medium-grained. Some of their tops are gradational, but all of their bases are sharp and some overlie surfaces with desiccation cracks. The top few metres of the cyclothem consist of red, clayey siltstones and sandy siltstones with abundant calcrete nodules, faint ripple-bedding and rare invertebrate burrows. The cyclothem terminates at the base of a thin sandstone, 10.5 m beneath the base of the St Maughans Formation. Directional structures throughout the cyclothem consistently indicate palaeoflow from the north.

Above the cyclothem are 20 m of red mudstone containing six calcrete profiles, the first three of which are particularly mature. The first and third profiles comprise massive limestone beds, and the second profile (the Psammosteus Limestone) is especially thick (9 m). It shows a gradual upward increase in the size and density of calcrete nodules from small, sporadic nodules at the base to a persistent horizon of closely packed, crudely prismatic nodules at the top. The Psammosteus Limestone and several other calcrete profiles show superb examples of pseudo-anticlines and nodule fans (Figure 5.48). The most prominent forms have a regularly undulose top in which cuspate or sharply rounded crests are separated by broadly rounded troughs on a scale of several metres. Within these 'folds', the calcrete nodules are arranged in fan-like arrays.

Figure 5.48 Calcrete profile in the Psammosteus Limestone at Lydney showing pseudo-anticlinal structure. Hammer (circled) for scale. (Photo: P.R. Wilby.)

The Anglo-Welsh Basin

Interpretation

The locality provides an excellent section of the strata at the boundary between the Raglan Mudstone Formation (Downtonian) and the St Maughans Formation (Dittonian). The boundary corresponds to an important tectonically induced diastem, represented by basin shut-down and major pedogenic carbonate development, and which more-or-less coincides with a major faunal and facies change (Allen and Tarlo, 1963; Allen, 1985; Dineley, 1999e). The Raglan Mudstone Formation was deposited on extensive coastal mudflats traversed by sparse channels (Allen, 1985). The section at Lydney has played an important part in assessing the nature, extent and timing of marine influence on these mudflats, and the nature of the transition from the marine environments of late Silurian times to the terrestrial, continental environments of early Devonian times (Allen, 1985). It is one of a handful of occurrences in the Welsh Borderland that has yielded a low-diversity, brackish to marine bivalve mollusc fauna (Allen, 1973a; King, 1934) at this level. These faunas, which are now known to extend up into the lowermost part of the St Maughans Formation (Barclay *et al.*, 1994), record short-lived, but probably quite extensive, brackish-water incursions on to the coastal mudflats (Allen, 1973a).

Further evidence for marine influence in the deposition of the Raglan Mudstone Formation is afforded by the lowermost facies of the cyclothem at Lydney. It consists of rapid alternations of siltstone and lenticular sandstone, which, based on modern analogues, suggests deposition in a tidal river channel in which the freshwater was periodically backed up by advancing tides (Allen, 1964a). The overlying part of the cyclothem was probably deposited in a non-tidal, meandering river that migrated across, and incised into, the deposits of an earlier channel system, with the finer topmost beds representing overbank facies. This non-tidal part is typical of cyclothems elsewhere in the Lower Old Red Sandstone and provides strong evidence for a fluvial origin of much of the sequence in the Anglo-Welsh Basin (Allen, 1964a).

The overlying calcrete profiles exhibit some of the best-developed pseudo-anticlines recorded from the Old Red Sandstone of Britain, comparable to those in Pembrokeshire. Lydney was one of five localities at which Allen (1974d) demonstrated a pedogenic origin for the limestones in the Old Red Sandstone of the Anglo-Welsh Basin. By comparison with modern examples, he concluded that they formed by replacive and displacive development of calcite in alluvial, floodplain muds. Pseudo-anticlines and related structures develop due to the swelling of the clay-rich sediment and the displacive growth of the calcite in response to seasonal wetting and drying (Allen, 1973b; 1974d, 1986). Allen (1974d, 1985) deduced that the climate in the Anglo-Welsh region during deposition of the Lower Old Red Sandstone was relatively warm (mean annual temperature of 16°–20°C), with low, seasonal rainfall averaging 100 mm to 500 mm per annum. He estimated that each calcrete profile records a period (and depositional break) of the general order of 10^4 years. Marriott and Wright (2004) provide a more recent discussion of calcrete formation, which may have taken tens of thousands to millions of years. The Psammosteus Limestone at Lydney, as at other localities, is particularly mature and is notable for its considerable thickness. This, together with the incoming of thick sandstones above the limestone, suggests that it records a regional, tectonically controlled event that resulted in changes to the source and rate of sediment supply to the basin (Allen, 1985).

Conclusions

The riverside cliffs at Lydney provide one of the best and most accessible sections in the uppermost part of the Přídolí–Lower Devonian (Downtonian) Raglan Mudstone Formation. The beds comprise a basal complex sandstone body of intertidal origin, and pass up through fluvial sediments into a stacked series of calcrete profiles, including the regionally important Psammosteus Limestone. The site is important because of its role in helping to model the depositional environments of the Lower Old Red Sandstone, and in determining the degree of marine influence during the transition into the overlying, wholly terrestrial Old Red Sandstone facies. In addition, it has played a key role in understanding the genesis of fossil calcretes and their palaeoenvironmental and palaeoclimatic significance. It is a site of continuing research, particularly into the composition of its fish faunas.

Albion Sands and Gateholm Island

ALBION SANDS AND GATEHOLM ISLAND, PEMBROKESHIRE (SM 771 074)

Potential ORS GCR site

W.J. Barclay

Introduction

This site was described in detail in the companion Silurian stratigraphy GCR volume (by Lane in Aldridge *et al.*, 2000) on account of the late Silurian (Ludlow–Přídolí) age of the lowermost beds of Old Red Sandstone red-bed facies. Only a brief summary is given here, with an updated interpretation based on recent work on the marine Silurian–Old Red Sandstone junction by Hillier (2000) and Hillier and Williams (2004). The site comprises the cliffs and foreshore at Albion Sands and on the adjacent Gateholm Island (Figure 5.49). It provides a magnificent section of almost 250 m of strata dipping steeply on the southern limb of the Marloes Anticline (Figure 5.50). Following early accounts by the [British]

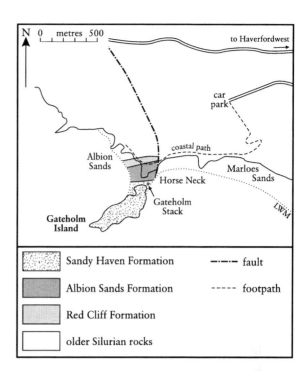

Figure 5.49 Location and simplified geology of Albion Sands and Gateholm Island. Based on Williams (1978) and Lane (2000c, fig. 6.20).

Figure 5.50 Oblique aerial view looking ENE to Albion Sands, the promontory of Horse Neck connecting with Gateholm Island, and Marloes Sands. (Photo: S. Howells.)

Geological Survey (Cantrill *et al.*, 1916), descriptions were given by Williams (1971, 1978), Allen and Williams (1978), Allen *et al.* (1981a), Bassett (1982b) and Williams *et al.* (1982). Hillier (2000) and Hillier and Williams (2004) have carried out a recent sedimentological study of the basal part of the succession. The Old Red Sandstone rests on the Gray Sandstone Group of Wenlock age, the nature of the junction continuing to generate debate (see Interpretation, below). The Old Red Sandstone comprises, in upward succession, the Red Cliff Formation, Albion Sands Formation (with a thin wedge of Lindsway Bay Formation) and the Sandy Haven Formation (Figure 5.51). All of the Old Red Sandstone strata are part of the Milford Haven Group (Allen and Williams, 1978).

Description

The following account summarizes that of Allen and Williams (1978) and Williams (1978). The Red Cliff Formation sharply overlies the Gray Sandstone Group at the northern end of the section. It comprises 51.6 m of interbedded red-brown mudstones and very fine- to fine-grained, bioturbated sandstones. Calcrete nodule horizons are present in a few of the mudstones. A Ludlow (late Ludfordian) spore assemblage has recently been recovered from the formation (Hillier and Williams, 2004). The Albion Sands Formation is named from this locality. It is over 100 m thick and consists mainly of thick, pale yellow to buff, multi-storey sandstones. Thin red-brown laminated mudstone interbeds occur throughout, with 11 m of mudstones with calcrete nodules lying at the top of the formation, as defined by Allen and Williams (1978). Calcrete nodules appear to be absent in the mudstone interbeds below, prompting Lane (in Aldridge *et al.*, 2000) to note that the topmost 11 m are much more like the calcrete-bearing mudstones in the overlying Sandy Haven Formation. Mudcracks in the laminated mudstones are desiccation cracks, and not synaeresis cracks as stated by Lane (in Aldridge *et al.*, 2000). The sandstones comprise individual bodies over 2 m thick and common multi-storey bodies up to 5 m thick. They have sharp tops and bases, the latter resting on erosion surfaces and commonly containing lenses of intraformational conglomerate with mudstone clasts up to cobble size. The sandstones also contain much igneous debris, and extraformational conglomerates also occur locally. The petrography of the sandstones and conglomerates indicates a westerly provenance. Thin airfall dust- and crystal-lithic tuffs occur sporadically, including a distinctive 1 m-thick, lilac, purple, red and yellow mottled dust tuff. A thin (7.42 m) wedge of conglomerates in the upper part of the formation outcrops on Horse Neck (SM 7718 0751) is correlated with the Lindsway Bay Formation. Here, it comprises petromictic conglomerates with exotic igneous pebbles interbedded with ill-sorted, granule-rich mudstones. The conglomerates have the greatest assortment of clasts of any petromict in the Old Red Sandstone of southern Britain.

The base of the Sandy Haven Formation is placed on Gateholm Stack (SM 7725 0746), where there is a conformable transition with the underlying Albion Sands Formation. About 85 m of beds comprise mainly bright red mudstones and thin pebbly sandstones. The mudstones are heavily calcretized, with mature soil carbonates and pseudo-anticlinal structures. The pebbly sandstones have yielded ?*Onchus wheathillensis*, *Pachytheca* and lingulids. Large-scale syn-sedimentary faults were recorded at Horse Neck by Hillier (2000).

Interpretation

The nature of the base of the Old Red Sandstone has generated much debate. A conformable contact with the underlying Wenlock Gray Sandstone Group was advocated by Cantrill *et al.* (1916), Sanzen-Baker (1972), Walmsley and Bassett (1976) and Hurst *et al.* (1978) and assigned a late Wenlock, or, at the latest, early Ludlow age. Allen and Williams (1978) preferred a disconformable contact, with the basal Old Red Sandstone being Downtonian (Přídolí) in age. Recent spore analysis has confirmed a late Wenlock (Homerian) age for the Gray Sandstone Group (Hillier and Williams, 2004). Hillier (2000) re-affirmed a conformable junction, with terrestrial red-bed sedimentation commencing here in the Ludlow. This is confirmed by the presence of late Ludfordian spores in the Red Cliff Formation (Hillier and Williams, 2004). Hillier (2000) and Hillier and Williams (2004) attribute the succession to the final of five episodes of incision and valley-fill in a structurally controlled basin. Transtensional movement on a series of NE-trending faults produced rifting on the southern margins of the Lower Palaeozoic Welsh Basin, producing local basins (the Skomer Basin in this case) fed with

Albion Sands and Gateholm Island

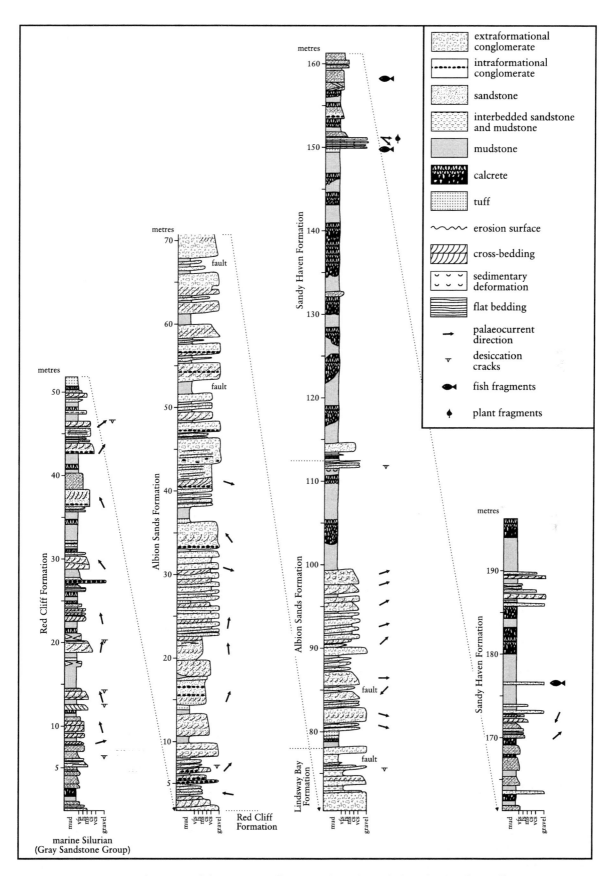

Figure 5.51 Vertical section of the strata at Albion Sands and Gateholm Island. After Williams (1978).

sediment derived from the Pretannia landmass to the south. Relative sea-level changes led to alternating marine flooding and fluvial incision. The Red Cliff Formation is interpreted by Hillier (2000) as the deposits of mudflats and tidal, fluvially influenced channels, the fluvial input being derived from the south (Hillier and Williams, 2004). However, a southerly source conflicts with the presence of white micas of northerly Laurentian (Scottish Highland) origin (Sherlock et al., 2002). Renewed transtensional activity produced tilt block topography, with the Albion Sands Formation being the fluvial deposits of a low-sinuosity, east-flowing, axial drainage system and the Lindsway Bay Formation representing gravel fans that prograded northwards from the uplifted fault scarps of the Pretannia landmass.

Conclusions

The magnificent cliffs and foreshore outcrops at this site expose a conformable succession of Old Red Sandstone sedimentary rocks of Silurian (late Ludfordian to Přídolí) age. The early age of the basal red rocks make them the oldest Old Red Sandstone facies in the Anglo-Welsh Basin, and only at Stonehaven in Scotland are there Old Red facies as old. The site is also the type locality of the Albion Sands Formation, in which sediment provenance indicators indicate, most importantly, a westerly source. Northerly palaeoflow patterns in the Lindsway Bay Formation are also of critical importance. The site is thus crucial to the interpretation of the sedimentary environments and sediment sources of late Silurian strata, and the implications for basin development and fault-controlled infill architecture at that time.

LITTLE CASTLE HEAD, PEMBROKESHIRE (SM 855 065)

Potential ORS GCR site

W.J. Barclay

Introduction

The Little Castle Head site (Figures 5.52, 5.53) is described in the companion Silurian stratigraphy GCR volume (Aldridge et al., 2000) on account of the late Přídolí age of the strata. Only a brief account is presented here. Little Castle Head lies on the north shore of Milford Haven, south Pembrokeshire. Formerly assigned to the Red Marls of Pembrokeshire (Cantrill et al., 1916), the strata belong now to the Sandy Haven Formation of the Milford Haven Group (Allen and Williams, 1978). These include two volcanic horizons, the Townsend Tuff Bed and Pickard Bay Tuff Bed, of which the former is the most regionally extensive, being known throughout Wales and the Welsh Borderland (Allen and Williams, 1981a; Parker et al., 1983). The section lies in the Winsle Block, one of several fault-bounded basins in south Pembrokeshire. The faults that define the block are the Benton Fault to the north and the Ritec Fault to the south, which acted as synsedimentary basin-margin faults and which were re-activated as

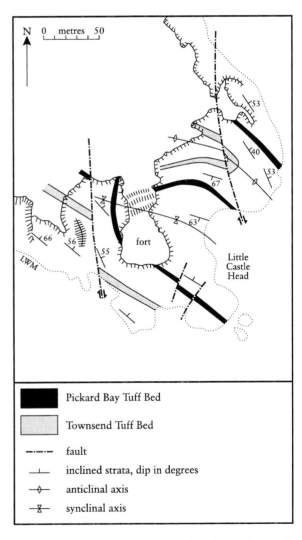

Figure 5.52 Geological map of Little Castle Head. Based on Allen (1980), Hancock et al. (1982) and Parker et al. (1983).

Little Castle Head

Figure 5.53 Oblique aerial view looking north-west to Little Castle Head. The strata are tightly folded and include the Townsend Tuff Bed and Pickard Bay Tuff Bed, which weather to slots and recesses. (Photo: S. Howells.)

thrusts in the Variscan Orogeny (Sanzen-Baker, 1972; Dunne, 1983; Powell, 1987, 1989; Marshall, 2000b). Variscan compression resulted in the tight folding present at the site.

Description

The Sandy Haven Formation (Allen and Williams, 1978) comprises about 850–900 m of predominantly red mudstones/siltstones, with minor quartz conglomerates and sandstones and a distinctive suite of airfall tuffs. The Little Castle Head site exposes beds in the middle part of the Sandy Haven Formation. This consists of bright red mudstones with abundant calcretes, minor purple and grey-green, lithic sandstones and a few granule- to pebble-grade conglomerates with vein quartz and lava clasts. It also contains the tuffs, the principal ones being the Townsend and Pickard Bay Tuff beds (Allen and Williams, 1978; 1981a). Of these tuffs (the magenta beds of Cantrill *et al.*, 1916), the Townsend Tuff Bed is the thicker, comprising three falls (A, B and C) in 2–4 m of beds (Figure 5.54).

Interpretation

The red calcrete-rich mudstones that dominate the succession are interpreted as the deposits of an extensive coastal mudflat, subjected to periodic marine influence and crossed by small, wet-season streams that were confined to channels, but also subject to flash sheet-flooding. These streams were probably largely sourced from the north, but some of the exotic clasts in the conglomerates may have been derived from the Pretannia landmass to the south (Cope and Bassett, 1987; Bluck *et al.*, 1992). Polygonal arrays of cracks in the mudstones are desiccation cracks, not synaeresis cracks, as stated by Lane (2000d). The well-developed calcretes at this level point to prolonged periods of subaerial exposure of the mudflats and the formation of thick carbonate soil profiles. The Townsend Tuff Bed is a regionally extensive marker bed of great value in correlation of the late Přídolí red beds throughout the Anglo-Welsh Basin. In the absence of faunas at this level, Allen and Williams (1981a) suggested that it could be used

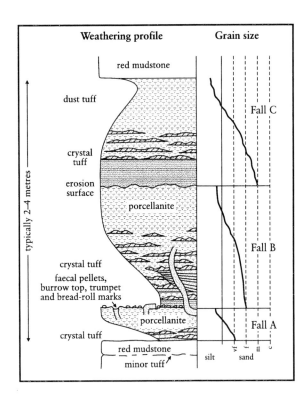

Figure 5.54 Schematic profile of the Townsend Tuff Bed illustrating its main sedimentological features. After Allen and Williams (1981a).

as the local Silurian–Devonian boundary. Palynomorph and thelodont assemblages suggest that this lies at a higher level (e.g. Richardson *et al.*, 2000), but sampling so far has been of suitable facies at a level when early Devonian species were already established. Radiometric dating of the tuffs may provide some degree of age precision. Allen and Williams (1981a) speculated that the tuffs were the products of Plinian eruptions. The thinnest of the Townsend tuffs (Fall A) was deposited by strong east–west winds from a centre 100–200 km away, but its location is unknown (Bevins in Stephenson *et al.*, 1999).

Conclusions

Little Castle Head provides a reference section for the Sandy Haven Formation of late Přídolí age. Red mudstones rich in soil carbonate (calcrete) horizons point to a coastal alluvial mudflat subjected to prolonged periods of emergence. The site also provides a reference section for the Townsend Tuff Bed, an important volcanic ashfall marker bed in the region, here in a succession tightly folded by Variscan compressive forces.

WEST ANGLE BAY (NORTH), PEMBROKESHIRE (SM 852 034)

P.R. Wilby

Introduction

The low cliffs at West Angle Bay (North), south Pembrokeshire, expose a continuous section from the Ridgeway Conglomerate Formation at the base, through the Skrinkle Sandstones Group, up into the Lower Limestone Shale Group (Figure 5.55). The age of the Ridgeway Conglomerate is uncertain, lacking diagnostic fossils, bounded by unconformities and lying in a fault-bounded block; late Early Devonian and Mid-Devonian ages have been suggested. The formation is one of only three in south Wales to have been sourced from the south. The age of the overlying Skrinkle Sandstones Group is well constrained, extending from the Late Devonian to the lowermost Carboniferous (Tournaisian).

The section is regionally important because it demonstrates the transitional nature of the boundary between the non-marine Old Red Sandstone facies and the marine deposits of the Lower Carboniferous strata in this part of the Anglo-Welsh Basin. It is also important in providing evidence with which to correlate the local Upper Old Red Sandstone succession with international biostratigraphical schemes based on spores and conodonts, and in allowing the location of the Devonian–Carboniferous boundary. In addition, the site is the stratotype for the West Angle Formation of the Skrinkle Sandstones Group, and was the first locality at which bedload-transported pedogenic mud aggregates were recognized in the Lower Old Red Sandstone of the Anglo-Welsh Basin (Ékes, 1993).

Description

Dixon (1921) was the first person to examine and log the succession at West Angle Bay in detail, following early work by De la Beche (1846). Williams (1964), Hassan (1966) and Marshall (1977, 2000a,b) have refined the lithostratigraphy of the section and studied its sedimentology. The section is described in several field guides (Williams, 1971; Marshall, 1978; Williams *et al.*, 1982). The nature and age of the transitional boundary between the Old

West Angle Bay (North)

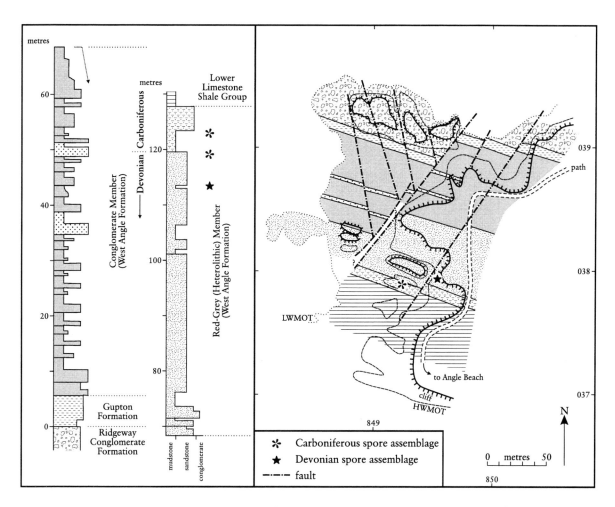

Figure 5.55 Geological sketch map of the north side of West Angle Bay and log of the Skrinkle Sandstones Group. After Williams *et al.* (1982).

Red Sandstone facies and the overlying Lower Limestone Shale Group have been discussed by numerous workers (e.g. De la Beche, 1846; Marshall, 1977, 2000a,b). Early faunal studies suggested that the Devonian–Carboniferous boundary coincided with the top of the Old Red Sandstone facies (at the top of the Skrinkle Sandstones Group) (Salter, 1863; Cantrill *et al.*, 1916; Dixon, 1921; George, 1970). However, spore and conodont studies (Dolby, 1971; Bassett and Jenkins, 1977) demonstrated that the Devonian–Carboniferous boundary lies at a lower level, within the top part of the Skrinkle Sandstones Group.

The faulted succession forms part of the northern limb of the Angle Syncline, a WNW-trending Variscan structure, the axis of which passes through the centre of the bay. The strata dip steeply (typically between 55° and 70°) and are repeated, but with some important lithological differences, on the south side of the bay (Dixon, 1921; Marshall, 2000b). The strata were deposited on the southern margin of the Wales–Brabant Massif, immediately to the south of the Ritec Fault, in the small (30 × 10 km) Tenby–Angle Basin that developed by intermittent subsidence during Devonian and Early Carboniferous times (Powell, 1987, 1989; Marshall, 2000a,b). The exposed succession is divided into three major lithostratigraphical units – Ridgeway Conglomerate Formation, Skrinkle Sandstones Group and Lower Limestone Shale Group.

The Ridgeway Conglomerate Formation consists of red-brown and purple, coarse, polymict, extraformational conglomerates 1–3 m thick, interbedded with thicker (1–25 m) red-brown siltstone-dominated units. The conglomerates contain subangular to well-rounded clasts, generally 1–5 cm in diameter, with some up to 20 cm, in a sandstone matrix.

The clasts are mainly of quartzite, vein quartz, lithic greywacke, siltstone, felsite and green phyllite. Some of the conglomerates are patchily cemented by calcite and most have sharp tops and bases, the latter resting on erosion surfaces. Most are poorly sorted and internally massive, although some show clast imbrication and poorly developed trough cross-bedding. Some contain sandstone interbeds. The interbedded siltstone-dominated units typically comprise a series of stacked, 0.5–2 m-thick fining-up cycles, each resting on a scoured surface and commonly commencing with a thin intraformational conglomerate. Typically, basal planar laminated or low-angle cross-laminated sandstones are succeeded by trough cross-bedded or ripple cross-laminated siltstones. Many of the cycles are capped by mudstones 20–40 cm thick with abundant trace fossils, including the giant form *Beaconites antarcticus*, in their upper parts. In addition, desiccation cracks and immature nodular calcrete profiles also occur. At West Angle Bay, the Ridgeway Conglomerate Formation is 120 m thick, but thins rapidly to the south-east away from the Ritec Fault (Dixon, 1921; Williams in Owen *et al.*, 1971). Its top is marked by calcretized red-brown mudstones and by a distinctive, 1.25 m-thick, grey-green, medium-grained, cross-bedded sandstone. Palaeocurrent data indicate that the formation was derived from the south (Williams, 1964; 1971; in Owen *et al.*, 1971).

The overlying Skrinkle Sandstones Group is of Late Devonian (Upper Old Red Sandstone) age, established by the presence of *Holoptychius* sp. (Dixon, 1921). The group is lithologically very variable and is divided into the Gupton and West Angle formations (Marshall, 1977). At West Angle Bay, the Gupton Formation is represented only by the Stackpole Sandstone Member (Figure 5.56). The West Angle Formation is subdivided into the Conglomerate and Red-Grey

Figure 5.56 Strata in West Angle Bay; view looking east (SM 8503 0380). The beds dip steeply and young southwards (left to right). The Ridgeway Conglomerate Formation is succeeded by the Stackpole Sandstone Member of the Gupton Formation, which in turn is overlain by the basal beds of the Conglomerate Member of the West Angle Formation. (Photo: P.R. Wilby.)

members. The Gupton Formation is 6 m thick and its base is sharp, slightly uneven and marked by a re-entrant in the cliff face (SM 8503 0380). The basal 1.25 m are dull red, silty mudstones and very finely laminated sandstone interbeds, arranged in centimetre- to decimetre-scale packages with abundant *Skolithos*-like and *Chondrites*-like burrows at some horizons. Decimetre-thick, hard, pink, quartzose sandstones form 90% of the upper part of the formation. These form horizontally bedded and low-angle cross-laminated lenticular sheets, or trough cross-bedded, channelized bodies, commonly with pebbly layers, resting on erosion surfaces. Palaeocurrents indicate southerly flow away from the Ritec Fault.

The Conglomerate Member of the West Angle Formation is 63 m thick and comprises a stacked succession of fining-upward cycles in which red and locally grey or green, conglomerate- and sandstone-based beds up to 2.5 m thick are capped by thicker beds of siltstone. The base of the member is a cross-bedded conglomerate about 1 m thick, with a sharp, scalloped base. Three beds preserve excellent examples of lateral accretion sets (e.g. SM 8502 0379, SM 8498 0373). The conglomerates are cross-bedded and contain a varied clast suite, including vein quartz, acid igneous rocks, mylonite, purple quartzite and intraformational fragments. They are generally of pebble grade and moderately well-rounded to subangular. The bases of the conglomerates and sandstones are sharp, but not obviously erosion surfaces. Most of the siltstones are red and structureless, and several contain immature, nodular calcrete horizons. Disturbed, dark grey mudstone laminae are preserved in a grey-green siltstone approximately 20 m above the base (SM 8498 0373). Palaeoflow data indicate south-westerly sediment dispersal.

The base of the overlying Red-Grey Member (Heterolithic Member of Marshall, 2000a) of the West Angle Formation coincides with a malachitic horizon (SM 8494 0375) first noted by Dixon (1921). The member is 59 m thick and dominated by red, mauve and green-grey, decimetre-scale beds of quartzitic sandstone and silty mudstone, commonly in fining-upward units capped by mottled mudstones and nodular calcretes. The sandstones are fine- to coarse-grained, commonly cross-bedded and sharp based. Some rest on erosion surfaces, particularly those towards the top of the member, where their bases are defined by thin (6–7 cm) basal intraformational granule layers.

The most characteristic feature of the Red-Grey Member are five so-called 'marine' beds (Dixon, 1921). These consist of grey mudstone, limestone or calcareous sandstone and contain bivalves, ostracods, plant debris, root traces and, at one level, a thin coal. Two of the 'marine' beds are particularly distinctive. The lower one lies above a 0.4 m-thick nodular calcrete, approximately 33 m above the base of the member and is most easily examined on a large southward-facing bedding plane (SM 8494 0370). Its base consists of 15 cm of grey mudstone that has yielded bivalves, plant debris (Dixon, 1921) and abundant fish teeth. Above is a burrowed, grey sandstone that fines upwards over 0.4 m into grey mudstone, then red-grey mottled mudstone and then red beds again. A few metres to the south the higher 'marine' bed is exposed in a crevice in the cliff. At its base is a 0.45 m-thick, green-grey, fining-upward, cross-bedded quartzitic sandstone with calcretes. This is overlain by 0.3 m of grey, ferruginous interbedded mudstone and sandstone with abundant *Planolites* burrows, overlain in turn by 0.5 m of calcareous, brown and grey, trough cross-bedded sandstone. This is capped by a 10 cm-thick, rubbly, bioclastic limestone.

The top of the Red-Grey Member is best exposed on the foreshore (c. SM 8492 0369). Here, the characteristic red-grey interbedding is replaced about 7 m beneath the top of the member by the abrupt appearance of interbedded grey shales and thin (generally less than 15 cm), yellow-weathering, dolomitic sandstones. Dixon (1921) recorded fish fragments, bivalves (*Modiola lata*), crinoid ossicles and brachiopods (*Lingula* sp.) from these beds and an orthocone nautiloid (*Orthoceras* sp.) from the red siltstone immediately beneath. These beds are succeeded by about 3 m of cross-bedded sandstone containing bryozoa and ooliths, which mark the top of the Skrinkle Sandstones Group. At West Angle Bay the total thickness of the Skrinkle Sandstones Group is 130 m, but it thickens dramatically to the south (Marshall, 1977). It is succeeded by grey mudstones and calcareous sandstones of the Lower Limestone Shale Group, which contains abundant sedimentary slump folds near its base (SM 8491 0362) (Kelling and Williams, 1966).

The Anglo-Welsh Basin

Interpretation

During Early and Late Devonian times, sedimentation in south Pembrokeshire was strongly influenced by a series of southerly dipping, NW-trending growth faults that controlled the formation of a series of small, fault-bounded basins, each with its own distinct lithostratigraphy (Allen and Williams, 1978; Powell, 1989). The rocks exposed along the foreshore of West Angle Bay were deposited in the southernmost of these basins. The succession is confined immediately to the north of the GCR site by the Ritec Fault, and to the south by a series of major faults on the northern margin of the former so-called 'Bristol Channel Landmass' (Dunne, 1983; Tunbridge, 1986; Brooks *et al.*, 1988). Syndepositional movement on the faults resulted in considerable lateral facies and thickness variations in the succession across the basin (Powell, 1989).

The succession is dominated by continental deposits. The Ridgeway Conglomerate Formation has been interpreted as a proximal alluvial-fan with braided stream deposits (Williams, 1971; Williams *et al.*, 1982; Cope and Bassett, 1987). Braided stream deposition is suggested by the upward-fining cyclothems at West Angle Bay, as well as the presence of *Beaconites antarcticus*, which is believed to be the trace of an animal that inhabited the banks of active river channels (Allen and Williams, 1981b). At least some of the siltstones may have been transported as pedogenic aggregates of mud pellets and deposited from stream bedload (Ékes, 1993). Based on palaeocurrent trends, structural data and the predominance of exotic Lower Palaeozoic sedimentary/metasedimentary clasts, the Ridgeway Conglomerate Formation is widely believed to have been sourced from the 'Bristol Channel Landmass' to the south (Dixon, 1921; Williams, 1964; Williams *et al.*, 1982; Tunbridge, 1986; Cope and Bassett, 1987). This is consistent with the northwards decrease in the average particle size of the unit and with its dramatic thickening towards the Ritec Fault. Lacking diagnostic fossils and bounded by unconformities, the age of the Ridgeway Conglomerate is uncertain. It has generally been assigned to the late Early Devonian (Lower Old Red Sandstone) (Dixon, 1921; Thomas, 1978; Williams, 1978; Allen, 1979; Powell, 1987, 1989; Ékes, 1993) on tectonic (post-Acadian deformation) and lithological grounds, although it may extend into the Mid-Devonian (Middle Old Red Sandstone) (Williams, 1964, 1971; Allen, 1965b, 1977; Allen *et al.*, 1967; Tunbridge, 1986).

The Skrinkle Sandstones Group was deposited following major tectonic inversion in Mid-Devonian times. This event completely changed the palaeogeography of the region (Allen, 1974a) and switched the sediment source of the Tenby–Angle Basin from the south to the north (Allen, 1965b). Continued movement on the basin-bounding faults during Late Devonian time (Powell, 1989) imparted a southerly tilt on the basin and a reversal in the sense of thickening of the sequence. Allen (1965b, 1974a) interpreted the Skrinkle Sandstones Group as a large-scale, fining-upward system comprising lower braided stream deposits and upper heterolithic coastline barrier deposits. Allen (1986) suggested that the Gupton Formation may be of shallow marine origin. More recently, however, Marshall (1977, 2000a,b) interpreted the lower, laminated part of the Stackpole Sandstone Member of the Gupton Formation as the lacustrine deposits of a lake that may have drained to the south or east, and the overlying quartzose sandstones as high-energy, sandy braidplain ephemeral stream and sheet-flood deposits.

SE-directed palaeocurrents and the textural maturity of the Gupton Formation indicate axial basin-fill, parallel to the bounding faults (Marshall 2000a,b). In contrast, the succeeding Conglomerate Member of the West Angle Formation records the influx of immature, locally derived fluvial sediments sourced from the north-east, across the Ritec Fault (Marshall, 1978). Once established, this sediment transport path persisted throughout the deposition of the remainder of the West Angle Formation. The varied suite of clasts in the conglomerates, which is very different to those in the Ridgeway Conglomerate Formation, suggests a source in Carmarthen or north Pembrokeshire (Marshall, 1977) and confirms the presence of a major depositional break between the Ridgeway Conglomerate Formation and Skrinkle Sandstones Group.

The overlying Red-Grey Member of the West Angle Formation is interpreted as the deposits of a coastal barrier/lagoon complex (Allen, 1965b; Marshall, 1978; 2000a,b). Marshall (2000b) recognized a complex of tidal-flat, lagoonal and washover fan deposits, culminating in the topmost barrier sandstone. The Red-Grey Member is laterally equivalent to the Shirehampton Beds of the Bristol district (see **Portishead** GCR site report, this chapter). Both record the start of the northward marine transgression of the Lower Carboniferous sea on to the Wales–Brabant Massif ('St George's Land'), culminating in deposition of the fully marine Lower Limestone Shale Group. Spore and conodont data from West Angle Bay indicate that the Devonian–Carboniferous boundary is located 8–15 m below the top of the Skrinkle Sandstones Group (Dolby, 1971; Bassett and Jenkins, 1977).

Conclusions

The low cliffs at West Angle Bay (North) expose a continuous section encompassing the Ridgeway Conglomerate Formation of possible Mid-Devonian age, the Late Devonian to Early Carboniferous Skrinkle Sandstones Group, and the succeeding Lower Limestone Shale Group. The section exhibits a wide range of lithologies, representing depositional environments in a small, fault-bounded basin.

The site is important because:

- it shows the transition from the non-marine Upper Old Red Sandstone facies, through a number of marginal marine grey beds at the top of the Skrinkle Sandstones Group, to the fully marine Lower Limestone Shale Group.
- microfossil evidence indicates that the Devonian–Carboniferous boundary lies within Old Red Sandstone facies, 8–15 m below the top of the Skrinkle Sandstones Group.
- it provides evidence to show that deposition during Devonian times in this region was tectonically controlled and confined to fault-bounded sub-basins. Correlation between these sub-basins is difficult, but detailed spore analysis at West Angle Bay has enabled the Upper Old Red Sandstone lithostratigraphy of the Tenby–Angle area to be correlated with an international biostratigraphical scheme based principally on marine sequences.
- in addition to being the stratotype for the West Angle Formation, the locality is critical to the understanding of the palaeoenvironmental setting of the Skrinkle Sandstones Group. The site offers continued opportunities to collect Late Devonian fish material.

FRESHWATER WEST, PEMBROKESHIRE (SR 884 996–SR 887 988)

Potential ORS GCR site

W.J. Barclay and B.P.J. Williams

Introduction

Foreshore and cliff exposures at Freshwater West (Figure 5.57) in the Pembroke peninsula south of Milford Haven provide a magnificent, continuously exposed, easily accessible section (depending on tides) of the entire Old Red Sandstone succession. The basal Lower Old Red Sandstone rocks rest unconformably on marine rocks of Wenlock age at the north end of the section, the strata dipping steeply and younging southwards. At the southern end of the section, Upper Old Red Sandstone rocks are conformably overlain by the Dinantian Lower Limestone Shale Group. The section was first mapped in detail and described by the [British] Geological Survey (Dixon, 1921, 1933a,b). Field guides by Williams (1971, 1978), Allen *et al.* (1981b) and Williams *et al.* (1982), and detailed study by Marshall (1977, 2000a,b) have highlighted the importance of the section. Together with the presence of the Townsend Tuff Bed and the Chapel Point Calcretes Member, the recent discoveries of an early Devonian microflora (Higgs, 2004) and of exceptionally well-preserved burrowing traces (*Beaconites barretti*; Morrissey and Braddy, 2004) have added to that importance. This is the Old Red Sandstone section most visited by geologists and students in southern Britain because of its completeness and the superb exposure of all of the formations. Varied sedimentary structures and trace fossils can be studied in detail on wave-polished surfaces. The site provides an easily demonstrable picture of the evolving nature of sandbody architecture in the Lower Old Red Sandstone (Williams and Hillier, 2004). It also provides the most proximal, complete and accessible section through the unconformity-bounded Ridgeway Conglomerate Formation, and is important in having yielded fragments of the late Devonian fish *Holoptychius* from the basal beds of the Skrinkle Sandstones Group (Dixon, 1921). Furthermore, the Flimston Bay Fault, which transects the entire rock sequence with increasing amount of throw to the south, provides an additional, spectacular feature.

The Anglo-Welsh Basin

Figure 5.57 Aerial oblique view of Freshwater West looking south to Great Furzenip headland. Steep, south-dipping beds of the Freshwater West Formation are cut by the Flimston Bay Fault. The Skrinkle Sandstones Group crops out on Great Furzenip. (Photo: S. Howells.)

Description

Descriptions of the section have been given by Dixon (1921, 1933a,b), Williams (1971, 1978), Allen *et al.* (1981b) and Williams *et al.* (1982). Marshall (2000a,b) described the Upper Old Red Sandstone in detail. Freshwater West lies in the Pembroke peninsula south of Milford Haven and south of the Ritec Fault (Figure 5.58). The strata at Freshwater West occupy the southern limb of the Castlemartin Corse Anticline (Hancock, 1973), dip steeply (50°–80°) southwards and are locally overturned. The southern third of the section, in the Upper Old Red Sandstone, lies within the Castlemartin Tank Range, and permission to visit must be obtained in advance from the Commandant, Merrion Camp, Castlemartin.

The section can be reached from the B4319. From this road above Little Furzenip (SR 885 994) there is a magnificent view of the steeply dipping succession, exposed in a rock platform between high-tide and low-tide levels. The Old Red Sandstone comprises most of the section (Figure 5.59), outcrops being confined in the north to the foreshore between Little Furzenip and the main beach of Freshwater West (SR 8872 9900). To the south, the Ridgeway Conglomerate Formation and Skrinkle Sandstones Group crop out in cliffs about 35 m high. The north-trending Flimston Bay Fault, a Variscan dextral wrench fault, bisects the entire section, displacing the strata about 120 m in the middle of the outcrop. This fault also isolates the Little Furzenip stack from the headland and is marked by a shatter zone on Great Furzenip headland (SR 887 987). It dies out through a myriad of small splay faults immediately north of Little Furzenip and does not affect the rock sequence on the northern limb of the Castlemartin Corse Anticline (Hancock, 1973).

At the northern (Little Furzenip) end of the section, the basal Old Red Sandstone rests unconformably on marine Silurian rocks of Wenlock age, the Gray Sandstone Group (Williams, 1971, 1978; Bassett, 1974; Allen and Williams, 1979b; Allen *et al.*, 1981b;

Freshwater West

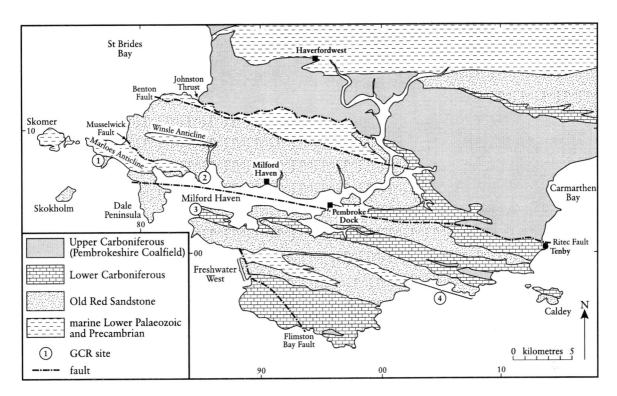

Figure 5.58 Geological map of the Pembroke peninsula and location of the potential GCR site at Freshwater West. Locations of GCR Old Red Sandstone sites on the peninsula are also shown. (1 – Albion Sands and Gateholm Island; 2 – Little Castle Head; 3 – West Angle Bay (North); 4 – Freshwater East–Skrinkle Haven.)

Williams *et al.*, 1982). The Gray Sandstone Group consists of conglomerates, green fine- to medium-grained sandstones, thin limestones and dark grey mudstones. The group is truncated by an irregular scoured, channelled surface that is overlain by the basal Old Red Sandstone rocks, the Freshwater East Formation. This is 18.45 m thick and comprises conglomerates in its lower part, with finer-grained sandstones above. The conglomerates are green to dark grey, of cobble- and pebble-grade, and are interbedded with fine- to medium-grained sandstones and green-grey mudstones. They are framework supported and include clasts up to 450 mm of quartzitic and lithic sandstones, vein quartz and olive-green mudstone in a coarse-grained sandstone matrix, some of the clasts matching lithologies in the underlying Silurian strata. The conglomerates are massive or cross-bedded with internal scour surfaces and overlie erosion surfaces. The upper, finer-grained beds contain lingulids, plant remains, arthropod tracks, *Pachytheca* and ostracoderm fragments (Dixon, 1921; Richardson and Lister, 1969; Williams, 1978).

The Freshwater East Formation is succeeded by the Moor Cliffs Formation. This is 119.56 m thick and consists predominantly of red and green mudstones in which calcrete glaebules and tubules are abundant, commonly arranged in pseudo-anticlinal structures. Subordinate lithologies include very fine- to fine-grained sandstones, exotic and intraformational conglomerates and airfall tuffs. The Townsend Tuff Bed is the thickest (2.81 m) of the tuffs. It crops out on the foreshore immediately north of Little Furzenip (SR 8847 9946) and is visible when the sand cover is removed. It lies 37.89 m above the base of the Moor Cliffs Formation (in contrast to 300 m on the northern limb of the same fold) and is succeeded by 13.53 m of beds in which there are five thin (0.01–0.11 m), pink-purple dust tuffs (markers B, C, E, F and G of Allen and Williams, 1982; Marker D is apparently absent here because of erosion). These are overlain by the Pickard Bay Tuff Bed, which is 1.25 m thick (Allen and Williams, 1982). A medium- to coarse-grained, cross-bedded litharenite lies 56 m above the Townsend Tuff Bed and forms a prominent rib on the foreshore and Little Furzenip headland. The sandstone rests on an erosion surface and the cross-bedding is in four to six tabular sets. The highest beds of the

The Anglo-Welsh Basin

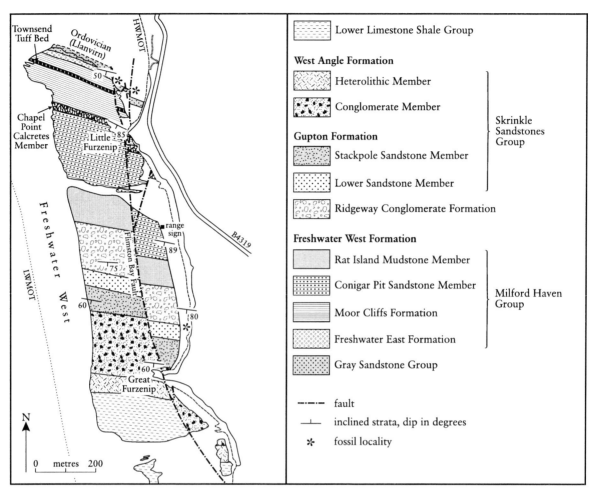

Figure 5.59 Geological map of Freshwater West. Based on Williams (1978), Allen *et al.* (1981b) and Williams *et al.* (1982).

formation crop out on Little Furzenip headland (SR 8851 9939), where red mudstones, in beds up to 5.7 m thick, contain abundant calcrete, mainly in the form of large concretions and rods, which locally show a preferred orientation parallel to cleavage or are arranged in pseudo-anticlinal festoons. The calcretes are arranged in stacked profiles and are correlated with the Chapel Point Calcretes Member of Pembrokeshire and the Psammosteus Limestone (Bishop's Frome Limestone) of the Anglo-Welsh Basin.

The abrupt junction between the Moor Cliffs Formation and the overlying Freshwater West Formation is well exposed on the southern side of Little Furzenip headland. The latter is 345.78 m thick and can be examined in great detail on the wave-polished foreshore section between the headland and the base of the Ridgeway Conglomerate Formation to the south (Williams, 1971, 1978). The lower part of the formation (the Conigar Pit Sandstone Member) is sandstone-dominated (e.g. Williams and Hillier, 2004), whereas mudstone predominates in the upper part (the Rat Island Mudstone Member). The Conigar Pit Sandstone comprises intraformational conglomerates, very fine- to medium-grained sandstones and calcrete-bearing mudstones, commonly arranged in upward-fining cycles (Figure 5.60). The conglomerates contain calcrete and mudstone clasts, mainly of pebble grade, and overlie low-relief erosion surfaces. They and the coarser sandstones contain fish fragments at four horizons, including plates, spines and scales of pteraspids, onchids and traquairaspids. The member can be examined in detail near the range warning sign (SR 8857 9917). The sandstones are mainly fine grained, red and green, parallel or cross-bedded

Freshwater West

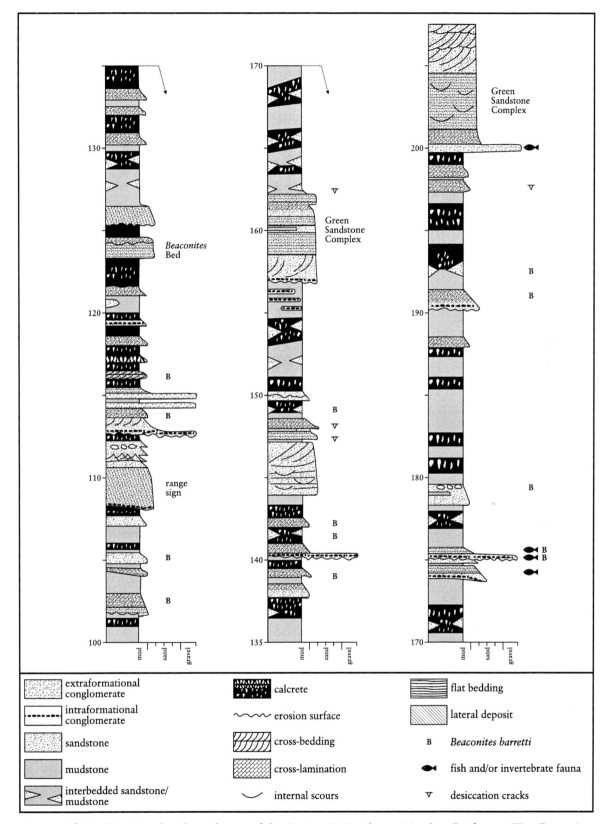

Figure 5.60 Sedimentary log through part of the Conigar Pit Sandstone Member, Freshwater West Formation at Freshwater West. Thicknesses are in metres from the base of the member. Based on Allen *et al.* (1981b) and Williams *et al.* (1982).

(Figure 5.61), and fine upwards into ripple cross-laminated units. Some of the sandstone–conglomerate bodies are stacked into thick complexes with lateral accretion surfaces. Many of the sandstones and some mudstones are bioturbated. Concentrations of burrowing activity occur at two horizons and include exceptionally preserved examples of *Beaconites barretti* (Morrissey and Braddy, 2004). Plant debris occurs in some sandstones, and spores (including two new species) have been recovered from green mudstone layers within green sandstones 112 m and 160 m from the base of the member (Higgs, 2004). The microflora is assigned to the middle part of the *Emphanisporites-micrornatus–Streelispora-newportensis* (MN) Zone of mid-Lochkovian age.

Figure 5.61 Sandstones of the Conigar Pit Sandstone Member, Freshwater West, 114 m above the base of the member. The sequence youngs from left to right. Planar cross-bedded sandstone overlies flat-bedded sandstone and is overlain by sandstone/mudstone and massive calcrete. The sandstones are interpreted as sheet-flood deposits. (Photo: B.P.J. Williams.)

The mudstones of the Conigar Pit Sandstone Member are laminated or massive, blue-mottled and contain small burrows. Some very fine-grained sandstone–mudstone complexes contain desiccation cracks and arthropod tracks. Some soft-sediment deformation structures in the form of sandstone balls in mudstone occur locally. Calcrete is common in the mudstones, including nodules, massive limestone and veins, and in pseudo-anticlinal fans. The top of the Conigar Pit Sandstone is placed at the top of the highest thick (5.18 m), green sandstone complex, which has a rare extraformational conglomerate at its base.

The Rat Island Mudstone Member consists mainly of thick, red mudstones that are blue-mottled, laminated and contain calcrete in the form of nodules, pseudo-anticlinal fans and some massive horizons. There are a few very fine-grained, ripple-laminated sandstones with *Beaconites antarcticus*. There are also some prominent thick (2.05 m), coarse, pebble intraformational conglomerates with burrows and fish scales.

The disconformable junction between the Rat Island Mudstone and the overlying Ridgeway Conglomerate Formation is placed at the base of an intraformational conglomerate–mudstone complex. The mudstones in the Ridgeway Conglomerate are different from those in the underlying Rat Island Mudstone in being brighter red and ill-sorted, with scattered granules and small pebbles of vein quartz and carbonate. The sandstones are also different in that they are phyllarenitic in composition. The Ridgeway Conglomerate Formation (Williams, 1971, 1978) is a broadly upward-coarsening sequence of red, sandy mudstones with calcrete and ill-sorted mudrocks, fine- to coarse-grained litharenites rich in metamorphic and sedimentary rocks, and exotic, mainly clast-supported, petromict conglomerates (Figure 5.62). The formation is 115 m thick and contains eleven conglomerate bodies ranging from 0.6 m to 9 m thick, which make up 30 m of the formation. These overlie erosion surfaces and are massive or have weak internal cross-bedding or horizontal bedding, with some clast imbrication. The conglomerates are mainly framework-supported, with clasts ranging from 4 mm to 64 mm (Figure 5.63), but a 9 m-thick unit 14 m below the top of the formation contains vein quartz boulders up to

Freshwater West

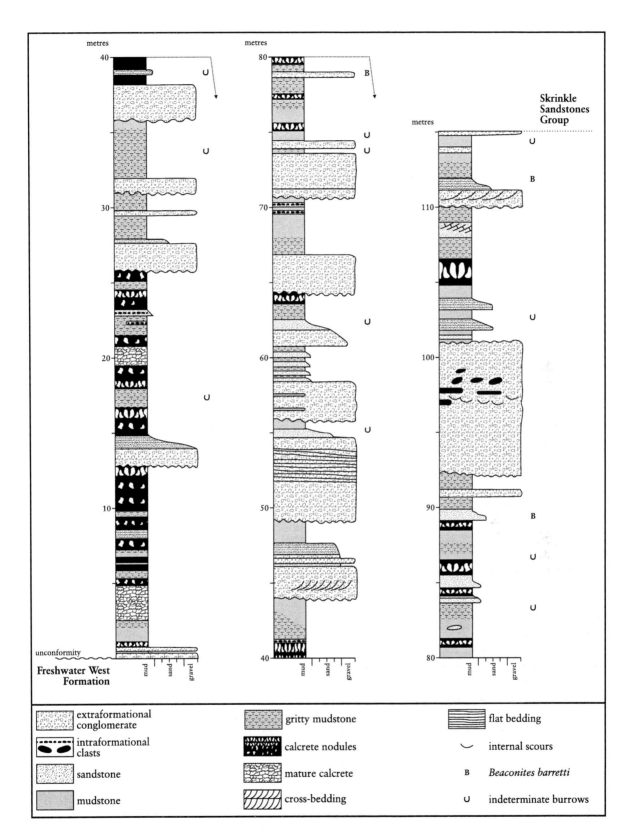

Figure 5.62 Sedimentary log of the Ridgeway Conglomerate Formation at Freshwater West. Based on Williams *et al.* (1982) and Allen *et al.* (1981b).

Figure 5.63 Conglomerate in the Ridgeway Conglomerate Formation, Freshwater West, 73 m above base of the formation. The sequence youngs from left to right. A pronounced pebble fabric is aligned parallel to the foresets of a large-scale cross-stratified bedset. (Photo: B.P.J. Williams.)

30 cm, the largest recorded anywhere in the formation. The clasts are of protoquartzite, lithic greywacke, siltstone and vein quartz, with a marked increase in abundance of phyllite debris in the higher conglomerates. The matrix is coarse-grained, lithic sandstone with rare carbonate cement. The thick conglomerates are poorly to fairly sorted, with some clast imbrication. The thinner units are matrix-supported and contain irregular gravel seams and lenses, outsize clasts and isolated gravel-filled scours.

The sandstones in the formation are fine- to coarse-grained litharenites, many of them rich in phyllite fragments. They are massive or graded, parallel bedded or cross-bedded and locally burrowed. The mudstones are mainly thick and bright red and are ill-sorted, with quartz grains and granules, as well as seams of vein quartz, carbonate debris and exotic granule- and fine pebble-grade detritus. Calcretes occur as nodular, prismatic, massive and vein horizons.

The Ridgeway Conglomerate Formation is overlain with no local angular discordance, but with regional unconformity, by the Skrinkle Sandstones Group. The abrupt facies change is well exposed in the cliff and foreshore (SR 8873 9891). Fish scales (*Holoptychius*) are recorded from a very fine-grained sandstone in the cliff immediately above the junction (Dixon, 1921).

The Skrinkle Sandstones Group comprises two formations, the Gupton Formation and the West Angle Formation (Marshall, 1977, 1978, 2000a,b). The Gupton Formation is subdivided into the Lower Sandstone Member and overlying Stackpole Sandstone Member. Marshall (e.g. 2000a) carried out a detailed analysis of the facies present and their associations. The Lower Sandstone Member is 55 m thick and crops out in the cliffs and foreshore (SR 8876 9885), where it consists of thick, lenticular and tabular sandstones and pebbly sandstones that coarsen and thicken upwards. These show an upward change through three facies associations from lowermost small multi-storey sandstones in a background of mudstone and siltstone to thicker single-storey sandstones and then to stacked, pebbly, sandstone-based fining-upward units. The Stackpole Sandstone Member is exposed on the foreshore and in the cliff near the headland of Great Furzenip (SR 8875 9878). It comprises a lower heterolithic, mudstone-dominated unit that includes a laminated facies of thinly bedded sandstones and mudstones passing up into fining-upward couplets of rippled sandstones and mudstones. An upper sandstone-dominated unit comprises pale yellow-white quartz arenites arranged in wedging and mutually eroded bodies. The sandstones are mainly parallel-laminated, and scour-and-fill structures are abundant. Where the ground rises to meet the headland of Great Furzenip, the white sandstones of the Stackpole Sandstone Member are overlain by the West Angle Formation. The lower part of this formation (the Conglomerate Member) consists of red sandstones, conglomerates and mudstones, and is markedly different in texture and composition from the underlying beds. The sandstones and conglomerates are rich in lithic debris and there is an abundance of calcrete in the mudstones, the facies being arranged in fining-upward couplets. Thin sheets of intraformational conglomerate and overlying rippled sandstone occur within the mudstones.

The upper part of the West Angle Formation (Red-Grey Member) is predominantly red, but is characterized by the incoming of sporadic grey-green sandstones and pebbly sandstones with well-preserved plant fragments. The sandstones occur at the bases of fining-upward sequences capped by calcretized red mudstones. Grey mudstones with freshwater microfossils, coalified plant debris and lingulids appear higher in the member, along with a few sheets of phosphatized pebble conglomerate. The topmost beds of the formation are interbedded red and grey sandstones with mudstone interbeds capped by a mature, calcareous sandstone with well-rounded quartz pebbles and sparse marine brachiopods and bryozoa.

Interpretation

The unconformity between the Moor Cliffs Formation and the Gray Sandstone Group at the base of the section represents a gap of at least 6 Ma, from the late Wenlock (Homerian) to the Lochkovian. This contrasts with the situation at Albion Sands, where the basal Old Red Sandstone facies has a gradational relationship with the Gray Sandstone Group (Hillier, 2000; Hillier and Williams, 2004). The lower part of the Freshwater East Formation is interpreted as a proximal alluvial deposit. The sheet-like, heterolithic units in the upper part represent deposition in an alluvial coastal-plain environment, with wave ripples and the presence of lingulids suggesting episodic marine influence, perhaps in estuarine tidal-flats or as storm washovers (Williams and Hillier, 2004).

The Moor Cliffs Formation is dominated by red mudstones that were traditionally referred to overbank deposition of a high-sinuosity fluvial systems in an alluvial coastal-plain setting (Allen and Williams, 1978; Williams *et al.*, 1982). Recently, however, detailed examination of the mudrocks of the Lower Old Red Sandstone has revealed a range of possible origins (Marriott and Wright, 1996, 2004; Marriott *et al.*, 2005). These authors propose a depositional environment similar to a modern tropical dryland river system in which moderately sinuous, anastomosing ephemeral rivers reworked muddy floodplain sediment during seasonal flooding. The floodplain sediments were pedified to varying degrees as calcic vertisols (calcretes) and mud soil aggregates that were reworked by wind and/or water as sand- or silt-sized pelleted aggregates. Infrequent major flood events caused extensive stripping of sediment from the floodplain surface and the deposition of sand sheets. Ephemeral lakes formed in depressions on the floodplain after floods, trapping aeolian dust. Marriott and Wright (2004) recognize four mudrock facies:

- Heterolithic facies dominated by red mudstone with minor interbedded intraformational conglomerates (Type B of Allen and Williams, 1979b) and extraformational conglomerates (Type A of Allen and Williams, 1979b) and sandstones. These occur as low-angle inclined or horizontal units. The former are interpreted as lateral accretion deposits of the point bars of meandering channels, with the mudstone deposited from bedload as pedogenic mud aggregates. The horizontal units may be the products of rapid, crevassing events.
- Pedified purplish red to brownish red mudstones with calcrete nodules, pseudo-anticlines and slickensided surfaces.
- Brownish red burrowed mudstone units with *Beaconites*- and *Arenicolites*-type burrows. These commonly contain wave-rippled bedforms and are interpreted as the deposits of shallow, semi-permanent floodplain lakes, although bioturbation, desiccation and pedogenesis largely destroyed any fine lacustrine lamination.
- Massive, brick-red mudstones, interpreted as fluvial deposits of silt- and sand-sized pedogenic aggregates, although aeolian reworking of some of the deposits may have also taken place.

Marriott and Wright (1993, 1996, 2004) suggested a dynamic environment in which stable periods with minor flood events produced steady aggradation of the floodplains. These were interrupted occasionally by major flood events when vast amounts of sediment were stripped and deposition in river channels took place.

The Anglo-Welsh Basin

The alluvial architecture of the sandstones of the Moor Cliffs Formation has been intensively studied (e.g. Williams and Hillier, 2004), particularly in the interval immediately above the Townsend Tuff Bed (Allen and Williams, 1982; Love, 1993; Love and Williams, 2000; Love *et al.*, 2004). The sheet-like finer-grained sandstones were probably the products of unconfined sheet flooding over the floodplain. The coarser, cross-bedded sandstone 56 m above the Townsend Tuff Bed is an amalgamation of units produced by sheet-flood events, its erosive incision reflecting a drop in sea level that was possibly induced tectonically (Williams and Hillier, 2004).

The common presence of calcrete nodules in the formation is evidence of repeated, seasonal wetting and drying of the floodplain in a semi-arid climate. The thick succession of stacked, mature calcretes (Stage III of Machette, 1985) of the Chapel Point Calcretes Member points to a prolonged period of non-deposition, sediment starvation and basin shut-down. Each mature profile is estimated to have taken 10 000 years (e.g. Allen, 1974d, 1986), although Marriott and Wright (2004) note that rates of carbonate formation are highly variable, depending on factors including the porosity of the parent sediment and the carbonate content of the percolating water. Hundreds of thousands to millions of years may have been required for Stage III calcrete formation. The regional extent of these calcretes demonstrates 'shut-down' of the Anglo-Welsh Basin, after which sediment provenance, fish fauna and ichnofacies all changed.

The change in basin architecture instigated deposition of the more proximal Freshwater West Formation. The common cyclic arrangement of the facies in fining-upward sequences in the lower part (the Conigar Pit Sandstone Member) points locally to a fluvial origin in high-sinuosity meandering streams, as evidenced by common lateral accretion surfaces (Williams, 1978; Allen *et al.*, 1982). However, the basin infill architecture is variable, with unconfined sheet-flood facies also being abundant (Williams and Hillier, 2004). The basal part of the member here is more mud-rich than elsewhere in Pembrokeshire, perhaps due to its elevated position in the hanging-wall of the Ritec/Benton Fault (Williams and Hillier, 2004). A northerly provenance is indicated by palaeocurrent directions, in contrast to the southerly derived Ridgeway Conglomerate Formation above. The Rat Island Mudstone Formation is of muddy floodplain origin similar to that of the Moor Cliffs Formation (Marriott and Wright, 1993, 1996, 2004; Marriott *et al.*, 2005).

The Ridgeway Conglomerate Formation is interpreted as the deposits of an alluvial-fan system that prograded northwards over fringing alluvial mudflats or playas from an emergent southern source (Williams, 1971; Williams *et al.*, 1982; Tunbridge, 1986; Ékes, 1993). The source area may have formed as a result of uplift in the footwall of a fault in what is now the Bristol Channel during Acadian deformation (Marshall, 2000a). At least some of the red mudrock interbeds in the formation owe their origin to deposition from bedload of pedogenic mud aggregates reworked from muddy floodplains (Ékes, 1993). Other mudrock facies represent mudflows on the fan surface (Williams *et al.*, 1982).

The Skrinkle Sandstones Group is interpreted as the fill of a post-Acadian synrift, fault-bounded basin in which alluvial-fan, lacustrine, braidplain delta and high-energy braidplain deposition took place successively in two superimposed sequences, the first an axial basin-fill, the second transverse fill (Marshall, 2000a,b). The Ritec Fault bounded the basin to the north, the succession thickening southwards in the hanging-wall of a fault in the present-day Bristol Channel. The first phase of the axial basin filling commenced with deposition of the Lower Sandstone Member of the Gupton Formation. Marshall (2000a,b) interpreted the three facies associations of the member respectively as small sheet-flood systems in a lacustrine or floodplain setting; isolated channel-fills; and meandering channel deposits. The coarsening- and thickening-upward nature of the succession suggested to Marshall a distal to proximal change and a SE-prograding terminal fan system. The lower part of the Stackpole Sandstone Member (mudstone-rich heterolithic association) is interpreted as lacustrine and lake-delta deposition, the upper (quartz arenite association) as high-energy, fluvial braidplain deposition. The textural maturity of the sandstones suggests long axial transport paths rather than local sourcing from the footwall of the Ritec Fault.

The West Angle Formation represents the second, transverse, basin-fill succession. Palaeocurrent directions in the lower part (the Conglomerate Member) indicate south and south-west drainage. Lateral accretion sets indicate point-bar formation and meandering river channels, with thin sheets of intraformational conglomerate and overlying rippled sandstone in the mudstone floodplain facies probably representing sheet-flood events. The upper part of the West Angle Formation (Red-Grey Member) is interpreted by Marshall (2000b) as recording the start of the Carboniferous transgression. The sporadic green beds and plant fragments point to a higher water-table on the alluvial floodplain. The progressive incoming of grey mudstones with freshwater microfossils, coalified plant debris and lingulids higher in the member suggests lacustrine deposition. The sheets of phosphatized pebble conglomerate contain bryozoa and sharks' teeth and indicate sporadic marine inundation prior to the main Carboniferous transgression. The calcareous sandstone at the top of the sequence has north-directed palaeocurrents and is interpreted by Marshall as a coastal barrier sand, with the underlying beds being lagoonal and washover fan deposits.

Conclusions

The importance of this site lies in the magnificent exposure of the entire Old Red Sandstone succession and the ease of accessibility of most of the rocks, which attracts large numbers of students and researchers. It is the classic section of the Lower Old Red Sandstone in southern Britain. The site has played a crucial role in the understanding of Old Red Sandstone–basement relationships in the Anglo-Welsh Basin, and of the tectonic and sedimentary history of the basin. The detailed sedimentological studies carried out here have vastly increased our knowledge of the terrestrial depositional environments of the Devonian. Also of prime importance is the part the site has played in the elucidation of the effects of contemporaneous faulting in controlling deposition. Along with the **West Angle Bay (North)** and **Freshwater East–Skrinkle Haven** sites, this site provides a complete picture of the transition into the overlying Carboniferous marine succession. Recent palynological study has added to the stratigraphical importance of the site by yielding an Early Devonian plant spore assemblage. Exceptionally well-preserved trace fossils provide an insight into the early animals that inhabited dry land for the first time.

FRESHWATER EAST–SKRINKLE HAVEN, PEMBROKESHIRE (SS 021 981–SS 082 974)

W.J. Barclay

Introduction

The sea cliffs and foreshore of south Pembrokeshire, south Wales offer the best exposures of the Old Red Sandstone in southern Britain. This 8 km-long, well-exposed along-strike section, also known as 'Tenby Cliffs', extends from Freshwater East in the west to Skrinkle Haven in the east (Figure 5.64). It gives a complete transect through the Old Red Sandstone, from the junction with the Silurian Gray Sandstone Group at Freshwater East, through the Lower Devonian Red Marls, the ?Middle Devonian Ridgeway Conglomerate Formation, the Upper Devonian Skrinkle Sandstones Group and the junction of the Skrinkle Sandstones Group with the overlying Carboniferous Lower Limestone Shale Group at Skrinkle Haven. The strata lie on the southern limb of the Pembroke Syncline (or northern limb of the Freshwater East Anticline), dipping very steeply to the NNE or being almost vertical. With the strata striking almost parallel to the coast, individual beds can be traced laterally for long distances, allowing detailed measurement of sandbody sizes and confirmation of the lateral persistence of marker beds such as calcretes and ash-fall tuffs. It should be noted, however, that large parts of the cliffs are inaccessible and dangerous.

Of particular interest is the group of eight closely spaced marker tuff beds within the Moor Cliffs Formation, the most distinctive of which are the Townsend Tuff Bed, Pickard Bay Tuff Bed and Rook's Cave Tuff. Their identification, combined with the extensive exposures, has allowed a detailed analysis of the alluvial architecture and sedimentary environments of parts of the succession, notably of that between the Townsend and the Pickard Bay Tuff beds (Allen and Williams, 1982) and of the Moor Cliffs Formation as a whole (Love, 1993; Love and Williams, 2000). The Townsend Tuff Bed and Rook's Cave Tuff between Manorbier Bay and Presipe preserve a unique set of trace fossils at their junctions with the underlying floodplain mudstones (Allen and Williams, 1981a; Morrissey and Braddy, 2004). The Ridgeway Conglomerate is important because of its

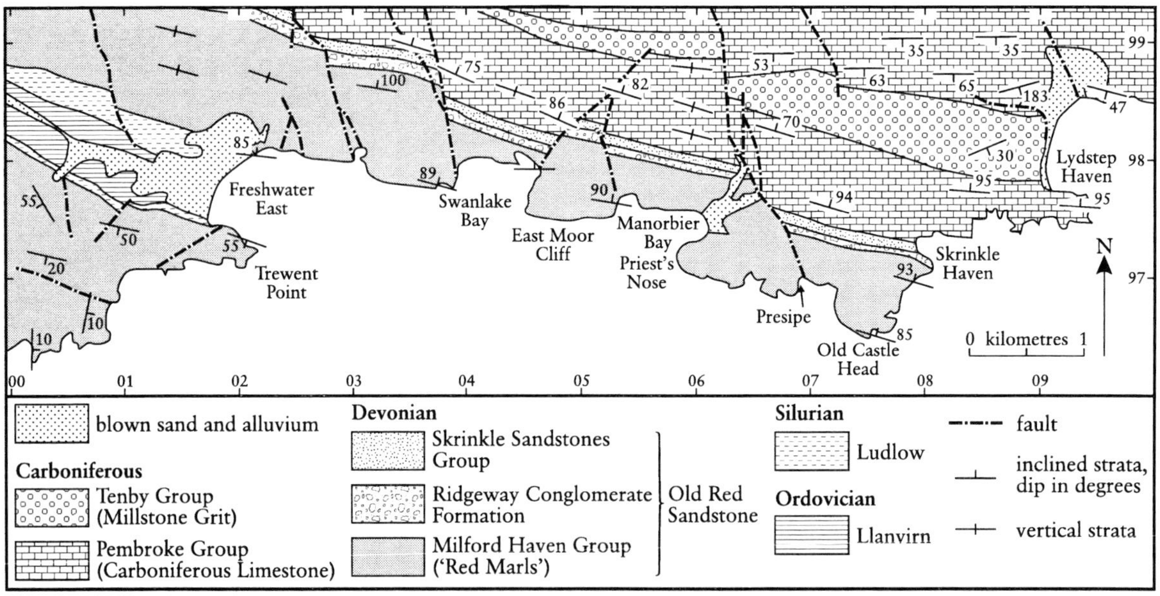

Figure 5.64 Geological map of the Freshwater East–Skrinkle Haven coast section. After British Geological Survey 1:50 000 sheets 244 and 245 (England and Wales), Pembroke and Linney Head (1983).

derivation from a local southerly source and the implications for Mid-Devonian tectonics and palaeogeography. The site also includes the type locality of the Upper Devonian Skrinkle Sandstones Group. The section east of Manorbier was included in a field guide by Williams *et al.* (1982).

Description

The foreshore (depending on extent of sand cover) and cliffs on the north-east side of Freshwater East bay expose the Basement Group (of Dixon, 1921) and the lower part of the Moor Cliffs Formation (the Lower Marl Group of Dixon). The Basement Group lies disconformably on fossiliferous sandstones and mudstones of the Gray Sandstone Group of Wenlock age. Dixon (1921, p. 41) gave a section and Bassett (1982b) described the succession. A basal conglomerate–sandstone 1.4 m thick is overlain by the lower part of the mudstone-dominated Moor Cliffs Formation, in which there are four complexes of green mudstones and green or pale grey, micaceous sandstones. The lowest green beds yielded a spore assemblage (Richardson and Lister, 1969). These green beds contain sporadic lingulids and ostracoderm fragments, together with the vascular plants *Cooksonia*, *Hostinella*, and *Tautilicaulis* (Edwards, 1979b). Malachite occurs on joint faces. There are fine examples of calcretes in the intervening red mudstones, as well as some airfall tuffs. Hancock *et al.* (1982) gave details of the structures.

West Moor Cliff provides exposures of the lower part of the Moor Cliffs Formation, Swanlake Bay exposes the upper part. Allen and Williams (1982) included the Swanlake Bay section in their detailed sedimentological analysis of the section between the Townsend Tuff Bed and the overlying Pickard Bay Tuff Bed. Allen (1974e) analysed cycles in the overlying Freshwater West Formation at Manorbier Bay (SS 057 976; SS 059 973) and at Swanlake Bay (SS 043 981). Marriott and Wright (1993, 2004) gave a detailed description of the mudrocks of the Moor Cliffs Formation.

Love and Williams (2000, fig. 4) and Love *et al.* (2004) gave detailed lithofacies profiles of part of the Moor Cliffs Formation at its type locality of East Moor Cliff and from Priest's Nose to Rook's Cave in the cliff east of Manorbier Bay (Figure 5.65). The section analysed extends from about 25 m below to 50–60 m above the Rook's Cave Tuff, which lies about 90 m below the Townsend Tuff Bed. Marriott and Wright (1993) examined 17 m of strata below, and 69 m above the Rook's Cave Tuff, 40 m west of Rook's Cave.

The cliffs to the east of Priest's Nose expose a magnificent strike section of the Moor Cliffs

Freshwater East–Skrinkle Haven

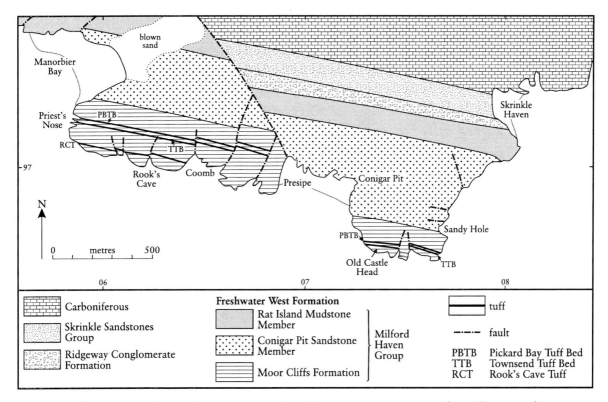

Figure 5.65 Geological map of the Manorbier–Skrinkle Haven coast section. After Williams *et al.* (1982).

Formation (e.g. Marriott and Wright, 2004). The sequence is dominated by calcretized, red mudstones/siltstones, many showing pseudo-anticlinal structures (Figure 5.66). There are a few, thin, fine- to medium-grained sandstones, as well as some thin lenses and lenticular sheets of intraformational conglomerate (Allen and Williams, 1979b) and a very few thin, extraformational conglomerate lenses. Conspicuous, laterally persistent slots occur where softer airfall tuffs have been weathered out. The Rook's Cave Tuff buried an irregular surface with shallow depressions and channels, probably occupied by water, separated by inter-pond, slightly eroded tracts of mud (Williams *et al.*, 1982). Organic burrow structures ('trumpet marks' of Allen and Williams, 1981a) and arthropod tracks (Morrissey and Braddy, 2004) occur at the base of the tuff (Allen and Williams, 1981a). Love and Williams (2000) record six tuffs in the 60 m of beds above the Rook's Cave Tuff.

Figure 5.66▶ Calcrete forming pseudo-anticlinal structure, Moor Cliffs Formation, Presipe. View looking west, near-vertical sequence younging left to right. (Photo: W.J. Barclay.)

The Anglo-Welsh Basin

At Old Castle Head (Figure 5.67) a deep slot marks the outcrop of the Townsend Tuff Bed (Figure 5.68), which is visible only at very lowest water-level. Details of the Townsend Tuff Bed at East Moor Cliff (SS 0456 9762), Priest's Nose (SS 0586 9723) and Old Castle Head (SS 0741 9664) were given by Allen and Williams (1981a). It comprises a complex of three superimposed, normally graded airfall tuffs (falls A, B and C). Thickness of the complex ranges from 2.44 m at Priest's Nose to 3.39 m at Old Castle Head. Crystal tuffs at the bases of the falls grade up into porcellanous dust tuffs. The basal fall (A) is the thinnest (0.17–0.47 m) and has a sharp, non-erosional top strewn with faecal debris. Organic structures on its top at East Moor Cliff include those described by Allen and Williams (1981a) as 'burrow tops', and 'trumpet' and 'bread roll' marks, possibly related to *Beaconites*. Fall B ranges from 1.17 m at Priest's Nose to 1.72 m at Old Castle Head. At East Moor Cliff it is 1.42 m thick and comprises crystal and dust tuffs interbedded at mainly centimetre scale. Fall C ranges from 1.10 m at Priest's Nose to 1.15 m at East Moor Cliff and 1.2 m at Old Castle Head. At all three localities it has an irregular base overlying an erosion surface.

Williams *et al.* (1982) gave details of the section at Presipe (SS 0690 9707) near a major dextral wrench fault. The west cliff of the bay exposes a fine mudstone/siltstone-dominated section below the Rook's Cave Tuff, which lies high up in the north-west corner of the bay 5.85 m below a prominent fine- to medium-grained, cross-bedded sandstone. The section shows a magnificent development of calcretized mudstones and calcretes, with nodule fans displaying pseudo-anticlinal structure (Figure 5.66; Love, 1993; Love and Williams, 2000; Love *et al.*, 2004). Bedding planes in the section in the central part of the bay preserve a range of trace fossils, including arthropod tracks and large scour marks and *Beaconites* burrows (Morrissey and Braddy, 2004). Mud-filled

Figure 5.67 Oblique aerial view looking east of Old Castle Head. Vertical strata of the Moor Cliffs Formation. The strata young to the north (right to left) and a fault displaces them dextrally. Tuffs are marked by slots in the cliffs (RCT – Rook's Cave Tuff; TT – Townsend Tuff Bed; PBT – Pickard Bay Tuff Bed; CPC– Chapel Point Calcretes Member). Other slots probably mark unnamed tuffs. (Photo: S. Howells.)

Freshwater East–Skrinkle Haven

Figure 5.68 The Townsend Tuff Bed, Old Castle Head. View looking east, sequence younging right to left. (Photo: B.P.J. Williams.)

desiccation polygons and rippled surfaces are also seen. The beds are arranged in upward-fining cycles and some of the sandstones contain fine examples of soft-sediment deformation.

The Chapel Point Calcretes Member, the local name for the regionally developed Psammosteus Limestone, forms the top of the Moor Cliffs Formation. It crops out near King's Quoit (SS 0596 9733) and comprises 10–15 m of massive, purple careously weathered calcrete, which form a pronounced notch on the foreshore (Figure 5.69) and alcove in the cliff. Immediately overlying it is a very distinctive conglomerate 1.5–2.2 m thick and containing angular quartz pebbles (Williams and Hillier, 2004). The calcrete is at least 11.2 m thick at Conigar Pit to the east (Williams et al., 1982).

The overlying beds comprise the Conigar Pit Sandstone Member at the base of the Freshwater West Formation. Conigar Pit (SS 0724 9696) provides a magnificent type section of the Conigar Pit Sandstone Member (Williams et al., 1982), with the junction with the underlying Chapel Point Calcretes Member exposed at low-water mark (SS 0722 9680). The Conigar Pit Sandstone mainly comprises a stacked succession of upward-fining cycles of intraformational conglomerates, sandstones and mudstones with calcretes. The conglomerates are of small pebble grade, massive or cross-bedded and contain ostracoderm fragments. The sandstones are fine- to medium-grained and ripple cross-laminated and contain *Beaconites* burrows and mud-filled desiccation cracks. The most notable feature of the section is the occurrence of eleven sandstone-mudstone lateral accretion complexes. They range from 0.7 m to 3.09 m thick, the epsilon cross-bedding being magnificently seen and showing southerly flow direction. North- and west-directed sets are also present (Williams et al. 1982).

The calcareous siltstones of the Rat Island Mudstone Member of the Freshwater West Formation are exposed at Manorbier (Wright and Marriott, 1996) and on the headland south of Skrinkle Haven (Marriott and Wright, 1993, 1996, 2004). Their junction with the vertical Ridgeway Conglomerate Formation (SS 0800 9718) is erosional, although the basal conglomerate appears to pass laterally into red and green siltstones high in the face. However, a small fault at this point may complicate the relationship (Williams, 1971).

Williams (1971) described the Ridgeway Conglomerate. It is attenuated to 47 m at Skrinkle Haven and consists of interbedded siltstones, sandstones and extraformational conglomerates (Williams, 1971, fig. 5). Four conglomerate beds comprise 9 m of the formation. One 4.5 m-thick, green conglomerate lies 24 m above the base of the formation. Imbricated pebbles are set in a coarse-grained sandstone matrix with a carbonate cement. A thick (4 m) mature, massive calcrete overlies the conglomerate. Ékes (1993) noted abundant, metre-thick, massive calcretes showing pseudo-anticlinal 'gilgai'-type structures. The pebbles of the Ridgeway Conglomerate have not yet been matched with older formations (Allen, 1974a), but palaeocurrent evidence and decreasing pebble size northwards indicate a southerly derivation (Williams, 1971; Allen, 1974a; Cope and Bassett, 1987; Powell, 1989; Ékes, 1993).

The Anglo-Welsh Basin

Figure 5.69 Near-vertical basal beds of the Freshwater West Formation. View looking WSW at southern end of Manorbier Bay. The Chapel Point Calcretes Member forms the notch on the left, the sequence younging left to right. (Photo: W.J. Barclay.)

Skrinkle Haven is the type locality of the Skrinkle Sandstones Group, which unconformably overlies the Ridgeway Conglomerate. Williams (1971) provided a description. It is 91 m thick and includes a lower quartzite unit (15.5 m) and upper conglomerate unit 54 m thick. The upper unit contains a 0.35 m-thick, grey and red mottled siltstone containing *Lingula* sp. and bivalves including *Modiola* sp.. The top of the Skrinkle Sandstones Group is sharp, where green sandstones are overlain by grey shales of the Lower Limestone Shale Group.

Interpretation

Allen and Williams (1982) interpreted the succession between the Townsend and Pickard Bay Tuff beds as having been deposited in an extensive mudflat environment, with sporadic fluvial channels. Widespread calcretes point to periodic subaerial exposure of the mudflats and precipitation of carbonate in the vadose zone. The ubiquitous destruction of sedimentary lamination is attributed to organic bioturbation.

Allen and Williams (1982) were unclear as to whether the mudflats were subjected to tidal influence (an interpretation that they favoured), as suggested by the strong currents that affected the ashfall tuffs, although the evidence of variability of palaeocurrents is lacking. Marriott and Wright (1993, 1996) ruled out tidal influence during deposition of the Moor Cliffs Formation, favouring an unstable, flashy alluvial-floodplain system with ephemeral stream channels subject to common erosional events that reworked earlier soil horizons and the floodplain sediments. Love and Williams (2000) also discounted any marine influence. The alluvial-floodplain facies associations indicate prolonged periods of stability, non-deposition and soil formation. The calcretes indicate a seasonally wet, monsoonal, sub-tropical climate. Love and Williams (2000) and Marriott and Wright (1996) favoured an active, channelized depositional floodplain setting for much of the mudstones and siltstones, rather than a strictly overbank model favoured by earlier workers (e.g. Allen and Williams, 1981b). The floodplain channels were ephemeral, extremely broad and with low relief and flow, depositing mud and silt as bedload aggregates. Love (1993) and Love and Williams (2000) emphasize the role of land-plant colonization, with the primitive, shallow-rooted vegetation being unable to protect the floodplain from the rapid removal of sediments and calcrete soils during wet-season flooding.

Love and Williams (2000) and Love *et al.* (2004) interpret the sedimentary architecture of the Moor Cliffs Formation as the product of deposition by sporadic, large, broad, sandy channel systems with highly variable discharges and ephemeral flow, and by ephemeral, flashy, shallow streams on extensive interfluvial areas. Marriott and Wright (2004) envisage an environment analogous to a modern dryland system in which mud-dominated, moderately sinuous ephemeral rivers reworked floodplain muds during seasonal flooding. There is little evidence to support the conventional view that the mudstones/siltstones were deposited by suspension from standing water. For the heterolithic mudstone, sandstone–conglomerate mudrock facies and the massive brick-red mudrocks, Marriott and Wright (1996, 2004) favour active deposition of mud/silt aggregates from bedload. Rippled bodies and laminated sheets were deposited in through-draining larger channels and lumps of silt/mud aggregates and calcrete clasts were transported as coarse sand and gravel bedload in the shallow interdistributary channels and deposited as bar forms. Units of burrowed mudrocks are attributed to lacustrine deposition in floodplain lakes, the pedified and desiccation-cracked, calcrete-prone mudrocks to steady aggradation on the floodplain during relatively stable periods, occasionally interrupted by flood events (see **Freshwater West** GCR site report, this chapter, for a fuller account of Marriott and Wright's important work). The strong cyclicity of the succession and of the calcrete palaeosols, which show increasing upward maturity, is attributed by Love and Williams (2000) to autocyclic processes of fluvial avulsion and aggradation.

The Conigar Pit Sandstone is interpreted by Hillier and Williams (2004) to represent a range of fluvial environments. Multi-storey sandstone bodies occupied the main channel belts and were mainly laterally accreted channel sand-bodies. Also present are heterolithic sandstone/ mudstone beds representing ephemeral channel deposition and sheet sands deposited from unconfined sheet floods. The Rat Island Mudstone comprises four facies, as detailed by Marriott *et al.* (2005) at a section (SS 0580 9775) on the west side of Manorbier Bay. These authors compare the environment to the Channel Country of central Australia, where mud sheet-flood deposits on a muddy braidplain border anastomosing fluvial systems. Intraformational conglomerates with mudstone and calcrete clasts were deposited as gravels in slightly sinuous, ephemeral channels, and heterolithic mudstone/sandstone units were deposited as accretionery benches in the channels. Pedified, purplish red mudstones with carbonate nodules are palaeo-vertisols (Stage II or III calcretes of Machette, 1985). Brick-red, pelleted mudstones were probably deposited as fine sand-sized aggregates in bedload during sheet-flood events. Subaerial exposure between flows was sufficiently long to allow pedogenic processes to form immature calcrete vertisols. Alternatively, the upper parts of the beds of pelleted mudrocks may have been deposited as aeolian dust. The facies are stacked in fining-upward units, 2.5m–5 m thick, representing ephemeral channel-zone deposition in which fluvial reworking of the palaeo-vertisols as clay pellet bedload took place.

The age of the Ridgeway Conglomerate Formation remains unknown. Unconformities at its base and top constrain it to between late Early and Mid-Devonian age (Marshall, 2000a,b). Allen (1974a) suggested that its base marks an important depositional break that extended into Mid-Devonian, and perhaps even into Late Devonian, times. Williams (1971) favoured a Mid-Devonian age, but later (Williams, 1978) suggested a late Early Devonian age. Powell (1989) gave a Siegenian (Early Devonian) age, but provided no evidence on which this is based. It is now clear that the formation post-dates the Acadian deformation (B.P.J. Williams, pers. comm.).

Allen (1974a) compared the Ridgeway Conglomerate Formation to the sediments of modern semi-arid playa-basins and alluvial-fans, with the conglomerates probably being of local origin; an alluvial-fan–braided river setting has also been suggested by several authors (e.g. Williams, 1971; Williams *et al.*, 1982; Powell, 1989; Ékes, 1993). The sediments were probably sourced from Lower Palaeozoic or

?Precambrian outcrops in the Bristol Channel Landmass to the south. Tunbridge (1986) and Cope and Bassett (1987) suggested that Mid-Devonian tectonics in the Bristol Channel area produced this positive area.

The Skrinkle Sandstones Group is of Late Devonian to Early Carboniferous age. It was deposited in a fault-bounded basin (the Tenby–Angle Basin), of which the Ritec Fault marked the northern limit (Powell, 1989; Marshall, 2000a,b). No detailed sedimentological work has been carried out at Skrinkle Haven, but Marshall interpreted the succession in **West Angle Bay (North)** to the west (see GCR site report, this chapter) as the product of alluvial-fan, alluvial-plain and lacustrine deposition. This was controlled by movements of the basin-bounding faults, within an overall transgressive regime as subsidence on the southern margin of the Wales–Brabant Massif was matched by rising sea level in early Carboniferous times.

Conclusions

The magnificent, cliff exposures along this 6 km-long section of coast reveal the entire succession of Old Red Sandstone strata in south Pembrokeshire. The high quality and lateral persistence of the extensive exposures of near-vertical strata allow detailed analyses of facies and their architecture. The modern sedimentological research of the section has been instrumental in the elucidation of the alluvial environments of the Old Red Sandstone, ranging from the distal floodplains of Late Silurian and Early Devonian times to the more proximal alluvial-fan deposition of the southerly sourced Early–Mid-Devonian Ridgeway Conglomerate Formation and the Upper Devonian Skrinkle Sandstones Group. Of special interest are the marker ashfall tuffs, of which the Townsend Tuff Bed, Rook's Cave Tuff and Pickard Bay Tuff Bed are the principal ones. These preserve a unique set of trace fossils, including faecal pellets, burrow-fills and arthropod tracks at their junctions with the floodplain mudstones. The magnificent exposures of calcretes, showing varying stages of maturity and gilgai-type pseudo-anticlinal structures, as well as the facies and architecture of the sandbodies and mudrocks, have been utilized to model the contemporaneous sub-tropical, semi-arid, seasonally wet climate of Late Silurian and Devonian times.

LLANSTEFFAN, CARMARTHENSHIRE (SN 350 100)

W.J. Barclay

Introduction

The site is a coastal section exposing beds traditionally referred to the Red Marl Group (Lower Old Red Sandstone) of Pembrokeshire (Strahan *et al.*, 1909). The presence of the Chapel Point Calcretes Member (Psammosteus Limestone) at the site, in a magnificent development of stacked, mature, pedogenic calcrete profiles, allows subdivision of the succession, with the beds above the limestones correlated with the lowermost part of the Freshwater West Formation of south Pembrokeshire (Allen, 1978c; Allen *et al.*, 1981b; Williams *et al.*, 1982). These beds include large, fluvial, sandbodies deposited in large rivers, as well as the deposits of small, muddy, interfluvial distributaries with little sand, but much intraclast gravel. This site has provided important evidence towards the understanding of Early Devonian alluvial-plain geomorphology and drainage patterns. Apart from the type locality of the Chapel Point Calcretes Member on Caldey Island, the limestones here are some of the best examples of mature Old Red Sandstone calcrete profiles seen in the Anglo-Welsh Basin, all completely exposed in extensive cliffs. The site has been instrumental in the understanding, interpretation and significance of these calcretes.

Description

The GCR site lies immediately south of the village of Llansteffan and extends along the coast of the Towy estuary from 100 m north-east of Llansteffan Castle (SN 3526 1026) south-westwards to near St Anthony's Cottage (SN 3470 0085) in Scott's Bay (Figure 5.70). Murchison (1839) noted that the section exposed 'the finest example of a limestone of the Old Red Sandstone System in Carmarthenshire'. Strahan (in Strahan *et al.*, 1909) was the first to describe the section. More recently, Allen (1978c, 1986), Allen *et al.* (1981b), Cope (1982), Marriott and Wright (1996), Wright and Marriott (1996) and Jenkins (1998) examined the section, providing details and sedimentological analyses. Morrissey and Braddy (2004) describe animal burrows and trails in the Freshwater West Formation.

Llansteffan

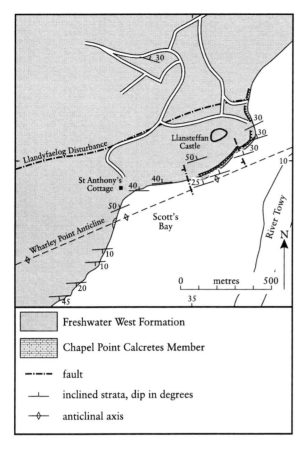

Figure 5.70 Geological sketch map of the Llansteffan area. After British Geological Survey 1:10 560 manuscript map Carmarthenshire 45SE (1906).

The site lies on the northern limb of the Wharley Point (or St Anthony's) Anticline, a NE-trending anticline, which lies in the hanging-wall of the Llandyfaelog Disturbance (Strahan in Strahan *et al.*, 1909). This major structure has a similar trend to the anticline and crops out close to the northern end of the site. The axis of the anticline crosses the coast south-west of the western end of the site, 230 m south of St Anthony's Cottage, the Llansteffan succession being repeated and higher strata exposed towards Wharley Point.

The Chapel Point Calcretes Member is the lowest bed exposed in the site, cropping out along the axis of the anticline (Figure 5.71) and being continuously exposed from east of Llansteffan Castle (SN 353 101) to south-west of the castle (SN 350 099). Correlated with the regionally developed Psammosteus Limestone, it forms the uppermost part of the Moor Cliffs Formation of the Pembroke peninsula. Figure 5.72 is a graphic log of the near-strike succession exposed in the section, on the northern limb of the Wharley Point Anticline. The following description is based largely on that given by Strahan (in Strahan *et al.*, 1909), Allen (1978c) and Allen *et al.* (1981b). The succession comprises three parts, in ascending order: thick calcretes (the Chapel Point Calcretes Member); interbedded calcretes, mudstones and intraformational conglomerates; and sandstones and mudstones in upward-fining sequences. The cliffs from below the castle for 350 m south-westwards expose 15 m comprising six closely spaced calcrete profiles. Strahan (in Strahan *et al.*, 1909) referred to the beds as massive limestone composed of concretions perpendicular to bedding. The calcretes are grey to green and rubbly to massive, and show an upward increase of maturity, with increasing carbonate content and a gradation from discrete nodules, which increase upwards in size and coalesce into rubbly and massive limestones. Irregular, horizontally laminated limestones are present locally, but the profiles have a mainly prismatic, bedding-perpendicular fabric (Allen, 1974d, 1978c, 1986; Allen *et al.*, 1981b). Allen (1974d, 1986) differentiated the calcretes on the basis of their maturity into three types. Type A profiles comprise scattered nodules (glaebules) with locally more intense concentrations (Stage I of Machette, 1985); in Type B, the glaebules are larger and closely packed at intermediate to upper levels, giving a crude prismatic fabric (Stage II of Machette). Type C profiles represent the most mature profiles in which closely packed glaebules coalesce (Stage III of Machette) and contain laminated carbonate layers (Stage IV of Machette). Most of the calcretes in this section are of Type B (stages II to III).

Overlying the calcretes is a mudstone-dominated succession, which marks the base of the Freshwater West Formation. It contains some calcretes and three types of sandstone bodies (Allen, 1980):

- thin, mainly very fine-grained sandstones with lateral accretion structures;
- thick, mainly fine-grained, cross-bedded or parallel-laminated sandstones; and
- lenticular intraformational conglomerates.

The conglomerates are bounded above and below by mudstones, locally cross-bedded and preserve dune or bar forms and internal channelling (Allen, 1978c; Allen and Williams, 1979b). The conglomerates occur at the base of

Figure 5.71 The Chapel Point Calcretes Member, Llansteffan. The topmost calcrete of a stacked succession of calcretes overlies the careously weathered top of the one below. (Photo: W.J. Barclay.)

the succession and overlie erosion surfaces (Allen, 1978c; Allen *et al.*, 1981b). The calcretes, which lie within the mudstones exposed in and to the south-west of a slight embayment south-west of the Chapel Point Calcretes Member cliff, show varying degrees of maturity (but are mostly Type A) and well-developed pseudo-anticlines (Allen, 1973b).

Marriott and Wright (1996) described a section of 6 m of strata lying 10 m above the base of the Freshwater West Formation. They interpreted some of the red mudstones in the succession as originating as sand-sized mud aggregates or pellets produced in soils and deposited from bedload in small, sinuous channels, similar to the aggregates described by Ékes (1993) in the Ridgeway Conglomerate Formation of West Angle Bay.

There are several metre-scale, mainly cross-laminated, very fine-grained sandstones with lateral accretion surfaces. Between 29 m and 35 m above the base of the section is a complex of upward-fining and upward-thinning intraformational conglomerates interbedded with mudstones. The conglomerate bodies have sharp bases resting on downcutting erosion surfaces and are either cross-bedded or horizontally bedded. The higher conglomerates are thin, of granule grade and lack quartz sand (Type B of Allen and Williams, 1979b). A conglomerate near the top of this part of the section contains plant fragments and *Pachytheca*.

The upper (third) part of the succession is exposed in the cliff that extends towards St Anthony's Cottage (SN 346 099–SN 347 099). Two thick, green sandstone bodies lie 80 m and 95 m above the base of the section. They are coarser and thicker than the underlying laterally accreted sandstones and contain no lateral accretion structures. The lower one is 5.2 m thick, has a thin, fish-bearing intraformational conglomerate resting on a scoured surface, and comprises parallel-laminated sandstones overlain by cross-bedded, fine-grained sandstones. At the top, 0.8 m of cross-laminated, very fine-grained sandstone passes up into calcretized mudstone. Internal erosion surfaces at the bases of sets are strewn with intraformational clasts. The higher sandstone is 7.2 m thick and comprises a complex of trough cross-bedded,

Llansteffan

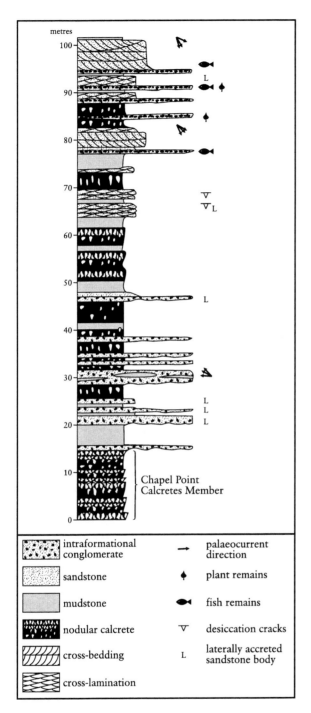

Figure 5.72 Vertical section of the strata exposed at Llansteffan. Based on Allen (1978c) and Allen *et al.* (1981b).

parallel-laminated and cross-laminated sandstones resting on a thick basal intraformational conglomerate containing ostracoderm fragments. The burrow trace fossil *Beaconites* is very common towards the top. Allen *et al.* (1981b) referred it to *Beaconites antarcticus*, Morrissey and Braddy (2004) to *Beaconites barretti*. Arthropod trackways and foraging traces occur in mudstones near Wharley Point (Morrissey and Braddy, 2004).

Interpretation

The succession of six stacked calcrete profiles comprising the Chapel Point Calcretes Member at the base of the section correlates with the regionally developed Psammosteus Limestone. It represents a prolonged period of geomorphological stability, with little subsidence of the alluvial floodplain or fluvial incision, allowing subaerial exposure and pedogenic carbonate formation, each profile representing periods that may have ranged between tens of thousands to millions of years (Allen, 1974d, 1986; Allen and Williams, 1979b; Marriott and Wright, 1993, 2004). Some of the mudstones/siltstones have been interpreted as originating as sand-sized pedogenic mud aggregates (pellets) produced in floodplain soils and deposited from bedload in small, sinuous channels (Marriott and Wright, 1996, 2004).

The cross-bedded, intraformational conglomerates in the overlying beds suggest deposition in broad, shallow channels. Their distribution, many as bar and dune forms within mudstones, may represent ephemeral drainage systems in interfluvial areas between the main distributary channels (Allen and Williams, 1979b; Allen *et al.*, 1981b). The cross-laminated, very fine-grained sandstones with lateral accretion surfaces may represent high-sinuosity channels with sluggish water flows. The higher conglomerates are thin sheets, also lacking in quartz sand, and suggesting deposition of reworked muds (including pedogenic mud aggregates) and calcretes in ephemeral streams and sheet floods (Marriott and Wright, 1996). The green, coarser-grained, thicker sandstones 80 m and 95 m above the base of the section are interpreted as the deposits of larger, higher-energy, low-sinuosity streams, suggesting the establishment of major rivers in the area (Allen, 1978c; Allen *et al.* 1981b). The inception of major fluvial sedimentation above the Chapel Point Calcretes Member was a regional event recognized throughout the Anglo-Welsh Basin, marking the establishment of generally south-flowing rivers from a source nearer than that of the streams which deposited the strata below the limestone (Allen and Crowley, 1983).

The Anglo-Welsh Basin

Conclusions

The site is of regional and national importance in providing the best onshore and most easily accessible exposure of a condensed succession of stacked Old Red Sandstone mature fossil soil carbonate (calcrete) profiles, representing the best development of the Chapel Point Calcretes Member (Psammosteus Limestone) in the Anglo-Welsh Basin. It provides important information on the contemporaneous climate, which, by analogy with the occurrence of modern calcretes, was semi-arid, seasonally wet and tropical.

Detailed sedimentological analyses at the site have been instrumental in building models of Old Red Sandstone alluvial sedimentation. Analyses of the sandstone and intraformational conglomerate bodies have provided an insight into a range of fluvial environments, including major, sand-filled distributary channels in the higher parts of the section, and more localized shallow, ephemeral flashy streams in the lower parts. A further point of interest lies in some of the mudstones above the Chapel Point Calcretes Member, which may have originated as pedogenic mud aggregates deposited from stream bedload or sheet floods.

PORTISHEAD, NORTH SOMERSET (ST 461 770)

P.R. Wilby

Introduction

The Portishead GCR site exposes Lower and Upper Old Red Sandstone beds and provides the best exposure on the southern side of the Bristol Channel of the Lower–Upper Devonian boundary as preserved in continental red-bed facies. It also exposes the biostratigraphically important Woodhill Bay Fish Bed. The site has played an important role in identifying the provenance of the Old Red Sandstone in southern Britain, and in demonstrating its thickening southwards across the Anglo-Welsh Basin. The importance of the site also lies in its unique fish fauna, for which it is described separately in the fossil fishes GCR volume (Dineley and Metcalf, 1999).

The low cliffs of Woodhill Bay, Kilkenny Bay and Black Nore Point to the south of Portishead (Figure 5.73) expose a continuous, 354 m-thick section through the upper part of the Black Nore Sandstone Formation and the lower part of the overlying Portishead Formation. The latter includes the lithostratigraphically important Woodhill Bay Conglomerate at its base and, 27 m higher, the Woodhill Bay Fish Bed. The Black Nore Sandstone Formation is unfossiliferous (Kellaway and Welch, 1955), but is correlated with the Lower Devonian Brownstones Formation of south Wales (Kellaway and Welch, 1955; Green, 1992). The Portishead Formation is correlated with the Quartz Conglomerate and Tintern Sandstone formations of the Forest of Dean (Kellaway and Welch, 1955). It is mainly of Late Devonian age, but probably extends into the earliest Carboniferous (Neves and Dolby, 1967; Utting and Neves, 1970).

Description

The site is located on the southern limb of an ENE-trending anticline that forms a prominent ridge that extends from Portishead 5 km south-westwards to Clevedon. The anticline plunges WSW at its south-western end and ENE around Portishead. Consequently, the lowest part of the sequence is exposed at the southern end of the section between Kilkenny Bay and Black Nore Point. The rocks dip at a moderate angle (typically 20°) to the south-east and are cut by a number of NW-trending faults, many of which show post-Triassic movement (Wallis, 1927; Pick, 1964b). Most of the faults are minor, with throws of the order of 10 m to the south-west. At various points along the section, the Old Red Sandstone is blanketed by a discontinuous cover of dolomitic conglomerate, a marginal facies of the Triassic Mercia Mudstone Group.

The sedimentology and stratigraphy of the Old Red Sandstone at Portishead was described by Reynolds and Greenly (1923), Wallis (1927, 1928), Kellaway and Welch (1948, 1955, 1993), Butler *et al.* (1972) and Dodd (1986). The site is included in two field guides (Reynolds, 1921; Williams and Hancock, 1977). Wallis (1927)

Figure 5.73▶ Location and geology of the Portishead GCR site. (a) Geological map (after British Geological Survey 1:50 000 Sheet 264 (England and Wales), Bristol (2003)); (b) detailed map of section in southern Kilkenny Bay (based on Pick (1964a) and Williams and Hancock (1977)). Letters in (b) are 'formations' of Pick (1964a).

Portishead

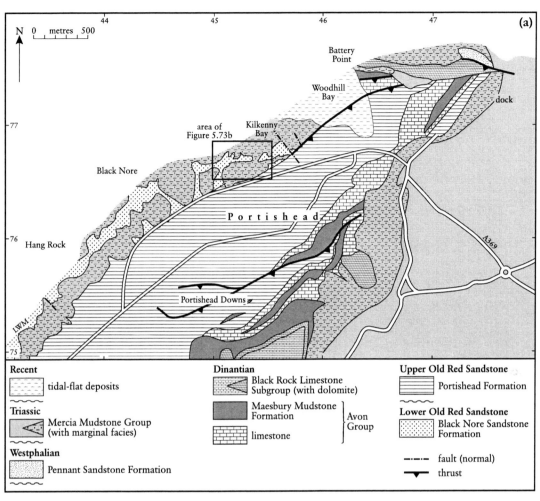

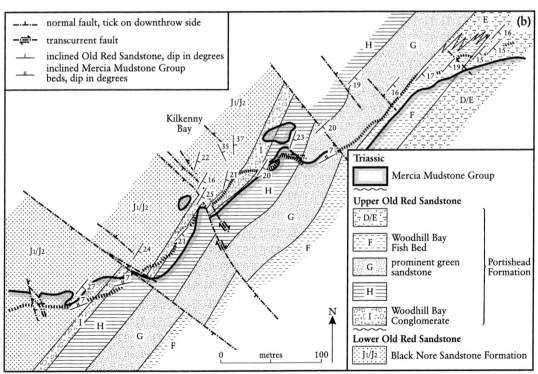

investigated the provenance of the sediments and discussed their depositional environment. The most detailed study was by Pick (1964a,b). He subdivided the Portishead Formation into nine formations (A to I) and the Black Nore Sandstone Formation into three (J1 to J3). However, only the Woodhill Bay Conglomerate and Woodhill Bay Fish Bed (Pick's formations I and F respectively) justify separate names, the remaining units representing laterally impersistent facies variations that are not mappable over a significant distance (Williams and Hancock, 1977). Also, Dodd (1986) suggested that formations A to C, which crop out south of the GCR site near Clevedon, may, in fact, be the same beds that are exposed at Portishead, repeated by faulting, with Pick's formations C and G being the same unit. The structural setting of the region was described by Reynolds and Greenly (1924). The diverse fish fauna from the Woodhill Bay Fish Bed, which is unique in southern Britain and includes the only British record of the placoderm *Groenlandaspis*, is described in the fossil fishes GCR volume (Dineley, 1999g).

The succession (Figure 5.74) is most easily examined in descending stratigraphical order by traversing the section from north-east to south-west. For convenience, it is described using Pick's (1964a,b) 'formations'. The highest unit ('formation' D), which forms part of the Portishead Formation, crops out near the top of the cliff adjacent to the seawall at the southern end of Woodhill Bay. It is approximately 3 m thick and consists of centimetre- to decimetre-scale beds of soft, red and green, cross-laminated, sandy siltstones and interbedded fine-grained sandstones that are harder, red, micaceous and parallel-laminated. Some of the sandstones rest on erosion surfaces, fill minor channels and contain thin, discontinuous intraformational mud-flake conglomerates (Williams and Hancock, 1977). Some are bioturbated, some preserve desiccation cracks and some contain fish fragments (Wallis, 1928). Beneath this unit are 7.5–17 m of mauve-red and green-grey, trough cross-bedded, fine- to medium-grained, metre-scale beds of sandstone (Pick's 'formation' E). The lowermost sandstones contain dispersed vein quartz pebbles (averaging 1.3 cm in length), mudstone rip-up clasts and rare eurypterid remains (Simpson, 1951). The base of this unit is marked just to the north-east of an indentation in the cliff (ST 4602 7695) by an erosion surface above which lies a conglomerate and channelized, cross-bedded sandstone, each about 1 m thick. Together, 'formations' E and D form a fining-upward sequence that fills a large channel.

The underlying Woodhill Bay Fish Bed ('formation' F of Pick, 1964a,b; beds 5–10 of Wallis, 1927, 1928) is seen best in the lower two thirds of the cliff just to the south-west of the indentation. At Woodhill Bay it is 10 m thick, but it is not laterally extensive, being cut out north-eastwards by an erosion surface. It consists predominantly of red, micaceous siltstones and sandy siltstones, but also contains some decimetre-scale beds of green and red, fine-grained sandstone and green siltstone (Pick, 1964a; Williams and Hancock, 1977). Most of the siltstone beds are 1–5 cm thick and contain small, simple horizontal and vertical burrows (Dodd, 1986). The sandstones increase in abundance in the middle and upper part of the unit and four prominent beds (up to 1 m thick) can be traced across the outcrop. They are parallel-laminated, planar cross-bedded or trough cross-bedded, and contain rare pebbles and abundant intraclasts. Some bedding surfaces are rippled or display primary current lineation, and the base of the thickest sandstone preserves load casts. The important ichthofauna, which is dominated by abundant disarticulated scales and teeth of the crossopterygian fish *Holoptychius* and *Glyptopomus*, is concentrated in the thickest sandstones (beds 7, 8 and 10 of Wallis). It indicates a Late Devonian (mid-Famennian) age for the bed (Kellaway and Welch, 1993) and includes the only British record of the otherwise widespread form *Groenlandaspis* (Ritchie, 1975; Dineley, 1999g). In addition, fragmentary carbonized plant debris occurs locally and Wallis (1927) recorded a single, poorly preserved brachiopod.

The base of the Woodhill Bay Fish Bed is sharp and defined by a gently scoured, pebble-strewn surface. Beneath is a 14 m-thick, pale green sandstone unit (Pick's 'formation' G), the upper part of which is particularly distinctive and of probable aeolian origin (Dodd, 1986). It is fine grained, moderately well- to very well-sorted, and contains large-scale planar and trough cross-sets. In contrast to the units above and below, intraclasts and quartz pebbles are absent, and the cross-bedding indicates transport predominantly from the south-east.

'Formation' H consists of 13.5 m of red and green-grey, cross-laminated and cross-bedded,

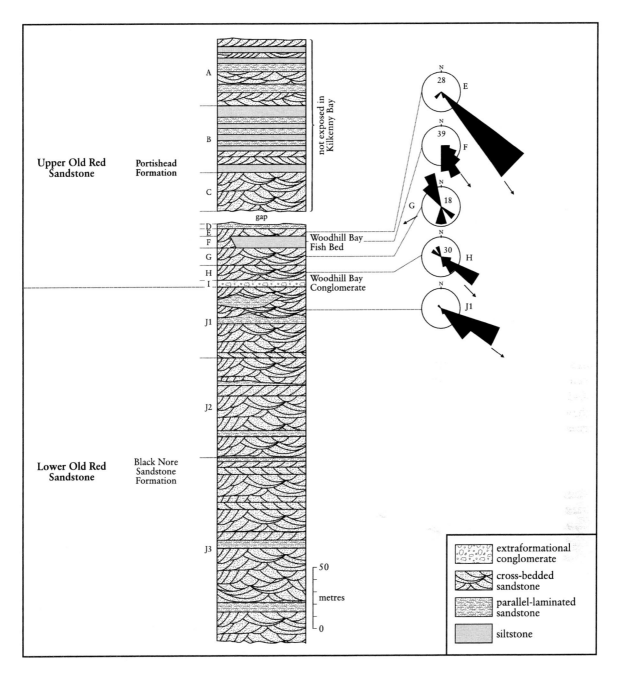

Figure 5.74 Generalized vertical section of the Old Red Sandstone at Portishead, with selected palaeocurrent rose diagrams. The rose diagrams show preferred palaeocurrent directions grouped in 20° classes and plotted as number frequency percent. The circles mark the 20% frequency level. Small arrows show the vector means of the cross-bedding dip azimuths. Based on Pick (1964a) and Williams and Hancock (1977). Letters A, B, etc. are 'formations' of Pick (1964a).

metre-scale beds of fine- to coarse-grained sandstone and pebbly sandstone with subordinate thin siltstones (up to 1.2 m thick) and minor (15–25 cm thick) conglomerates. Many of the sandstones are channelized, an excellent example of which occurs just above the base of the unit at the bottom of the cliff (ST 4585 7688).

Underlying these beds is the Woodhill Bay Conglomerate (Pick's 'formation' I), designated as the lowest unit of the Portishead Formation. Although thin (4–4.5 m), the bed is laterally extensive (Kellaway and Welch, 1955). It consists of well-rounded pebbles (averaging 2 cm) and cobbles (up to 15 cm), predominantly of vein quartz and dark red-brown quartzite, set in a coarse-grained sandstone matrix cemented by quartz and calcite. In addition to quartz and quartzite, Wallis (1927) recorded exotic clasts of chert, jasper, lithic sandstone, quartz schist, mica schist and spilite. The conglomerate is generally unsorted, although in places the largest clasts are concentrated near the base of the unit and there is locally a preferred orientation of the pebbles parallel to bedding. Rare lenses up to 30 cm thick of yellow, coarse-grained, cross-bedded sandstone also occur. Palaeocurrents throughout the Portishead Formation are consistently from the north and north-west, except for most of those in the upper part of 'formation' G, which are from the south-east.

The junction between the Woodhill Bay Conglomerate and the underlying Black Nore Sandstone Formation is seen best in the cliff (ST 4585 7686) and near some steps (ST 4576 7681). It is a sharp, irregular surface with scouring up to 1 m depth and a locally well-developed calcrete. A pipe-like carbonate body (ST 4579 7681), 2.5 m wide at its base and 2.5 m high, extends from the Black Nore Sandstone Formation up into the Woodhill Bay Conglomerate (Figure 5.75). This is an unusual calcrete, apparently controlled in part by a joint in the conglomerate, and unrelated to any contemporaneous pedogenesis at the top surface of the Black Nore Sandstone Formation.

The underlying Black Nore Sandstone Formation consists predominantly of stacked, commonly lenticular (decimetre- to metre-thick) beds of red and pale greenish grey, fine- to medium-grained, quartzitic sandstones. There are 292 m of beds at Portishead, the base of the formation not being seen. Some sandstones have a carbonate cement and some contain beds rich in muscovite. Almost all of the sandstones rest on erosion surfaces, their bases commonly marked by intraformational conglomerates containing angular mudstone clasts and rounded calcrete pebbles. Nodular calcretes are common, particularly in the uppermost part of the formation (Williams and Hancock, 1977). There are also a few extraformational conglomerates with quartz pebbles and a very few siltstone interbeds. The sandstones are either planar bedded or cross-bedded and indicate sediment transport from the north-west. Pick (1964a) subdivided the formation into three 'formations' (J1, J2 and

Figure 5.75 Conglomerates of the Woodhill Bay Conglomerate above sandstones of the Black Nore Sandstone Formation, Kilkenny Bay (ST 4579 7681). A pipe-like carbonate body extends across the junction and carbonate nodules occur in the Black Nore Sandstone Formation. (Photo: BGS No. A10737, reproduced with the permission of the Director, British Geological Survey, © NERC.)

J3), J2 being distinguished on the high concentration of calcrete clasts in its intraformational conglomerates. Thickness variations of the three units suggested to Pick that they wedge out laterally.

Interpretation

The section at Portishead is the stratotype for the Old Red Sandstone in the Bristol and Mendip Hills region (Kellaway and Welch, 1948, 1955, 1993; Green, 1992). The succession is characterized by a scarcity of fossils and a patchy outcrop distribution, making it difficult to correlate with adjacent areas. Lithologically, the Black Nore Sandstone Formation resembles the Lower Devonian Brownstones Formation of south Wales and the Welsh Borderland (Kellaway and Welch, 1955; Green, 1992). North of Bristol, the Upper Devonian strata are correlated with the Quartz Conglomerate and Tintern Sandstone formations of the Forest of Dean (Kellaway and Welch, 1993). However, these two formations cannot be separately identified at Portishead and are grouped together as the Portishead Formation, the numerous conglomerates and pebbly beds of the Portishead Formation being the distal correlatives of the Quartz Conglomerate.

The Black Nore Sandstone Formation and the Portishead Formation were deposited during two periods of sedimentation separated by a major (Mid-Devonian) orogenic phase. Most workers (e.g. Kellaway and Welch, 1955, 1993; Pick, 1964a; Allen, 1965b) placed the break at the base of the Woodhill Bay Conglomerate, although there is no angular discordance. Tunbridge (1986) suggested that the conglomerate may be of late Early or early Mid-Devonian age. Williams and Hancock (1977) questioned the significance of the Woodhill Bay Conglomerate–Black Nore Sandstone Formation junction, suggesting that the main unconformity may lie a few metres higher, at the top of the Woodhill Bay Conglomerate.

The Black Nore Sandstone Formation consists predominantly of alluvial channel-fill facies (Pick, 1964a; Williams and Hancock, 1977). Aeolian sandstones have been reported at Clevedon (Dodd, 1986), just to the south of the GCR site, the only occurrence of such facies at this level in the Anglo-Welsh Basin. The formation was one of the first in which the limestones of the Old Red Sandstone were interpreted as calcretes, and their significance in terms of understanding the palaeoclimatic and depositional setting of the Old Red Sandstone was recognized (Pick, 1964a).

The succeeding Portishead Formation is also mainly fluvial, but includes important interfluvial lacustrine or mudflat deposits (Pick, 1964a; Allen, 1965b). Dodd (1986) interpreted the upper part of 'formation' G as aeolian, probably deposited as relatively small barchan or barchanoid dunes. He tentatively linked the development of the dune field to deflation of the underlying sandflats in response to a eustatic regression.

The fish fauna of the Portishead Formation is unique in southern Britain and of considerable biostratigraphical and palaeoecological significance (Dineley, 1999g). In particular, the presence of *Groenlandaspis*, the only British record of this widely dispersed genus, suggests that freshwater migration routes were probably open between this region and other parts of Laurasia and Gondwana at that time. In addition, the occurrence of *Coccosteus* in the Woodhill Bay Fish Bed suggests that the Portishead Formation may pass laterally into the Pickwell Down Sandstone of north Devon and west Somerset (Kellaway and Welch, 1993).

The probable source area for the sediment of the Portishead Formation can be suggested from its petrology (Reynolds and Greenly, 1924; Wallis, 1927; Allen, 1965b) and palaeocurrent data (Wallis, 1927; Pick, 1964a; Allen, 1965b). Most workers (e.g. Reynolds and Greenly, 1924; Wallis, 1927; Allen, 1965b) concluded that the source was a Precambrian massif lying to the north-west, with a similar composition to the green schists of the Gwna Melange in the Mona Complex of Anglesey. However, based on tectonic criteria and on the regional distribution of similar conglomerates, Tunbridge (1986) and Cope and Bassett (1987) suggested that the Woodhill Bay Conglomerate may have been sourced from a periodically exposed landmass to the west in the area of the present Bristol Channel.

In the Bristol area, the continental Portishead Formation passes up through a transitional sequence into fully marine Carboniferous rocks. At Portishead, the transitional sequence (the Shirehampton Formation) is concealed beneath Recent tidal-flat sediments. It was logged nearby by Reynolds and Greenly (1923) at Portishead Pier station, and by Butler *et al.* (1972) in a temporary exposure at Woodhill (ST 4689 7733). At these localities it consists of intercalated red fluvial sandstones and mudstones, and grey, marine limestones and mudstones (Butler *et al.*, 1972; Kellaway and Welch, 1993) with a 'mixed' Devonian–Carboniferous fauna. Traditionally, the Devonian–Carboniferous boundary was placed at the base of the Shirehampton Formation in this region (see Kellaway and Welch, 1955). However, palynological evidence suggests that the boundary may lie in the upper part of the Portishead Formation (Neves and Dolby, 1967; Utting and Neves, 1970).

Conclusions

The cliffs to the south of Portishead between Woodhill Bay and Black Nore Point are the type locality for the Lower Devonian Black Nore Sandstone Formation and the overlying Upper Devonian to Lower Carboniferous Portishead Formation. The Portishead Formation includes important stratigraphical marker beds – the Woodhill Bay Conglomerate and the Woodhill Bay Fish Bed, the latter containing a diverse fauna that is unique in southern Britain and which provides a valuable insight into the palaeoecology of Devonian fish and their migration routes across the Old Red Sandstone continent. The site has been the subject of sustained research. It is regularly visited by undergraduate parties, and its eurypterid fauna is currently being studied. The nature of the Lower–Upper Old Red Sandstone junction, in particular the unusual calcrete development, requires detailed study to resolve conflicting interpretations. Aeolian sandstones reported in both the Black Nore Sandstone Formation and Portishead Formation also require further investigation, those in the former being unknown elsewhere in the Anglo-Welsh Basin at that stratigraphical level.

GLENTHORNE, DEVON (SS 795 499–SS 805 495)

P.R. Wilby

Introduction

The coastal outcrops at Glenthorne (Figure 5.76), which extend from Giant's Rib (SS 795 499) to The Caves (SS 805 495), provide the most representative section of the Hangman Sandstone Formation (*sensu* Edwards, 1999), the only substantial continental deposit of northerly derivation and proved Mid-Devonian age in southern Britain. The site is nationally important because it provides evidence of the southerly progradation of continental facies into the mainly marine area of north Devon and west Somerset when most of the Wales–Brabant Massif to the north was being eroded. It includes abundant examples of unconfined sheet-flood sandstones, which have not been widely recognized elsewhere in the Old Red Sandstone of Britain until recently

Description

The Hangman Sandstone Formation conformably overlies the shallow marine Lynton Formation (Simpson, 1964; Evans, 1983; Edmonds *et al.*, 1985) and is succeeded by the fully marine Ilfracombe Slates (Holwill, 1962; Edmonds *et al.*, 1985). Although largely unfossiliferous (see summary by Edwards, 1999), it is believed to be of mid-Eifelian age (Goldring *et al.*, 1978; Edmonds *et al.*, 1985), perhaps extending into early Givetian times (Evans, 1922; Knight, 1990; Edwards, 1999). A formal lithostratigraphical subdivision of the Hangman Sandstone proposed by Tunbridge (1978) broadly follows that of earlier workers (e.g. Lane, 1965), but the units are difficult to trace inland away from the well-exposed coast. The modified classification proposed by Edwards (1999) is therefore adopted in the following account. Estimates of the thickness of the Hangman Sandstone Formation vary considerably, principally because of structural complexities (Evans, 1922), ranging between 1097 m and 1658 m (Lane, 1965; Tunbridge, 1980b, 1981b, 1984, 1986). The Trentishoe Member makes up most of the formation (perhaps over 80%) and forms the entire section at Glenthorne. Its sedimentology has been studied in some detail, the most important contributions being by Tunbridge (1978, 1980b, 1981b, 1983a, 1984), Jones (1995) and Edwards (1999). The source of the sediments was discussed by Tunbridge (1983b, 1986) and Cope and Bassett (1987), and the nature of the upper and lower boundaries of the formation were described by Webby (1965a, 1966a) and Tunbridge (1983a). Descriptions of the sections at Glenthorne and nearby exposures were given by Tunbridge (1978), Jones (1995) and Edwards (1996, 1999). Details of the sequences preserved in the nearby inliers

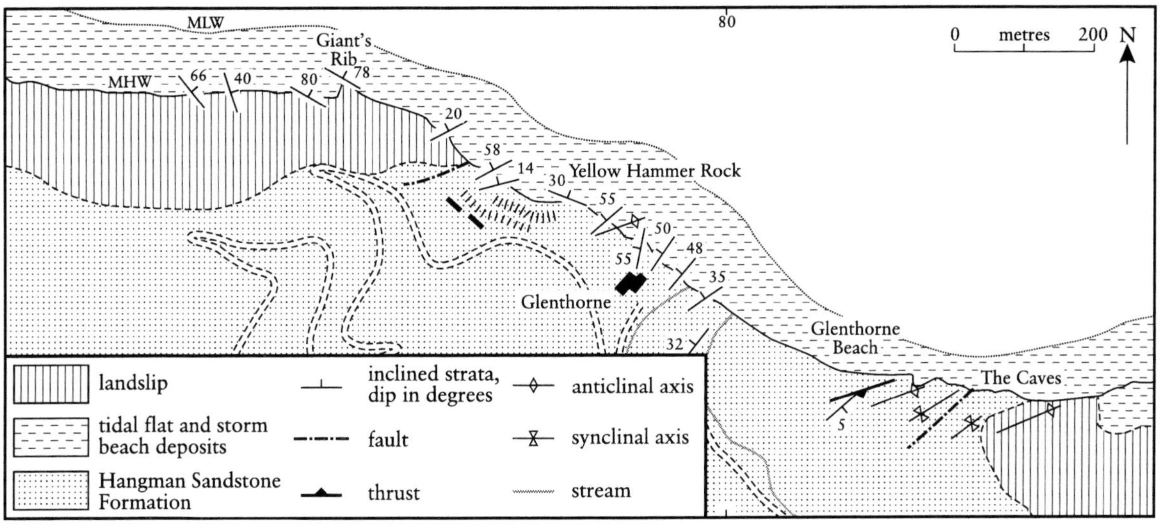

Glenthorne

Figure 5.76 Geological map of the Glenthorne area, north Devon. Based on British Geological Survey 1:10 560 Sheet SS 74NE (1983) and 1:10 000 manuscript map 84NW (1993).

forming the Brendon Hills and Quantock Hills were given by Webby (1965b, 1966b).

Glenthorne is situated in a complex structural zone on the northern limb of the Lynton Anticline (Shearman, 1967; Sanderson and Dearman, 1973; Edwards, 1999). The exposed sequence is intensely folded and cut by numerous subvertical faults striking north-west or ENE. Most of the folds are open, plunge to the north-east and ENE at 10° to 30° and have axial surfaces dipping about 40° to the south-east. Locally, an axial planar fracture cleavage is developed in the more argillaceous units, and some of the sandstones are affected by bedding-parallel shear. Because of the structural complexity, it is not possible to examine a continuous sequence through the Trentishoe Member at the site, or to assess major vertical facies variations. However, the locality exposes the most important constituent facies of the unit (Tunbridge, 1984; Jones, 1995; Edwards, 1999) (Figure 5.77). Estimates of the abundance of each of these facies in the Minehead district by Jones (1995) appear to be representative of the sequence exposed at Glenthorne and are quoted below.

About 75% of the succession at Glenthorne comprises a variable suite of grey, purple and red, fine- to medium-grained, metre-scale beds of well-sorted, hard, lithic, quartzose sandstones and subordinate siltstones (Tunbridge, 1986; Strong, 1995). Approximately 40% of the succession consists of distinctive sheet-like sandstones, in which two facies are distinguished by differences in sandstone thickness (Tunbridge, 1984; Jones, 1995; Edwards, 1999). The thicker sheet sandstones (facies 4 of Edwards, 1999) form laterally persistent, 0.1–1.0 m-thick beds that are traceable over tens of metres with no obvious thinning. They have sharp, planar bases that rest on erosion surfaces and locally contain intraformational mudstone and siltstone clasts. Internally, they are dominated by low-angle planar or trough cross-bedding that commonly passes up into cross-lamination. Current- and wave-ripple sets occur in the topmost parts of some of the beds, and convolute lamination, climbing ripples and primary current lineation occur locally. In places, these thick sheet sandstones are stacked to form composite units, particularly good examples of which are exposed at Giant's Rib and Yellow Hammer Rock (Figure 5.78), and in the cliffs between the eastern end of Glenthorne Beach and The Caves.

The thinner sheet sandstone facies (facies 5 of Edwards, 1999) consists of laminae and thin beds (up to 0.1 m thick) of very fine- to fine-grained sandstone with sharp bases not characterized by erosion surfaces. Most of these sheets persist laterally for only a few metres. Many of them exhibit unidirectional cross-lamination and a fining-upward trend, or more rarely, wave ripple cross-lamination. Typically, this facies is interbedded with mudstones containing abundant desiccation cracks. It is best exposed on the foreshore at the base of the cliff approximately

Facies	Main characters	Interpretation
1. Single-storey channel sandstone	erosion surface-based sandbodies, up to 3.5 metres thick	high-energy river channels
2. Multi-storey channel sandstone	erosively based, vertically stacked sandbodies, up to 12 metres thick	high-energy river channels
3. Laterally accreted channel sandstone	erosively based sandbodies, up to 1 metre thick	low-energy minor river channels
4. Thick sheet sandstone	sharp or erosively based beds of sandstone in laterally persistent sheets, up to 1 metre thick	high-energy sheet-floods deposited on subaerial mudflats
5. Thin sheet sandstone	laminae to thin beds of sandstone up to 0.1 metres thick, typically interbedded with facies 7	distal or weak sheet-floods deposited into lakes; later modification by emergence and desiccation
6. Massive to laminated mudstone	mudstones in beds up to 2 metres thick; local carbonate nodules	deposition from suspension in perennial lakes
7. Desiccated and remobilized mudstone	mudstones, with common desiccation cracks; typically interbedded with facies 5; abundant bioturbation and local carbonate nodules	deposition from suspension in ephemeral lakes; emergence caused drying of sediment surface; folded and convoluted laminae caused by water escape
8. Mudstone with extraformational pebbles	mudstones with scattered quartz pebbles	cohesive, subaerial debris flows

Figure 5.77 Summary of the sedimentary facies of the Hangman Sandstone Formation and their interpretation. After Edwards (1999).

midway between Giant's Rib and Yellow Hammer Rock.

Interbedded with the sheet sandstone facies, or separated from them by thin mudstone- and siltstone-dominated units, are single-storey, channelized sandstones, each up to 3.5 m thick (facies 1 of Edwards, 1999). Most of these sandstones rest on erosion surfaces and their bases are commonly marked by laterally impersistent intraformational conglomerates. Internally, they are dominated by low-angle, unidirectional, planar or trough cross-bedding, and their upper surfaces are generally sharp and locally preserve asymmetrical ripple forms. Convolute bedding is also common and thin mudstone beds are present locally. This facies comprises approximately 20% of the sequence and is best exposed on the eastern side of Yellow Hammer Rock just above a modern storm beach ridge (Figures 5.78, 5.79). At several points, the channelized sandstones are stacked to form composite, multi-storey units (facies 2 of Edwards, 1999). These comprise approximately 15% of the succession and some are over 5 m thick. Some (e.g. SS 7963 4980) display convolute bedding and de-watering pipes.

In addition to the sandstone facies described above, Tunbridge (1984, fig. 17b) logged a 1.75 m-thick laterally accreted channel sandstone at the site (facies 3 of Edwards, 1999). Such sandstones are volumetrically unimportant at Glenthorne and only form about 1% of the sequence in the Minehead district. All of the sandstone facies show a consistent southerly palaeocurrent flow.

Interbedded with the sandstone facies, in units generally 0.1–2.5 m thick, are two finer-grained, mudstone-dominated facies. The first (facies 6 of Edwards, 1999) comprises approximately 9% of the succession and consists of red, purple and grey-green mudstones, silty mudstones and siltstones, some in interbeds mainly less than 5 cm thick. Some siltstones are internally cross-laminated. The mudstones are massive or laminated, and some contain simple, vertical burrows. The second facies (facies 7 of Edwards, 1999) comprises about 15% of the sequence and is characterized by an abundance of desiccation and remobilization structures. These include folds, diapirs, pipes, upcurled mudstone laminae and sand-filled desiccation cracks up to 6 cm deep, many of which penetrate multiple laminae.

Glenthorne

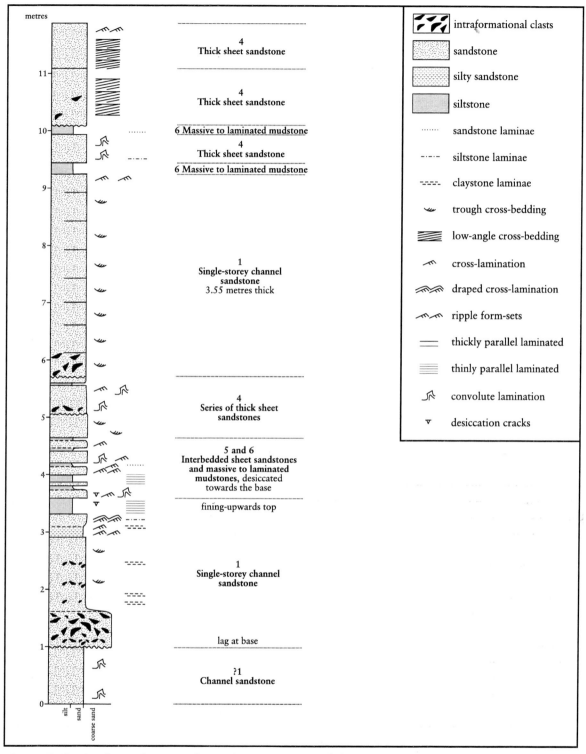

Figure 5.78 Graphic sedimentary log and interpretation of the Hangman Sandstone Formation at Yellow Hammer Rock (SS 7980 4975). After Edwards (1999).

Locally, these structures are so common that the sediment appears comprehensively brecciated or even homogenized. Sand- and mud-filled burrows, probably attributable to *Beaconites* and *Planolites*, are also common at certain levels.

Two other fine-grained facies (mudstones with pedogenic calcretes and mudstones with

The Anglo-Welsh Basin

Figure 5.79 Sedimentary features of the Hangman Sandstone Formation at Yellow Hammer Rock (SS 7980 4970). A single-storey channel sandstone with an undulatory erosive base (Facies 1) overlies sharp-based sheet sandstones (Facies 4). The rucksack (circled) is 0.5 m high. (Photo: BGS No. GS480, reproduced with the permission of the Director, British Geological Survey, © NERC.

extraformational pebbles; Jones, 1995; Edwards, 1999) together comprise less than 2% of the succession in the Minehead district and are not important in the Glenthorne section.

Interpretation

The Mid-Devonian Epoch was marked across much of southern Britain by an episode of uplift and erosion corresponding to the final (Acadian) stage of the Caledonian Orogeny. Only in the predominantly marine basin of south-west England did sedimentation continue uninterrupted from Early Devonian up into Mid- and Late Devonian times (Dineley, 1999g). North Devon and west Somerset occupied a position between the mainly continental upland area of the Wales–Brabant Massif to the north and the wave-dominated shoreline to the south (Webby, 1965a, 1966a; Edmonds *et al.*, 1975). Shifts in the position of the shoreline in response to fluctuations in sea level (House, 1975) are recorded by the alternation of shallow marine and terrestrial facies. The Hangman Sandstone Formation is dominated by the terrestrial Trentishoe and Rawn's members, but includes important shallow marine units at its base (the Hollowbrook Member of Tunbridge, 1983a) and top (the Sherrycombe and Little Hangman members; Goldring *et al.*, 1978; Edmonds *et al.*, 1985).

The Trentishoe Member prograded over a low-lying coastal plain on the northern margin of the Rheic Ocean (Evans, 1922) from a source to the north during a period of eustatic sea-level rise (Tunbridge, 1983a, 1986), the sediment supply keeping pace with rising sea level. The source of the sediment has been the subject of debate. Petrographic and heavy-mineral analyses show a close, but not identical match with the Lower Devonian Brownstones Formation of south Wales and the Welsh Borderlands. This unit, having then been recently deposited, is likely to have been poorly consolidated and particularly susceptible to rapid erosion during Acadian uplift (Tunbridge, 1984). However, garnet, which is the dominant heavy mineral in the Brownstones Formation, is rare and generally highly etched in the Trentishoe Member. This led investigators to discount south Wales as the source of the Trentishoe Member in favour of a landmass, either in the Bristol Channel area (Dewey, 1982) or in some other, as yet, unidentified lateral source (Hallam, 1934). The rarity of garnet in the Trentishoe Member has been interpreted more recently as a diagenetic phenomenon and the Brownstones Formation has been re-instated as the most likely source (Tunbridge, 1981b, 1984, 1986; Edwards, 1999), which is entirely consistent with the palaeocurrent evidence. In marked contrast, the overlying Rawn's Member consists of much coarser and more immature facies, including abundant conglomerates, that contain exotic, angular clasts which do not match any possible source rocks in south Wales. It records a dramatic switch in sediment source and is widely believed to have been derived from the intermittently exposed Bristol Channel landmass (Tunbridge, 1984; Cope and Bassett, 1987).

Tunbridge (1984), following Evans (1922), but contrary to Holwill *et al.* (1969), interpreted the Trentishoe Member as an extensive sandy ephemeral stream and clay playa complex. Three facies associations represent a down-slope transition from a system of low sinuosity channels, through an ephemeral flood complex, into a playa lake. The proximal sequence is represented by the multi-storey channelized sandstone facies which were deposited by a network of small, sandy, low-sinuosity, ephemeral streams, each channel being cut and filled during a single flood event, resulting in avulsion and the incision of new courses. The medial facies association is represented by cyclical, upward-coarsening units several metres thick containing the sheet sandstone facies. These thicken and coarsen upwards, some stacking into multiple bodies and some being incised and cut out by erosion surfaces underlying channelized sandstones. Tunbridge (1981b, 1984) proposed that the dimensions of these sheet sandstones and the abundance of parallel lamination are consistent with deposition from unconfined flood waters fed by ephemeral streams during peak flood periods, with the cyclicity being a response to progradation of lobes during consecutive floods. The distal facies association comprises mainly the desiccated mudstone and sandstone facies, but also contains rare thin sheet sandstones. It represents the final stage in the down-slope decline in flow velocity and the main site of muddy flood deposition.

Jones (1995) and Edwards (1999) supported Tunbridge's (1981b, 1984) tripartite division of the facies, but suggested that it does not necessarily reflect down-slope changes. Instead, they proposed that it may be due to variations in channel flux, fan retrogradation, and/or lateral shifts in fan position. Consequently, they interpreted the entire sequence as having been deposited on the distal part of an alluvial-fan complex, with periodic development of ephemeral mudflats. This, they suggested, is more consistent with the relative abundance of both sheet and channel sandstones, which imply continual proximity to a sediment source. In this model, periods of high and frequent discharge resulted in the formation of sandflats, whereas mudflat–playa-lake environments formed during periods of low discharge. Ephemeral lakes may also have formed after sheet flooding.

Tunbridge (1984) recognized some problems with his model and conceded that the presence of a sandy coastline to the south (Tunbridge, 1983a) showed that not all of the sands could have been deposited inland of the clay playas. Similarly, he recognized that his most distal facies could represent periods of greater aridity, or may have been deposited within local depressions (cf. Jones, 1995; Edwards, 1999).

Conclusions

The cliffs at Glenthorne Beach provide excellent exposures of the only substantial Middle Devonian Old Red Sandstone facies development in southern Britain – the Trentishoe Member of the Hangman Sandstone Formation. Although the section is intensely folded and faulted, it affords an excellent opportunity to examine representative examples of the most important constituent lithologies of the member, one of which (sheet-flood sandstone) is poorly represented elsewhere in the Old Red Sandstone. The site is of national importance because of the evidence that it provides of the southerly advance of continental Old Red Sandstone facies into the mainly marine area of north Devon and west Somerset at a time when the rest of the Anglo-Welsh Basin to the north was undergoing erosion. This has important implications for the palaeogeography of the basin, and for the relationship between sea-level changes and the rates of source erosion and sediment transport during Mid-Devonian times.

References

In this reference list the arrangement is alphabetical by author surname for works by sole authors and dual authors. Where there are references that include the first-named author with others, the sole-author works are listed chronologically first, followed by the dual author references (alphabetically) followed by the references with three or more authors listed *chronologically*. Chronological order is used within each group of identical authors.

Aldridge, R.J., Siveter, David J., Siveter, Derek J., Lane, P.D., Palmer, D. and Woodcock, N.H. (2000) *British Silurian Stratigraphy*, Geological Conservation Review Series, No. 19, Joint Nature Conservation Committee, Peterborough, 542 pp.

Allen, J.R.L. (1961) The highest Lower Old Red Sandstone of Brown Clee Hill, Shropshire. *Proceedings of the Geologists' Association*, **72**, 205–19.

Allen, J.R.L. (1962) Intraformational conglomerates and scoured surfaces in the Lower Old Red Sandstone of the Anglo-Welsh cuvette. *Geological Journal*, **3**, 1–20.

Allen, J.R.L. (1963a) The classification of cross-stratified units, with notes on their origin. *Sedimentology*, **2**, 93–114.

Allen, J.R.L. (1963b) Depositional features of Dittonian rocks: Pembrokeshire compared with the Welsh Borderland. *Geological Magazine*, **100**, 385–400.

Allen, J.R.L. (1964a) Studies in fluviatile sedimentation: six cyclothems from the Lower Old Red Sandstone, Anglo-Welsh Basin. *Sedimentology*, **3**, 163–98.

Allen, J.R.L. (1964b) Primary current lineation in the Lower Old Red Sandstone (Devonian), Anglo-Welsh Basin. *Sedimentology*, **3**, 89–108.

Allen, J.R.L. (1965a) The sedimentation and palaeogeography of the Old Red Sandstone of Anglesey, North Wales. *Proceedings of the Yorkshire Geological Society*, **35**, 139–85.

Allen, J.R.L. (1965b) Upper Old Red Sandstone (Farlovian) paleogeography in South Wales and the Welsh Borderland. *Journal of Sedimentary Petrology*, **35**, 167–95.

Allen, J.R.L. (1970) Studies in fluviatile sedimentation: a comparison of fining-upwards cyclothems, with special reference to coarse-member composition and interpretation. *Journal of Sedimentary Petrology*, **40**, 298–323.

Allen, J.R.L. (1971) The sedimentation of the Old Red Sandstone in the Forest of Dean. In *Geological Excursions in South Wales and the Forest of Dean* (eds D.A. Bassett and M.G. Bassett), Geologists' Association South Wales Group, Cardiff, pp. 9–19.

Allen, J.R.L. (1973a) A new find of bivalve molluscs in the uppermost Downtonian (Lower Old Red Sandstone) of Lydney, Gloucestershire. *Proceedings of the Geologists' Association*, **84**, 27–9.

Allen, J.R.L. (1973b) Compressional structures (patterned ground) in Devonian pedogenic limestones. *Nature Physical Science*, **243**, 84–6.

References

Allen, J.R.L. (1974a) The Devonian rocks of Wales and the Welsh Borderland. In *The Upper Palaeozoic and post-Palaeozoic Rocks of Wales* (ed. T.R. Owen), University of Wales Press, Cardiff, pp. 47–84.

Allen, J.R.L. (1974b) Source rocks of the Lower Old Red Sandstone: exotic pebbles from the Brownstones, Ross-on-Wye, Hereford and Worcester. *Proceedings of the Geologists' Association*, **85**, 493–510.

Allen, J.R.L. (1974c) Sedimentology of the Old Red Sandstone (Siluro–Devonian) in the Clee Hills area, Shropshire, England. *Sedimentary Geology*, **12**, 73–167.

Allen, J.R.L. (1974d) Studies in fluviatile sedimentation: implications of pedogenic carbonate units, Lower Old Red Sandstone, Anglo-Welsh outcrop. *Geological Journal*, **9**, 181–208.

Allen, J.R.L. (1974e) Studies in fluviatile sedimentation: lateral variation in some fining-upwards cyclothems from the Red Marls, Pembrokeshire. *Geological Magazine*, **9**, 1–16.

Allen, J.R.L. (1975) Source rocks of the Lower Old Red Sandstone: Llanishen Conglomerate of the Cardiff area, South Wales. *Proceedings of the Geologists' Association*, **86**, 63–76.

Allen, J.R.L. (1977) Wales and the Welsh Borders. In *A Correlation of Devonian Rocks in the British Isles* (eds M.R. House, J.B. Richardson, W.G. Chaloner, J.R.L. Allen, C.H. Holland and T.S. Westoll), Geological Society of London Special Report, No. **8**, Scottish Academic Press, Edinburgh, pp. 40–54.

Allen, J.R.L. (1978a) Locality B2. Ross-on-Wye, Hereford and Worcester. In *A Field Guide to Selected Outcrop Areas of the Devonian of Scotland, the Welsh Borderland and South Wales: International Symposium on the Devonian System (P.A.D.S. 78)* (eds P.F. Friend and B.P.J. Williams), Palaeontological Association, London, p. 63.

Allen, J.R.L. (1978b) Locality B1. Lydney, Gloucestershire. In *A Field Guide to Selected Outcrop Areas of the Devonian of Scotland, the Welsh Borderland and South Wales: International Symposium on the Devonian System (P.A.D.S. 78)* (eds P.F. Friend and B.P.J. Williams), Palaeontological Association, London, pp. 61–3.

Allen, J.R.L (1978c) Locality B11. Llanstephan, Dyfed. In *A Field Guide to Selected Outcrop Areas of the Devonian of Scotland, the Welsh Borderland and South Wales: International Symposium on the Devonian System (P.A.D.S. 78)* (eds P.F. Friend and B.P.J. Williams), Palaeontological Association, London, pp. 79–82.

Allen, J.R.L. (1979) Old Red Sandstone facies in external basins, with particular reference to southern Britain. In *The Devonian System: A Palaeontological Association International Symposium* (eds M.R. House, C.R. Scrutton and M.G. Bassett), *Special Papers in Paleontology*, No. **23**, Palaeontological Association, London, pp. 65–80.

Allen, J.R.L. (1980) Institute of Geological Sciences field seminar on the Lower Old Red Sandstone in South Wales and the Welsh Borders, unpublished.

Allen, J.R.L. (1983a) Studies in fluviatile sedimentation: bars, bar-complexes and sandstone sheets (low sinuosity braided streams) in the Brownstones (L. Devonian), Welsh borders. *Sedimentary Geology*, **33**, 237–93.

Allen, J.R.L. (1983b) Gravel overpassing on humpback bars supplied with mixed sediment: examples from the Lower Old Red Sandstone, southern Britain. *Sedimentology*, **30**, 285–94.

Allen, J.R.L. (1985) Marine to freshwater: the sedimentology of the interrupted environmental transition (Ludlow–Siegenian) in the Anglo-Welsh region. *Philosophical Transactions of the Royal Society of London*, **B309**, 85–104.

Allen, J.R.L. (1986) Pedogenic calcretes in the Old Red Sandstone facies (late Silurian–early Carboniferous) of the Anglo Welsh area, southern Britain. In *Palaeosols: Their Recognition and Interpretation* (ed. V.P. Wright), Blackwell Scientific Publications, Oxford, pp. 58–86.

Allen, J.R.L. and Crowley, S.F. (1983) Lower Old Red Sandstone fluvial dispersal systems in the British Isles. *Transactions of the Royal Society of Edinburgh: Earth Sciences*, **74**, 61–8.

Allen, J.R.L. and Tarlo, L.B. (1963) The Downtonian and Dittonian facies of the Welsh Borderland. *Geological Magazine*, **100**, 129–55.

Allen, J.R.L. and Williams, B.P.J. (1978) The sequence of the earlier Lower Old Red Sandstone (Siluro–Devonian), north of Milford Haven, southwest Dyfed Wales. *Geological Journal*, **13**, 113–36.

References

Allen, J.R.L. and Williams, B.P.J. (1979a) Old Red Sandstone facies and Wenlock stratigraphy and palaeogeography in Wales and the Welsh Borderland. *Proceedings of the Geologists' Association*, 90, 229–30.

Allen, J.R.L. and Williams, B.P.J. (1979b) Interfluvial drainage on Siluro–Devonian alluvial plains in Wales and the Welsh Borders. *Journal of the Geological Society of London*, 136, 361–6.

Allen, J.R.L. and Williams, B.P.J. (1981a) Sedimentology and stratigraphy of the Townsend Tuff Bed (Lower Old Red Sandstone) in South Wales and the Welsh Borders. *Journal of the Geological Society of London*, 138, 15–29.

Allen, J.R.L. and Williams, B.P.J. (1981b) *Beaconites antarcticus*: a giant channel-associated trace fossil from the Lower Old Red Sandstone of South Wales and the Welsh Borders. *Geological Journal*, 16, 255–69.

Allen, J.R.L. and Williams, B.P.J. (1982) The architecture of an alluvial suite: rocks between the Townsend Tuff and Pickard Bay Tuff Beds (early Devonian) southwest Wales. *Philosophical Transactions of the Royal Society of London*, B297, 51–89.

Allen, J.R.L., Dineley, D.L. and Friend, P.F. (1967) Old Red Sandstone basins of North America and Northwest Europe. In *Proceedings of the International Symposium on the Devonian System* (ed. D.H. Oswald), Alberta Association of Petroleum Geologists, Calgary, 1, pp. 69–98.

Allen, J.R.L., Halstead, L.B. and Turner, S. (1968) Dittonian ostracoderm fauna from the Brownstones of Wilderness Quarry, Mitcheldean, Gloucestershire. *Proceedings of the Geological Society of London*, No. 1649, 141–53.

Allen, J.R.L., Bassett, M.G., Hancock, P.L., Walmsley, V.G. and Williams, B.P.J. (1976) Stratigraphy and structure of the Winsle Inlier, southwest Dyfed, Wales. *Proceedings of the Geologists' Association*, 87, 221–9.

Allen, J.R.L., Thomas, R.G. and Williams, B.P.J. (1981a) Field Meeting: the facies of the Lower Old Red Sandstone, north of Milford Haven, southwest Dyfed, Wales. *Proceedings of the Geologists' Association*, 92, 251–67.

Allen, J.R.L., Elliott, T. and Williams, B.P.J. (1981b) Old Red Sandstone and Carboniferous fluvial sediments in South Wales. In *Field Guides to Modern and Ancient Fluvial Systems in Britain and Spain. Proceedings of the Second International Conference on Fluvial Sediments: Modern and Ancient Fluvial Systems: Sedimentology and Processes, University of Keele, 21–25 September 1981* (ed. T. Elliott), University of Keele, Keele, pp. 1.1–1.39.

Allen, J.R.L., Thomas, R.G. and Williams, B.P.J. (1982) The Old Red Sandstone north of Milford Haven. In *Geological Excursions in Dyfed, South-west Wales* (ed. M.G. Bassett), National Museum of Wales, Cardiff, for the Geologists' Association South Wales Group, pp. 123–49.

Allen, P.A. (1981a) Sediments and processes on a small stream-flow dominated, Devonian alluvial fan, Shetland Islands. *Sedimentary Geology*, 29, 31–66.

Allen, P.A. (1981b) Devonian lake margin environments and processes, SE Shetland, Scotland. *Journal of the Geological Society of London*, 138, 1–14.

Allen, P.A. and Marshall, J.E.A. (1981) Depositional environments and palynology of the Devonian south-east Shetland Basin. *Scottish Journal of Geology*, 17, 257–73.

Almond, J. (1983) The sedimentology of the earliest Lower Old Red Sandstone, south central Wales. Unpublished PhD thesis, University of Bristol.

Almond, J., Williams, B.P.J. and Woodcock, N.H. (1993) The Old Red Sandstone of the Brecon Beacons to Black Mountain area. In *Geological Excursions in Powys, Central Wales* (eds N.H. Woodcock and M.G. Bassett), National Museum of Wales and University of Wales Press, Cardiff, for the Geologists' Association South Wales Group, pp. 311–30.

Anderson, J.G.C. (1944) The Dalradian rocks of Arran. *Transactions of the Geological Society of Glasgow*, 20, 274.

Anderson, J.G.C. (1954) "Hutton's Unconformity". *Geological Magazine*, 91, 85–6

Anderson, L.I. and Trewin, N.H. (2003) An Early Devonian arthropod fauna from the Windyfield cherts, Aberdeenshire, Scotland. *Palaeontology*, 46, 467–509.

Anderson, L.I., Crighton, W.R.B. and Hass, H. (2004) A new crustacean with 'cladoceran' affinities from the early Devonian (Pragian) Rhynie chert hot-spring complex. *Transactions of the Royal Society of Edinburgh: Earth Sciences*, 94, 355–69.

References

Anderton, R., Bridges, P.H., Leeder, M.R. and Selwood, B.W. (1979) *A Dynamic Stratigraphy of the British Isles*, Allen and Unwin, London, 301 pp.

Andrews, J.E., Turner, M.S., Nabi, G. and Spiro, B. (1991) The anatomy of an early Dinantian floodplain: palaeo-environment and early diagenesis. *Sedimentology*, 38, 271–87.

Archer, R. (1978) The Old Red Sandstone outliers of Gamrie and Rhynie, Aberdeenshire. Unpublished PhD thesis, University of Newcastle-upon-Tyne.

Armstrong, H.A. and Owen, A.W. (2000) Age and provenance of limestone clasts in Lower Old Red Sandstone conglomerates: implications for the geological history of the Midland Valley Terrane. In *New Perspectives on the Old Red Sandstone* (eds P.F. Friend and B.J.P. Williams), *Geological Society Special Publication*, No. 180, Geological Society, London, pp. 459–71.

Armstrong, M. (1977) The Old Red Sandstone of Easter Ross and the Black Isle. In *The Moray Firth Area – Geological Studies* (ed. G. Gill), Inverness Field Club, Inverness, pp. 25–34.

Armstrong, M. and Paterson, I.B. (1970) The Lower Old Red Sandstone of the Strathmore Region. *Report of the Institute of Geological Sciences*, 70/12, 24 pp.

Armstrong, M., Donovan, R.N. and Mykura, W. (1978a) Western Moray Firth and Caithness. In *A Field Guide to Selected Outcrop Areas of the Devonian of Scotland, the Welsh Borderland and South Wales: International Symposium on the Devonian System (P.A.D.S. 78)* (eds P.F. Friend and B.P.J. Williams), Palaeontological Association, London, pp. 32–7.

Armstrong, M., Donovan, R.N. and Paterson, I.B. (1978b) Angus and Kincardine. In *A Field Guide to Selected Outcrop Areas of the Devonian of Scotland, the Welsh Borderland and South Wales: International Symposium on the Devonian System (P.A.D.S. 78)* (eds P.F. Friend and B.P.J. Williams), Palaeontological Association, London, pp. 13–22.

Armstrong, M., Paterson, I.B. and Browne, M.A.E. (1985) *Geology of the Perth and Dundee District*, Memoir of the British Geological Survey, sheets 48W, 48E and 49 (Scotland), HMSO, London, 108 pp.

Arthurton, R.S. and Wadge, A.J. (1981) *Geology of the Country Around Penrith*, Memoir of the Geological Survey of Great Britain, Sheet 24 (England and Wales), HMSO, London, 177 pp.

Astin, T.R. (1982) The Devonian geology of the Walls Peninsula, Shetland. Unpublished PhD thesis, University of Cambridge.

Astin, T.R. (1985) The palaeogeography of the Middle Devonian Lower Eday Sandstone, Orkney. *Scottish Journal of Geology*, 21, 353–75.

Astin, T.R. (1990) The Devonian lacustrine sediments of Orkney, Scotland; implications for climatic cyclicity, basin structure and maturation history. *Journal of the Geological Society of London*, 147, 141–51.

Astin, T.R. and Rogers, D.A. (1991) "Subaqueous shrinkage cracks" in the Devonian of Scotland reinterpreted. *Journal of Sedimentary Petrology*, 61, 850–9.

Astin, T.R. and Rogers, D.A. (1992) "Subaqueous shrinkage cracks" in the Devonian of Scotland reinterpreted. Discussion. *Journal of Sedimentary Petrology*, 62, 923–4.

Astin, T.R. and Rogers, D.A. (1993) "Subaqueous shrinkage cracks" in the Devonian of Scotland reinterpreted. Discussion. *Journal of Sedimentary Petrology*, 63, 566–7.

Balin, D.F. (1993) Upper Old Red Sandstone sedimentation in the eastern Midland area, Scotland. Unpublished PhD thesis, University of Cambridge.

Balin, D.F. (2000) Calcrete morphology and karst development in the Upper Old Red Sandstone at Milton Ness, Scotland. In *New perspectives on the Old Red Sandstone* (eds P.F. Friend and B.P.J. Williams), *Geological Society Special Publication*, No. 180, Geological Society, London, pp. 485–501.

Ball, W.H. and Dineley, D.L. (1961) The Old Red Sandstone of Brown Clee Hill and the adjacent area. I. Stratigraphy. *Bulletin of the British Museum (Natural History)*, A5, 177–242.

Banks, H. (1980) Floral assemblages in the Siluro–Devonian. In *Biostratigraphy of Fossil Plants: Successional and Paleoecological Analyses* (eds D.L. Dilcher and T.N. Taylor), Dowden, Hutchinson and Ross, Stroudsberg, pp. 1–24.

Barclay, W.J. (1975) *Annual Report for 1974*. Institute of Geological Sciences, London, p. 21.

Barclay, W.J. (1989) *Geology of the South Wales Coalfield: Part II: The Country Around Abergavenny*, 3rd edn, Memoir of the British Geological Survey, Sheet 232 (England and Wales), HMSO, London, 147 pp.

References

Barclay, W.J. and Wilby, P.R. (2003) *Geology of the Talgarth District – A Brief Explanation of the Geological Map*, Sheet Explanation of the British Geological Survey, Sheet 214 (England and Wales), British Geological Survey, Keyworth, 22 pp.

Barclay, W.J., Taylor, K. and Thomas, L.P. (1988) *Geology of the South Wales Coalfield: Part V: The Country Around Merthyr Tydfil*, Memoir of the British Geological Survey, Sheet 231 (England and Wales), 3rd edn, HMSO, London, 52 pp.

Barclay, W.J., Glover, B.W. and Mendum, J.R. (1993) "Sub-aqueous shrinkage cracks" in the Devonian of Scotland reinterpreted. Discussion. *Journal of Sedimentary Petrology*, **63**, 564–5.

Barclay, W.J., Rathbone, P.A., White, D.E. and Richardson, J.B. (1994) Brackish water faunas from the St. Maughans Formation: the Old Red Sandstone section at Ammons Hill, Hereford and Worcester, UK, re-examined. *Geological Journal*, **29**, 369–79.

Barclay, W.J., Davies, J.R., Humpage, A., Waters, R.A., Wilby, P.R., Williams, M. and Wilson, D. (in press) *Geology of the Country Around Brecon*, Sheet Explanation of the British Geological Survey, Sheet 213 (England and Wales), British Geological Survey, Keyworth.

Bassett, M.G. (1974) Review of the stratigraphy of the Wenlock Series in the Welsh Borderland and South Wales. *Palaeontology*, **17**, 745–77.

Bassett, M.G. (1982a) Ordovician and Silurian sections in the Llangadog–Llandilo area. In *Geological Excursions in Dyfed, South-west Wales* (ed. M.G. Bassett), National Museum of Wales, Cardiff, for the Geologists' Association South Wales Group, pp. 271–87.

Bassett, M.G. (1982b) Silurian rocks of the Marloes and Pembroke peninsulas. In *Geological Excursions in Dyfed, South-west Wales* (ed. M.G. Bassett), National Museum of Wales, Cardiff, for the Geologists' Association South Wales Group, pp. 103–22.

Bassett, M.G. and Jenkins, T.B.H. (1977) Tournaisian conodont and spore data from the uppermost Skrinkle Sandstones of Pembrokeshire, South Wales. *Geologica et Palaeontologica*, **11**, 121–34.

Bassett, M.G., Lawson, J.D. and White, D.E. (1982) The Downton Series as the fourth Series of the Silurian System. *Lethaia*, **15**, 1–24.

Bassett, M.G., Bluck, B.J., Cave, R., Holland, C.H. and Lawson, J.D. (1992) Silurian. In *Atlas of Palaeogeography and Lithofacies* (eds J.C.W. Cope, J.K. Ingram and P.F. Rawson), *Geological Society Memoir*, No. **13**, Geological Society, London, pp. 37–56.

Benton, M.J., Cook, E. and Turner, P. (2002) *Permian and Triassic Red Beds and the Penarth Group of Great Britain*, Geological Conservation Review Series, No. 24, Joint Nature Conservation Committee, Peterborough, 337 pp.

Blieck, A. (1981) Le genre *Protopteraspis* Leriche (Vertébrés, Hétérostracés) du Dévonien inférieur nord Atlantique. *Palaeontographica, Abteilung A*, **173**, 141–59.

Blieck, A. (1984) *Les Hétérostracés Pteraspidiformes, Agnathans du Silurien Dévonien du Continent Nord-Atlantique et des Blocs Avoisines: Revision Systematique, Phylogenie, Biostratigraphie, Biogeographie, Cahiers de Paléontologie: Section vertébrés*, Éditions du CNRS, Paris, 200 pp.

Blieck, A. (1985) Paléoenvironments des Hétérostracés Vertébrés agnathes Ordoviciens à Dévoniens. *Bulletin du Muséum National d'Histoire Naturelle, Paris: Science de la Terre*, **7**, 143–55.

Blieck, A. and Janvier, P. (1989) Vertébré Agnathes du Dévonien inférieur de l'Artois (Pas-de-Calais, France): implications biostratigraphiques. *Ann Paléont. (Vert–Invert)*, **75**(3), 125–67.

Blieck, A. and Cloutier, R. (2000) Biostratigaphical correlations of Early Devonian vertebrate assemblages of the Old Red Sandstone Continent. *Courier Forschungsinstitut Senckenberg*, **223**, 223–69.

Bluck, B.J. (1967). Deposition of some Upper Old Red Sandstone conglomerates in the Clyde area: A study in the significance of bedding. *Scottish Journal of Geology*, **3**, 139–67.

Bluck, B.J. (1978) Sedimentation in a late orogenic basin: the Old Red Sandstone of the Midland Valley of Scotland. In *Crustal Evolution in Northwestern Britain and Adjacent Regions: Proceedings of an International Conference held in Glasgow University, 5–8 April 1977* (eds D.R. Bowes and B.E. Leake), *Geological Journal Special Issue*, No. **10**, Seel House Press, Liverpool, pp. 249–78.

References

Bluck, B.J. (1980a) Evolution of a strike-slip fault-controlled basin, Upper Old Red Sandstone, Scotland. In *Sedimentation in Oblique-slip Mobile Zones* (eds P.F. Ballance and H.G. Reading), *International Association of Sedimentologists Special Publication*, No. 4, Blackwell Scientific Publications, Oxford, pp. 63–78.

Bluck, B.J. (1980b) Structure, generation and preservation of upward fining, braided stream cycles in the Old Red Sandstone of Scotland. *Transactions of the Royal Society of Edinburgh: Earth Sciences*, 71, 29–46.

Bluck, B.J. (1984) Pre-Carboniferous history of the Midland Valley of Scotland. *Transactions of the Royal Society of Edinburgh: Earth Sciences*, 75, 275–95.

Bluck, B.J. (1986) Upward coarsening sedimentation units and facies lineages, Old Red Sandstone, Scotland. *Transactions of the Royal Society of Edinburgh: Earth Sciences*, 77, 251–64.

Bluck, B.J. (1992) Upper Old Red Sandstone of the Firth of Clyde. In *Geological Excursions around Glasgow and Girvan* (eds J.D. Lawson and D.S. Weedon), Geological Society of Glasgow, Glasgow, pp. 220–9.

Bluck, B.J. (2000) Old Red Sandstone basins and alluvial systems of Midland Scotland. In *New Perspectives on the Old Red Sandstone* (eds P.F. Friend and B.P.J. Williams), *Geological Society Special Publication*, No. 180, Geological Society, London, pp. 417–37.

Bluck, B.J. (2001) Caledonian and related events in Scotland. *Transactions of the Royal Society of Edinburgh: Earth Sciences*, 91, 375–404.

Bluck, B.J. (2002) The Midland Valley terrane. In *The Geology of Scotland*, 4th edn (ed. N.H. Trewin), The Geological Society, London, pp. 149–66.

Bluck, B.J., Cope, J.C.W. and Scrutton, C.T. (1992) Devonian. In *Atlas of Palaeogeography and Lithofacies* (eds J.C.W. Cope, J.K. Ingham and P.F. Rawson), *Geological Society Memoir*, No. 13, Geological Society, London, pp. 57–66.

Braddy, S.J. (2000) Eurypterids from the Early Devonian of the Midland Valley of Scotland. *Scottish Journal of Geology*, 36, 115–22.

Brandon, A. (1989) *Geology of the Country Between Hereford and Leominster*, Memoir of the British Geological Survey, Sheet 198 (England and Wales), HMSO, London, 62 pp.

Brandon, A. and Hains, B.A. (1979) Geology of Hereford City and its environs. *Institute of Geological Sciences, Open File Report*, 1979/1.

Bridge, J.S. and Diemer, J.A. (1983) Quantitative interpretation of an evolving ancient river system. *Sedimentology*, 30, 599–623.

British Geological Survey (1987) *Arran. Scotland Special Sheet. Solid. 1:50 000*, Ordnance Survey, Southampton, for the British Geological Survey.

British Geological Survey (1990) *Greenock. Scotland Sheet 30. Solid. 1:50 000*, British Geological Survey, Keyworth.

British Geological Survey (1993) *Alford. Scotland Sheet 76W. Solid. 1:50 000*, British Geological Survey, Keyworth.

British Geological Survey (1996) *Foyers. Scotland Sheet 7SE. Solid. 1:50 000*, British Geological Survey, Keyworth.

British Geological Survey (1999) *Orkney Islands. Scotland Special Sheet. Solid and Drift. 1:100 000*, British Geological Survey, Keyworth.

British Geological Survey (2003) *Appleby. England and Wales Sheet 30. Solid. 1:50 000*, British Geological Survey, Keyworth.

British Geological Survey (in press) *Brecon. England and Wales Sheet 213. 1:50 000*, British Geological Survey, Keyworth.

Brooks, M., Trayner, P.K. and Trimble, T.J. (1988) Mesozoic reactivation of Variscan thrusting in the Bristol Channel area, UK. *Journal of the Geological Society of London*, 145, 439–45.

Browne, M.A.E. (1999) Balmerino to Wormit, Sheriffmuir Road to Menstrie Burn; Craig Rossie; Tillicoultry. In *Caledonian Igneous Rocks of Great Britain* (D. Stephenson, R.E. Bevins, D. Millward, A.J. Highton, I. Parsons, P. Stone and W.J. Wadsworth), Geological Conservation Review Series, No. 17, Joint Nature Conservation Committee, Peterborough, pp. 531–42.

Browne, M., Hargreaves, R. and Smith, I. (1985) *The Upper Palaeozoic Basins of the Midland Valley of Scotland, Investigation of the Geothermal Potential of the UK*, British Geological Survey, Keyworth, 48 pp.

Browne, M.A.E., Dean, M.T., Hall, I.H.S., McAdam, A.D., Munro, S.K. and Chisholm, J.I. (1999) A lithostratigraphical framework for the Carboniferous rocks of the Midland Valley of Scotland. Version 2. *British Geological Survey, Research Report*, **RR/99/07**.

References

Browne, M.A.E., Smith, R.A. and Aitken, A.M. (2002) Stratigraphical framework for the Devonian (Old Red Sandstone) rocks of Scotland south of a line from Fort William to Aberdeen. *British Geological Survey, Research Report*, **RR/01/04**.

Bucheit, A.K., Hargrove, U.S. and Donovan, R.N. (1997) Hutton's first unconformity revisited, Newton Point, Arran, Scotland. *American Association of Petroleum Geologists Bulletin*, **81**, 1347.

Burchette, T.P. (1981) The Lower Limestone Shales. In *A Field Guide to the Carboniferous Limestone Around Abergavenny* (eds V.P. Wright, M. Raven and T.P. Burchette), Department of Geology, University of Wales, Cardiff, pp. 13–27.

Burgess, I.C. (1961) Fossil soils of the Upper Old Red Sandstone of South Ayrshire. *Transactions of the Geological Society of Glasgow*, **24**, 138–53.

Burgess, I.C. and Wadge, A.J. (1974) *Classical Areas of British Geology: The Geology of the Cross Fell Inlier: Explanation of 1:25 000 Geological Special Sheet Comprising Parts of Sheets NY53, 62, 63, 64, 71, 72, 73*, HMSO, London, 92 pp.

Burgess, N.D. and Richardson, J.B. (1995) Late Wenlock to early Přídolí cryptospores and miospores from south and southwest Wales, Great Britain. *Palaeontographica, Abteilungen B*, **236**, 1–44.

Butler, M., Williams, B.P.J. and Bradshaw, R. (1972) A new exposure of the Old Red Sandstone–Lower Limestone Shale transition at Portishead, Somerset. *Proceedings of the Bristol Naturalists' Society*, **32**, 151–5.

Cameron, I.B. and Stephenson, D. (1985) *British Regional Geology: The Midland Valley of Scotland*, 3rd edn, HMSO, London, 172 pp.

Campbell, R. (1913) The geology of southeastern Kincardineshire. *Transactions of the Royal Society of Edinburgh*, **48**, 923–60.

Cantrill, T.C., Dixon, E.E.L., Thomas, H.H. and Jones, O.T. (1916) *The Geology of the South Wales Coalfield: Part XII: The Country Around Milford*, Memoir of the Geological Survey of Great Britain, Sheet 227, HMSO, London, 185 pp.

Capewell, J.G. (1954) The basic intrusions and an associated vent near Little Mell Fell, Cumberland. *Transactions of the Leeds Geological Association*, **6**, 243–8.

Capewell, J.G. (1955) The post-Silurian pre-marine Carboniferous sedimentary rocks of the eastern side of the English Lake District. *Quarterly Journal of the Geological Society of London*, **111**, 23–46.

Carroll, S. (1991) Terrestrial, fluvial and marginal lacustrine ecosystems in the Old Red Sandstone of the Orcadian basin. Unpublished PhD thesis, University of Aberdeen.

Carroll, S. (1995a) Geology of the Stonehaven District, 1:10 000 sheets NO88NW, NO88NE, NO88SW and NO88SE (South of the Highland Boundary Fault). *British Geological Survey, Onshore Geology Series, Technical Report*, **WA/94/19**.

Carroll, S. (1995b) Geology of the Inverbervie and Catterline District, 1:10 000 sheets NO87NW, NO87NE and NO87SW. *British Geological Survey, Onshore Geology Series, Technical Report*, **WA/94/20**.

Carroll, S. (1995c) Geology of the Fettercairn District, 1:10 000 sheets NO67NW, NO67NE, NO67SW, NO67SE and NO68SE (South of the Highland Boundary Fault). *British Geological Survey, Onshore Geology Series, Technical Report*, **WA/95/91**.

Carroll, S. and Trewin, N.H. (1995) *Cornulatichnus*: a new trace fossil from the Old Red Sandstone of Orkney. *Scottish Journal of Geology*, **31**, 37–41.

Channell, J.E.T., McCabe, C. and Woodcock, N.H. (1992) Early Devonian (pre-Acadian) magnetization directions in Lower Old Red Sandstone of south Wales (UK). *Geophysical Journal International*, **108**, 883–94.

Cherns, L. (1988) Faunal and facies dynamics in the Upper Silurian of the Anglo-Welsh Basin. *Palaeontology*, **31**, 451–502.

Chisholm, J.I. and Dean, J.M. (1974) The Upper Old Red Sandstone of Fife and Kinross: a fluviatile sequence with evidence of marine incursion. *Scottish Journal of Geology*, **10**, 1–30.

Christie-Blick, N. and Biddle, K.T. (1985) Deformation and basin formation along strike-slip faults. In *Strike-slip Deformation and Sedimentation* (eds N. Christie-Blick and K.T. Biddle), *Society of Economic Paleontologists and Mineralogists Special Publication*, No. **37**, SEPM, Tulsa, pp. 1–34.

Clarke, B.B. (1952) A note on the geology of the Deepwell tufa at Moccas and Brobury Scar. *Transactions of the Woolhope Naturalists' Field Club*, **33**, 48.

References

Clarke, B.B. (1955) The Old Red Sandstone of the Merbach Ridge, with an account of the Middlewood Sandstone, a new fossiliferous horizon 500 feet below the Psammosteus Limestone. *Transactions of the Woolhope Naturalists' Field Club*, **34**, 195–218.

Clarke, P. (1990) Sedimentological studies in Lower Old Red Sandstone basins, Northern Scotland. Unpublished PhD thesis, University of Queens, Belfast.

Clarkson, E.N.K. (1986) Pease Bay to Cove. In *Lothian Geology: An Excursion Guide* (eds A.D. McAdam and E.N.K. Clarkson), Scottish Academic Press, Edinburgh, pp. 140–5.

Cleal, C.J. and Thomas, B.A. (1995) *Palaeozoic Palaeobotany of Great Britain*, Geological Conservation Review Series, No. 9, Chapman and Hall, London, 295 pp.

Cleal, C.J., Thomas, B.A., Bevins, R.E. and Wimbledon, W.A. (2001) Deciding on a new world order. *Earth Heritage*, **16**, 10–13.

Collins, A.G. and Donovan, R.N. (1977) The age of two Old Red Sandstone sequences in southern Caithness. *Scottish Journal of Geology*, **13**, 53–7.

Collinson, J.D. (1978) Alluvial sediments. In *Sedimentary Environments and Facies*, 2nd edn (ed. H.G. Reading), Blackwell Scientific Publications, Oxford, pp. 15–60.

Conybeare, W.D. and Phillips, W. (1822) *Outlines of the Geology of England and Wales*, W. Phillips, London.

Cope, J.C.W. (1979) Early history of the southern margin of the Towy Anticline in the Carmarthen area, South Wales. In *The Caledonides of the British Isles, Reviewed* (eds A.L. Harris, C.H. Holland and B.E. Leake), Geological Society of London Special Publication, No. 8, Scottish Academic Press, Edinburgh, pp. 527–32.

Cope, J.C.W. (1981) The Swansea Valley Fault, Wales. *Geological Magazine*, **118**, 309–10.

Cope, J.C.W. (1982) The geology of the Llanstephan peninsula. In *Geological Excursions in Dyfed, South-west Wales* (ed. M.G. Bassett), National Museum of Wales, Cardiff, for the Geologists' Association South Wales Group, pp. 259–69.

Cope, J.C.W. and Bassett, M.G. (1987) Sediment sources and Palaeozoic history of the Bristol Channel area. *Proceedings of the Geologists' Association*, **98**, 315–30.

Cope, J.C.W., Ingham, J.K. and Rawson, P.F. (1992) Atlas of Palaeogeography and Lithofacies, *Geological Society Memoir*, No. 13, Geological Society, London, 153 pp.

Cossey, P.J., Adams, A.E., Purnell, M.A., Whiteley, M.J., Whyte, M.A. and Wright, V.P. (2004) *British Lower Carboniferous Stratigraphy*, Geological Conservation Review Series, No. 29, Joint Nature Conservation Committee, Peterborough, 617 pp.

Craig, G.Y. (1960). Grantshouse, Siccar Point, Catcraig. In *Edinburgh Geology: An Excursion Guide* (eds G.H. Mitchell, E.K. Walton and D. Grant), Oliver and Boyd, Edinburgh, pp. 89–101.

Craig, G.Y. (1986). Siccar Point: Hutton's classic unconformity. In *Lothian Geology: An Excursion Guide* (eds A.D. McAdam and E.N.K. Clarkson), Scottish Academic Press, Edinburgh, pp. 146–51.

Craig, G.Y. (1991) *Geology of Scotland*, 3rd edn, The Geological Society of London, London, 612 pp.

Craig, G.Y., McIntyre, D.B. and Waterston, C.D. (1978) *James Hutton's Theory of the Earth: the Lost Drawings*, Scottish Academic Press in association with the Royal Society of Edinburgh and the Geological Society of London, Edinburgh, 67 pp.

Crampton, C.B. and Carruthers, R.G. (1914) *The Geology of Caithness*, Memoir of the Geological Survey of Great Britain, sheets 110, 116 with parts of 109, 115 and 117 (Scotland), HMSO, Edinburgh, 194 pp.

Croft, W.N. (1953) Breconian: a stage name of the Old Red Sandstone. *Geological Magazine*, **90**, 429–32.

Croft, W.N. and George, E.A. (1959) Blue-green algae from the Middle Devonian of Rhynie, Aberdeenshire. *Bulletin of the British Museum (Natural History) Geology*, **3**, 341–53.

Dakyns, J.R., Tiddeman, R.H. and Goodchild, J.G. (1897) *The Geology of the Country between Appleby, Ullswater and Haweswater*, Memoir of the Geological Survey of Great Britain, Quarter-sheet 102SW (England and Wales), HMSO, London, 102 pp.

Dare-Edwards, A.J. (1984) Aeolian clay deposits of south-eastern Australia; parna or loessic clay? *Transactions of the Institute of British Geographers*, **9**, 337–44.

Davidson, R.G. and Trewin, N.H. (1999) Unusual soft tissue preservation in Middle Devonian fish-bearing nodule beds. *Palaeontological Association 43rd Annual Meeting, University of Manchester, Abstracts.*

References

Davidson, R.G. and Trewin, N.H. (2002) Guts and gizzard stones; unusual preservation in Scottish Middle Devonian fishes. *Palaeontological Association 46th Annual Meeting, University of Manchester, Abstracts.*

Davies, A., McAdam, A.D. and Cameron, I.B. (1986) *Geology of the Dunbar District*, Memoir of the British Geological Survey, Sheet 33E and part of Sheet 41 (Scotland), HMSO, London, 69 pp.

Dean, M.T. (2001) Faunas from and biostratigraphy of various exposures in the Appleby district. *British Geological Survey Report*, **IR/01/143**.

De la Beche, H.T. (1846) *On the Formation of the Rocks of South Wales and South Western England*, Memoir of the Geological Survey of the United Kingdom, **1**, HMSO, London, 296 pp.

De Souza, H.A.F. (1982) Abstract NDS 165. In *Numerical Dating in Stratigraphy* (ed. G.S. Odin), John Wiley and Sons, Chichester, pp. 847–8.

Denison, R.H. (1956) A review of the habitat of the earliest vertebrates. *Fieldiana: Geology*, **16**, 359–457.

Dewey, J.F. (1982) Plate tectonics and the evolution of the British Isles. *Journal of the Geological Society of London*, **139**, 371–414.

Dewey, J.F. and Strachan, R.A. (2003) Changing Silurian–Devonian relative plate motion in the Caledonides: sinistral transpression to sinistral transtension. *Journal of the Geological Society of London*, **160**, 219–29.

Dineley, D.L. (1999a) Mid-Devonian fossil fishes sites of Scotland. In *Fossil Fishes of Great Britain* (D.L. Dineley and S.J. Metcalf), Geological Conservation Review Series, No. 16, Joint Nature Conservation Committee, Peterborough, pp. 167–222.

Dineley, D.L. (1999b) Silurian fossil fishes sites of Scotland. In *Fossil Fishes of Great Britain* (D.L. Dineley and S.J. Metcalf), Geological Conservation Review Series, No. 16, Joint Nature Conservation Committee, Peterborough, pp. 31–62.

Dineley, D.L. (1999c) Early Devonian fossil fishes sites of Scotland. In *Fossil Fishes of Great Britain* (D.L. Dineley and S.J. Metcalf), Geological Conservation Review Series, No. 16, Joint Nature Conservation Committee, Peterborough, pp. 142–4.

Dineley, D.L. (1999d) Late Devonian fossil fishes sites of Scotland. In *Fossil Fishes of Great Britain* (D.L. Dineley and S.J. Metcalf), Geological Conservation Review Series, No. 16, Joint Nature Conservation Committee, Peterborough, pp. 241–62.

Dineley, D.L. (1999e) Late Silurian fossil fishes sites of the Welsh Borders. In *Fossil Fishes of Great Britain* (D.L. Dineley and S.J. Metcalf), Geological Conservation Review Series, No. 16, Joint Nature Conservation Committee, Peterborough, pp. 63–106.

Dineley, D.L. (1999f) Early Devonian fossil fishes sites of the Welsh Borders. In *Fossil Fishes of Great Britain* (D.L. Dineley and S.J. Metcalf), Geological Conservation Review Series, No. 16, Joint Nature Conservation Committee, Peterborough, pp. 107–44.

Dineley, D.L. (1999g) Mid- and Late Devonian fossil fishes sites of England and Wales. In *Fossil Fishes of Great Britain* (D.L. Dineley and S.J. Metcalf), Geological Conservation Review Series, No. 16, Joint Nature Conservation Committee, Peterborough, pp. 223–39.

Dineley, D.L. and Metcalf, S.J. (1999) *Fossil Fishes of Great Britain*, Geological Conservation Review Series, No. 16, Joint Nature Conservation Committee, Peterborough, 675 pp.

Dixon, E.E.L. (1921) *Geology of the South Wales Coalfield: Part XIII: The Country Around Pembroke and Tenby*, Memoir of the Geological Survey of Great Britain, sheets 244 and 245, HMSO, London, 220 pp.

Dixon, E.E.L. (1933a) Some recent stratigraphical work and its bearing on south Pembrokeshire problems. *Proceedings of the Geologists' Association*, **44**, 217–25.

Dixon, E.E.L. (1933b) Notes on the geological succession in south Pembrokeshire. *Proceedings of the Geologists' Association*, **44**, 402–11.

Dodd, C. (1986) The sedimentology of fluvio-aeolian interactions with geological examples from the British Isles. Unpublished PhD thesis, Plymouth Polytechnic.

Dolby, G. (1971) Spore assemblages from the Devonian–Carboniferous transition measures in south-west Britain and southern Eire. In *Colloque sur la Stratigraphie du Carbonifère* (eds M. Streel and R.H. Wagner), *Les Congrès et Colloques de l'Université de Liège*, No. **55**, Université de Liège, Liège, pp. 267–74.

References

Donovan, R.N. (1970) The geology of the coastal tract near Wick, Caithness. Unpublished PhD thesis, University of Newcastle-upon-Tyne.

Donovan, R.N. (1975) Devonian lacustrine limestones at the margin of the Orcadian Basin, Scotland. *Journal of the Geological Society of London*, **131**, 489–510.

Donovan, R.N. (1978) Caithness. In *A Field Guide to Selected Outcrop Areas of the Devonian of Scotland, the Welsh Borderland and South Wales: International Symposium on the Devonian System (P.A.D.S. 78)* (eds P.F. Friend and B.P.J. Williams), Palaeontological Association, London, pp. 37–53.

Donovan, R.N. (1980) Lacustrine cycles, fish ecology and stratigraphic zonation in the Middle Devonian of Caithness. *Scottish Journal of Geology*, **16**, 35–50.

Donovan, R.N. and Foster, R.J. (1972) Subaqueous shrinkage cracks from the Caithness Flagstone Series (Middle Devonian) of Northeast Scotland. *Journal of Sedimentary Petrology*, **42**, 309–17.

Donovan, R.N., Foster, R.J. and Westoll, T.S. (1974) A stratigraphical revision of the Old Red Sandstone of north-eastern Caithness. *Transactions of the Royal Society of Edinburgh*, **69**, 167–201.

Donovan, R.N., Peacock, J.D. and Mykura, W. (1978) South of the Moray Firth. In *A Field Guide to Selected Outcrop Areas of the Devonian of Scotland, the Welsh Borderland and South Wales: International Symposium on the Devonian System (P.A.D.S. 78)* (eds P.F. Friend and B.P.J. Williams), Palaeontological Association, London, pp. 22–7.

Donovan, R.N., Bucheit, A.K. and Hargrove, U.S. (1998) Carbonate pedogenesis at Newton Point, Isle of Arran, Scotland; Hutton's first unconformity revisited. *Abstracts with Programs – Geological Society of America*, **30**, part 3, p. 5.

Dubinin, V.B. (1962) Class Acaromorpha: mites or gnathosomic chelicerate arthropods. In *Fundamentals of Palaeontology* (ed. B.B. Rodendorf), Academy of Science USSR, Moscow, pp. 447–73.

Dunlop, J.A., Anderson, L.I., Kerp, H. and Hass, H. (2003) Preserved organs of Devonian harvestmen. *Nature*, **425**, 916.

Dunlop, J.A., Anderson, L.I., Kerp, H. and Hass, H. (2004) A harvestman (Arachnida: Opiliones) from the Early Devonian Rhynie cherts, Aberdeenshire, Scotland. *Transactions of the Royal Society of Edinburgh: Earth Sciences*, **94**, 341–54.

Dunne, W.M. (1983) Tectonic evolution of SW Wales during the Upper Palaeozoic. *Journal of the Geological Society of London*, **140**, 257–65.

Durant, G. (1999a). Port Schuchan to Dunure Castle; Culzean Harbour; Turnberry Lighthouse to Port Murray. In *Caledonian Igneous Rocks of Great Britain* (D. Stephenson, R.E. Bevins, D. Millward, A.J. Highton, I. Parsons, P. Stone and W.J. Wadsworth), Geological Conservation Review Series, No. 17, Joint Nature Conservation Committee, Peterborough, pp. 542–52.

Durant, G. (1999b). South Kerrera. In *Caledonian Igneous Rocks of Great Britain* (D. Stephenson, R.E. Bevins, D. Millward, A.J. Highton, I. Parsons, P. Stone and W.J. Wadsworth), Geological Conservation Review Series, No. 17, Joint Nature Conservation Committee, Peterborough, pp. 489–92.

Edmonds, E.A., Mckeown, M.C. and Williams, M. (1975) *British Regional Geology: Southwest England*, 4th edn, HMSO, London, 136 pp.

Edmonds, E.A., Whittaker, A. and Williams, B.J. (1985) *Geology of the Country Around Ilfracombe and Barnstaple*, Memoir of the British Geological Survey, sheets 277 and 293 (England and Wales), HMSO, London, 96 pp.

Edwards, D. (1979a) The early history of vascular plants based on late Silurian and early Devonian floras of the British Isles. In *The Caledonides of the British Isles, Reviewed* (eds A.L. Harris, C.H. Holland and B.E. Leake), *Geological Society of London Special Publication*, No. **8**, Scottish Academic Press, Edinburgh, pp. 405–10.

Edwards, D. (1979b) A late Silurian flora from the Lower Old Red Sandstone of south-west Dyfed. *Palaeontology*, **22**, 23–52.

Edwards, D. (1980) Early land floras. In *The Terrestrial Environment and the Origin of Land Vertebrates* (ed. A.L. Panchen), *Systematics Association Special Volume*, **15**, Academic Press, London, pp. 55–85.

Edwards, D. and Fanning, U. (1985) Evolution and environment in the late Silurian to early Devonian: the rise of the pteridophytes. *Philosophical Transactions of the Royal Society of London*, **B309**, 147–65.

References

Edwards, D. and Richardson, J.B. (1978) Brecon Beacons Quarry (Storey Arms), Powys. In *A Field Guide to Selected Outcrop Areas of the Devonian of Scotland, the Welsh Borderland and South Wales: International Symposium on the Devonian System (P.A.D.S. 78)* (eds P.F. Friend and B.P.J. Williams), Palaeontological Association, London, p. 75.

Edwards, D. and Richardson, J.B. (2004) Silurian and Lower Devonian plant assemblages from the Anglo-Welsh Basin: a palaeobotanical and palynological synthesis. *Geological Journal*, **39**, 375–402.

Edwards, D., Richardson, J.B. and Thomas, R.G. (1978) Heol Senni Quarry, Powys. In *A Field Guide to Selected Outcrop Areas of the Devonian of Scotland, the Welsh Borderland and South Wales: International Symposium on the Devonian System (P.A.D.S. 78)* (eds P.F. Friend and B.P.J. Williams), Palaeontological Association, London, pp. 77–8.

Edwards, D.S. (1986) *Aglaophyton major*, a non-vascular land-plant from the Devonian Rhynie chert. *Botanical Journal of the Linnean Society*, **93**, 173–204.

Edwards, D.S. and Lyon, A.G. (1983) Algae from the Rhynie chert. *Botanical Journal of the Linnean Society*, **86**, 37–55.

Edwards, R.A. (1996) Selected geological locality details for the Minehead district. *British Geological Survey, Onshore Geology Series, Technical Report*, **WA/96/32**.

Edwards, R.A. (1999) *The Minehead District – a Concise Account of the Geology*, Memoir of the British Geological Survey, Sheet 278 and part of Sheet 294 (England and Wales), The Stationery Office, London, 128 pp.

Ékes, C. (1993) Bedload-transported pedogenic mud aggregates in the Lower Old Red Sandstone in southwest Wales. *Journal of the Geological Society of London*, **150**, 469–71.

El Saadawy, W. and Lacey, W.S. (1979a) The sporangia of *Horneophyton lignieri* (Kidston and Lang) Barghoorn and Darrah. *Review of Palaeobotany and Palynology*, **28**, 137–44.

El Saadawy, W. and Lacey, W.S. (1979b) Observations on *Nothia aphylla* Lyon ex Høeg. *Review of Palaeobotany and Palynology*, **27**, 119–47.

Ellis, N.V. (ed.), Bowen, D.Q., Campbell, S., Knill, J.L., McKirdy, A.P., Prosser, C.D., Vincent, M.A. and Wilson, R.C.L. (1996) *An Introduction to the Geological Conservation Review*, Geological Conservation Review Series, No. 1, Joint Nature Conservation Committee, Peterborough, 131 pp.

Engel, M.S. and Grimaldi, D.A. (2004) New light shed on the oldest insect. *Nature*, **427**, 627–30.

Evans, J.W. (1922) The geological structure of the country round Combe Martin, north Devon. *Proceedings of the Geologists' Association*, **33**, 201–28.

Evans, K.M. (1983) Note on the age and fauna of the Lynton Beds (Lower Devonian) of north Devon. *Geological Journal*, **18**, 297–305.

Eyles, V.A. (1953) West Midlands. *Summary of Progress for the Geological Survey of Great Britain*, for 1952, 25–6.

Fannin, N.G.T. (1969) Stromatolites from the Middle Old Red Sandstone of Western Orkney. *Geological Magazine*, **106**, 77–88.

Fannin, N.G.T. (1970) The sedimentary environment of the Old Red Sandstone of Western Orkney. Unpublished PhD thesis, University of Reading.

Fayers, N.S. and Trewin, N.H. (2003) A new crustacean from the Early Devonian Rhynie chert, Aberdeenshire, Scotland. *Transactions of the Royal Society of Edinburgh: Earth Sciences*, **93**, 355–82

Fayers, N.S. and Trewin, N.H. (2004) A review of the palaeoenvironments and biota of the Windyfield chert. *Transactions of the Royal Society of Edinburgh: Earth Sciences*, **94**, 325–39.

Fayers, N.S. and Trewin, N.H. (in press) A hexapod from the Early Devonian Windyfield chert, Rhynie, Scotland. *Palaeontology*.

Fayers, N.S., Dunlop, J.A. and Trewin, N.H. (in press) A new early Devonian trigonotarbid arachnid from the Windyfield chert, Rhynie, Scotland. *Journal of Systematic Palaeontology*.

Fletcher, T.P., Auton, C.A., Highton, A.J., Merritt, J.W., Robertson, S. and Rollin, K.E. (1996) *Geology of Fortrose and Eastern Inverness District*, Memoir of the British Geological Survey, Sheet 84W (Scotland), HMSO, London, 137 pp.

Foster, R.J. (1972) The solid geology of northeast Caithness. Unpublished PhD thesis, University of Newcastle-upon-Tyne.

Francis, E.H., Forsyth, I.H., Read, W.A., and Armstrong, M. (1970) *The Geology of the Stirling District*, Memoir of the Geological Survey of Great Britain, Sheet 39 (Scotland), HMSO, Edinburgh, 357 pp.

References

Friend, P.F. and Williams, B.P.J. (1978) *A Field Guide to Selected Outcrop Areas of the Devonian of Scotland, the Welsh Borderland and South Wales: International Symposium on the Devonian System (P.A.D.S. 78)*, Palaeontological Association, London, pp. 106.

Friend, P.F. and Williams, B.P.J. (2000) *New perspectives on the Old Red Sandstone, Geological Society Special Publication*, No. **180**, Geological Society, London, 623 pp.

Friend, P.F., Williams, B.P.J., Ford, M. and Williams, E.A. (2000) Kinematics and dynamics of Old Red Sandstone basins. In *New perspectives on the Old Red Sandstone* (eds P.F. Friend and B.P.J. Williams), *Geological Society Special Publication*, No. **180**, Geological Society, London, pp. 29–60.

Fyfe, J.A. (1985) The sedimentology of the Old Red Sandstone section from Pease Bay to Horse Roads Rock, Berwickshire. Unpublished thesis, University of St Andrews.

Gatliff, R.W., Richards, P.C., Smith, K., Graham, C.C., McCormac, M., Smith, N.J.P., Long, D., Cameron, T.D.J., Evans, D., Stevenson, A.G., Bulat, J. and Ritchie, J.D. (1994) *United Kingdom Offshore Regional Report: The Geology of the Central North Sea*, HMSO, London, for the British Geological Survey, 118 pp.

Geikie, A. (1879) On the Old Red Sandstone of Western Europe. Part 1. *Transactions of the Royal Society of Edinburgh*, **28**, 345–453.

Geikie, A. (1897) *The Ancient Volcanoes of Great Britain*, Macmillan, London.

George, T.N. (1970) *British Regional Geology: South Wales*, 3rd edn, HMSO, London, 152 pp.

Gibling, M.R., Nanson, G.C. and Maroulis, J.C. (1998) Anastomosing river sedimentation in the Channel Country of central Australia. *Sedimentology*, **45**, 595–619.

Gillen, C. (1987). Huntly, Elgin and Lossiemouth. In *Excursion Guide to the Geology of the Aberdeen Area* (eds N.H. Trewin, B.C. Kneller and C. Gillen), Scottish Academic Press, Edinburgh, pp. 149–60.

Gillen, C. and Trewin, N.H. (1987) Dunnottar to Stonehaven and the Highland Boundary Fault. In *Excursion Guide to the Geology of the Aberdeen Area* (eds N.H. Trewin, B.C. Kneller and C. Gillen), Scottish Academic Press, Edinburgh, pp. 265–73.

Goldring, R., Tunbridge, I.P., Whittaker, A. and Williams, B.J. (1978) 2: North Devon. In *A Field Guide to Selected Areas of the Devonian of South-west England: International Symposium on the Devonian System (P.A.D.S. 78)* (ed. C.T. Scrutton), Paleontological Association, London, pp. 8–27.

Gould, D. (1997) *Geology of the Country Around Inverurie and Alford*, Memoir of the British Geological Survey, sheets 76E and 76W (Scotland), The Stationery Office, London, 116 pp.

Green, G.W. (1992) *British Regional Geology: Bristol and Gloucester Region*, 3rd edn, HMSO, London.

Green, J.F.N. (1918) The Mell Fell Conglomerate. *Proceedings of the Geologists' Association*, **29**, 117–25.

Greenly, E. (1919) *The Geology of Anglesey*, Memoir of the Geological Survey of Great Britain, HMSO, London.

Greig, D.C. (1988) *Geology of the Eyemouth District*, Memoir of the British Geological Survey, Sheet 34 (Scotland), HMSO, London, 78 pp.

Greig, D.C. and Davies, A. (1978) Berwickshire Coast. In *A Field Guide to Selected Outcrop Areas of the Devonian of Scotland, the Welsh Borderland and South Wales: International Symposium on the Devonian System (P.A.D.S. 78)* (eds P.F. Friend and B.P.J. Williams), Palaeontological Association, London, pp. 6–12.

Greig, D.C., Wright, J.E., Hains, B.A. and Mitchell, G.H. (1968) *Geology of the Country Around Church Stretton, Craven Arms, Wenlock Edge and Brown Clee*, Memoir of the Geological Survey of Great Britain, Sheet 166 (England and Wales), HMSO, London, 378 pp.

Gunn, W. (1903) In *The Geology of North Arran, South Bute and the Cumbraes with parts of Ayrshire and Kintyre* (W. Gunn, A. Geikie, B.N. Peach and A. Harker), Memoir of the Geological Survey of Great Britain, Sheet 21 (Scotland), HMSO, Edinburgh, pp. 51–3.

Habgood, K.S. (2000) Integrated approaches to the cycling of primary productivity in early terrestrial ecosystems. Unpublished PhD thesis, Cardiff University.

Hall, A.J. and Donovan, R.N. (1978) Origin of complex sulphide nodules related to diagenesis of lacustrine sediments of Middle Devonian age from the Shetland Islands. *Scottish Journal of Geology*, **14**, 289–99.

Hall, I.H.S. and Chisholm, J.I. (1987) Aeolian sediments in the late Devonian of the Scottish Midland Valley. *Scottish Journal of Geology*, **23**, 203–8.

References

Hall, I.H.S., Taylor, K. and Thomas, L.P. (1973) The stratigraphy of the Upper Old Red Sandstone in South Breconshire. *Bulletin of the Geological Survey of Great Britain*, No. 44, 45–62.

Hall, I.H.S., Browne, M.A.E. and Forsyth, I.H. (1998) *Geology of the Glasgow District*, Memoir of the British Geological Survey, Sheet 30E (Scotland), The Stationery Office, London, 117 pp.

Hallam, A.D. (1934) The geology of the Hangman Grits of the Quantock Hills. *Geological Magazine*, 71, 433–46.

Halliday, A.N., McAlpine, A. and Mitchell, J.G. (1982) $^{40}Ar/^{39}Ar$ age of the Hoy lavas, Orkney. In *Numerical Dating in Stratigraphy* (ed. G.S. Odin), John Wiley and Sons, Chichester, pp. 928–30.

Halstead, L.B. (1985) The vertebrate invasion of fresh water. *Philosophical Transactions of the Royal Society*, B309, 243–58.

Hamilton, R.F.M. and Trewin, N.H. (1988) Environmental controls on fish faunas of the Middle Devonian Orcadian Basin. In *Devonian of the World: Proceedings of the Second International Symposium on the Devonian, Calgary, Canada* (eds N.J. McMillan, A.F. Embry and D.J. Glass), Canadian Society of Petroleum Geologists, Calgary, 3, pp. 589–600.

Hancock, P.L. (1973) Structural zones in Variscan Pembrokeshire. *Proceedings of the Ussher Society*, 2, 509–20.

Hancock, P.L., Dunne, W.M. and Tringham, M.E. (1982) Variscan structures in south-west Dyfed. In *Geological Excursions in Dyfed, South-west Wales* (ed. M.G. Bassett), National Museum of Wales, Cardiff, for the Geologists' Association South Wales Group, pp. 215–48.

Hanken, N.M. and Størmer, L. (1975) The trail of a large Silurian eurypterid. *Fossils and strata*, 4, 255–70.

Harker, A. and Marr, J.E. (1891) The Shap Granite and the associated igneous and metamorphic rocks. *Quarterly Journal of the Geological Society of London*, 47, 266–328.

Harper, J.C. and Hartley, J.J. (1938) The Silurian inlier of Lisbellow, Co. Fermanagh, with a note on the age of the Fintona Beds. *Proceedings of the Royal Irish Academy*, 45B, 73–87.

Hassan, A.M. (1982) Palynology, stratigraphy and provenance of the Lower Old Red Sandstone of the Brecon Beacons (Powys) and Black Mountains (Gwent and Powys), South Wales. Unpublished PhD thesis, University of London.

Hassan, T.H. (1966) The petrology of the Skrinkle Sandstone and contiguous deposits of south Pembrokeshire. Unpublished PhD thesis, Chelsea College of Science and Technology.

Haughton, P.D.W. (1988) A cryptic Caledonian flysch terrane in Scotland. *Journal of the Geological Society of London*, 145, 685–703.

Haughton, P.D.W. (1989) Structure of some Lower Old Red Sandstone conglomerates, Kincardineshire, Scotland: deposition from late-orogenic antecedent streams? *Journal of the Geological Society of London*, 146, 509–25.

Haughton, P.D.W. and Bluck, B.J. (1988) Diverse alluvial sequences from the Lower Old Red Sandstone of the Strathmore region, Scotland – implications for the relationship between late Caledonian tectonics and sedimentation. In *Devonian of the World: Proceedings of the Second International Symposium on the Devonian, Calgary, Canada* (eds N.J. McMillan, A.F. Embry and D.J. Glass), Canadian Society of Petroleum Geologists, Calgary, 2, pp. 269–93.

Haughton, P.D.W. and Farrow, C.M. (1989) Compositional variation in Lower Old Red Sandstone detrital garnets from the Midland Valley of Scotland and the Anglo-Welsh Basin. *Geological Magazine*, 126, 373–96.

Haughton, P.D.W., Rogers, G. and Halliday, A.N. (1990) Provenance of Lower Old Red Sandstone conglomerates, SE Kincardineshire: evidence of the timing of Caledonian terrane accretion in central Scotland. *Journal of the Geological Society of London*, 147, 105–20.

Hawley, D. (1991) *Pwll-Y-Wrach Nature Reserve Geology Trail*, Brecknock Wildlife Trust, Brecon.

Heard, A. (1926) *Psilophyton breconensis. Report of the British Association for the Advancement of Science*, Southampton, 311–12.

Heard, A. (1927) On the Old Red Sandstone plants showing structure from Brecon (South Wales). *Quarterly Journal of the Geological Society of London*, 83, 195–207.

Hickling, G. (1908) The Old Red Sandstone of Forfarshire. *Geological Magazine, New Series*, 5, 396–408.

References

Hickling, G. (1912) On the geology and palaeontology of Forfarshire. *Proceedings of the Geologists' Association*, **23**, 302–11.

Higgs, K.T. (2004) An early Devonian (Lochkovian) microflora from the Freshwater West Formation, Lower Old Red Sandstone, southwest Wales. *Geological Journal*, **39**, 359–74.

Hillier, R.D. (2000) Silurian marginal marine sedimentation and the anatomy of the marine–Old Red Sandstone transition in Pembrokeshire, SW Wales. In *New Perspectives on the Old Red Sandstone* (eds P.F. Friend and B.P.J. Williams), *Geological Society Special Publication*, No. **180**, Geological Society, London, pp. 343–54.

Hillier, R.D. and Williams, B.P.J. (2004) Sedimentation and tectonics: the marine Silurian–basal Old Red Sandstone transition in southwest Wales. *Geological Journal*, **39**, 237–56.

Hillier, R.D. and Williams, B.P.J. (in press) Chapter 5. The alluvial Old Red Sandstone basins. In *The Geology of England and Wales* (eds P.J. Brenchley and P.F. Rawson).

Hirst, S. (1923) On some arachnid remains from the Old Red Sandstone (Rhynie chert bed, Aberdeenshire). *Annals and Magazine of Natural History (Series 9)*, **12**, 455–74.

Hirst, S. and Maulik, S. (1926a) On some arthropod remains from the Rhynie chert (Old Red Sandstone). *Geological Magazine*, **63**, 69–71.

Hirst, S. and Maulik, S. (1926b) (Untitled contribution to 'Announcements and Inquiries' section). *Geological Magazine*, **63**, 288.

Holland, C.H. and Lawson, J.D. (1963) Facies patterns in the Ludlovian of Wales and the Welsh Borderland. *Liverpool and Manchester Geological Journal*, **3**, 269–88.

Holwill, F.J.W. (1962) The succession of limestones within the Ilfracombe Beds (Devonian) of north Devon. *Proceedings of the Geologists' Association*, **73**, 281–93.

Holwill, F.J.W., House, M.R., Gauss, G.A., Hendriks, E.M.L. and Dearman, W.R. (1969) Summer (1966) field meeting in Devon and Cornwall, report of the directors. *Proceedings of the Geologists' Association*, **80**, 43–62.

Horne, J. (1923) The *Geology of the Lower Findhorn and Lower Strath Nairn*, Memoir of the Geological Survey of Great Britain, Sheet 84 and part of 94 (Scotland), HMSO, Edinburgh, 128 pp.

Horne, J. and Hinxman, L.W. (1914) *The Geology of the Country Around Beauly and Inverness: Including a Part of the Black Isle*, Memoir of the Geological Survey of Great Britain, Sheet 83 (Scotland), HMSO, Edinburgh, 108 pp.

House, M.R. (1975) Facies and time in Devonian tropical areas. *Proceedings of the Yorkshire Geological Society*, **40**, 233–88.

House, M.R. (1977) Subdivision of the marine Devonian. In *A Correlation of Devonian Rocks in the British Isles* (eds M.R. House, J.B. Richardson, W.G. Chaloner, J.R.L. Allen, C.H. Holland and T.S. Westoll), *Geological Society of London Special Report*, No. **8**, Scottish Academic Press, Edinburgh, pp. 4–10.

House, M.R. and Kirchgasser, W.T. (1993) Devonian goniatite biostratigraphy and timing of facies movements in the Frasnian of eastern North America. In *High Resolution Stratigraphy* (eds E.A. Hailwood and R.B. Kidd), *Geological Society Special Publication*, No. **70**, The Geological Society, London, pp. 267–92.

House, M.R., Richardson, J.B., Chaloner, W.G., Allen, J.R.L., Holland, C.H. and Westoll, T.S. (1977) *A Correlation of Devonian Rocks in the British Isles*, Geological Society of London Special Report, No. **8**, Scottish Academic Press, Edinburgh, 110 pp.

Hurst, J.M., Hancock, N.J. and McKerrow, W.S. (1978) Wenlock stratigraphy and palaeogeography in Wales and the Welsh Borderland. *Proceedings of the Geologists' Association*, **89**, 197–226.

Hurst, J.M., Hancock, N.J. and McKerrow, W.S. (1979) Old Red Sandstone facies and Wenlock stratigraphy and palaeogeography in Wales and the Welsh Borderland; reply by the authors. *Proceedings of the Geologists' Association*, **90**, 231–2.

Hutton, J. (1795) *Theory of the Earth with proofs and illustrations*, **1**, 429–30.

Hutton, J. (1899) *Theory of the Earth with proofs and illustrations* (ed. Sir A. Geikie), **3**, 235–37, 262–3.

Institute of Geological Sciences (1973) *Cromarty. Scotland Sheet 94. Solid. 1:50 000*, Ordnance Survey, Southampton, for the Institute of Geological Sciences.

Institute of Geological Sciences (1977) *Ammanford. England and Wales Sheet 230. Solid. 1:50 000*, Ordnance Survey, Southampton, for the Institute of Geological Sciences.

Institute of Geological Sciences (1979) *Merthyr Tydfil. England and Wales Sheet 231. Solid. 1:50 000*, Ordnance Survey, Southampton, for the Institute of Geological Sciences.

References

James, D.M.D. (1987) Tectonics and sedimentation in the Lower Palaeozoic back-arc basin of South Wales, UK. Some quantitative aspects of basin development. *Norsk Geologisk Tidsskrift*, **67**, 419–26.

Jameson, R. (1800) *Mineralogy of the Scottish Isles*, **1**, 77–9.

Jameson, R. (1805) *A Mineralogical Description of the County of Dumfries*, Bell and Bradgate, Edinburgh.

Jameson, R. (1821) *Manual of Mineralogy*, A. Constable and Co., Edinburgh, 491 pp.

Janaway, T.M. and Parnell, J. (1989) Carbonate production within the Orcadian Basin, Northern Scotland: a petrographic and geochemical study. *Palaeogeography, Palaeoclimatology and Palaeoecology*, **70**, 89–105.

Jenkins, G. (1998) An investigation into marine influence during deposition of the Lower Old Red Sandstone, Anglo-Welsh Basin, UK. Unpublished PhD thesis, University of Wales, Cardiff.

Johnstone, G.S. and Mykura, W. (1989) *British Regional Geology: The Northern Highlands of Scotland*, 4th edn, HMSO, London, 219 pp.

Jones, N.S. (1995) The sedimentology of the Devonian Hangman Sandstone Formation and Permo–Triassic strata from the Minehead sheet 278. *British Geological Survey, Technical Report*, **WH/95/119R**.

Kellaway, G.A. and Welch, F.B.A. (1948) *British Regional Geology: Bristol and Gloucester District*, 2nd edn, HMSO, London.

Kellaway, G.A. and Welch, F.B.A. (1955) The Upper Old Red Sandstone and Lower Carboniferous rocks of Bristol and the Mendips compared with those of Chepstow and the Forest of Dean. *Bulletin of the Geological Survey of Great Britain*, **9**, 1–21.

Kellaway, G.A. and Welch, F.B.A. (1993) *Geology of the Bristol District*, Memoir of the British Geological Survey, Bristol Special Sheet (England and Wales), HMSO, London, 199 pp.

Kelling, G. and Williams, B.P.J. (1966) Deformation structures of sedimentary origin in the Lower Limestone Shales (basal Carboniferous) of south Pembrokeshire, Wales. *Journal of Sedimentary Petrology*, **36**, 927–39.

Kellock, E. (1969) Alkaline basic igneous rocks in the Orkneys. *Scottish Journal of Geology*, **5**, 140–53.

Kelman, R., Feist, M., Trewin, N.H. and Hass, H. (2004) Charophyte Algae from the Rhynie Chert. *Transactions of the Royal Society of Edinburgh: Earth Sciences*, **94**, 443–53.

Kerp, H., Hass, H. and Mosbrugger, V. (2001) New data on *Nothia aphylla* Lyon, 1964 ex El Saadawy et Lacey, 1979: a poorly known plant from the Lower Devonian Rhynie chert. In *Plants Invade the Land: Evolutionary and Environmental Perspectives* (eds P.G. Gensel and D. Edwards), Columbia University Press, New York, pp. 52–82.

Kerp, H., Trewin, N.H. and Hass, H. (2004) New gametophytes from the Early Devonian Rhynie chert. *Transactions of the Royal Society of Edinburgh: Earth Sciences*, **94**, 409–26.

Kidston, R. and Lang, W.H. (1917) On Old Red Sandstone plants showing structure, from the Rhynie chert bed, Aberdeenshire. Part I. *Rhynia gwynne-vaughanii* Kidston and Lang. *Transactions of the Royal Society of Edinburgh*, **51**(24), 761–84.

Kidston, R. and Lang, W.H. (1920a) On Old Red Sandstone plants showing structure, from the Rhynie chert bed, Aberdeenshire. Part II. Additional notes on *Rhynia Gwynne-vaughanii* Kidston and Lang; with descriptions of *Rhynie major*, n.sp., and *Hornia lignieri*, n.g., n.sp. *Transactions of the Royal Society of Edinburgh*, **52**(24), 603–27.

Kidston, R. and Lang, W.H. (1920b) On Old Red Sandstone plants showing structure, from the Rhynie chert bed, Aberdeenshire. Part III. *Asteroxylon mackiei*, Kidston and Lang. *Transactions of the Royal Society of Edinburgh*, **52**(26), 643–80.

Kidston, R. and Lang, W.H. (1921) On Old Red Sandstone plants showing structure, from the Rhynie chert bed, Aberdeenshire. Part V. The Thallophyta occurring in the peat-bed; the succession of the plants throughout a vertical section of the bed, and the conditions of accumulation and preservation of the deposit. *Transactions of the Royal Society of Edinburgh*, **52**(33), 855–902.

Kimber, R.N. and Johnson, G.A.L. (1984) Lake District highlands and islands during the Upper Palaeozoic. *Proceedings of the Cumberland Geological Society*, **4**, 377–90.

King, L.M. (1994) Subsidence analysis of Eastern Avalonian sequences: implications for Iapetus closure. *Journal of the Geological Society of London*, **151**, 647–57.

References

King, W.W. (1934) The Downtonian and Dittonian strata of Great Britain and north-western Europe. *Quarterly Journal of the Geological Society of London*, **90**, 526–67.

Knight, R.R.W. (1990) The Devonian of north Devon – a palynologist's (or 'conodontologist's') dream or nightmare? *Proceedings of the Ussher Society*, **7**, 306.

Kokelaar, B.P. (1982) Fluidization of wet sediments during the emplacement and cooling of various igneous bodies. *Journal of the Geological Society of London*, **139**, 21–34.

Knudsen, T.L. (2000) The provenance of Devonian sandstones from Shetland: a Sm–Nd and trace element study. *Scottish Journal of Geology*, **36**, 61–72.

Lagios, E. (1983) A gravity study of the eastern Berwickshire Devonian basins, SE Scotland. *Scottish Journal of Geology*, **19**, 189–203.

Lamont, A. (1952) Ecology and correlations of the Pentlandian, a new division of the Silurian System in Scotland. *Report XVIII, International Geological Congress, London*, Part X, 27–32.

Lane, P.D. (2000a) The Přídolí Series. In *British Silurian Stratigraphy* (R.J. Aldridge, David J. Siveter, Derek J. Siveter, P.D. Lane, D. Palmer and N.H. Woodcock), Geological Conservation Review Series, No. 19, Joint Nature Conservation Committee, Peterborough, pp. 427–58.

Lane, P.D. (2000b) Capel Horeb Quarry. In *British Silurian Stratigraphy* (R.J. Aldridge, David J. Siveter, Derek J. Siveter, P.D. Lane, D. Palmer and N.H. Woodcock), Geological Conservation Review Series, No. 19, Joint Nature Conservation Committee, Peterborough, pp. 440–4.

Lane, P.D. (2000c) Albion Sands and Gateholm Island. In *British Silurian Stratigraphy* (R.J. Aldridge, David J. Siveter, Derek J. Siveter, P.D. Lane, D. Palmer and N.H. Woodcock), Geological Conservation Review Series, No. 19, Joint Nature Conservation Committee, Peterborough, pp. 447–51.

Lane, P.D. (2000d) Little Castle Head. In *British Silurian Stratigraphy* (R.J. Aldridge, David J. Siveter, Derek J. Siveter, P.D. Lane, D. Palmer and N.H. Woodcock), Geological Conservation Review Series, No. 19, Joint Nature Conservation Committee, Peterborough, pp. 446–7.

Lane, R. (1965) The Hangman Grits – an introduction and stratigraphy. *Proceedings of the Ussher Society*, **1**, 166–7.

Lang, W.H. (1927) Contributions to the study of the Old Red Sandstone flora of Scotland. VI. *Transactions of the Royal Society of Edinburgh*, **55**, 443–52.

Le Maitre, R.W. (1989) *A Classification of Igneous Rocks and Glossary of Terms: Recommendations of the International Union of Geological Sciences Subcommission on the Systematics of Igneous Rocks*, Blackwell Scientific Publications, Oxford, 193 pp.

Leeder, M.R. (1973) Sedimentology and palaeogeography of the Upper Old Red Sandstone in the Scottish Border Basin. *Scottish Journal of Geology*, **9**, 117–44.

Leeder, M.R. (1974) Origin of the Northumberland Basin. *Scottish Journal of Geology*, **10**, 283–96.

Leeder, M.R. (1975) Pedogenic carbonates and flood sediment accumulation rates: a quantitative model for alluvial arid-zone lithofacies. *Geological Magazine*, **112**, 257–70.

Leeder, M.R. (1976) Palaeogeographic significance of pedogenic carbonates in the topmost Upper Old Red Sandstone of the Scottish Border Basin. *Geological Journal*, **11**, 21–8.

Leeder, M.R. and Bridges, P.H. (1978) Upper Old Red Sandstone near Kirkbean, Dumfries and Galloway. *Scottish Journal of Geology*, **14**, 267–72.

Loeffler, E.J. and Thomas, R.G. (1980) A new pteraspid ostracoderm from the Devonian Senni Beds Formation of South Wales and its stratigraphical significance. *Palaeontology*, **23**, 287–96.

Love, S.E. (1993) Floodplain deposits as indicators of sandbody geometry and reservoir architecture. Unpublished PhD thesis, University of Aberdeen.

Love, S.E. and Williams, B.P.J. (2000) Sedimentology, cyclicity and floodplain architecture in the Lower Old Red Sandstone of SW Wales. In *New Perspectives on the Old Red Sandstone* (eds P.F. Friend and B.P.J. Williams), *Geological Society Special Publication*, No. **180**, Geological Society, London, pp. 371–88.

Love, S.E., Davies, S. and Gould, S. (2004) Inter-tuff correlation and architecture within the Moor Cliffs Formation, Lower Old Red Sandstone, southwest Wales: a multidisciplinary approach for reservoir characterisation. *Geological Journal*, **39**, 299–314.

References

Lovell, R.W.W. (1978a) The sedimentology and stratigraphy of the Upper Old Red Sandstone and Lower Limestone Shales of the South Wales Coalfield. Unpublished PhD thesis, University of Bristol.

Lovell, R.W.W. (1978b) Abercriban, Powys. In *A Field Guide to Selected Outcrop Areas of the Devonian of Scotland, the Welsh Borderland and South Wales: International Symposium on the Devonian System (P.A.D.S. 78)* (eds P.F. Friend and B.P.J. Williams), Palaeontological Association, London, pp. 72–3.

Lumsden, G.I., Tulloch, W., Howells, M.F. and Davies, A. (1967) *The Geology of the Neighbourhood of Langholm*, Memoir of the Geological Survey of Great Britain, Sheet 11 (Scotland), HMSO, Edinburgh, 255 pp.

Lyon, A.G. (1962) On the fragmentary remains of an organism referable to the nematophytales, from the Rhynie chert, "*Nematoplexus rhyniensis*" gen. et. sp. nov. *Transactions of the Royal Society of Edinburgh*, **65**, 79–87.

Lyon, A.G. (1964) Probable fertile region of *Asteroxylon mackiei* Kidston and Lang. *Nature*, **203**, 1082–3.

Lyon, A.G. and Edwards, D. (1991) The first zosterophyll from the Lower Devonian Rhynie Chert, Aberdeenshire. *Transactions of the Royal Society of Edinburgh: Earth Sciences*, **82**, 323–32.

Machette, M.N. (1985) Calcic soils of the south-western United States. In *Soils and Quaternary Geology of the South West United States* (ed. D.L. Weide), *Geological Society of America Special Paper*, No. **203**, Geological Society of America, Boulder, pp. 363–81.

MacDonald, R. and Walker, B.H. (1985) Geochemistry and tectonic significance of the Lower Carboniferous Cockermouth lavas, Cumbria. *Proceedings of the Yorkshire Geological Society*, **45**, 141–6.

MacGregor, A.G. and Eckford, R.J.A. (1946) The Upper Old Red and Lower Carboniferous sediments of Teviotdale and Tweedside, and the stones of the abbeys of the Scottish Borderland. *Transactions of the Edinburgh Geological Society*, **14**, 230–52.

MacGregor, A.R. (1968) Edzell and Glen Esk. In *Fife and Angus Geology: An Excursion Guide* (ed. A.R. MacGregor), Scottish Academic Press, Edinburgh, pp. 85–97.

MacGregor, A.R. (1996a) Excursion 1. Arbroath, Crawton and Stonehaven. In *Fife and Angus Geology: An Excursion Guide*, 3rd edn (ed. A.R. MacGregor), Pentland Press, Edinburgh, pp. 79–92.

MacGregor, A.R. (1996b) Excursion 2. Edzell and Glen Esk. In *Fife and Angus Geology: An Excursion Guide*, 3rd edn (ed. A.R. MacGregor), Pentland Press, Edinburgh, pp. 93–108.

Macintyre, R.M., Cliff, R.A. and Chapman, N.A. (1981) Geochronological evidence for phased volcanic activity in Fife and Caithness vents, Scotland. *Transactions of the Royal Society of Edinburgh: Earth Sciences*, **72**, 1–7.

Mackie, A. (1980) Sandstone quarrying in Angus – some thoughts on an old craft. *The Edinburgh Geologist*, **8**, 14–25.

Mader, D. and Yardley, M.J. (1985) Migration, modification and merging in aeolian systems and the significance of depositional mechanisms in Permian and Triassic dune sands of Europe and North America. *Sedimentary Geology*, **43**, 85–218.

Malcolmson, J. (1859) On the relations of the different parts of the Old Red Sandstone in which organic remains have been discovered in the counties of Moray, Banff, Nairn and Inverness. *Quarterly Journal of the Geological Society of London*, **15**, 336–51.

Marr, J. (1899) Note on a conglomerate at Melmerby. *Quarterly Journal of the Geological Society of London*, **55**, 11–15.

Marriott, S.B. and Wright, V.P. (1993) Palaeosols as indicators of geomorphic stability in two Old Red Sandstone alluvial suites, South Wales. *Journal of the Geological Society of London*, **150**, 1109–20.

Marriott, S.B. and Wright, V.P. (1996) Sediment recycling on Siluro-Devonian floodplains. *Journal of the Geological Society of London*, **153**, 661–4.

Marriott, S.B. and Wright, V.P. (2004) Mudrock deposition in an ancient dryland system: Moor Cliffs Formation, Lower Old Red Sandstone of southwest Wales, UK. *Geological Journal*, **39**, 277–98.

Marriott, S.B., Wright, V.P. and Williams, B.P.J. (2005) A new evaluation of fining upward sequences in a mud-rock dominated succession of the Lower Old Red Sandstone of South Wales, UK. In *Fluvial Sedimentology VII* (eds M.D. Blum, S.B. Marriott and S. LeClair), *International Association of Sedimentologists Special Publication*, No. **35**, Blackwell, London, pp. 517–29.

References

Marshall, J.D. (1977) Sedimentology of the Skrinkle Sandstones Group (Devonian–Carboniferous), south-west Dyfed. Unpublished PhD thesis, University of Bristol.

Marshall, J.D. (1978) West Angle Bay, Dyfed. In *A Field Guide to Selected Outcrop Areas of the Devonian of Scotland, the Welsh Borderland and South Wales: International Symposium on the Devonian System (P.A.D.S. 78)* (eds P.F. Friend and B.P.J. Williams), Palaeontological Association, London, pp. 99–101.

Marshall, J.D. (2000a) Sedimentology of a Devonian fault-bounded braidplain and lacustrine fill in the lower part of the Skrinkle Sandstones, Dyfed, Wales. *Sedimentology*, 47, 325–42.

Marshall, J.D. (2000b) Fault-bounded basin-fill: fluvial response to tectonic controls in the Skrinkle Sandstones of SW Pembrokeshire, Wales. In *New Perspectives on the Old Red Sandstone* (eds P.F. Friend and B.P.J. Williams), *Geological Society Special Publication*, No. 180, Geological Society, London, pp. 401–16.

Marshall, J.E.A. (1991) Palynology of the Stonehaven Group, Scotland: evidence for a mid Silurian age and its geological implications. *Geological Magazine*, 128, 283–6.

Marshall, J.E.A. (2000) Devonian (Givetian) miospores from the Walls Group, Shetland. In *New Perspectives on the Old Red Sandstone* (eds P.F. Friend and B.P.J. Williams), *Geological Society Special Publication*, No. 180, Geological Society, London, pp. 473–83.

Marshall, J.E.A. and Hewett, A.J. (2003) Chapter 6. Devonian. In *The Millennium Atlas: Petroleum Geology of the Central and Northern North Sea* (eds and co-ordinators D. Evans, C. Graham, A. Armour and P. Bathurst), Geological Society of London, London, pp. 65–81.

Marshall, J.E.A. and House, M.R. (2000) Devonian stage boundaries in England, Wales and Scotland. *Courier Forschungsinstitut Senckenberg*, 225, 83–90.

Marshall, J.E.A., Haughton, P.D.W. and Hillier, S.J. (1994) Vitrinite reflectivity and the structure and burial history of the Old Red Sandstone of the Midland Valley of Scotland. *Journal of the Geological Society of London*, 151, 425–38.

Marshall, J.E.A., Rogers, D.A. and Whiteley, M.J. (1996) Devonian marine incursions into the Orcadian Basin. *Journal of the Geological Society of London*, 153, 451–66.

Marston, R.J. (1971) The Foyers granitic complex, Inverness-shire, Scotland. *Quarterly Journal of the Geological Society of London*, 126, 331–68.

M'Caw, L.S. (1936) The Black Mountains: A physical, agricultural and geographical survey 1932–1936. Unpublished MA thesis, University of Manchester.

McAdam, A.D. and Clarkson, E.N.K. (1986) *Lothian Geology: An Excursion Guide*, Scottish Academic Press, Edinburgh.

McAlpine, A. (1978) The Upper Old Red Sandstone of Orkney, Caithness and neighbouring areas. Unpublished PhD thesis, University of Newcastle-upon-Tyne.

McCormac, M. (2001) The Upper Palaeozoic rocks and Quaternary deposits of the Shap and Penrith district, Cumbria. *British Geological Survey Report*, RR/01/10.

McGarvie, D.W. (1999a) Ben Nevis and Allt A'Mhuilinn. In *Caledonian Igneous Rocks* (D. Stephenson, R.E. Bevins, D. Millward, A.J. Highton, I. Parsons, P. Stone and W.J. Wadsworth), Geological Conservation Review Series, No. 17, Joint Nature Conservation Committee, Peterborough, pp. 492–7.

McGarvie, D.W. (1999b) Stob Dearg and Cam Ghleann. In *Caledonian Igneous Rocks* (D. Stephenson, R.E. Bevins, D. Millward, A.J. Highton, I. Parsons, P. Stone and W.J. Wadsworth), Geological Conservation Review Series, No. 17, Joint Nature Conservation Committee, Peterborough, pp. 510–13.

McKerrow, W.S., Dewey, J.F. and Scotese, C.R. (1991) The Ordovician and Silurian development of the Iapetus Ocean. In *The Murchison Symposium: Proceedings of an International Conference on the Silurian* System (eds M.G. Bassett, P.D. Lane and D. Edwards), *Special Papers in Palaeontology*, No. 44, Palaeontological Association, London, pp. 165–78.

Mechie, J. and Brooks, J. (1984) A seismic study of deep geological structure in the Bristol Channel area, SW Britain. *Geophysical Journal of the Royal Astronomical Society*, 78, 661–89.

Melvin, J. (1976) Sedimentological studies in Upper Palaeozoic sandstones near Bude, Cornwall and Walls, Shetland. Unpublished PhD thesis, University of Edinburgh.

Melvin, J. (1985) Walls Formation, Western Shetland: distal alluvial plain deposits within a tectonically active Devonian basin. *Scottish Journal of Geology*, 21, 23–40.

References

Miall, A.D. (1996) *The Geology of Fluvial Deposits: Sedimentary Facies, Basin Analysis and Petroleum Geology*, Springer-Verlag, Berlin, 582 pp.

Miles, R.S. (1968) The Old Red Sandstone antiarchs of Scotland: Family Bothriolepidae. *Monograph of the Palaeontographical Society*, **122**, No. 552, 1–130.

Miller, H. (1841) *The Old Red Sandstone; Or, New Walks in an Old Field*, J. Johnstone, Edinburgh.

Millward, D. (2003) Little Mell Fell Quarry. In *Carboniferous and Permian Igneous Rocks of Great Britain* (D. Stephenson, S.C. Loughlin, D. Millward, C.N. Waters and I.T. Williamson), Geological Conservation Review Series, No. 27, Joint Nature Conservation Committee, Peterborough, pp. 122–5.

Millward, D., McCormac, M., Hughes, R.A., Entwisle, D.C., Butcher, A. and Raines, M.G. (2003) *Geology of the Appleby District – A Brief Explanation of the Geological Map*, Sheet Explanation of the British Geological Survey, Sheet 30 (England and Wales), British Geological Survey, Keyworth, 30 pp.

Milne, D. (1843) Geological Account of Roxburghshire. *Transactions of the Royal Society of Edinburgh*, **15**, 433–502.

Morrissey, L.B. and Braddy, S.J. (2004) Terrestrial trace fossils from the Lower Old Red Sandstone southwest Wales. *Geological Journal*, **39**, 315–36.

Morrissey, L.B., Braddy, S.J., Bennett, J.P., Marriott, S.B. and Tarrant, P.R. (2004) Fish trails from the Lower Old Red Sandstone of Tredomen Quarry, Powys, southeast Wales. *Geological Journal*, **39**, 337–58.

Morton, D.J. (1976) Lower Old Red Sandstone sedimentation in the north-west Midland Valley and north Argyll areas of Scotland. Unpublished PhD thesis, University of Glasgow.

Mould, D.D.C.P. (1946) The geology of the Foyers 'granite' and the surrounding country. *Geological Magazine*, **83**, 249–65.

Muir, R.O. and Ridgway, J.M. (1975) Sulphide mineralization of the continental Devonian sediments of Orkney (Scotland). *Mineralium Deposita*, **10**, 205–15.

Murchison, R.I. (1834) *Field Notebook, Volume 13, 1834 June*. Murchison papers, LDGSL 839, M/N 69, Geological Society archives, London, 27–28.

Murchison, R.I. (1839) *The Silurian System*, John Murray, London.

Mykura, W. (1976) *British Regional Geology: Orkney and Shetland*, HMSO, Edinburgh, 149 pp.

Mykura, W. (1982) The Old Red Sandstone east of Loch Ness, Inverness-shire. *Report of the Institute of Geological Sciences*, **82/13**, 26 pp.

Mykura, W. (1991) Old Red Sandstone. In *Geology of Scotland*, 3rd edn (ed. G.Y. Craig), The Geological Society of London, London, pp. 297–346.

Mykura, W. and Phemister, J. (1976) *The Geology of Western Shetland*, Memoir of the Geological Survey of Great Britain, Sheet 127 and parts of 125, 126 and 128 (Scotland), HMSO, Edinburgh, 304 pp.

Neves, R. and Dolby, G. (1967) Assemblage of miospores from the Portishead Beds (Upper Old Red Sandstone) of the Mendip Hills, England. *Pollen et Spores*, **9**, 607–14.

Nicol, J. (1847) On the geology of Roxburghshire. *Transactions of the Highland and Agricultural Society of Scotland*, **10**, 35–83.

Oldham, R.D. (1900) The Basal (Carboniferous) Conglomerate of Ullswater and its mode of origin. *Geological Magazine*, **7**, 564.

Owen, G. (1995) Senni Formation of the Devonian Old Red Sandstone, Dyfed, Wales: anatomy of a semi-arid floodplain. *Sedimentary Geology*, **95**, 221–35.

Owen, G. and Hawley, D. (2000) Depositional setting of the Lower Old Red Sandstone at Pantymaes Quarry, central South Wales: new perspectives on the significance and occurrence of 'Senni Beds' facies. In *New Perspectives on the Old Red Sandstone* (eds P.F. Friend and B.P.J. Williams), *Geological Society Special Publication*, No. **180**, Geological Society, London, pp. 389–400.

Owen, T.R., Bloxham, T.W., Jones, D.G., Walmsley, V.G. and Williams, B.P.J. (1971) Summer (1968) field meeting in Pembrokeshire, South Wales. *Proceedings of the Geologists' Association*, **82**, 17–60.

Parker, A., Allen, J.R.L. and Williams, B.P.J. (1983) Clay mineral assemblages of the Townsend Tuff Bed (Lower Old Red Sandstone), South Wales and the Welsh Borders. *Journal of the Geological Society of London*, **140**, 769–79.

Parnell, J. (1987) Mineralized magadi-type cherts in the Devonian of Scotland: support for a model of sulfide deposition in low-temperature alkaline conditions. *Economic Geology*, **82**, 1053–6.

References

Parnell, J., Marshall, J. and Astin, T.R. (1990) *Field Guide to Lacustrine Deposits of the Orcadian Basin, Scotland*, British Sedimentological Research Group, Cambridge, 45 pp.

Parry, S.F. (2004) Age and underlying cause of hot-spring activity at Rhynie, Aberdeenshire, Scotland. Unpublished PhD thesis, University of Aberdeen.

Paterson, I.B. and Hall, I.H.S. (1986) Lithostratigraphy of the late Devonian and early Carboniferous rocks in the Midland Valley of Scotland. *Report of the British Geological Survey*, **18/3**, 14 pp.

Paterson, I.B., Browne, M.A.E. and Armstrong, M. (1976) Letters to the Editors: Upper Old Red Sandstone palaeogeography. *Scottish Journal of Geology*, **12**, 89–91.

Paterson, I.B., Hall, I.H.S. and Stephenson, D. (1990) *Geology of the Greenock District*, Memoir of the British Geological Survey, Sheet 30W and part of Sheet 29E (Scotland), HMSO, London, 69 pp.

Paton, R.L. (1981) *A Catalogue of Fossil Vertebrates in the Royal Scottish Museum, Part 6, Placodermi, Royal Scottish Museum Information Series: Geology*, No. 9, Royal Scottish Museum, Edinburgh, 70 pp.

Peach, B.N. and Horne, J. (1903) The Canonbie Coalfield: its geological structure and relations to the Carboniferous rocks of the north of England and central Scotland. *Transactions of the Royal Society of Edinburgh*, **40**, 835–77.

Peach, C.W. (1868) On fossil fishes of the Old Red Sandstone of Caithness and Sutherland, with notices of some new to these comments. *Report of the British Association for the Advancement of Science*, Dundee, **37**, 72.

Peacock, D.P.S. (1961) The Old Red Sandstone rocks of the Edzell District. Unpublished honours thesis, University of St Andrews.

Peacock, J.D., Berridge, N.G., Harris, A.L. and May, F. (1968) *The Geology of the Elgin District*, Memoir of the Geological Survey of Great Britain, Sheet 95 (Scotland), HMSO, Edinburgh, 165 pp.

Pennington, J.J. (1975) The geology of the Argyll Field. In *Petroleum and the Continental Shelf of North-west Europe* (ed. A.W. Woodland), Applied Science Publishers, Barking, pp. 285–91.

Phillips, E.R. and Aitken, A.M. (1998) The petrology and composition of the Lower Old Red Sandstone exposed in the Aberfoyle area (Sheet 38E), Central Scotland. *British Geological Survey, Mineralogy and Petrology Series, Technical Report*, **WG/98/10**.

Phillips, E.R. and Carroll, S. (1995) The petrology and provenance of the basal Lower Old Red Sandstone exposed between Ruthery Head and Dunnicaer, Stonehaven, Scotland. *British Geological Survey, Technical Report* **WG/95/21**.

Phillips, E.R., Smith, R.A. and Carroll, S. (1998) Strike-slip, terrane accretion and the pre-Carboniferous evolution of Midland Valley of Scotland. *Transactions of the Royal Society of Edinburgh: Earth Sciences*, **89**, 209–24.

Phillips, W. (1818) *Geology of England and Wales*, London.

Pick, M.C. (1964a) The stratigraphy and sedimentary features of the Old Red Sandstone, Portishead coastal section, N.E. Somerset. *Proceedings of the Geologists' Association*, **75**, 199–221.

Pick, M.C. (1964b) The Triassic Dolomitic Conglomerate and structure of the Old Red Sandstone, Portishead coastal section, N.E. Somerset. *Proceedings of the Bristol Naturalists' Society*, **30**, 445–50.

Playfair, J. (1805) Biographical account of the late Dr James Hutton, FRS. Edin. *Transactions of the Royal Society of Edinburgh*, **5** (Part 3), 39–99.

Plummer, P.S. and Gostin, V.A. (1981) Shrinkage cracks: desiccation or synaeresis? *Journal of Sedimentary Petrology*, **51**, 1147–56.

Potter, J.F. and Price, J.H. (1965) Comparative sections through rocks of Ludlovian–Downtonian age in the Llandovery and Llandeilo districts. *Proceedings of the Geologists' Association*, **76**, 379–401.

Powell, C.L., Edwards, D. and Trewin, N.H. (2000a) A new vascular plant from the Lower Devonian Windyfield chert, Rhynie, NE Scotland. *Transactions of the Royal Society of Edinburgh: Earth Sciences*, **90**, 331–49.

Powell, C.L., Trewin, N.H. and Edwards, D. (2000b) Palaeocology and plant succession in a borehole through the Rhynie cherts, Lower Old Red Sandstone, Scotland. In *New Perspectives on the Old Red Sandstone* (eds P.F. Friend and B.P.J. Williams), *Geological Society Special Publication*, No. **180**. Geological Society, London, pp. 439–57.

Powell, C.M. (1987) Inversion tectonics in S.W. Dyfed. *Proceedings of the Geologists' Association*, **98**, 193–203.

References

Powell, C.M. (1989) Structural controls on Palaeozoic basin evolution and inversion in southwest Wales. *Journal of the Geological Society of London*, **146**, 439–46.

Powrie, J. (1861) Cephalaspids of Forfar. *The Geologist*, **4**, 137–90.

Powrie, J. (1864) The Scottish *Pteraspis*. *The Geologist*, **7**, 172.

Powrie, J. (1870) On the connection of the Lower, Middle and Upper Old Red Sandstones of Scotland. *Transactions of the Edinburgh Geological Society*, **1**, 115–32.

Ramos, A. and Friend, P.F. (1982) Upper Old Red Sandstone sedimentation near the unconformity at Arbroath. *Scottish Journal of Geology*, **18**, 297–315.

Rayner, D.H. (1963) The Achanarras Limestone of the Middle Old Red Sandstone, Caithness, Scotland. *Proceedings of the Yorkshire Geological Society*, **34**, 1–44.

Rayner, D.H. (1983) New observations on *Sawdonia ornata* from Scotland. *Transactions of the Royal Society of Edinburgh*, **74**, 79–93.

Rayner, D.H. (1984) New finds of *Drepanophycus spinaeformis* Goppert from the Lower Devonian of Scotland. *Transactions of the Royal Society of Edinburgh*, **75**, 353–63.

Rayner, R.J. (1995) Auchensail Quarry. In *Palaeozoic Palaeobotany of Great Britain* (C.J. Cleal and B.A. Thomas), Geological Conservation Review Series, No. 9, Chapman and Hall, London, pp. 78–80.

Read, H.H. (1923) *The Geology of the Country Around Banff, Huntly and Turriff*, Memoir of the Geological Survey of Scotland, sheets 86 and 96, HMSO, Edinburgh, 240 pp.

Remy, W. and Hass, H. (1991a) *Langiophyton mackiei nov. gen., nov. spec.*, ein Gametophyt mit Archegoniophoren aus dem Chert von Rhynie (Unterdevon Schottland). *Argumenta Palaeobotanica*, **8**, 69–117.

Remy, W. and Hass, H. (1991b) Ergänzende Beobachtungen an *Lyonophyton rhyniensis*. *Argumenta Palaeobotanica*, **8**, 1–27.

Remy, W. and Hass, H. (1991c) *Kidstonophyton discoides nov. gen. nov. spec.*, ein Gametophyt aus dem Chert von Rhynie (Unterdevon, Schottland). *Argumenta Palaeobotanica*, **8**, 29–45.

Remy, W. and Remy, R. (1980) *Lyonophyton Rhyniensis n. gen. et nov. spec.*, ein Gametophyt aus dem Chert von Rhynie (Unterdevon, Schottland). *Argumenta Palaeobotanica*, **6**, 37–72.

Remy, W., Gensel, P.G. and Hass, H. (1993) The gametophyte generation of some early Devonian land plants. *International Journal of Plant Science*, **154**, 35–58.

Remy, W., Taylor, T.N. and Hass, H. (1994) Early Devonian fungi: A blastocladalean fungus with sexual reproduction. *American Journal of Botany*, **81**, 690–702.

Repcheck, J. (2003) *The Man Who Found Time: James Hutton and the Discovery of the Earth's Antiquity*, Simon & Schuster, London, 247 pp.

Retallack, G.J. (1990) *Soils of the Past: An Introduction to Paleopedology*, Unwin Hyman, Boston, 520 pp.

Reynolds, S.H. (1921) *A Geological Excursion Handbook for the Bristol District*, 2nd edn, Arrowsmith, Bristol, 224 pp.

Reynolds, S.H. and Greenly, E. (1923) The Old Red Sandstone and Carboniferous Limestone of the Portishead–Clevedon area. *Proceedings of the Bristol Naturalists' Society*, **6**, 92–7.

Reynolds, S.H. and Greenly, E. (1924) The geological structure of the Portishead–Clevedon area. *Quarterly Journal of the Geological Society of London*, **80**, 447–67.

Rice, C.M. and Ashcroft, W.A. (2004) The geology of the northern half of the Rhynie Basin, Aberdeenshire, Scotland. *Transactions of the Royal Society of Edinburgh*, **94**, 299–308.

Rice, C.M. and Trewin, N.H. (1988) Lower Devonian gold-bearing hot spring system near Rhynie, Scotland. *Transactions of the Institution of Mining and Metallurgy*, **97**, B141–144.

Rice, C.M., Ashcroft, W.A., Batten, D.J., Boyce, A.J., Caulfield, J.B.D., Fallick, A.E., Hole, M.J., Jones, E., Pearson, M.J., Rogers, G., Saxton, J.M., Stuart, F.M., Trewin, N.H. and Turner, G.A. (1995) A Devonian auriferous hot spring system, Rhynie, Scotland. *Journal of the Geological Society of London*, **152**, 229–50.

Rice, C.M., Trewin, N.H. and Anderson, L.I. (2002) Geological setting of the Early Devonian Rhynie cherts, Aberdeenshire, Scotland: an early terrestrial hot spring system. *Journal of the Geological Society of London*, **159**, 203–14.

Richardson, J.B. (1967) Some British Lower Devonian spore assemblages and their stratigraphic significance. *Review of Palaeobotany and Palynology*, **1**, 111–29.

Richardson, J.B. and Lister, T.R. (1969) Upper Silurian and Lower Devonian spore assemblages from the Welsh Borderlands and south Wales. *Palaeontology*, **12**, 201–52.

References

Richardson, J.B. and MacGregor, D.C. (1986) *Silurian and Devonian Spore Zones of the Old Red Sandstone Continent and Adjacent Regions*, Geological Survey of Canada Bulletin, No. 364, Geological Survey of Canada, Ottawa, 79 pp.

Richardson, J.B., Rasul, S.M. and Al-Ameri, T. (1981) Acritarchs, miospores and correlation of the Ludlovian–Downtonian and Silurian–Devonian boundaries. *Review of Palaeobotany and Palynology*, 34, 209–24.

Richardson, J.B., Ford, J.H., and Parker, J. (1984) Miospores, correlation and age of some Scottish Lower Old Red Sandstone sediments from the Strathmore region (Fife and Angus). *Journal of Micropalaeontology*, 3, 109–24.

Richardson, J.B., Rodriguez, R.M. and Sutherland, S.J.E. (2000) Palynology and recognition of the Silurian/Devonian boundary in some British terrestrial sediments by correlation with Cantabrian and other European marine sequences – a progress report. *Courier Forschungsinstitut Senckenberg*, 220, 1–7.

Richmond, L.K. and Williams, B.P.J. (2000) A new terrane in the Old Red Sandstone of the Dingle Peninsula, SW Ireland. In *New Perspectives on the Old Red Sandstone* (eds P.F. Friend and B.P.J. Williams), Geological Society Special Publication, No. 180, Geological Society, London, pp. 147–83.

Ridgway, J.M. (1974) Sedimentology and palaeogeography of the Eday Group, Middle Old Red Standstone, Orkney. Unpublished PhD thesis, University of London.

Ritchie, A. (1975) *Groenlandaspis* in Antarctica, Australia and Europe. *Nature*, 254, 569–73.

Robertson, J.D. (1957) The habitat of the early vertebrates. *Biological Reviews, Cambridge*, 32, 256–87.

Robertson, S. (1987) Early sinistral transpression in the Lower Old Red Sandstone of Kincardineshire, Scotland. *Scottish Journal of Geology*, 23, 261–8.

Robertson, T. (1927) *The Geology of the South Wales Coalfield: Part II: The Country Around Abergavenny*, 2nd edn, Memoir of the Geological Survey of Great Britain, Sheet 232 (England and Wales), HMSO, London, 145 pp.

Robertson, T. (1932) The *Geology of the South Wales Coalfield: Part V: The Country Around Merthyr Tydfil*, 2nd edn, Memoir of the Geological Survey of Great Britain, Sheet 231 (England and Wales), HMSO, London, 283 pp.

Robinson, R.A.J., Rennie, C.A. and Oliver, G.J.H. (1998) Palaeocurrent data, source terrains and palaeogeographic setting of the Dalradian block: the Stonehaven–Dunnottar Groups revisited. *Tectonic Studies Group Annual General Meeting* (St Andrews).

Rock, N.M.S. and Rundle, C.C. (1986) Lower Devonian age for the 'Great (basal) Conglomerate', Scottish Borders. *Scottish Journal of Geology*, 22, 285–8.

Rogers, D.A. (1987) Devonian correlations, environments and tectonics across the Great Glen Fault. Unpublished PhD thesis, University of Cambridge.

Rogers, D.A. and Astin, T.R. (1991) Ephemeral lakes, mud pellet dunes and wind-blown sand and silt: reinterpretations of Devonian lacustrine cycles in north Scotland. In *Lacustrine Facies Analysis* (eds P. Anadon, Ll. Cabrera and K. Kelts), *International Association of Sedimentologists Special Publication*, No. 13, Blackwell Scientific Publications, Oxford, pp. 201–23.

Sallomey, J.T. and Piper, J.D.A. (1973) Palaeomagnetic studies in the British Caledonides IV. Lower Devonian lavas of the Strathmore region. *Geophysical Journal of the Royal Astronomical Society*, 58, 47–68.

Salter, J.W. (1863) On the Upper Old Red Sandstone and Upper Devonian rocks. *Quarterly Journal of the Geological Society of London*, 19, 474–96.

Salter, T.M. (1992) Facies, geometrical and palaeocurrent analysis of well-exposed alluvial reservoir analogues and applications to subsurface studies. Unpublished PhD thesis, University of Leeds.

Sanderson, D.J. and Dearman, W.R. (1973) Structural zones of the Variscan fold belt in SW England, their location and development. *Journal of the Geological Society of London*, 129, 527–36.

Sanzen-Baker, I. (1972) Stratigraphical relationships and sedimentary environments of the Silurian–early Old Red Sandstone of Pembrokeshire. *Proceedings of the Geologists' Association*, 83, 139–64.

Scotese, C.R. (2001) *Atlas of Earth History*, PALEOMAP Project, Arlington, Texas, 52 pp.

Scott, A.C., Edwards, D. and Rolfe, W.D.I. (1976) Fossiliferous Lower Old Red Sandstone near Cardross, Dunbartonshire. *Proceedings of the Geological Society of Glasgow*, 117, 4–5.

References

Scourfield, D.J. (1926) On a new type of crustacean from the Old Red Sandstone (Rhynie chert bed, Aberdeenshire) – *Lepidocaris rhyniensis*, gen. et sp. nov. *Philosophical Transactions of the Royal Society, London (Series B)*, **214**, 153–87.

Scourfield, D.J. (1940) Two new and nearly complete specimens of young stages of the Devonian fossil crustacean *Lepidocaris rhyniensis*. *Proceedings of the Linnean Society*, **152**, 290–8.

Sedgwick, A. and Murchison, R.I. (1839) On the classification of the older rocks of Devon and Cornwall. *Proceedings of the Geological Society of London*, **3**, 121–3.

Selden, P.A., Shear, W.A. and Bonamo, P.M. (1991) A spider and other arachnids from the Devonian of New York and reinterpretations of Devonian Araneae. *Palaeontology*, **34**, 241–81.

Serrane, M. (1992) Devonian extensional tectonics versus Carboniferous inversion in the northern Orcadian basin. *Journal of the Geological Society of London*, **149**, 27–37.

Shear, W.A., Selden, P.A., Rolfe, W.D.I., Bonamo, P.M. and Grierson, J.D. (1987) New terrestrial arachnids from the Devonian of Gilboa, New York (Arachnida, Trigonotarbida). *American Museum Novitates*, **2901**, 1–74.

Shear, W.A., Jeram, A.J. and Selden, P.A. (1998) Centipede legs (Arthropoda, Chilopoda, Scutigeromorpha) from the Silurian and Devonian of Britain and the Devonian of North America. *American Museum Novitates*, **3232**, 1–16.

Shearman, D.J. (1967) On Tertiary fault movements in north Devonshire. *Proceedings of the Geologists' Association*, **78**, 555–66.

Sherlock, S.A., Jones, K.A. and Kelley, S.P. (2002) Fingerprinting polyorogenic detritus using the $^{40}Ar/^{39}Ar$ ultraviolet laser microprobe. *Geology*, **30**, 515–18.

Shotton, F.W. (1935) The stratigraphy and tectonics of the Cross Fell Inlier. *Quarterly Journal of the Geological Society of London*, **91**, 639–701.

Simon, J.B. and Bluck, B.J. (1982) Palaeodrainage of the southern margin of the Caledonian mountain chain in the northern British Isles. *Transactions of the Royal Society of Edinburgh: Earth Sciences*, **73**, 11–15.

Simpson, B. (1971) The Palaeozoic succession in the Black Mountains between Pontardawe and Llandeilo. In *Geological Excursions in South Wales and The Forest of Dean* (eds D.A. Bassett and M.G. Bassett), The Geologists' Association South Wales Group, Cardiff, pp. 143–54.

Simpson, S. (1951) A new eurypterid from the Upper Old Red Sandstone of Portishead. *The Annals and Magazine of Natural History*, **4**, 849–61.

Simpson, S. (1959) Fascicule 3aVI – Devonian. *Lexique Stratigraphique International. Volume 1: Europe* (eds W.F. Whittard and S. Simpson), Congrès Géologique International. Centre National De La Recherche Scientifique, Paris, 131 pp.

Simpson, S. (1964) The Lynton Beds of north Devon. *Proceedings of the Ussher Society*, **1**, 121–2.

Siveter, D.J. (2000) Sawdde Gorge. In *British Silurian Stratigraphy* (R.J. Aldridge, David J. Siveter, Derek J. Siveter, P.D. Lane, D. Palmer and N.H. Woodcock), Geological Conservation Review Series, No. 19, Joint Nature Conservation Committee, Peterborough, pp. 403–7.

Siveter, D.J., Owens, R.M. and Thomas, A.T. (1989) *Silurian Field Excursions: A Geotraverse across Wales and the Welsh Borderlands*, National Museum of Wales Geological Series, No. 10, National Museum of Wales in conjunction with The Geologists' Association, Cardiff, 133 pp.

Smith, A., Braddy, S.J., Marriott, S.B. and Briggs, D.E.G. (2003). Arthropod trackways from the Early Devonian of South Wales: a functional analysis of producers and their behaviour. *Geological Magazine*, **140**, 63–72.

Smith, D.B. (1980) *Summary of Geology. Ross-on-Wye Special 1:10,000 Sheet*, British Geological Survey, Keyworth.

Smith, R.A. (1995) The Siluro–Devonian evolution of the southern Midland Valley of Scotland. *Geological Magazine*, **132**, 503–13.

Smith, R.A. (1996) Geology of the Gass Water area, explanation of 1:10 000 Sheet NS62SE. *British Geological Survey, Onshore Geology Series, Technical Report*, **WA/96/22**.

Smith, R.A. (1999a) Scurdie Ness to Usan Harbour; Black Rock to East Comb. In *Caledonian Igneous Rocks of Great Britain* (D. Stephenson, R.E. Bevins, D. Millward, A.J. Highton, I. Parsons, P. Stone and W.J. Wadsworth), Geological Conservation Review Series, No. 17, Joint Nature Conservation Committee, Peterborough, pp. 525–31.

References

Smith, R.A. (1999b) Crawton Bay. In *Caledonian Igneous Rocks of Great Britain* (D. Stephenson, R.E. Bevins, D. Millward, A.J. Highton, I. Parsons, P. Stone and W.J. Wadsworth). Geological Conservation Review Series, No. 17, Joint Nature Conservation Committee, Peterborough, pp. 522–5.

Smith, T.E. (1967) A preliminary study of sandstone sedimentation in the Lower Carboniferous of the Tweed Basin. *Scottish Journal of Geology*, 3, 282–305.

Smith, T.E. (1968) The Upper Old Red Sandstone–Carboniferous junction at Burnmouth, Berwickshire. *Scottish Journal of Geology*, 4, 349–54.

Soper, N.J. and Woodcock, N.H. (2003) The lost Lower Old Red Sandstone of England and Wales: a record of post-Iapetan flexure or Early Devonian transtension? *Geological Magazine*, 140, 627–47.

Soper, N.J., Strachan, R.A., Holdsworth, R.E., Gayer, R.A. and Greiling, R.O. (1992) Sinistral transpression and the Silurian closure of Iapetus. *Journal of the Geological Society of London*, 149, 871–80.

Squirrell, H.C. and Downing, R.A. (1969) *Geology of the South Wales Coalfield: Part I: The Country Around Newport (Mon.)*, 3rd edn, Memoir of the Geological Survey of Great Britain, Sheet 249 (England and Wales), HMSO, London, 333 pp.

Squirrell, H.C. and White, D.E. (1978) Stratigraphy of the Silurian and Old Red Sandstone of the Cennen Valley and adjacent areas, south-east Dyfed, Wales. *Report of the Institute of Geological Sciences*, 78/6, 45 pp.

Stamp, L.D. (1923) The base of the Devonian, with special reference to the Welsh Borderlands. *Geological Magazine*, 60, 276–82, 331–6, 367–72, 385–410.

Steel, R.J. (1974) Cornstone (fossil caliche) – its origin, stratigraphic, and sedimentological importance in the New Red Sandstone, Western Scotland. *Journal of Geology*, 82, 351–69.

Stephenson, D. (1972) Middle Old Red Sandstone alluvial fan and talus deposits at Foyers, Inverness-shire. *Scottish Journal of Geology*, 8, 121–7.

Stephenson, D. (1977) Intermontane basin deposits associated with an early Great Glen feature in the Old Red Sandstone of Inverness-shire. In *The Moray Firth Area – Geological Studies* (ed. G. Gill), Inverness Field Club, Inverness, pp. 35–45.

Stephenson, D. (1999) Pettico Wick to St Abb's Harbour. In *Caledonian Igneous Rocks of Great Britain* (D. Stephenson, R.E. Bevins, D. Millward, A.J. Highton, I. Parsons, P. Stone and W.J. Wadsworth), Geological Conservation Review Series, No. 17, Joint Nature Conservation Committee, Peterborough, pp. 552–6.

Stephenson, D. and Gould, D. (1995) *British Regional Geology: The Grampian Highlands*, HMSO, London, for the British Geological Survey, 261 pp.

Stephenson, D., Bevins, R.E., Millward, D., Highton, A.J., Parsons, I., Stone, P. and Wadsworth, W.J. (1999) *Caledonian Igneous Rocks*, Geological Conservation Review Series, No. 17, Joint Nature Conservation Committee, Peterborough, 648 pp.

Stephenson, D., Loughlin, S.C., Waters, C.N. and Williamson, I.T. (2003) *Carboniferous and Permian Igneous Rocks of Great Britain north of the Variscan Front*, Geological Conservation Review Series, No. 27, Joint Nature Conservation Committee, Peterborough, 374 pp.

Strahan, A. and Gibson, W. (1900) *The Geology of the South Wales Coalfield: Part II: The Country Around Abergavenny*, Memoir of the Geological Survey of Great Britain, Sheet 232 (England and Wales), HMSO, London, 145 pp.

Strahan, A.H., Gibson, W. and Cantrill, T.C. (1904) *The Geology of the South Wales Coalfield: Part V: The country Around Merthyr Tydfil*, Memoir of the Geological Survey of Great Britain, Sheet 231 (England and Wales), HMSO, London, 132 pp.

Strahan, A., Cantrill, T.C., Dixon, E.E.L., and Thomas, H.H. (1907) *The Geology of the South Wales Coalfield: Part VII: The Country Around Ammanford*, Memoir of the Geological Survey of Great Britain, Sheet 230 (England and Wales), HMSO, London, 246 pp.

Strahan, A., Cantrill, T.C., Dixon, E.E.L. and Thomas, H.H. (1909) *The Geology of the South Wales Coalfield: Part X: The Country Around Carmarthen*, Memoir of the Geological Survey of Great Britain, Sheet 229 (England and Wales), HMSO, London, 177 pp.

Straw, S.H. (1930) The Siluro–Devonian boundary in south-central Wales. *Journal of the Manchester Geological Association*, 1 (for 1929), 79–102.

References

Streel, M., Loboziak, S., Steemans, P. and Bultynck, P. (2000) Devonian miospore stratigraphy and correlation with the global stratotype sections and points. *Courier Forschungsinstitut Senckenberg*, **220**, 9–23.

Strong, G.E. (1995) Petrography of Permo-Triassic and Devonian rock specimens from the Minehead area, Somerset. *British Geological Survey, Mineralogy and Petrology Series, Technical Report*, **WG/95/10**.

Sweet, I.P. (1985) Sedimentology of the Lower Old Red Sandstone near New Aberdour, Grampian Region. *Scottish Journal of Geology*, **21**, 239–59.

Syba, E. (1989) The sedimentation and provenance of the Lower Old Red Sandstone Greywacke Conglomerate, Southern Midland Valley, Scotland. Unpublished PhD thesis, University of Glasgow.

Tarrant, P.R. (1991) The ostracoderm *Phialaspis* from the Lower Devonian of the Welsh Borderland and South Wales. *Palaeontology*, **34**, 399–438.

Taylor, K. (1972) New fossiliferous localities in the Upper Old Red Sandstone of the Ystradfellte–Cwm Taff district of Breconshire. *Bulletin of the Geological Survey of Great Britain*, **38**, 11–14.

Taylor, K. and Thomas, L.P. (1974) Field meeting in the Upper Old Red Sandstone of south Breconshire. *Proceedings of the Geologists' Association*, **85**, 423–32.

Taylor, K. and Thomas, L.P. (1975) Geological Survey boreholes in south Breconshire and their bearing on the stratigraphy of the Upper Old Red Sandstone and the Carboniferous–Old Red Sandstone boundary east of the Afon Hepste. *Bulletin of the Geological Survey of Great Britain*, **54**, 1–39.

Taylor, T.N., Remy, W. and Hass, H. (1992a) Fungi from the Lower Devonian Rhynie Chert: Chytridiomycetes. *American Journal of Botany*, **79**, 1233–41.

Taylor, T.N., Hass, H. and Remy, W. (1992b) Devonian fungi: interactions with the green alga *Palaeonitella*. *Mycologia*, **84**, 901–10.

Taylor, T.N., Remy, W., Hass, H. and Kerp, H. (1995) Fossil arbuscular mycorrhizae from the Early Devonian. *Mycologia*, **87**, 560–73.

Taylor, T.N., Hass, H. and Kerp, H. (1997) A cyanolichen from the Lower Devonian Rhynie Chert. *American Journal of Botany*, **84**, 992–1004.

Taylor, T.N., Hass, H. and Kerp, H. (1999) The oldest fossil ascomycetes. *Nature (London)*, **399**, 648.

Thirlwall, M.F. (1979) The petrochemistry of the British Old Red Sandstone volcanic province. Unpublished PhD thesis, University of Edinburgh.

Thirwall, M.F. (1981) Implications for Caledonian plate tectonic models of chemical data from volcanic rocks of the British Old Red Sandstone. *Journal of the Geological Society of London*, **138**, 123–38.

Thirlwall, M.F. (1983) Isotope geochemistry and origin of calc-alkaline lavas from a Caledonian continental margin volcanic arc. *Journal of Volcanology and Geothermal Research*, **18**, 589–631.

Thirlwall, M.F. (1988) Geochronology of Late Caledonian magmatism in northern Britain. *Journal of the Geological Society of London*, **145**, 951–67.

Thirlwall, M.F. (1989) Movement on proposed terrane boundaries in northern Britain: constraints from Ordovician–Devonian igneous rocks. *Journal of the Geological Society of London*, **146**, 373–6.

Thomas, R.G. (1978) The stratigraphy, palynology and sedimentology of the Lower Old Red Sandstone Cosheston Group, S.W. Dyfed, Wales. Unpublished PhD thesis, University of Bristol, 552 pp.

Tomkeieff, S.I. (1953) "Hutton's Unconformity", Isle of Arran. *Geological Magazine*, **90**, 404–8.

Torsvik, T.H., Smethurst, M.A., Meert, J.G., Van der Voo, R., Trench, A., Abrahamsen, N. and Halvorsen, E. (1992) Baltica: a synopsis of Vendian–Permian palaeomagnetic data and their palaeotectonic implications. *Earth Science Reviews*, **40**, 133–52.

Torsvik, T.H., Smethurst, M.A., Meert, J.G., Van der Voo, R. and McKerrow, W.S. (1996) Continental break-up and collision in the Neoproterozoic and Palaeozoic – a tale of Baltica and Laurentia. *Earth Science Reviews*, **33**, 133–52.

Trench, A. and Haughton, P.D.W. (1990) Palaeomagnetic and geochemical evaluation of a terrane-linking ignimbrite: evidence for the relative position of the Grampian and Midland Valley terranes in late Silurian times. *Geological Magazine*, **127**, 241–357.

Trench, A. and Torsvik, T.H. (1992) The closure of the Iapetus Ocean and Tornquist Sea; new palaeomagnetic constraints. *Journal of the Geological Society of London*, **149**, 867–70.

References

Trewin, N.H. (1976) Correlation of the Achanarras and Sandwick Fish Beds, Middle Old Red Sandstone, Scotland. *Scottish Journal of Geology*, **12**, 205–8.

Trewin, N.H. (1986) Palaeoecology and sedimentology of the Achanarras fish bed of the Middle Old Red Sandstone, Scotland. *Transactions of the Royal Society of Edinburgh: Earth Sciences*, **77**, 21–46.

Trewin, N.H. (1987a) Pennan, unconformity within the Old Red Sandstone. In *Excursion Guide to the Geology of the Aberdeen Area* (eds N.H. Trewin, B.C. Kneller and C. Gillen), Scottish Academic Press, Edinburgh pp. 127–30.

Trewin, N.H. (1987b) Crawton: lavas and conglomerates of the Lower ORS. In *Excursion Guide to the Geology of the Aberdeen Area* (eds N.H. Trewin, B.C. Kneller and C. Gillen), Scottish Academic Press, Edinburgh, pp. 259–64.

Trewin, N.H. (1987c) Devonian of St Cyrus and Milton Ness. In *Excursion Guide to the Geology of the Aberdeen Area*, (eds N.H. Trewin, B.C. Kneller and C. Gillen), Scottish Academic Press, Edinburgh, pp. 251–8.

Trewin, N.H. (1992) 'Subaqueous shrinkage cracks' in the Devonian of Scotland re-interpreted – Discussion. *Journal of Sedimentary Petrology*, **62**, 921–2.

Trewin, N.H. (1993) The Old Red Sandstone of Caithness. In *Excursion Guide to the Geology of East Sutherland and Caithness* (eds N.H. Trewin and A. Hurst), Scottish Academic Press, Edinburgh, pp. 123–66.

Trewin, N.H. (1994) Depositional environment and preservation of biota in the Lower Devonian hot-springs of Rhynie, Aberdeenshire, Scotland. *Transactions of the Royal Society of Edinburgh: Earth Sciences*, **84**, 433–42.

Trewin, N.H. (1996) The Rhynie Cherts: an early Devonian ecosystem preserved by hydrothermal activity. In *Evolution of Hydrothermal Ecosystems on Earth (and Mars?)* (eds G.R. Bock and J.A. Goode), Ciba Foundation Symposium, No. 202, Wiley, Chichester, pp. 131–49.

Trewin, N.H. and Davidson, R.G. (1996) An early Devonian lake and its associated biota in the Midland Valley of Scotland. *Transactions of the Royal Society of Edinburgh: Earth Sciences*, **86**, 233–46.

Trewin, N.H. and Davidson, R.G. (1999) Lake-level changes, sedimentation and faunas in a Middle Devonian basin-margin fish bed. *Journal of the Geological Society of London*, **156**, 535–48.

Trewin, N.H. and Kneller, B.C. (1987a) Old Red Sandstone and Dalradian of Gamrie Bay. In *Excursion Guide to the Geology of the Aberdeen Area* (eds N.H. Trewin, B.C. Kneller and C. Gillen), Scottish Academic Press, Edinburgh, pp. 113–26.

Trewin, N.H. and Kneller, B.C. (1987b) The Lower Old Red Sandstone of New Aberdour. In *Excursion Guide to the Geology of the Aberdeen Area* (eds N.H. Trewin, B.C. Kneller and C. Gillen), Scottish Academic Press, Edinburgh, pp. 131–41.

Trewin, N.H. and Kneller, B.C. (1987c) Old Red Sandstone and Dalradian of Quarry Haven. In *Excursion Guide to the Geology of the Aberdeen Area* (eds N.H. Trewin, B.C. Kneller and C. Gillen), Scottish Academic Press, Edinburgh, pp. 143–7.

Trewin, N.H. and Knoll, A.H. (1999) Preservation of Devonian chemotrophic filamentous bacteria in calcite veins. *Palaios*, **14**, 288–94.

Trewin, N.H. and Rice, C.M. (1992) Stratigraphy and sedimentology of the Devonian Rhynie chert locality. *Scottish Journal of Geology*, **28**, 37–47.

Trewin, N.H. and Thirlwall, M.F. (2002) Old Red Sandstone. In *The Geology of Scotland*, 4th edn (ed. N.H. Trewin), The Geological Society, London, pp. 213–49.

Trewin, N.H. and Wilson, E. (2004) Correlation of the Early Devonian Rhynie chert beds between three boreholes at Rhynie, Aberdeenshire. *Scottish Journal of Geology*, **40**, 73–81.

Trewin, N.H., Fayers, S.R. and Kelman, R. (2003) Subaqueous silicification of the contents of small ponds in an Early Devonian hot-spring complex, Rhynie, Scotland. *Canadian Journal of Earth Sciences*, **40**, 1697–712.

Trotter, F.M. (1942) *Geology of the Forest of Dean Coal and Iron Ore Field*, Memoir of the Geological Survey of Great Britain, HMSO, London, 95 pp.

Tunbridge, I.P. (1978) The sedimentology of late Lower and early Middle Devonian rocks of the Bristol Channel region, southern England. Unpublished PhD thesis, University of Reading.

Tunbridge, I.P. (1980a) Possible Devonian uplift on the Swansea Valley Fault, Wales. *Geological Magazine*, **117**, 497–8.

References

Tunbridge, I.P. (1980b) The Yes Tor Member of the Hangman Sandstone Group (north Devon). *Proceedings of the Ussher Society*, **5**, 7–12.

Tunbridge, I.P. (1981a) Old Red Sandstone sedimentation – an example from the Brownstones (highest Lower Old Red Sandstone) of south central Wales. *Geological Journal*, **16**, 111–24.

Tunbridge, I.P. (1981b) Sandy high-energy flood sedimentation – some criteria for recognition, with an example from the Devonian of S.W. England. *Sedimentary Geology*, **28**, 79–95.

Tunbridge, I.P. (1983a) The Middle Devonian shoreline in north Devon, England. *Journal of the Geological Society of London*, **140**, 147–58.

Tunbridge, I.P. (1983b). The Bristol Channel landmass (abstract). *Proceedings of the Ussher Society*, **5**, 490.

Tunbridge, I.P. (1984) Facies model for a sandy ephemeral stream and clay playa complex; the Middle Devonian Trentishoe Formation of north Devon, U.K. *Sedimentology*, **31**, 697–715.

Tunbridge, I.P. (1986) Mid-Devonian tectonics and sedimentation in the Bristol Channel. *Journal of the Geological Society of London*, **143**, 107–15.

Turner, S. (1973) Siluro-Devonian thelodonts from the Welsh Borderland. *Journal of the Geological Society of London*, **129**, 557–84.

Turner, S., Vergoossen, J.M.J. and Williams, R.B. (1995) Early Devonian microvertebrates from Pwll-Y-Wrach, Talgarth, South Wales. *Geobios*, **19**, 377–82.

Tyrrell, G.W. (1928) *The Geology of Arran*, Memoir of the Geological Survey of Great Britain, Arran Map including parts of sheets 13 and 21 (Scotland), HMSO, Edinburgh, 292 pp.

Utting, J. and Neves, R. (1970) Palynology of the Lower Limestone Shale Group (basal Carboniferous Limestone Series) and Portishead Beds (Upper Old Red Sandstone) of the Avon Gorge, Bristol, England. In *Colloque sur la Stratigraphie du Carbonifère* (eds M. Streel and R.H. Wagner), *Les Congrès et Colloques de l'Université de Liège*, No. **55**, Université de Liège, Liège, pp. 411–22.

Vergoossen, J.M.J. (2000) Acanthodian and chondrichthyan microremains in the Siluro-Devonian of the Welsh Borderland, Great Britain. *Courier Forschungsinstitut Senckenberg*, **223**, 175–99.

Wadge, A.J. (1978) Devonian. In *The Geology of the Lake District* (ed. F. Moseley), *Yorkshire Geological Society Occasional Publication*, No. 3, Yorkshire Geological Society, Leeds, pp.164–7.

Walker, T.R. (1967) Formation of red beds in modern and ancient sediments. *Bulletin of the Geological Society of America*, **78**, 353–68.

Wallace, T. (1880) The structural geology of Strathnairn. *Transactions of the Edinburgh Geological Society*, **3**, 204–19.

Wallis, F.S. (1927) Notes on sections of Old Red Sandstone in the Bristol district. *Proceedings of the Bristol Naturalists' Society*, **6**, 400–5.

Wallis, F.S. (1928) The Old Red Sandstone of the Bristol District. *Quarterly Journal of the Geological Society of London*, **83**, 760–89.

Walmsley, V.G. (1962) Upper Silurian–Devonian contacts in the Welsh Borderland and South Wales. In *Symposiums-Band: 2: Internationalen Arbeitstagung über die Silur/Devon-Grenze und die Stratigraphie von Silur und Devon, Bonn-Bruxelles 1960* (H.K. von Erben), E. Schweizerbart'sche, Stuttgart, pp. 288–95.

Walmsley, V.G. and Bassett, M.G. (1976) Biostratigraphy and correlation of the Coralliferous Group (Silurian) of Pembrokeshire, Wales. *Proceedings of the Geologists' Association*, **87**, 191–220.

Ward, J.C. (1876) *The Geology of the Northern Part of the English Lake District*, Memoir of the Geological Survey of the United Kingdom, Quarter sheet 101SE, HMSO, London, 132 pp.

Waters, R.A. and Lawrence, D.J.D. (1987) *Geology of the South Wales Coalfield: Part III: The Country Around Cardiff*, 3rd edn, Memoir of the British Geological Survey, Sheet 263 (England and Wales), HMSO, London, 114 pp.

Watson, D.M.S. (1935) Fossil fishes of the Orcadian Old Red Sandstone. In *The Geology of the Orkneys* (G.V. Wilson, W. Edwards, J. Knox, R.C.B. Jones and J.V. Stephens), Memoir of the Geological Survey of Great Britain, sheets 117 to 122 (Scotland), HMSO, Edinburgh, pp. 157–69.

Webby, B.D. (1965a) The Middle Devonian marine transgression in north Devon and west Somerset. *Geological Magazine*, **102**, 478–88.

References

Webby, B.D. (1965b) The stratigraphy and structure of the Devonian rocks in the Brendon Hills, west Somerset. *Proceedings of the Geologists' Association*, **76**, 39–60.

Webby, B.D. (1966a) Middle–Upper Devonian paleogeography of north Devon and west Somerset, England. *Palaeogeography, Palaeoclimatology, Palaeoecology*, **2**, 27–46.

Webby, B.D. (1966b) The stratigraphy and structure of the Devonian rocks in the Quantock Hills, west Somerset. *Proceedings of the Geologists' Association*, **76**, 321–43.

Welch, F.B.A. and Trotter, F.M. (1961) *Geology of the Country Around Monmouth and Chepstow*, Memoir of the Geological Survey of Great Britain, sheets 233 and 250 (England and Wales), HMSO, London, 142 pp.

Wellman, C.H. (1993) A land plant microfossil assemblage of Mid Silurian age from the Stonehaven Group, Scotland. *Journal of Micropalaeontology*, **12**, 47–66.

Wellman, C.H. (1994) Palynology of the 'Lower Old Red Sandstone' at Glen Coe, Scotland. *Geological Magazine*, **131**, 563–6.

Wellman, C.H. (2004) Palaeoecology and palaeophytogeography of the Rhynie chert plants: evidence from integrated analysis of *in situ* and dispersed spores. *Proceedings of the Royal Society of London, Series B*, **271**, 985–92.

Wellman, C.H., Thomas, R.G., Edwards, D. and Kenrick, P. (1998) The Cosheston Group (Lower Old Red Sandstone) in southwest Wales: age, correlation and palaeobotanical significance. *Geological Magazine*, **135**, 397–412.

Wellman, C.H., Habgood, K., Jenkins, G. and Richardson, J.B. (2000) A new plant assemblage (microfossil and megafossil) from the Lower Old Red sandstone of the Anglo-Welsh Basin: its implications for the palaeoecology of early terrestrial ecosystems. *Review of Palaeobotany and Palynology*, **109**, 161–96.

Westoll, T.S. (1948) The vertebrate-bearing strata of Scotland. *18th International Geological Congress, London, Abstracts*, **18**, 74–5.

Westoll, T.S. (1951) A new cephalaspid fish from the Downtonian of Scotland, with notes on the structure and classification of ostracoderms. *Transactions of the Royal Society of Edinburgh*, **61**, 341–57.

Westoll, T.S. (1977) Northern Britain. In *A Correlation of Devonian Rocks of the British Isles* (eds M.R. House, J.B. Richardson, W.G. Chaloner, J.R.L. Allen, C.H. Holland, and T.S. Westoll), *Geological Society of London Special Report*, No. **8**, Scottish Academic Press, Edinburgh, pp. 66–92.

White, D.E. and Lawson, J.D. (1989) The Přídolí Series in the Welsh Borderland and south-central Wales. In *A Global Standard for the Silurian System* (eds C.H. Holland and M.G. Bassett), National Museum of Wales Geological Series, No. **9**, National Museum of Wales, Cardiff, pp. 131–41.

White, E.I. (1935) The ostracoderm *Pteraspis* Kner and the relationships of the agnathous vertebrates. *Philisophical Transactions of the Royal Society, London*, **B225**, 381–457.

White, E.I. (1938) New pteraspids from South Wales. *Quarterly Journal of the Geological Society of London*, **94**, 85–155.

White, E.I. (1946) The Genus *Phialaspis* and the Psammosteus Limestones. *Quarterly Journal of the Geological Society of London*, **101**, 207–42.

White, E.I. (1950) The vertebrate faunas of the Lower Old Red Sandstone of the Welsh Borders. *Bulletin of the British Museum (Natural History)*, **A1**, 57–67.

Whitehead, T.H. and Pocock, R.W. (1947) *Dudley and Bridgnorth*, Memoir of the Geological Survey of Great Britain, Sheet 167 (England and Wales), HMSO, London, 226 pp.

Williams, B.P.J. (1964) The stratigraphy, petrology and sedimentation of the Ridgeway Conglomerate and associated formations in south Pembrokeshire. Unpublished PhD thesis, University of Wales (Swansea).

Williams, B.P.J. (1971) Sedimentary features of the Old Red Sandstone and Lower Limestone Shales of south Pembrokeshire, south of the Ritec Fault. In *Geological Excursions in South Wales and the Forest of Dean* (eds D.A. Bassett and M.G. Bassett), The Geologists' Association South Wales Group, Cardiff, pp. 222–39.

Williams, B.P.J. (1978) The Old Red Sandstone of the Welsh Borderland and South Wales. In *A Field Guide to Selected Outcrop Areas of the Devonian of Scotland, the Welsh Borderland and South Wales: International Symposium on the Devonian System (P.A.D.S. 78)* (eds P.F. Friend and B.P.J. Williams), Palaeontological Association, London, pp. 55–106.

Williams, B.P.J. and Hancock, P.L. (1977) The sedimentology and structure of the Upper Palaeozoic rocks at Portishead. In *Geological Excursions in the Bristol District* (ed. R.J.G. Savage), University of Bristol, Bristol, pp. 29–46.

References

Williams, B.P.J. and Hillier, R.D. (2004) Variable alluvial sandstone architecture within the Lower Old Red Sandstone, southwest Wales. *Geological Journal*, **39**, 257–75.

Williams, B.P.J., Allen, J.R.L. and Marshall, J.D. (1982) Old Red Sandstone facies of the Pembroke Peninsula, south of the Ritec Fault. In *Geological Excursions in Dyfed, South-west Wales* (ed. M.G. Bassett), National Museum of Wales, Cardiff, for the Geologists' Association South Wales Group, pp. 151–74.

Williams, E.A., Friend, P.A. and Williams, B.P.J. (2000) A review of Devonian time scales: databases, construction and new data. In *New Perspectives on the Old Red Sandstone* (eds P.F. Friend and B.P.J. Williams), *Geological Society Special Publication*, No. **180**, Geological Society, London, pp. 1–21.

Wills, L.J. (1948) *The Palaeogeography of the Midlands*, University Press of Liverpool, Liverpool, 147 pp.

Wills, L.J. (1950) *The Palaeogeography of the Midlands*, 2nd edn, University Press of Liverpool, Liverpool, 147 pp.

Wills, L.J. (1952) *A Palaeogeographical Atlas of the British Isles and Adjacent Parts of Europe*, Reprint, Blackie and Son, London, 64 pp.

Wilson, H.M. and Anderson, L.I. (2004) Morphology and taxonomy of Paleozic millipedes (Diplopoda: Chilognatha: Archipolypoda) from Scotland. *Journal of Palaeontology*, **78**, 169–84.

Wilson, G.V., Edwards, W., Knox, J., Jones, R.C.B. and Stephens, J.V. (1935) *The Geology of the Orkneys*, Memoir of the Geological Survey of Great Britain, sheets 117 to 122 (Scotland), HMSO, Edinburgh, 205 pp.

Wood, S. and Norman, D. (1991) Tynet Burn: on the trail of Orcadian fish. *Earth Science Conservation*, **29**, 15–17.

Woodcock, N.H. (2000a) Devonian sedimentation and volcanism of the Old Red Sandstone continent. In *Geological History of Britain and Ireland* (eds N.H. Woodcock and R. Strachan), Blackwell Science, Oxford, pp. 207–23.

Woodcock, N.H. (2000b) Introduction to the Silurian. In *British Silurian Stratigraphy* (R.J. Aldridge, David J. Siveter, Derek J. Siveter, P.D. Lane, D. Palmer and N.H. Woodcock), Geological Conservation Review Series, No. 19, Joint Nature Conservation Committee, Peterborough, pp. 1–22.

Woodcock, N.H. and Gibbons, W. (1988) Is the Welsh Borderlands Fault System a terrane boundary? *Journal of the Geological Society of London*, **145**, 915–23.

Woodcock, N.H. and Strachan, R. (2000) *Geological History of Britain and Ireland*, Blackwell Science, Oxford, 423 pp.

Wright, V.P. and Marriott, S.B. (1996) A quantitative approach to soil occurrence in alluvial deposits and its application to the Old Red Sandstone of Britain. *Journal of the Geological Society of London*, **153**, 907–13.

Yang, C.X. (1997) Quaternary sedimentation, parna, landforms, and soil landscapes of the Wagga Wagga 1:100 000 map sheet, southeastern Australia. *Australian Journal of Soil Research*, **35**, 643–68.

Young, V.T. (1995) Micro-remains from Early and Middle Devonian acanthodian fishes from the U.K. and their biostratigraphic possibilities. *Ichthyolith Issues, Special Publication*, **1**, 65–8.

Glossary

This glossary provides brief explanations of the technical terms used in the introductions to the chapters and in the 'conclusions' sections of the site reports. These explanations are not rigorous scientific definitions, but are intended to help the general reader. Detailed stratigraphical terms are omitted as they are given context within the tables and figures. Words in **bold** type indicate an internal reference to another glossary entry.

Abrasion: the process of wearing away parts of **fossils** or rocks by **sediment**-laden water or air. The process produces an increasingly smoothed and rounded outline shape.

Acadian Orogeny: the Early to **Mid-Devonian** phase of mountain building (late Caledonian) along a collision zone between **Avalonia** and **Laurentia**, after the **subduction** of the **Iapetus Ocean** and the production of a range of mountains stretching south-westwards from Scandinavia, through northern Britain and Ireland, Greenland and North America.

Acanthodian: any member of the extinct (**Silurian–Permian**) **class** Acanthodii (**phylum** Chordata). These primitive jawed fish, the so-called 'spiny sharks' with spine-supported fins and a covering of small scales, occupied both marine and fresh waters.

Accretion: the build up of **sediment** by physical processes such as water and wind.

Acme: the point of maximum sea or lake level (**highstand**) during a marine or **lacustrine transgression**.

Adhesion wart: a small, irregular, wart-like **sand** acccumulation formed by wind that changes direction rapidly whilst blowing over a moist sand surface.

Aeolian: descriptive of **sediments** carried and deposited by the wind.

Age: a geological time unit (cf. **chronostratigraphy**), usually taken to be the smallest standard division of geological time.

Agglomerate: a **pyroclastic** rock with predominantly rounded **clasts** greater than 64 mm in diameter.

Aggradation: the building upwards of a river valley or floodplain by accumulation of **fluvial** deposits; can also be applied to material deposited by other agencies, such as wind or sea.

Aggregate: a mass or body of rock fragments and/or mineral grains.

Agnathan: any member of the **class** Agnatha (**phylum** Chordata). Primitive jawless vertebrates which have existed since the Cambrian Period, they include a large number of extinct marine and freshwater groups, but are now reduced to two: hagfishes and lampreys.

Algae (sing. **alga**): a large and diverse division of the plant kingdom, consisting of mainly aquatic organisms. Simple plants that have no true stems, roots or leaves, they contain chlorophyll and therefore can **photosynthesize**. They range from microscopic single cells to very large multi-cellular structures.

Allochthonous: descriptive of **fossils** or rocks that lived or formed elsewhere to their current position.

Glossary

Alluvial: a term applied to the environments, action and products of rivers or streams. Alluvial deposits are composed of **clastic** material deposited in river channels and floodplains.

Alluvial fan: a cone-shaped deposit of water-laid deposits at the confluence of a tributary stream with the main river.

Alluvium: **sediment** deposited by rivers.

Ammonoid: any member of the extinct subclass Ammonoidea (**class** Cephalopoda). Important **zone fossils** for the **Palaeozoic** and Mesozoic **eras**.

Anastomosing: descriptive of a system which branches or contains a network; for example the channel pattern of a **braided river**, or the veins on a leaf which form a netlike pattern.

Andesite (adj. **andesitic**): a fine-grained volcanic (**extrusive**) rock containing more than 53 wt% SiO_2. Intermediate in composition between a **basalt** and a **rhyolite**.

Anglo-Welsh Basin: a Devonian sedimentary basin that occupied a large part of what is now Wales and England, in which the **Old Red Sandstone** was deposited.

Anoxic: literally 'without oxygen'; often used to describe an anaerobic environment.

Antecedent: of a river or drainage system controlled by a pre-exisiting system.

Anticline: an arch-shaped upfold of rocks produced by **tectonic** activity with younger **strata** on the outermost part of the arch and older rock in the core (cf. **syncline**).

Arenite (adj. **arenaceous**): a general term for a detrital, **clastic sedimentary rock** made of **sand**-sized particles.

Argillite (adj. **argillaceous**): a general term for a fine-grained, **clay**-rich, **clastic sedimentary rock**.

Arkose (adj. **arkosic**): an **arenaceous** rock that contains at least 25% **feldspar**.

Arthropod: any member of the **phylum** Arthropoda, the largest and most diverse phylum of the animal kingdom. These invertebrate animals are characterized by a segmented body and paired antennae, wings or legs. Examples include insects, **crustaceans** and arachnids.

Ash (volcanic): unconsolidated deposit consisting of **pyroclastic** material (glass shards, crystals etc.) less than 2 mm in size. In consolidated rocks the term is commonly used to denote the size of individual **volcaniclastic** fragments (e.g. coarse-ash grains and fine-ash grains).

Autochthonous: descriptive of **fossils** or rocks that lived or formed in their current positions.

Avalonia: a small early **Palaeozoic** crustal plate consisting of parts of the maritime states of North America, England, Wales, south-east Ireland and part of western Europe, which split from **Gondwana** early in **Ordovician** times and moved northwards, colliding with **Laurentia** during the **Silurian Period**.

Avulsion: the process when **aggradation** on a floodplain diverts a river channel to a new course, at a lower elevation on the floodplain.

Axial drainage: a river drainage system that flows along the axis of a valley or **basin**.

Ball-and-pillow structure: a sedimentary structure with a hemispherical shape resembling balls and pillows. Caused by **sand** sinking into fluidized muddy **sediment**.

Baltica: an early **Palaeozoic** crustal plate - consisting of much of present-day northwestern Europe, including Scandinavia, European Russia and parts of Central Europe. The plate formed the south-eastern continental margin of the **Iapetus Ocean** and amalgamated with **Avalonia** before moving northwards from Late **Ordovician** times and colliding with **Laurentia** to form the Caledonian mountain belt.

Bar: a lenticular deposit, usually **sandy**, formed in and at the margins of a river channel, or offshore of a beach.

Barrier deposits: **sand** deposits that accumulate in shallow, offshore waters by wave action.

Back-barrier deposits: shallow-water deposits formed on the landward side of a barrier.

Basalt (adj. **basaltic**): a fine-grained, usually dark-coloured, basic, volcanic (**extrusive**) **igneous rock**. It usually occurs as a **lava** or **dyke**.

Basement: the oldest rocks recognized in a given area; a complex of **metamorphic** and/or **igneous rocks** that underlies all the sedimentary **formations**.

Basin: an area of **subsidence**, or depression, usually of considerable size, in which **sediments** and/or volcanic **strata** accumulate.

Basin-fill: the sedimentary succession deposited in a **basin**.

Glossary

Basin inversion: the **tectonic** process in which **basin-fill** is **uplifted**.

Batholith: a very large (over 100 km^2) igneous **intrusion**, most commonly **granite**, that extends to great (unknown) depth in the Earth's crust.

Bed: in **lithostratigraphy**, a subdivision of either a **member** or a **formation**; the smallest unit within the scheme of formal lithostratigraphical classification. Also used informally to indicate a **stratum** within a **sedimentary rock** succession.

Bedding plane: a planar feature in **sedimentary rocks** representing an original surface of deposition. Conspicuous bedding planes may indicate a short interruption in, or change in character of, **sediment** deposition.

Beyrichiacea: named after Beyrich, a German palaeontologist, an extinct group (late Ordovician?–early Carboniferous) of marine **ostracods**, which have **biostratigraphical** use.

Bioclast (adj. **bioclastic**): a **sediment** grain consisting of comminuted **fossil** remains.

Biofacies: a **facies** defined by its characteristic **fossil** assemblage, and reflecting a specific set of environmental conditions.

Biogenic: produced by living organisms or biological processes.

Biostratigraphy: the **stratigraphical** subdivision, classification and **correlation** of **sedimentary rocks** based on their **fossil** content.

Biota: the **flora** and **fauna** of a particular place; or the faunal and floral assemblage of a **bed** or other **stratigraphical** unit.

Bioturbation: the physical disturbance of unconsolidated **sediment**, such as by burrowing and feeding, caused by the organisms living on or in it. These disturbances are often preserved as **trace fossils** in ancient **sediments**.

Biozone: in **biostratigraphy**, a restricted unit of **sedimentary rock** defined by its **fossil** content, most usefully by **species** of narrowly defined temporal, but wide spatial, range, and named after one or more abundant or characteristic species.

Bivalve: any member of the **order** Bivalvia (**phylum** Mollusca (*see* **mollusc**)). These marine invertebrates are characterized by bodies enclosed in two, hinged, often mirror-image, shells (valves). Modern examples include cockles and mussels.

Boudinage: a structure in strongly deformed **sedimentary** or **metamorphic rocks**, in which an original competent layer between less competent layers has been stretched, thinned and broken at regular intervals into bodies resembling boudins or sausages.

Brachiopod: any member of the **phylum** Brachiopoda. These marine invertebrates are superficially similar to **bivalves** but with a different anatomy and two hinged shells that are typically dissimilar.

Braided river: a network of converging and diverging (**anastomosing**) streams.

Braidplain: the floodplain of a **braided river** system.

Breccia: a rock composed of angular broken fragments greater than 2 mm in diameter; can be **pyroclastic**, sedimentary or **fault**-related.

Breconian Stage: a local **chronostratigraphical** division of the **Old Red Sandstone** of the **Devonian Period** for the **Anglo-Welsh Basin**. Roughly equivalent to the more widely known **Emsian** and **Pragian stages**, it follows the **Dittonian Stage**.

Bryozoan: any member of the **phylum** Bryozoa. These very small, moss-like aquatic organisms often form permanent colonies, linked by their box-like skeletons of **calcium carbonate**.

Calc-alkaline: of a suite of **igneous rocks** characterized chemically by the steady increase in iron content relative to silica during evolution of the magma.

Calcareous: containing large quantities, or composed, of **calcium carbonate**.

Calci-: prefix indicating containing/composed of **calcium carbonate**.

Calcite: the most common, rock-forming crystalline form of **calcium carbonate**; the main constituent of **limestone** and the shells of many **brachiopods**, echinoderms and other invertebrates.

Calcium carbonate ($CaCO_3$): a colourless or white crystal compound, which occurs naturally as **limestone**, marble and chalk.

Calcrete: a soil rich in **calcium carbonate**, indicative of arid or semi-arid environments. In the geological record, the term is used for a fossilized soil (**palaeosol**) rich in **carbonate** that is now a **limestone** or **dolomite**.

Calcretized: of a **sediment** that has been altered wholly or partly to **calcrete**.

Glossary

Caldera: a circular, **basin**-shaped depression, usually many times greater than the size of any individual volcanic vent, caused by collapse of the roof of an underlying magma chamber following an eruption; also refers to the underlying volcanic structure.

Caledonian Orogeny: a major period of mountain building that took place about 430 million years ago, associated with the closure of the ancient **Iapetus Ocean** that was situated between Scotland and the rest of present-day Britain.

Caledonides: the **Palaeozoic** mountain chain that extended in a NE–SW-direction from Spitsbergen, eastern Greenland, Scandinavia, Scotland, northern Ireland, the Lake District of England, Wales into eastern Canada and the USA, which resulted from the closure of the **Iapetus Ocean**.

Caliche: *see* **hardpan**.

Carbonaceous: containing carbon.

Carbonate: a mineral salt of carbonic acid, usually referring to the common sedimentary form of **calcium carbonate** in **limestones** and invertebrate shells, but also encompassing other minerals, notably **dolomite**.

Carboniferous Period: a geological time division (cf. **chronostratigraphy**). Ranging from 362 to 290 million years ago, it precedes the **Permian Period**.

Carious weathering: differential **weathering** that produces honeycomb outcrop.

Cataclasis: the deformation of rock by fracture and rotation of **aggregates** and mineral grains.

Cataclasite: a rock containing angular fragments formed by **cataclasis**.

Cement: the mineral 'glue' that holds particles together in **sedimentary rocks**.

Cementstone: **argillaceous limestone** and **dolostone**.

Cephalaspid: any member of the extinct **order** Cephalaspida (**class** Agnatha (*see* **agnathan**)) with both an exoskeleton and endoskeleton of bone. Also referred to as **Osteostracans**. Characterized by a solid bony head-shield and broad **cornual** processes.

Chalcedony: a variety of **quartz** that is composed of microscopic crystals or fibres.

Channelized: of **fluvial** deposits formed in a river channel.

Chert: microcrystalline silica (**quartz** and **chalcedony**), which may be of organic or inorganic origin. It occurs as layers or nodules in **sedimentary rocks** (mainly chalk and **limestone**).

Chlorite: a green mineral, an anhydrous silicate of magnesium and alumina.

Chloritize: the conversion of a mineral to **chlorite**.

Chronostratigraphy: the subdivision and **correlation** of rock units on the basis of relative age. The hierarchy of principal chronostratigraphical units to which layers of **sedimentary rock** are allocated through the study and interpretation of their **stratigraphy** is erathem, **system**, **series** and **stage**, which are related, respectively to the geological time units of **era**, **period**, **epoch** and **age**. Rocks of the Devonian System (a chronostratigraphical unit) were laid down in the **Devonian Period** (a geological time unit).

Class: a category used in the taxonomic classification of organisms, which consists of one or several related **orders**. Similar classes are grouped into a **phylum**.

Clast: (adj. **clastic**): a fragment of a pre-existing rock. *See also* **bioclast**.

Clay: an extremely fine-grained **sediment** (grain-size less than 0.004 mm) composed of so-called 'clay minerals'.

Claystone: indurated **clay**, consisting predominantly of fine material and clay minerals.

Cleavage: a plane of incipient parting in a rock, produced by the alignment of platy crystals such as mica in response to confining pressure during deformation.

Coeval: belonging to or formed at the same time.

Concretion: a rounded or irregular mass of mineral matter concentrated around a nucleus and formed during **diagenesis** in a **sedimentary rock**.

Condensed deposit: a thin deposit formed where there was reduced sedimentation and/or **subsidence**.

Conglomerate: a **sedimentary rock** consisting of pebbles (cf. **breccia**).

Conodont (Conodonta): an extinct group of small eel-like marine animals, characterized by assemblages of paired tooth-like structures made of bone-like material. These 'teeth' have considerable use in **biostratigraphy**.

Contemporaneous: formed or occurring at the same time.

Contiguous: touching, in contact.

Continental: of **sediments** formed in a **terrestrial**, non-marine environment.

Glossary

Cornstone: a **concretionary limestone** characteristic of arid **terrestrial** environments (synonymous with **calcrete**).

Cornua (-ates): a horn or horn-like projection.

Correlation: the tracing and identification of a **stratigraphical** unit away from its **type area** by comparing **lithologies** and/or **fauna**.

Crevasse splay: a small fan-shaped accumulation of **sediment** formed when a river breaks through its banks or **levee**.

Cross-stratification: subsidiary bedding surfaces oblique to the upper and lower bounding surfaces of a particular **stratum** and representing ripples or dunes formed in the **sediment** by water currents (or wind). Large-scale features are named 'cross-bedding', small-scale features are known as 'cross-lamination'.

Crossopterygian: in some systems of classification any member of a mainly extinct **order** (Crossopterygii) of lobefin fish of the subclass Sarcopterygii (see **sarcopterygian**), **class Osteichthyes**.

Crustacean: any member of the **class** Crustacea (**phylum** Arthropoda (see **arthropod**)). These animals typically have two pairs of antennae, a pair of mandibles and often many other appendages, and are mainly aquatic. Examples include lobsters, barnacles and wood lice.

Cryptic: unidentified, hidden or unseen.

Crustal extension: expansion of the Earth's crust by **tectonic** forces.

Cyclothem: a succession of sedimentary layers, representing a sequence of depositional events that tend to be repeated; the result of cyclical sedimentation.

Dalradian: a large tract of (mainly) late **Precambrian** metamorphosed **sedimentary rocks** in the Grampian Highlands of Scotland and the north of Ireland.

Debris flow: a deposit formed by the down-slope movement of water- and air-bourne material.

Delta (adj. deltaic): a tract of **sediment**, typically fan-shaped, deposited where a river enters a lake or the sea.

Denudation: the combined processes of **weathering** and **erosion** that wear down landscapes.

Depocentre: the centre of (greatest) deposition.

Desiccation crack: a crack formed when wet **sediment** dries out.

Devonian Period: a geological time division (cf. **chronostratigraphy**). Ranging from 418 to 362 million years ago, it precedes the **Carboniferous Period** and is part of the **Palaeozoic Era**.

Dextral: of lateral, right-handed movement along a geological **fault**.

Diachronous: descriptive of a **lithological** unit, or **contiguous** rock body, that was deposited at different times in different locations and therefore differs in age from place to place.

Diagenesis: (adj. **diagenetic**): the post-depositional changes in mineralogy and texture of **sediments** and organisms that combine to produce rocks and **fossils**. The term excludes **metamorphic** alteration.

Diamict: a poorly or non-**sorted** non-**calcareous** terrigenous **sedimentary rock** or unconsolidated **sediment** that contains a wide range or particle sizes.

Disconformity (adj. disconformable): a break in continuity of deposition, (**unconformity**), where the **beds** above and below are parallel and therefore show no angular discordance.

Dissolution: the natural process of dissolving a solid; also known as solution.

Distal (adj. distally): far from the source.

Distributary channel: a river channel along which **sediment** has been transported.

Dittonian Stage: a local **chronostratigraphical** division of the **Old Red Sandstone** of the **Devonian Period** for the **Anglo-Welsh Basin**. Roughly equivalent to the more widely known **Lochkovian Stage**, it follows the **Downtonian Stage** and precedes the **Breconian Stage**.

Dolerite: a **mafic**, **igneous rock** that generally occurs in **dykes** and **sills**.

Dolomicrite: a **sedimentary rock** consisting of **clay**-sized **dolomite** crystals.

Dolomite ($CaMg(CO_3)_2$): a white or colourless mineral with a structure similar to **calcite** but with some calcium replaced by magnesium.

Dolostone: a **limestone** whose **carbonate** fraction contains more that 50% **dolomite**.

Downtonian Stage: a local **chronostratigraphical** division of the **Old Red Sandstone** for the **Anglo-Welsh Basin**. Of **Silurian** age it is roughly equivalent to the **Přídolí Series**, and precedes the **Dittonian Stage**.

Dyke: a sheet of **igneous rock** that has 'intruded' or 'cut through' pre-existing rocks.

Glossary

Early Devonian Epoch: a geological time division (cf. **chronostratigraphy**) of the **Devonian Period**. Ranging from 418 to 394 million years ago it includes the Lochkovian, Pragian and Emsian ages.

Ecosystem: habitat or environment inhabited by a group of animals and/or plants.

Eifelian Stage: a **chronostratigraphical** division of the **Middle Devonian Series**, comprising the rocks deposited during the Eifelian Age. The fourth **stage** of the **Devonian Period** it is dated to approximately 394–387.5 Ma and is preceded by the **Emsian Stage** and followed by the **Givetian Stage**.

Emsian Stage: a **chronostratigraphical** division of the **Lower Devonian Series**, comprising the rocks deposited during the Emsian Age. The third **stage** of the **Devonian Period** it is dated to approximately 409.5–394 Ma and is preceded by the **Pragian Stage** and followed by the **Eifelian Stage**.

Ephemeral: short-lived, intermittent.

Epichnial cast: a **trace fossil** cast that stands proud of the **sediment** or **bedding plane** surface.

Epoch: a geological time unit (cf. **chronostratigraphy**), of shorter duration than a **period** and itself divisible into **ages** (e.g. the Late Triassic Epoch).

Epsilon cross-bedding: **cross-bedding** formed by **lateral accretion** on a migrating **point bar** in a meandering stream.

Equigranular: a texture in which all the crystals are approximately the same size.

Era: a major geological time unit (cf. **chronostratigraphy**), which is divided into **periods** (e.g. the **Palaeozoic Era**).

Erosion: the wearing away of the land's surface by mechanical processes such as the flow of water, ice or wind.

Euramerica: the continental mass of northwestern Europe and North America, formed when the **Iapetus Ocean** was subducted during the **Caledonian Orogeny**.

Eurypterid: any member of the extinct (**Ordovician–Permian**) **order** Eurypterida (**class** Merostomata, **phylum** Arthropoda (*see* **arthropod**)). These large aquatic organisms (up to 2 m in length) superficially resemble scorpions.

Eutrophic: a body of water with high levels of plant nutrients, with correspondingly high productivity.

Evaporite (adj. **evaporitic**): a **sediment** or mineral grown from a saline solution by evaporation of water, which may be marine or **continental** in origin.

Exotic (**extraformational**): a rock or block unrelated to the rocks with which it is now associated, which has been moved from its place of origin; or of a pebble or **clast** from outwith a depositional **basin**, or of a **conglomerate** composed of such pebbles or clasts.

Extramontane: outwith a mountain belt.

Extrusive: descriptive of **igneous rocks** that have been extruded onto the Earth's surface, rather than being intruded beneath the surface.

Facies: the sum total of a rock's **lithological** and gross **faunal/floral** characteristics that together reflect the particular environment in which it formed.

Famennian Stage: a **chronostratigraphical** division of the **Upper Devonian Series**, comprising the rocks deposited during the Famennian Age. The sixth and final **stage** of the **Devonian Period**, it is dated to approximately 376.5–362 Ma and is preceded by the **Frasnian Stage**.

Family: a category used in the taxonomic classification of organisms, which consists of one or several related **genera**. Similar families are grouped into an **order**.

Fanglomerate: a **sedimentary rock** containing angular rock fragments cemented in a finer-grained groundmass; formed of coarse material in an **alluvial fan**.

Farlovian Stage: a local **chronostratigraphical** division of the **Old Red Sandstone** of the **Devonian Period** for the **Anglo-Welsh Basin**. Roughly equivalent to the more widely known **Famennian Stage**, it extends into the beginning of the Carboniferous System.

Fault: an approximately planar fracture surface in rock along which there has been some movement of one side relative to the other.

Fauna: animals; often referring to the characteristic animal assemblage of a region/time period.

Feldspar: a widespread, ubiquitous group of rock-forming silicate minerals that are the essential constituents of many **igneous rocks**. Variations in composition divide the group into two series – the 'alkali feldspars' with end-members albite (Na-rich) and orthoclase (K-rich), and the 'plagioclase feldspars', with end members albite and anorthite (Ca-rich).

Glossary

Feldspathic: a rock rich in **feldspar**.

Felsic: a pale-coloured rock poor in iron and magnesium minerals and rich in **quartz** and **feldspar**; the opposite of **mafic**.

Felsite: a field term for glassy and fine-grained **felsic igneous rocks**. Also called **felstone**.

Felstone: *see* **felsite**.

Fenestrae: cavity structures, which are generally infilled with crystals, also called 'birdseye structures'.

Fissile: descriptive of a **sedimentary rock** that contains very thin bedding or **cleavage laminae** along which the rock splits into thin sheets.

Flagstone: a hard, thin-bedded **sandstone**, firm **shale**, or other rock that splits easily along **bedding planes** or **joints** into flat slabs.

Flashy: of streams that flow as a result of sudden, heavy rainfall in tropical and sub-tropical areas.

Flexural subsidence: **subsidence** of the Earth's crust caused by loading in association with mountain building.

Flora: plants; often referring to the characteristic plant assemblage of a region/time period.

Flute mark/cast: a structure formed by small eddies that carve depressions in the surface of a **sediment** deposit. The cavities are asymmetrical in outline, with the deeper or thicker part at the upstream end, and are generally preserved on the base of the overlying **bed** as a 'cast' of the depression.

Fluvial: relating to a river or river system.

Flysch: deposits of dark, fine-grained, thin-bedded **sandstone**, **shales**, and **clay**, typically syn-**orogenic** and thought to be deposited by **turbidity currents**.

Fold: a bend in rock **strata** produced by earth movements.

Foliation: the planar arrangement of minerals, or other textural or structural features in rocks.

Footwall: the upthrow side of a geological **fault**.

Foreset: the steeply dipping surface of **cross-bedded strata**.

Formation: a succession of **contiguous** rock **strata** that is distinctive enough in its **lithology** from the surrounding rocks to be mapped as a unit; the fundamental unit of **lithostratigraphy**.

Fossil: the preserved remains of an animal or plant. *See also* **trace fossil**.

Frasnian Stage: a **chronostratigraphical** division of the **Upper Devonian Series**, comprising the rocks deposited during the Frasnian Age. The fifth **stage** of the **Devonian Period**, it is dated to approximately 382.5–376.5 Ma and is preceded by the **Givetian Stage** and followed by the **Famennian Stage**.

Friable: descriptive of a rock that is crumbly or easily broken.

GCR: Geological Conservation Review, in which nationally important geological and geomorphological sites were assessed and selected with a view to their long-term conservation as SSSIs.

Gedinnian Stage: a **chronostratigraphical** divison of the **Lower Devonian Series** in Belgium, France and Germany. Roughly equivalent to the **Lochkovian Stage**.

Genus (pl: **genera**): a category used in the taxonomic classification of organisms, which consists of one or several related **species**. Similar genera are grouped together into a **family**.

Geochemistry: the chemistry of rocks.

Geopetal: a sedimentary fabric that records the way up at the time of deposition. Commonly found in cavity fills within **limestones**.

Geotectonic: large-scale **tectonic** deformation of the Earth's crust.

Gilgai: the microrelief of small depressions or valleys and ridges on a soil surface. Found in soils that contain large amounts of **clay**, which expand and contract during wetting and drying.

Givetian Stage: a **chronostratigraphical** division of the **Middle Devonian Series**, comprising the rocks deposited during the Givetian Age. The fourth **stage** of the **Devonian Period**, it is dated to approximately 387.5–382.5 Ma and is preceded by the **Eifelian Stage** and followed by the **Frasnian Stage**.

Glaebule: a **concretion** found in **palaeosol** profiles.

Gneiss: a coarse-grained, inhomogeneous rock, common in relatively high-grade **metamorphic terranes**, characterized by a coarse **foliation** or layering more widely spaced, irregular or discontinuous than that in a **schist**.

Gondwana: a grouping of the major southern continental plates of Africa, Australasia, Antarctica, South America, India, several smaller plates and fragments of what are now parts of Mediterranean Europe, which together formed a massive southern supercontinent straddling the South Pole in early **Palaeozoic** times and began to split up when **Avalonia** broke away in early **Ordovician** times.

Glossary

Gorstian Stage: the older **chronostratigraphical** division of the **Ludlow Series** of the **Silurian Period**.

Graben: a linear block of crust downthrown between two parallel **faults** to form a **rift** or trough-shaped valley.

Graded beds: **beds** that show a change in grain size through the bed. Normal graded bedding is a fining upwards sequence. In reverse graded bedding, the grain size coarsens upwards.

Grampian Orogeny: the first mountain-building event of the **Caledonian Orogeny**, in the **Ordovian Period** from about 477 to 442 million years ago.

Granite: a pale-coloured, coarse-grained, typically **plutonic (intrusive) igneous rock**, with a high SiO_2 content. Commonly found in **batholiths** and veins.

Granitoid: a general term used to encompass unspecified coarse-grained, **quartz**-rich **igneous rocks** that may include compositional types such as **granite**, tonalite and **granodiorite**.

Granodiorite: a coarse-grained **igneous rock** similar to **granite** in texture but containing slightly less silica.

Graptolite: an extinct marine colonial organism belonging to the **class** Graptolithina (**phylum** Hemichordata) and characterized by a cup- or tube-shaped, highly resistant exoskeleton made of collagen.

Graptoloid: any **graptolite** belonging to the **order** Graptoloidea. The graptoloid exoskeleton is characterized by relatively few branches (stipes), lacks strengthening rods (dissepiments) and is without thecal (cup) differentiation. Typically **planktonic**.

Great Glen Fault: a major geological **fault** that trends north-east and crops out from Fort William to Inverness in Scotland and extends to the Shetland Islands and beyond. It separates two **terranes** and was active in the **Devonian Period**.

Greenstone: general term for **massive** basic **igneous rocks** that have been (partly) metamorphosed.

Greywacke: a poorly **sorted, clastic sedimentary rock** composed of fragments of rocks and crystals and **sand** set in a **clay**-rich **matrix**.

Groove mark/cast: a structure thought to be formed by the movement of objects along a stream bed by a continous current. These straight narrow depressions are a few millimetres deep and serveral centimetres long and wide, and may be preserved on the base of the overlying **bed** as a 'cast' of the depression.

Group: in **lithostratigraphy**, a grouping of two or more **formations** with significant unifying **lithological** and/or genetic features.

Half-graben: an elongate trough bounded by a normal **fault** on one side only. *See also* **graben**.

Halite: the chemical name for sodium chloride, common salt (NaCl).

Hanging-wall: the downthrow side of a geological **fault**.

Hardpan: the indurated surface horizon of a soil, such as **calcrete** (caliche), formed in arid and semi-arid climates.

Heterolithic: of varied **lithologies**.

Heterostracan: any member of the extinct **order** Heterostraci (**class** Agnatha (*see* **agnathan**)). Characterized by extensive head armour of large plates and a pair of common branchial openings on either side.

Hexapoda: a superclass of the **phylum** Arthropoda (*see* **arthropod**) characterized by having six legs, and consisting of the **classes** Insecta (insects) and Parainsecta.

Hiatus: a break or gap in sedimentation.

High: a region or area that underwent little or no **subsidence** and sedimentation.

Highland Boundary Fault: a major **tectonic** and **terrane** boundary extending north-eastwards across Scotland separating the Scottish Highlands from the Midland Valley. It was active during the **Devonian** and **Carboniferous periods** and experiences minor **seismic** activity today.

Highstand: a period of highest lake or sea level.

Holotype: the single specimen (the so-called 'type specimen') selected to epitomize a particular named **species**.

Honeycomb weathering: a form of chemical **weathering** in which numerous pits occur on a rock exposure, causing the surface to look similar to a large honeycomb. It typically occurs in arid regions, affecting granular rocks such as **sandstones** and **tuffs**.

Horst: an upfaulted block of crustal rocks, often on either side of a **graben**.

Hydrocarbon: an organic compound comprising hydrocarbon, oxygen and carbon, the term is also commonly applied to the energy compounds oil, gas and coal.

Glossary

Hydrothermal: of heat generated from hot sub-surface rock or groundwater.

Hypabyssal: descriptive of an igneous **intrusion**, or its rock, emplaced at a depth intermediate between **plutonic** and volcanic.

Hypolimnion: the layer of water below the thermocline in a lake that is non-circulating or perpetually cold.

Iapetus Ocean: a former ocean that separated the early **Palaeozoic** crustal plates of **Laurentia** and **Baltica** plus **Avalonia** until the ocean floor was subducted in **Ordovician–Silurian** times during the **Caledonian Orogeny**. It divided the present British Isles and its trace is situated between what is now Scotland and the northern part of Ireland, and the rest of Britain.

Ichnofauna (ichnofossil): an assemblage of **trace fossils**.

Ichthyofauna: the fish of a particular region.

Igneous rock: a rock that has formed from the cooling of molten magma, either following volcanic (**extrusive**) activity or **intrusive** processes. It consists of interlocking crystals, the size of which depends on the rate of cooling of the magma.

Ignimbrite: a **volcaniclastic** rock, typically silica-rich and pumiceous, formed by deposition from a **pyroclastic** flow; may partly or wholly comprise welded **tuff**.

Imbrication (adj. imbricated): a sedimentary fabric typically displaying elongate fragments that are aligned in a preferred angle to the line of bedding.

Index fossil (index species): a particular **fossil** (or **species**) that gives its name to a **biozone**.

Inlier: an outcrop of older rocks surrounded by younger rocks commonly exposed by **erosion** (cf. **outlier**).

Intercalation (adj. intercalated): layering within a sedimentary sequence of different rock types.

Intraclast: a fragment of rock derived from **coeval** parent material rather than an 'older' (**extraformational**) source.

Intraformational: of **conglomerate** containing **clasts** derived from the local depositional area.

Intramontane (intermontane): within a mountain belt.

Intrusion: (adj. **intrusive**): an **igneous rock** that has formed as a body intruded into other rocks below the Earth's surface.

Joint: a fracture in a rock that exhibits no displacement across it (unlike a **fault**).

Karst (adj. karstic): descriptive of a distinctive **terrane** developed upon a soluble rock, typically **limestone**; characterized by caves, sinkholes and dry valleys.

Kinematics: the mechanical processes involved in sedimentary **basin** formation.

Lacustrine: relating to, formed within, or produced by, lakes.

Lag deposit: a concentrated layer of coarse grained **sediments**, such as pebbles or bones, which accumulates at the bottom of a channel or the sea floor during deposition from strong currents.

Lagerstätten: a rock containing exceptionally abundant or exceptionally well-preserved **fossils** that are of considerable intrinsic interest.

Lamina (pl. laminae, laminations): the finest layer within a **sedimentary rock**, typically less than 10 mm thick.

Laminated: descriptive of a **bed** with a fabric composed of **laminae**.

Laminites: thin layers of generally fine-grained **sediment**, reflecting rapidly fluctuating, often seasonal, changes in sediment supply or environmental conditions; characteristic of lakes and other shallow **basins** of deposition where there is a restricted bottom **fauna**.

Lapillus (pl. lapilli): a **pyroclastic** fragment of any shape with a mean diameter of 2–64 mm.

Late Devonian Epoch: a geological time division (cf. **chronostratigraphy**) of the **Devonian Period**. Ranging from 382.5 to 362 million years ago it includes the Frasnian and Famennian ages

Lateral accretion: the build-up of **sediment** by river currents flowing obliquely to the main drainage direction.

Laurentia: the major North American crustal plate in early **Palaeozoic** times that straddled the equator, prior to the **subduction** of the **Iapetus Ocean**; comprised mainly of the ancient **Precambrian** core of the Canadian Shield, Greenland, Scotland and north-west Ireland.

Laurussia: the amalgamated plates of North America and Russia, following the **subduction** of the **Iapetus Ocean**.

Lava: an **igneous rock** formed during volcanic eruptions (molten rock at the Earth's surface).

Glossary

Levee: a broad ridge alongside a river or stream, deposited by floodwaters when they overtop the channel banks.

Limestone: **sedimentary rock** composed of **calcium carbonate**, often partly derived from the shells of organisms.

Lingulid: any member of the **order** Lingulida, extant shellfish of the **phylum** Brachiopoda (*see* **brachiopod**). Much more abundant in the **Palaeozoic Era** than at present.

Litharenite: an **arenaceous** rock composed of **lithic clasts**.

Lithic: relating to a rock **clast** found within a **sedimentary rock**.

Lithification: the conversion of **sediment** into rock.

Lithoclast: a mechanically deposited rock fragment, normally greater than 2 mm in diameter, derived from any older, (pre-existing) **lithified** rock.

Lithofacies: a **facies** defined by **sedimentary rock** type (using, for example, colour, texture and mineral composition).

Lithology: descriptive of the constitution of a **sediment** or other rock, including composition, texture, colour and hardness.

Lithostratigraphy: the organization and division of **strata** into mainly mappable rock units and their **correlation**, based entirely upon their **lithological** characteristics. Units are named according to their rank in a formal hierarchy, namely supergroup, **group**, **formation**, **member** and **bed**.

Littoral: descriptive of the zone between high- and low-water marks on a shoreline.

Load cast (structure): a protuberance of **sand** or coarse **clastic** material that extends downwards into a finer-grained, softer underlying material such as wet **mud** or **clay**. Produced by downsinking and unequal settling and compaction of the overlying material.

Lochkovian Stage: a **chronostratigraphical** division of the **Lower Devonian Series**, comprising the rocks deposited during the Lochkovian Age. The first **stage** of the **Devonian Period**, it is dated to approximately 418–413.5 Ma and is followed by the **Pragian Stage**.

Lower Devonian Series: a **chronostratigraphical** division of the Devonian System comprising the rocks deposited during the **Early Devonian Epoch**. It includes the **Lochkovian**, **Pragian** and **Emsian** stages.

Lowstand: a period of lowest sea level or lake level.

Ludfordian Stage: the younger **chronostratigraphical** division of the **Ludlow Series** of the **Silurian Period**.

Ludlow Bone Bed: a thin, lenticular **marker horizon** of **phospatized** fish fragments in south Wales and the Welsh Borderland. It was formerly taken to mark the base of the **Old Red Sandstone** and marks the base of the **Přídolí Series** in the **Anglo-Welsh Basin**.

Ludlow Series: a **chronostratigraphical** division of the **Silurian Period**.

Macrofossil: a **fossil** that is easily seen by the naked eye.

Mafic: a rock rich in magnesium and/or iron minerals.

Magnafacies: a major continuous belt of deposits that is homogeneous in **lithological** and **palaeontological** characteristics and that extends obliquely across time planes or through several **chronostratigraphical** units.

Marker band/bed/horizon: a **bed** or layer within a rock succession with distinctive, easily recognizable characteristics that allow it to be traced for long distances or to serve as a reference or datum, and thereby enabling **correlation**.

Mass flow: the transport, downslope under the force of gravity, of large, coherent masses of **sediment**, tephra or rock; commonly assisted by the incorporation of water, ice or air.

Massif: a very large topographical or structural upland feature.

Massive: descriptive of a **bed** or layer of **sedimentary rock** with an apparently uniform structure and lacking bedding fabric or **lamination**.

Marl: a fine-grained **calcium carbonate**-rich **mud** or **clay**.

Matrix: the fine-grained **sediment** or crystalline **cement** that infills the spaces between larger grains.

Matrix-supported: a **conglomerate** in which the constituent **clasts** are not in contact with each other.

Medial: of the central parts of a depositional system.

Megafossil: macrofossil.

Megasequence: a very thick sequence of **sedimentary rocks**.

Member: in **lithostratigraphy** a subdivision of a **formation**.

Glossary

Meta-: a prefix commonly used to indicate that a rock has been affected by low-grade **metamorphism**, while still retaining many recognizable features of its origin (e.g. metasedimentary).

Metamorphic rock: a rock that has been altered by the action of heat and/or pressure, without melting.

Metamorphism (adj. **metamorphic**): the process of radical alteration of the mineralogical and/or physical nature of rocks as a result of pressure and/or temperature.

Micrite: microcrystalline **calcite**; typically a lime **mud**.

Microfauna: microscopic animals.

Microflora: microscopic plants

Microfossil: a microscopic **fossil**.

Microgranite: a type of **granite** with characteristically small crystals.

Microspar: a recrystallized component of **limestone** rocks comprising mosaics of small (4–50 microns) crystals of **calcium carbonate**.

Microvertebrate: the small **fossil** remains of vertebrates, whether they be of juveniles of a large **species** or a small species. Remains tend to be disarticulated teeth and bones, and are usually size-**sorted** and deposited together by the processes of transport and deposition, especially by water currents.

Mid-Devonian Epoch: a geological time division (cf. **chronostratigraphy**) of the **Devonian Period**. Ranging from 394 to 382.5 million years ago it includes the Eifelian and Givetian ages.

Middle Devonian Series: a **chronostratigraphical** division of the Devonian System comprising the rocks deposited during the **Mid-Devonian Epoch**. It includes the **Eifelian** and **Givetian stages**.

Midland Microcraton: a triangular area underlying the English Midlands that was a stable, cratonic area on the south-east flanks of the **Lower Palaeozoic Welsh Basin**.

Milankovitch cycles: periodic shifts in the Earth's orbital parameters, thought to be one of the driving mechanisms behind glacial events.

Miospore: a general term for any **fossil** plant spore smaller than 0.2 mm.

Modiolopsid: a **species** of **bivalve**.

Molasse: of a **basin** and its deposits that form after a period of mountain building, the deposits being derived from the mountain range.

Mollusc: any member of the **phylum** Mollusca, which comprises about 5000 **species**. These invertebrates are characterized by a fleshy soft body and, usually, a hard shell. They may be marine, freshwater or **terrestrial**, and examples include gastropods (snails, limpets), **bivalves** (oysters, mussels), and cephalopods.

Monocline: a **fold** in which **strata** of otherwise uniform dip are locally steepened.

Mud: a mixture of **clay** and **silt**.

Mudflat: an area of predominantly fine-grained **mud** deposition.

Mudstone (mudrock): a fine-grained **sedimentary rock**; **lithified mud**.

Myriapod: any member of the extant (**Silurian–Recent**) **class** Myriapoda of **terrestrial** uniramous **arthropods**, which includes the centipedes and millipedes, some of which were amongst the first land-living animals.

Nautiloid: a member of the subclass Nautiloidea (**class** Cephalopoda). These marine invertebrates possess a multi-chambered external shell of **calcium carbonate** which may be straight or coiled. Only one **genus**, the *Nautilus*, survives today.

Neoproterozoic: the youngest **period** of the Proterozoic Eon of the **Precambrian**, preceding the Cambrian Period.

New Red Sandstone: a sequence of red, largely desert and **fluvial sedimentary rocks**, which was formed in the **Permo–Triassic** period.

Nodule: a small **concretion**, generally roughly spherical or ellipsoidal.

Offlap: the successive build-up of **strata** in a prograding sequence in a shrinking sea or on the margin of a rising landmass.

Old Red Sandstone: a classic term applied to the **terrestrial**, largely **clastic facies** of late **Silurian** to earliest Carboniferous age in Britain; characterized by red **sandstones**, **mudstones**, **siltstones**, **conglomerates** and **calcretes**.

Oncolite (adj. **oncolitic**): a spherical or sub-spherical particle, up to 5 cm in diameter, which is produced by the **accretion** of sedimentary material on to a mobile grain through the action of **algae**.

Ophiolite: an ordered sequence of petrogenetically related **ultramafic** rocks, gabbros, sheeted **dykes** and **basalt lavas** that originated through the generation of oceanic crust, but were subsequently thrust (obducted) onto continental crust.

Glossary

Opilionid: an **order** of the **class** Arachnida (**phylum** Arthropoda (*see* **arthropod**)) commonly referred to as 'harvestmen' or 'daddy longlegs'.

Orcadian Basin: a Mid-Devonian depositional **basin** in north-east Scotland in which a thick succession of **lacustrine** and **fluvial** deposits accumulated.

Order: a category used in the taxonomic classification of organisms, which consists of one or several related **families**. Similar orders are grouped together in a **class**.

Ordovician Period: a geological time division (cf. **chronostratigraphy**). Ranging from 495 to 290 million years ago it is the second **period** of the **Palaeozoic Era** and precedes the **Silurian Period**.

Orogen: a mountain belt produced by continental collision and **uplift**.

Orogeny: a process of mountain building during which the rocks and **sediments** of a particular area of a continent are deformed and **uplifted** to form mountain belts.

Orthoconglomerate: a **conglomerate** with an intact gravel framework, characterized by a mineral **cement**. Associated with coarse-grained **cross-bedded sandstones**.

Osteichthyes: a **class** of the **phylum** Chordata, also referred to as 'bony fish'.

Osteolepid: any member of the extinct **family** Osteolepidae (**order** Osteolepiformes, subclass Sarcopterygii (*see* **Sarcopterygian**); lobe-finned fishes.

Osteostracan: *see* **cephalaspid**.

Ostracoderma: a popular name applied to the extinct division of jawless and armoured **agnathans**.

Ostracod: any member of the subclass Ostracoda (**class** Crustacea (*see* **crustacean**), **phylum** Arthropoda (*see* **arthropod**)). These small invertebrates are mostly less than 1 mm in size and consist of two **calcareous** valves ('shells'). They can be found in a wide range of aquatic environments, both in fresh- and salt-water.

Outlier: an outcrop of younger rocks surrounded by older rocks (cf. **inlier**).

Overstep: a relationship in which a younger series of sedimentary **strata** rests upon a progressively older series of strata, the older and younger series of strata being separated by a plane of **unconformity**.

Overbank deposit: fine-grained **sediment** (silt and clay) deposited from suspension on a floodplain by floodwaters from a stream channel.

Pachytheca: a primitive, non-**vascular** Early Devonian plant **fossil**.

Palaeo-: 'ancient' (occurring or formed in geological time).

Palaeocurrent: a river current in a geological, **fluvial** deposit.

Palaeoclimate: the climate at a particular geological time.

Palaeoecology: the relationship between organisms and their environments in the geological past.

Palaeokarst: ancient **karst** landform.

Palaeomagnetism: the magnetic alignment of iron minerals imparted to a rock at the time of its formation; also the science of the reconstruction of the Earth's magnetic field and continents in geological time.

Palaeontology: the study of **fossil fauna** and **flora**, including their evolution and the reconstruction of pre-existing environments.

Palaeosol: an ancient or 'fossilized' soil.

Palaeoslope: the depositional slope at a particular geological time.

Palaeozoic Era: a geological time division; the first major division of geological time characterized by abundant life. It precedes the Mesozoic Era and is subdivided into 'Lower' and 'Upper' divisions at the top of the **Devonian Period**.

Palyno-: prefix indicating 'pollen' or 'spores'.

Palynomorph: a microscopic, acid-resistant, organic-walled body studied in palynology.

Ped: a naturally formed unit of soil structure.

Pedogenic: of processes relating to the development of soil profiles.

Pelite: a general term for fine-grained, clay-rich, **clastic sedimentary rocks**; often applied to metamorphosed **mudstones**.

Penecontemporaneous: formed or existing at the same time.

Peneplain (adj. **peneplained**): a virtually flat and featureless landscape of considerable size, caused by prolonged **weathering** and **erosion**, especially mass-wasting and sheetwash.

Period: a geological time unit (cf. **chronostratigraphy**); of shorter duration than an **era** and itself divisible into **epochs**.

Permian Period: a geological time division (cf. **chronostratigraphy**). Ranging from about 290 until 250 million years ago, it follows the **Carboniferous Period** and precedes the **Triassic Period**.

Permo–Triassic: a period of geological time spanning the **Permian** and **Triassic periods**.

Glossary

Petrifaction: a process of fossilization whereby organic matter is converted into a stony substance by the infiltration of water containing dissolved inorganic matter, which replaces the original organic materials, thereby often retaining the structure of the materials.

Petrography: the study of the origin, mineral composition, texture and history of rocks; includes **petrology**.

Petrology: the study of the mineral composition, texture and systematic classification of rocks.

Petromictic: comprising a mixture of different rock types (normally with regard to **sediment** pebble composition).

Phosphate: a compound salt of phosphoric acid, a mineral commonly associated with the preservation of bones and shells.

Phosphatic: descriptive of a rock containing large quantities of **phosphate**.

Phosphatize: the process of becoming enriched with **phosphate**.

Photosynthesis: the process whereby green plants trap light in chlorophyll and use it to synthesize carbohydrates from carbon dioxide and water.

Phyllarenite: a **litharenite** composed mainly of fragmented **metamorphic rocks** such as slate, phyllite and **schist**.

Phyllite: a **metamorphic rock** rich in the ferromagnesian mineral mica and texturally intermediate between slate and **schist**.

Phylum (pl. **phyla**): a category used in the taxonomic classification of organisms, which consists of one or several related **classes**. The phyla are grouped together into two kingdoms, the Plantae (plants) and the Animalia (animals).

Phytoplankton: plant forms of **plankton**, for example diatoms. They are often microscopic and with limited powers of locomotion, so mainly dispersed by wind and tide.

Piedmont: a sloping area at the bottom of a mountain or mountain range.

Pisoid (**pisolith**): a large ooid with a diameter of more than 2 mm.

Pisolite: a **sedimentary rock** consisting manly of **pisoids**.

Placoderm: any member of the extinct **class** Placodermi. A group of primitive **Palaeozoic** jawed fishes, with a dermal armour in two parts, one covering the head, the other the trunk, which are sometimes articulated.

Plankton (adj. **planktonic**): minute aquatic organisms that drift with water movement.

Playa: the flat dry bottom of a desert **basin**, often the bed of an **ephemeral** lake and underlain by **evaporites**.

Pluton (adj. **plutonic**): an **intrusion** of **igneous rock** emplaced at depth in the Earth's crust.

Point bar: a lenticular deposit formed on the inside bend of a meandering river channel.

Polychaete: a group of annelid worms, some bearing bristles, tentacles and hard organic tooth structures (**scolecodonts**).

Polygonal: refers to arrays of **desiccation cracks** in patterned ground, and subaqueous shrinkage cracks formed in **mudrocks**.

Polymictic: descriptive of a **conglomerate** that contains **clasts** of many different rock types.

Pragian Stage: a **chronostratigraphical** division of the **Lower Devonian Series**, comprising the rocks deposited during the Pragian Age. The second **stage** of the **Devonian Period** it is dated to approximately 413.5–409.5 Ma and is preceded by the **Lochkovian Stage** and followed by the **Emsian Stage**.

Precambrian: a widely used term to encompass the time preceding the Cambrian Period. It extends from the formation of the earth (4600 Ma) to the beginning of the Phanerozoic Eon (540 Ma).

Přídolí Series: a **chronostratigraphical** division of the Silurian System comprising the rocks deposited during the Přídolí Epoch. The final **series** of the **Silurian Period** it is dated to approximately 419–418 Ma and immediately precedes the **Devonian Period**.

Progradation: the advance of a sedimentary **facies** or belt of facies as a result of a change in sea or lake level, or uplift of the source area.

Proximal: near to the source.

Psammite: an **arenite**.

Pseudomorph: a replacement product, composed either of a single mineral or an assemblage of minerals, that retains the distinctive overall shape of the parent crystal.

Pseudonodule: a primary sedimentary structure consisting of a ball-like mass of **sandstone** enclosed in **shale** or **mudstone**. It is the result of the settling of **sand** into underlying **clay** or **mud** that welled up between isolated sand masses. Characterized by a rounded base with upturned or inrolled edges.

Pteraspid: any member of the extinct (**Palaeozoic**) **order** Pteraspidiformes, also known as **heterostracans**, abundant in early **Devonian** times.

Glossary

Pull-apart basin: a sedimentary **basin** formed by the pulling apart of adjacent blocks by **transcurrent faults**.

Pyroclastic: descriptive of unconsolidated deposits (tephra) and rocks that form directly by explosive ejection from a volcano.

Quartz: a rock-forming mineral composed entirely of silica (SiO_2); one of the most common minerals of the Earth's crust.

Quartzite (adj. **quartzitic, quartzose**): both an **arenaceous** rock composed primarily of **quartz**, and a **metamorphic rock** formed of quartz.

Quaternary Period: a geological time division (cf. **chronostratigraphy**). The latest **period** of geological time and the last period of the Cainozoic Era, it ranges from 1.8 Ma to the present day.

Radiometric dating: methods of dating rocks or minerals using the relative abundances of radioactive and stable isotopes of certain elements, together with known rates of decay of radioactive elements. Radiocarbon dating can extend back to only 50 000 years, but other elements (potassium, lead, uranium) can be used to obtain dates of the order of tens to thousands of millions of years.

Red beds: a collective term applied to **continental** sedimentary successions that are predominantly red in colour due to the presence of iron oxides and hydroxides formed in a highly oxidizing environment.

Regolith: a layer of unconsolidated, **weathered**, broken rock debris that lies below the soil and above the bedrock below.

Regression: retreat or contraction of the sea as a result of a fall in sea level or **uplift** of the land.

Relict: descriptive of a geological feature surviving in its primitive form.

Reworking: the natural excavation and transportation of **sediment** or **fossil** material that is then re-deposited elsewhere.

Rheic (Ocean): an approximately E–W-oriented ocean that opened up at the same time as the closure of the **Iapetus Ocean** and separated part of **Gondwana** (in part, what is now Brittany and central Germany) from southern Britain, northern France and northern Germany during the **Silurian Period**; its closure resulted in the **Variscan Orogeny**.

Rhizocretion: a hollow, **concretion**-like mass that formed around the root of a living plant.

Rhizolith: a **concretion** (normally **calcitic**) replacing a root.

Rhyolite: a fine-grained, pale coloured, acid ($SiO_2 > 63$ wt%) volcanic (**extrusive**) rock, with the same chemical and mineralogical composition as **granite**.

Rhythmite: a unit of a rhythmic succession or of **beds** that were developed by rhythmic sedimentation.

Rift: a depressed area of continental crust produced by tensile stretching of the crust and down-faulting along parallel **faults**.

Rip-up clast: a fragment of **sediment** that was eroded from river banks and re-deposited in the succeeding sedimentary layers.

Rotliegendes: 'red layers', a German **stratigraphical** term applied to the largely **continental** deposits of Lower to Mid-Permian times, which are commonly reddened with iron oxide minerals.

Rudite (adj: **rudaceous**): a coarse-grained **sedimentary rock**, either consolidated as in a **conglomerate**, or unconsolidated as in a till.

Sabkha: a salt-encrusted **supratidal** surface or coastal flat bordering a lagoon. Inland forms frequently support sand dunes.

Sand: **sediment** particles typically between 0.625 mm and 2 mm in diameter.

Sandbar: a **bar** or low ridge of **sand** that borders the shore and is built up, or near, to the suface of the water by currents or wave action.

Sandstone: a **sedimentary rock** composed of **lithified sand** grains between 0.625 mm and 2 mm in diameter.

Sand volcano: an accumulation of **sand** resembling a minature volcano formed by the expulsion of liquefied sand to the **sediment** surface.

Sand wedge: a body of **sand** shaped like a vertical wedge, with the apex pointing downward, formed by the filling in of thermal contraction cracks.

Sarcopterygian: any member of the subclass Sarcopyterygii, **class Osteichthyes**. This group of bony fishes, characterized by paired 'fleshy fins' and internal nostrils includes the **crossopterygians** and the dipnoans.

Scandian Orogeny: the phase of the **Caledonian Orogeny** that affected what is now the Scandinavian countries, when **Baltica** and **Laurentia** collided in the Ludlow Epoch.

Glossary

Schist: a coarse-grained **metamorphic rock** that displays a strong **foliation** (schistosity) that is commonly defined by mica alignment.

Scolecodont: the fossilized jaw of an annelid.

Scoriaceous: descriptive of **lavas** that are very highly **vesiculated**, giving them a 'clinkery' appearance.

Sediment: granular material such as **sand** or **mud** derived from the **weathering** and **erosion** of pre-existing rocks, biological activity (e.g. shells and organic matter), or chemical precipitation (e.g. **evaporites**).

Sedimentary rock: a rock composed of **sediments**, deposited by water, wind or ice.

Sedimentology: the study of **sediments** and **sedimentary rocks**, including their deposition, structure and composition.

Seismic: of movements in the Earth's crust causing earthquakes. **Seismic reflection** is the technique whereby artificially created seismic waves are differentially transmitted by rocks of differing density and recorded on their return to the surface.

Series: a **chronostratigraphical** division comprising all of the rocks formed during an **epoch**; it can be divided into **stages**.

Shale: a **mudrock** that splits easily into layers.

Shearing: deformation of a rock body by the sliding of one part relative to another part, in a direction parallel to their plane of contact.

Sheet flood: a broad expanse of rapidly moving water and debris, not confined to a channel and usually of short duration due to rapid runoff in an arid area.

Shoreface deposits: the deposits formed on the shoreward part of a barrier in shallow sea or between the seaward limit of the shore and the horizontal surface of the offshore.

Siegenian Stage: a **chronostratigraphical** divison of the **Lower Devonian Series** in Belgium, France and Germany. Roughly equivalent to the **Pragian Stage**.

Silcrete: an indurated silica soil (or **palaeosol**); the term was originally used for a **conglomerate** of **sand** and gravel cemented into a hard mass by silica.

Silicification (adj. **silicified**): the process of conversion of a **sediment**, plant or animal to **chert** by **hydrothermal** and/or chemical processes.

Sill: a tabular body of **igneous rock** that is more-or-less concordant with the bedding or **foliation** of the host rocks.

Silt: a fine-grained **sediment** intermediate in grain size between **clay** and **sand**.

Siltstone: a rock made of **silt**.

Silurian Period: a geological time division (cf. **chronostratigraphy**). Ranging from 439 to 418 million years ago, it precedes the **Devonian Period**.

Siluro–Devonian: a period of geological time spanning the **Silurian** and **Devonian periods**.

Sinistral: of left-hand lateral movement along a geological fault.

Sinter: a chemical **sedimentary rock** deposited by precipitation from mineral waters of springs, lakes or streams.

Sorting: the ordered distribution of grain sizes in a **sediment** or **sedimentary rock**.

Sparite (sparry calcite): a **limestone** in which the sparite **cement** is more abundant than the **micrite matrix**.

Species: a category used in the taxonomic classification of organisms. Similar species are grouped together in a **genus**.

Spherulite: a spherical mass of acicular crystals, commonly **feldspar**, radiating from a central point; commonly found in glassy silicic volcanic rocks as a result of devitrification.

SSSI: Site of Special Scientific Interest; the designation of an area of land for statutory protection under the Wildlife and Countryside Act 1981.

Stage: a **chronostratigraphical** division comprising all of the rocks formed during an **age**, and usually taken to be the smallest standard unit.

Staurolite-grade: a medium grade of **metamorphic rock**, based on the amount of temperature/pressure it has been subjected to, this referring to the presence of the index mineral staurolite.

Stratigraphy: the study of the temporal and spatial relationships within a rock succession.

Stratotype: a sequence of **sedimentary rocks** at a particular locality chosen as the standard against which other sequences can be compared. Stratotypes are established for **lithostratigraphical** and **biostratigraphical** units, both regionally and internationally.

Stratum (pl. **strata**): a **bed** or single layer in a succession of rock.

Stratovolcano: a cone-shaped volcano with a layered internal structure.

Strike: the trend of a geological surface (e.g. a **bedding plane**) measured at right angles to the direction of maximum slope or dip.

Glossary

Strike-slip: a **tectonic** break in which the predominant displacement is lateral rather than vertical.

Stromatolite: a **laminated** or mounded structure composed of **carbonate** trapped by cyanobacteria. Known in rocks throughout the geological record, they form in warm, shallow tropical seas in the present day.

Subaerial: of environments that exist and processes that operate in the open air.

Subarkose: a **sandstone** with insufficient **feldspar** to be termed an **arkose**, intermediate in composition between an arkose and a pure **quartz** sandstone.

Subduction: the process of one crustal plate descending into the mantle beneath another during plate convergence and collision, with the release of energy in the form of earthquakes and often accompanied by volcanicity.

Subsidence: the sinking of a local or regional portion of the Earth's surface with respect to its surroundings.

Subtidal: formed, or occurring, below the tides.

Sulphide: a compound of sulphur with a metal or semi-metal, such as pyrite FeS_2.

Supratidal: above the tides.

Synaeresis crack: a subaqueous shrinkage **crack**, formed by the loss of pore water from **clays** due to changes in salinity of the surrounding water.

Syn-: prefix indicating 'together' or '**contemporaneous** with'.

Syncline: a downfold of rock produced by **tectonic** deformation; the youngest rocks occur in its core.

System: a **chronostratigraphical** division comprising all of the rocks formed during a **period**; can be divided into **series**.

Talus: an accumulation of rock litter at the foot of a slope, generally with a wide size-range (up to several metres) and ungraded. Also called scree.

Taxon (pl. **taxa**): a named group of organisms of any rank.

Tectonism (adj. **tectonic**): deformation of the Earth's crust and the consequent structural effects (e.g. **faults** and **folds**).

Terrane: a small crustal plate or **fault**-bounded fragment of a larger plate, with distinctive characteristics, which may have been displaced considerable distances from its original site and welded to another plate during plate **tectonic** movement.

Terrestrial: of or relating to the Earth or the Earth's dry land.

Tetrapoda: four-footed vertebrates including amphibians, reptiles and mammals.

Thelodont: a member of the extinct **order** Thelodonti, **class** Agnatha (*see* **agnathan**). Characterized by their shark-like dermal denticles, which are commonly fossilized as separate elements within the **sediment**.

Thermal subsidence: subsidence of the Earth's crust as a result of heating.

Throw: the amount of vertical displacement between the rocks on either side of a **fault**.

Thrust fault: a **fault** characterized by movement or rocks under lateral compression along a low-angle fault plane.

Trace fossil (ichnofossil): a **biogenic** sedimentary structure produced by activity of an organism within a substrate; examples include burrows and footprints.

Transcurrent fault (strike-slip fault): a **fault** in which the major displacement is horizontal and parallel to the **strike** of a vertical or subvertical fault plane. Localized zones of deformation due to pressures and tensions across the fault occur at bends in the fault and can give rise to conditions of **transtension**. The latter process may cause the formation of rhombic-shaped **basins**, **graben**, or marginal basins that may be the focus of **rift**-related volcanic activity.

Transgression (adj. **transgressive**): the inundation of the land due to sea-level or lake-level rise, and the resulting sedimentary deposits.

Transpression: crustal shortening as a result of oblique compression across a **transcurrent (strike-slip) fault** or shear zone.

Transtension: crustal extension as a result of oblique tension across a **transcurrent fault** or shear zone leading to localized **rifts** or **basins**.

Triassic Period: a geological time division (cf. **chronostratigraphy**). Ranging from 245 to 208 million years ago, it is the first **period** of the Mesozoic Era, and is preceded by the **Permian Period**.

Trigonotarbida: an extinct (Silurian–Carboniferous) **order** of primitive chelicerate **arthropods**, which includes some of the earliest land-living animals.

Trough (cross-) bedding: cross-bedding formed in three-dimensional dunes in which the **foresets** (trough sets) are concave upwards.

Glossary

Truncation: the cutting or breaking off of the top of a geological structure or landform.

Tuff (adj. **tuffaceous**): cemented and **lithified** volcanic **ash**, comprising rock and crystal fragments from an explosive eruption.

Turbidite: any **sediment** or rock transported and deposited by a **turbidity current**, generally characterized by **graded bedding**, large amounts of **matrix** and commonly exhibiting a Bouma sequence.

Turbidity current: a highly turbid, dense current carrying large quantities of **clay**, **silt** and **sand** in suspension which flows down a submarine slope through less dense sea water.

Type locality/area: the place where the **type section** (or **stratotype**) for a **stratigraphical** unit is located, or from where the **type specimen** of a **fossil** came.

Type section: *see* **stratotype**.

Type specimen: a single specimen designated as typifying a named **species** or subspecies. *See also* **holotype**.

Ultramafic: of an **igneous rock** in which dark-coloured (iron/magnesium) minerals comprise more than 90% of the rock.

Unconformity: the surface that separates two sedimentary sequences of different ages; it represents a gap in the geological record when there was **erosion**, and/or **tectonism** and/or no deposition. There is often an angular discordance between the two sequences.

Uplift: movements that raises or upthrusts the Earth's crust, as in a dome, arch or orogen.

Upper Devonian Series: a **chronostratigraphical** division of the Devonian System comprising the rocks deposited during the **Late Devonian Epoch**. It includes the **Frasnian** and **Famennian stages**.

Variscan Orogeny: the period of mountain building in Europe that occurred in the **Carboniferous** and **Permian periods** as a result of the closure of the **Rheic Ocean**.

Vascular: of plants with conducting tissue.

Varve: a sedimentary **bed**, layer, or sequence of layers deposited in a body of still water within a year, and usually during a season.

Vertisol: a mineral soil rich in calcium and magnesium and with 30% or more of **clay**. Deep wide cracks develop when the soil is dry.

Vesicle: a gas bubble cavity, usually in a consolidated **lava** or shallow **intrusion**.

Volcanic arc: a generally curved or arc-shaped, linear belt of volcanoes above a **subduction** zone along a continental margin.

Volcaniclastic: generally applied to a **clastic** rock containing mainly material derived from volcanic activity, but without regard for its origin or environment of deposition (includes **pyroclastic** rocks and **sedimentary rocks** containing volcanic debris).

Volcanism: volcanic activity.

Vug (adj. **vuggy**): a cavity in a rock, which may contain a lining of crystalline minerals.

Wacke: an impure **sandstone** that consists of a mixture of angular and unsorted or poorly **sorted** mineral and rock fragments, with an abundant **matrix** of **clay** and fine **silt**.

Wadi: a gorge-like valley formed in arid or semi-arid environments.

Weathering: the chemical alteration and physical breaking down of rocks through the effects of exposure to the weather.

Welsh Basin: the Lower **Palaeozoic** sedimentary, deep-water basin that occupied much of what is now Wales.

Wenlock Series: a **chronostratigraphical** division of the **Silurian** System. It is preceded by the Llandovery Series and succeeded by the **Ludlow Series** and extended from about 430 to 424 million years ago.

Zone: a **stratigraphical** unit in many categories of stratigraphical classification. In **biostratigraphy** a zone is defined on the occurrence of a defining **fossil** (**index fossil**), or by the entry of a particular fossil taxon (*see* **biozone**).

Index

Note: Page numbers in **bold** and *italic* type refer to **tables** and *figures* respectively

Abdon limestones 170, 218
Abercriban Oolite 259
Abercriban Quarries 7, *214*, 220, 258–62
  Brownstones Formation *258*
  Carboniferous Limestone *258*, 259
  Grey Grits Formation 7, 220, 258–62
  Lower Limestone Shale Group *258–62*
    Castell Coch Limestone Formation 259
    Plateau Beds Formation *258*, 259–60, 262
Aberlemno Quarry 6, 8, *127*, 161–3
  Arbuthnott–Garvock Group 161, 163
    Dundee Flagstone Formation 161–3
    Scone Sandstone Formation 161, 163
      Melgund Sandstone Member 161
  see also Tillywhandland Quarry *and* Turin Hill
Acadian Orogeny 5, *14*–16, 130, 137, 190, 211, 215, 218, 227, 228, 256, 268, 300, 307, 322
acanthodians **8**, 110–12, 161, 163, 166, *229*, 230, 277
Achanarras Fish Bed 22, 107, 116

Achanarras Hill 90, *91*
Achanarras Limestone Member 6, *24–5*, 90–3, 111
Achanarras Quarry 6, 8, 22, *24–5*, 27, 90–3
  Achanarras Limestone Member 6, 90–3
  Lower Caithness Flagstone Group 90–3
    Robbery Head Subgroup 90–3
  Upper Caithness Flagstone Group 90–3
    Latheron Subgroup 93
adhesion warts 99, *100*, 101, 176
Afforsk Fault 113
Afon Dulais 254
Afon Sawdde *240*, 243
  see also Sawdde Gorge
Afon y Waen 7, 9, *214*, 220, 262–5
  Afon y Waen Fish Bed 220, 262–5
  Lower Limestone Shale Group *262, 264*
  Plateau Beds Formation 7, 262–5
Afon y Waen Fish Bed 220, 262–5
Aglaophyton major 120
Albadesmus almondi 143
Albion Sands Formation *212–13*, 217, *281–4*

Albion Sands and Gateholm Island 7, **11**, *214*, 281–4, *293*, 299
  Gray Sandstone Group 282, *283*
  Milford Haven Group 282
    Albion Sands Formation *281–4*
    Lindsway Bay Formation 282, 284
    Red Cliff Formation *281–4*
    Sandy Haven Formation *281–4*
algae 34, 53–6, 75–8, 93, 117, 120, 162–3
Aller's Mill 185
Allt a'Mhuilinn Mudstone Formation 139
Althaspis leachi 219, 274–6
  A. senniensis 219, 251–3
Alves and Scaat Craig Beds *24–5*
amphibians 100, 180
anaspid 143
andesite 16, 28, 119, 121–4, 128, 139, 142, 147–8, 151, 153, 157, 176–7, 199
Angle Syncline 287
Anglesey 5, 16, 211, *212–13*, *214, 216*, 218, 221–8, 274, 317
Anglo-Welsh Basin 4, 5, 9, **10**, **11**, 15–16, 17–18, 163, 170, 178, 186, 211–323

373

Index

Annandale 199, 202
annelids 121
annulatus sextantii Biozone 179
Apiculiretusipora 252
araneae 121
Arbroath *127*, *134–5*, *167–8*, 199
Arbroath Paving Stone 161
Arbroath Sandstone Member *134–5*, 167–9
Arbuthnott–Garvock Group 128–35, 139, 151, *158*, 161, 163, 166–7, 170–1, 176
  Ashbank Sandstone *134–5*
  Buttergask Flagstone *134–5*
  Catterline Conglomerate Formation *134–5*, *146*, *152–3*
    Barras Conglomerate *134–5*
    Rouen Bay Sandstone *134–5*
    St John's Knapp Sandstone *134–5*
    Three Wells Sandstone *134–5*
  Craighall Conglomerate Formation *134–5*
  Deep Conglomerate Formation *134–5*
  Dundee Flagstone Formation 6, 8, *134–5*, 139, 161–4, 166, *167*
  Finavon Conglomerate *134–5*
  Hatton Conglomerate *134–5*
  Montrose Volcanic Formation 12, *134–5*, 139, *167*
  Ruchill Flagstone Formation *134–5*
    Callander Craig Conglomerate *134–5*
    Gartan Conglomerate *134–5*
    Inchmurrin Conglomerate *134–5*
  Scone Sandstone Formation 128, *134–5*, 161, 163, 166–8, 170–2, 176, 178
    Arbroath Sandstone Member *134–5*, *167–9*
    Auchmithie Conglomerate Member *134–5*, 167

Campsie Limestone Member 7, *134–5*, 170–2
  Dunblane Sandstone *134–5*
  Melgund Sandstone Member 161
  Red Head Sandstone Member *134–5*, *167*
  Sheriffmuir Sandstone Member *134–5*, 176–8
  Tannadice Sandstone *134–5*
Archaeoperisaccus 73
Archaeothrix contexta 120
  A. oscillatoriformis 120
Archanodon jukesi 221
Archidesmus sp. 143
  A. macnicoli 166
Ardmore–Gallanach 8
Arenicolites 157, 220, 299
Argyll Field 73, *132–3*
Arkelton Hill 202
Arnton Fell 202
Arran *see* Isle of Arran
Arthostigma sp. 157
arthropoda *incertae sedis* 121
arthropods 139, 218
  Aberlemno Quarry 6, 162
  Den of Findon 114–16
  Freshwater East–Skrinkle Haven 303–4, 308
  Freshwater West 293
  Llansteffan 211
  Milton Ness 161
  Pantymaes Quarry 7, 246–50
  Porth-y-Mor 224
  Rhynie 6, **10**, 117, 120–4
  Tarbet Ness 100
  The Toutties 6, 140, 143–5
  Tillywhandland Quarry 166
Ascomycete 120
Ascomycete fungi 117
Ashbank Sandstone *134–5*
Asterolepis orcadensis Zone 85
Asteroxylon 117
  A. mackiei 120
Auchenaspis 9
Auchensail Quarry 7, **10**, *127*, 178–80
  Strathmore Group 179
  Teith Sandstone Formation 7, 179–80
Auchmithie Conglomerate Member *134–5*, 167

Auchtitench Sandstone Formation 128, *136*
Avalonia 5, 13–15, 245
Avon Group *see* Lower Limestone Shale Group
Ayre of Huxter *31–3*
Ayrshire 7, 16, 128, 137–9, 204

bacteria 56, 109, 111–12
Badbea Breccia *24–5*
Baligill Quarry 74
Ballagan Formation *181*, 185–6, 199, 201, *203*
Ballanucater Farm **10**, 139, 178, 180
Balleich Lava *134–5*
Balmaha 131
Balmerino to Wormit **12**, 139
Balnagown Group 6, *24–5*, 96–102
  Gaza Formation *24–5*, 97–102
  Tarbet Ness Formation *24–5*, 97–102
Baltica 4, 13–15
Bannisdale Formation 206
Banniskirk Quarry **8**
Barras Conglomerate *134–5*
Barren or Basement Group *24–5*
Basal Breccia Member 142
Basement Complex 27
Basement Group 302
Bay of Berstane 6, *24–5*, 27, 28, 59, 71–4
  Eday Marl Formation 71–4
  Lower Eday Sandstone Formation 72
  Upper Eday Sandstone Formation 72
Bay of Sannick 84
Bay of Skaill **10**, *64*
Bay of the Stairs *68–9*
Beaconichnus 157
Beaconites 59, 60, 72, 114, 115, 159, 161, 180, 304–5, 311, 321
  B. antarcticus 244, 288, 290, 296, 311
  B. barretti 218, 291, 296, *295*, *297*, 311
Belgium 17
Bell Hill 199
Ben Nevis and Allt a'Mhuilinn **12**, 139

Index

Ben Nevis Volcanic Formation **12**
Benton Fault 215, 219, 284, *293*, 300
Berriedale Flagstone Formation *24–5*
Berriedale Sandstone Formation *24–5*
Besom Farm Quarry **9**
Bettws Fault *254*
Biggar Volcanic Formation *136*
Billia Croo 52, *53*
biostratigraphy 16–17
bioturbation 60–1, 72–3, 79, 96, 275, 313
  see also burrows
biozones
  fish 17–18, *212–13*
  miospores 17–18, *212–13*, 258
Birk Beck 201
Birkenia sp. 143
Birrenswark Volcanic Formation 199, 203–4
Bishop's Frome Limestone (Member) *213*–18, 227
  Cusop Dingle *234*–9
  Devil's Hole 228
  Oak Dingle 230
  The Scar 232
  see also Chapel Point Calcretes Member; Psammosteus Limestone
bivalves 220, 229–30, 241–3, 277–89, 290
Black Cock Formation *240*
Blackhall Limestone *173*
Black Isle Syncline 96, 106,
Black Mountains 211, 218, 219, 220, 234, 235
Black Nore Point 312, 318
Black Nore Sandstone Formation *212–13*, 219, 312–18
Black Park, Edderton **8**
Black Rock to East Comb **12**, 139
Blaen-gweche Farm *254*
Blaenavon 268
Blairgowrie–Edzell *134–5*
Blo Geo 49, 51
Blorenge 271
Bochruben Formation 104
Bodafon Formation 218, 221–2, 226

Boddin Point 160
Bofrishlie Burn Sandstone *134–5*
Boghole and Whitemire Beds *24–5*
Bogmore, Muckle Burn **8**
Bohemian 13, 17–18
Borrowdale Volcanic Group 205–7
Borwick 66
bostonite 69
Bothriolepis **9**, 174, 184, 185, 186, 220–1, 264–5, 266–7
  B. hicklingi 184
  B. hydrophila 174
Brachyacanthus scutuger 165
brachiopods 220, 241–3, 259, 264–5, 266–78, 299, 315
Bracklinn Falls Conglomerate *134–5*
Bradnor Hill Quarry **9**
breccia 22, *24–5*, 26, 74–8, 102–7
Breck Ness *53*, 55
Brecon Beacons 211, 219, 220, *234*, 254, 274
Brecon Beacons Quarry see Craig-y-Fro Quarry
Breconensis–vallatus Zone 18
Breconian 17, *212–13*, 215, 219, 220, 253
  Heol Senni Quarry 251–3
  Ross-on-Wye, Royal Hotel **7**, 271–4
Breinton 232
Brendon Hills 319
Bressay Flagstone Facies *24–5*
Brewin's Bridge/Canal **11**
Bridge of Allan 176–7
Brindister Flagstone Formation *24–5*, 47
Bring Fault 27
Bristol 4–5, 317
Bristol Channel 16, 301, 307
Bristol Channel Landmass 211, *216–17*, 290, 317, 322
Brobury Scar see The Scar
Brontoscorpio sp. *229*
Brown Clee Hill 273
Brown and Grey Sandstone Member 140–2, 144
Brownstones Formation *212–13*, 219, 322
  Abercriban Quarries 258

Caeras Quarry **7**, 253–6
  Craig-y-cwm *269*
  Craig-y-fro Quarry 257
  Duffryn Crawnon 265–6
  Heol Senni Quarry *251*, 253
  Ross-on-Wye, Royal Hotel **7**, 271–4
  Wilderness (Land Grove) Quarry **7**, 274–6
bryzoa 289, 299, 301
Buchlyvie Sandstone *134–5*
Buckie Beds *24–5*
Burnmouth 199
Burnside Sandstone Formation 129, *138–9*, *167–9*, 173–4
burrows 111, 190, 220, 230, 235–9, 248, 288, 290, 304–5
  see also bioturbation
Burton Cliff Formation *212–13*
Bute Conglomerate Formation *138*
Buttergask Flagstone *134–5*
Buttermere Formation *205*

Cabrach 28, 117
Caddroun Burn 202–3
Caeras Conglomerate Member *254–6*
Caeras Quarry **7**, *214*, 219, 253–6
  Brownstones Formation **7**, 253–6
  Caeras Conglomerate Member 254–6
  Lower Limestone Shale Group 254
  Millstone Grit *254*
Caithness Basin 3–6, **8**, **10**, 21–5, *24–5*, 28–30, 74–96
Caithness Flagstone Group 22
  South Stromness Coast Section 52–6
  Taracliff Bay to Newark Bay 56–9
  Yesnaby and Gaulton Coast Section *64*, 67
calc-alkaline lavas 134–8, 139
Caldey Island *293*, 308
Caledonian intrusions 3, 15–16, 22, 102–7, 127–30, 139
Caledonian Orogeny 14–16, 21–2, 106, 128, 182, 199, 211, 217, 221, 227–8, 255, 322

375

Index

Callander Craig Conglomerate *134–5*
Camarotoechia nucula 255
Campsie Limestone Member 7, *134–5*, *170–2*
camptonite 66
Canas Solais 97, 98
Capel Berach Beds 243
Capel Horeb Member *242*, *243*, 245
Capel Horeb Quarry **10**, **11**
Caradoc 4–5
Carboniferous 3, 13, *17*, 187, 199–*200*, *202*, 206, *212–13*, 265–8
Carboniferous Limestone 7, **10**, 221, *254*, *258*, 259, 265–8, *269*, 302–7, 317
Cardiff 217–18, 221
Carlin Craig 147
Carmarthen Fans 211, 220, 254
Carmel 254
Carn Mor 102
Carn Powell Member *240*
Carnoustie 167–8
Carreg Cennen Castle 254
Carreg Cennen Disturbance 217, 246, 250
Carreg Cennen–Llandyfaelog Fault 219, 256
Carrick Hills **12**
Carrick Volcanic Formation *136*
Carron Sandstone Formation *134–5*, 140–5, 145–6, 149–50
Carse of Gowrie 168
Castell Coch Limestone Formation 259, *265*, *266*
Castlehaven 100
Castle Haven Conglomerate Member *134–5*, *146–8*, 150–1
Castle of Cowie Member *134–5*, *141*, 142, 144
Castle of Qui Ayre 66
Castle of Old Wick 85
Castlemartin Corse Anticline 292
Castracollis wilsonae 121
Catterline Conglomerate Formation *134–5*, *146*, *152–3*
Cefn Esgair South Borehole 264
Cennen Valley 217, 219, 254
centipedes 117, 121
cephalaspids **8**, **9**, 143, 161, 163, 166, 176, 230

Cephalaspis 166, 219, *229*, 230, 252
 C. pagei 162, 166
 C. powriei 166
 C. scotica 177
Ceratiocaris sp. 143
Chapel Point Calcretes Member 203–7, *212*–17, 237, 305, *306*, 308–11
 see also Bishop's Frome Limestone; Psammosteus Limestone
Chapelton Burn Member *136*
Cheiracanthus 108–11
 C. latus 111, 112
 C. murchisoni 110, 112
Cheirolepis 108, 111
 C. trailli 110, 112
Cheviot Volcanic Formation 199
chilopods 121
Chondrites 289
Chonetes striatella 255
chronostratigraphy 13, 16–17
Cil-yr-ychen Quarry *254–6*
Clashbenny Formation *138*
Clee Hills 211, 218, 219, 221
Clee Sandstone Formation *212–13*, 219
Clevedon 312, 314, 317
Clevigarth *48*, 51
Clew Bay Fault Zone *132–3*
Cliff Farm 277–8
climate 15, 116, 121, 128, 160, 176, 204, 219, 227, 253, 255, 258, 280, 306–7, 311
climatiid *229*
Climatius reticulatus 166
Clocksbriggs (or Wemyss) Quarry 161
Clousta Volcanic Rocks **12**, *24–5*
Clun Forest 211
Clydach Valley 254
Clyth Subgroup *24–5*
Cnoc Fyrish Conglomerate *24–5*
Coccosteus 34, 108, 110–11, 220, 317
 C. cuspidatus 34, 55, 66, 110, 112
Coccosteus Bed *108*, 109, 111
Cockburnspath 137, *181*, 199, 202
Cockburnspath Formation 184
Cockermouth Lavas 208

Coldra Limestone 218
conglomerates see named units
Conglomerate Facies *24–5*
Conglomerate Member 287–90, *294*, 298, 300
Conigar Pit Sandstone Member *212–13*, 250, *294–6*, 300, *303*, 305, 307
Coniston Group 206
conodonts 286, 287, 291
Cooksonia **10**, 252, 257–8, 302
 C. caledonica **10**, 162
coprolites *109*, 111, 165
Cornulatichnus 59
 C. edayensis 60
Cornwall 3
Cosheston Group *212–13*, 215, 219, 220, 253
 Burton Cliff Formation *212–13*
 Lawrenny Cliff Formation *212–13*
 Llanstadwell Formation *212–13*
 Mill Bay Formation *212–13*, 253
 New Shipping Formation *212–13*
Cotti Geo *38*
Courceyan *17*, 128, 187, *205*, 208
Cove Harbour *181–2*
Cowie Harbour 140, 142, 143
Cowie Harbour Conglomerate Member *134–5*, 140–2, 144
Cowie Harbour Fault *140*
Cowie Harbour Fish Bed 140, *141*, 143–4
Cowie Harbour Siltstone Member *134–5*, 140–5
Cowie Sandstone Formation *134–5*, 140–6, 149
 Basal Breccia Member 142
 Brown and Grey Sandstone Member 140–2, 144
 Castle of Cowie Member *134–5*, *141*, 142, 144
 Cowie Harbour Conglomerate Member *134–5*, 140–2, 144
 Cowie Harbour Siltstone Member *134–5*, 140–5
 Purple Sandstone Member *141*, 142, 144

Index

Red Sandstone Member 140–2, 144
Cowiedesmus eroticopodus 143
Craig of Monievreckie Conglomerate Formation *134–5*
Craig na h-Iolaire 102, 104–5
Craig-y-cwm 7, *214*, 268–71
    Brownstones Formation *269*
    Carboniferous Limestone *269*
    Millstone Grit *269*
    Quartz Conglomerate Group 7, 268–71
        Craig-y-cwm Formation 268–71
        Garn-gofen Formation 268–71
        Wern Watkin Formation 268, 269, 271
    Senni Formation *269*
Craig-y-cwm Formation *212–13*, 221, 268–71
Craig-y-Fro Quarry 7, **10**, *214*, 257–8
    Brownstones Formation 257
    Senni Formation 7, 257
Craig-yr-Hafod 268
Craighall Conglomerate Formation *134–5*
Crawton Basin 130–1, 144, 149, 151, 154
Crawton Bay 6, **12**, *127*, 131, 151–5
    Catterline Conglomerate Formation *152–3*
    Crawton Volcanic Formation **12**, 151–4
    Whitehouse Conglomerate Formation 152
Crawton Ness 151–3
Crawton Volcanic Formation **12**, *134–5*, *146*, 151–4
Creag Dhearg 102, 105
Creag nan Clag 102–6
Crieff–Callander Basin 131
Crigiau Stream 234, *236*, 239
crinoids 241, 289
Cromlix Mudstone Formation 128, *134–5*, 155–7, *170–2*
Cross Fell 199, *200*
Cross Fell Inlier 206–, 208
Crossgates–Burnside 168, 178
Crovie Fault *113–14*
Crovie Group *24–5*, 29, 113–16
Cruaday Quarry **8**, 55, 66
Crussolum sp. 121
crustaceans 117, 121
Culzean Harbour **12**, 139
Cusop Dingle 7, *214*, 217–18, 234–9
    Raglan Mudstone Formation 7, *234–7*
        Bishop's Frome Limestone *234–9*
        Townsend Tuff Bed *234–6*, 238–9
    St Maughans Formation *234–5*
Cwm Mill **9**, 218
Cwmyniscoy Mudstone Formation *265*
Cwrt-yr-ala Formation *212–13*, 221
cyanobacteria 117, 120
Cyathaspis **9**
    C. banksi 244
Cymbularia carina 244
Cyrtospirifer verneuili 220, 264

Dalradian Supergroup *155*, 192–3, 195
    Southern Highland Group 193
Dalmary Sandstone *134–5*
Daren Cilau 271
Dawsonites sp. 180
Dawston Burn 202–*3*
Deep Conglomerate Formation *134–5*
Deepwell 232–3
Deerness 56–9, 70
Deerness Syncline 27
Deerness Volcanic Member **12**, 57
Deheubarthia splendens 257–8
Den of Findon 6, **8**, 28–9, 112–16
    Crovie Group 113–16
    Findon Group 112–16
        Gamrie Fish Bed 112, *113*, 116
Devil's Hole 7, **9**, *214*, 217, 228–9
    Raglan Mudstone Formation 228
        Bishop's Frome Limestone 228
    St Maughans Formation 228

Devon 3, *5*, 13, 211, 215, 317, 318–23
Dickmont's Den 169
Dictyocaris 162
    D. slimoni 143, 145
Dictyocaris Member *141*, 142, 145
Digedi Brook 238
Dingle Peninsula 64
Dinley Burn 202–*3*
Diplacanthus 108–10
    D. crassisimus 110
    D. (Rhadinacanthus) longispinus 112
    D. striatus 110, 112
    D. tenuistriatus 112
Diplichnites 114, 166, 218
    D. gouldi 249
Diplocraterion 114
diplopod 140
Dipple Brae **8**
Dipterus 51, 79, 83, *108–9*, 116
    D. valenciennesi **8**, 110
Dittonian 17, *212–13*, *213*–17, 219, 221, 277–8
    Aberlemno Quarry 162–3
    Cusop Dingle *212–13*, 217
    Wilderness (Land Grove) Quarry 274–6
Djuba Water Syncline *39*
dolerite dykes *170*, 189
dolomite 61–3, 65–7, 87–90, 163–6
Doolie Ness Conglomerate *134–5*
Dornoch Firth *29*, 96
Dow Craig 172, *174–6*
Downie Point 144, 145–51
Downie Point Conglomerate Member *134–5*, *146–7*, 150–1
Downie Point to Dunnottar Castle *see* Dunnottar Coast Section
Downton Castle area **9**
Downton Castle Sandstone Formation 9, 212, *212–13*, 217, 240, 244
Downton Group 9, 10, 17, *212–13*, 215–17, 235–40, 247, 284
Downton–Dittonian boundary **9**
Downtonian 7, 17, *212–17*, 235, 247, 284, 291–301
draas 99, 101

377

Index

Drepanophycus spinaeformis 180
Dryden Flags Formation 118–19, *122*, 124
   Lower Shales *118–20*, 122
   Rhynie Cherts Unit 117–120, 123
   Shales with Muddy Sandstones *118*–20, 122–3
   Upper Shales *118–20*, 122
   White Sandstones *118*–19, 122, 124
Duffryn Crawnon 7, *214*, 220, 265–8
   Avon (Lower Limestone Shale) Group 265
      Cwmyniscoy Mudstone Formation 265
      Castell Coch Limestone Formation 265, *266*
      Brownstones Formation 265–6
      Grey Grits Formation 265–6, 268
   Pembroke (Carboniferous Limestone) Group 265
   Plateau Beds Formation 7, 265–8
Dulas Bay 218, *222*
Dulas Brook *234*, 239
Dumfries *200*, 202
Dun Chia Hill (Loch Duntelchaig) 6, *24–5*, 28, *29*, 102–6
   Grampian Group 102–4
      Bochruben Formation 104
   Inverness Sandstone Group *103*
Dunbar 199
Dunblane Sandstone *134–5*
Duncansby Head 30, *82–4*
Dundarg Fault 115
Dundee Flagstone Formation 6, 8, *134–5*, 139, 161–4, 166, *167*
Duneaton Volcanic Formation *136*
Dungavel Hill Conglomerate Member *136*
Dunnet Head 28, 30, 69
Dunnet Head Sandstone Group *24–5*, 30, 69
Dunnicaer 147
Dunnottar Castle Conglomerate Formation *134–5*, 145–7, 149

   Castle Haven Conglomerate Member *134–5*, *146*–8, 150–1
   Downie Point Conglomerate Member *134–5*, *146*–7, 150–1
   Strathlethan Sandstone Member *134–5*, *146*–7, 150
Dunnottar Coast Section 6, *127*, 131, 139, 145–51
   Catterline Conglomerate Formation *146*
   Cowie Sandstone Formation 145–6, 149
   Dunnottar Castle Conglomerate Formation 145–7, 149
      Castle Haven Conglomerate Member *146*–8, 150–1
      Downie Point Conglomerate Member *146*–7, 150–1
      Strathlethan Sandstone Member *146*–7, 150
   Stonehaven Group 145, 149–50
      Carron Sandstone Formation 145–6, 149–50
   Tremuda Bay Volcanic Formation *146*, 148
   Whitehouse Conglomerate Formation *146*
Dunnottar–Crawton Group 128–31, *155*
   Craig of Monievreckie Conglomerate Formation *134–5*
   Crawton Volcanic Formation 12, *134–5*, *146*, 151–4
      Lintrathen Tuff Member 153–4, *155*, 158
   Dunnottar Castle Conglomerate Formation *134–5*, 145–7, 149
      Castle Haven Conglomerate Member *134–5*, *146*–8, 150–1
      Downie Point Conglomerate Member *134–5*, *146*–7, 150–1
      Strathlethan Sandstone Member *134–5*, *146*–7, 150

   Gourdon Sandstone Formation *134–5*
      Doolie Ness Conglomerate *134–5*
      Rob's Cove Conglomerate *134–5*
   Tremuda Bay Volcanic Formation *134–5*, 145, *146*, 148
   Whitehouse Conglomerate Formation *134–5*, 139, *146*, 152
Dura Den Fish Bed 174
Dura Den Member *138*

East Moor Cliff *302*, 304
East Scapa Fault 27
Easter Rova Head 6, *24–5*, *26*, 43–7
Eastern Hole *181*, 185
Eastern Hole Conglomerate 185, 187
Ebullitiocaris oviformis 121
Eday Group 6, *24–5*, 56, 58–9, 61, 83, 84
   Eday Flagstone Formation 6, *24–5*, 56–9, 61–3, *72*
   Eday Marl Formation *24–5*, 56–7, 59–61, 71–4
   Lower Eday Sandstone Formation *24–5*, 56–8, 61–3, *72*
   Middle Eday Sandstone Formation *24–5*, *60*, 61–3
   Upper Eday Sandstone Formation *24–5*, 60, *72*
Eday Flagstone Formation 6, *24–5*, 56–9, 61–3, *72*
   Deerness Volcanic Member 12, 57
Eday Marl Formation *24–5*, 56–7, 59–61, 71–4
Eday Syncline 27, 59, 61
Edzell 131
Edzell Mudstone Formation 156
Edzell Sandstone Formation 156
Eifelian 4–5, **12**, *17*, *24–5*, *26*, 30–7, 66, 108–12
   Den of Findon 6, 8, *28–9*, 112–17
   Tynet Burn 6, *29*–30, 107–12
   Wick Quarries 85–90

378

Index

Eifelian–Givetian **8**, *24–5*, *29–30*, *113*–17, *212–13*, 319
   Achanarras Quarry **8**
   Den of Findon **8**, *28*–9, 112
   Melby **8**
   South Stromness Coast Section 26–7, 52–6
   Tarbet Ness *29*–30
   Yesnaby and Gaulton Coast Section *24–5*
Elgin *28*–9
Ellen's Geo Conglomerate *24–5*, 94
Emphanisporites 251
Emphanisporites-micrornatus-Streelispora-newportensis Zone 295
Emsian **10**, *24–5*, *64*–7, 94, 129, 211, *212–13*, 218
   Auchensail Quarry **10**, 178–80
   Melby 37
   Sarclet 94
   Yesnaby and Gaulton Coast Section *24–5*, *64*–7
Eoarthropleura sp. 121
Eoarthropleurids 121
Eophalangium sheari 121
Erieopterus 162
Esgryn Brook 234, *236*, 237, 239
Eshaness Coast **12**
'Ethie' lavas **12**
eurypterids 144, 162–3, 166
Euthacanthus sp. 166
   E. macnicoli 166
euthycarcinoid 117, 121
Exnaboe *see* The Cletts
Exnaboe Fish Bed 51–2
Eycott Volcanic Group 205
Eyemouth 137, *181*, 199
Eyemouth Volcanic Formation 199

Fairlie Sandstone Formation *138*
Fairtide Rock 277–8
Famennian *4–5*, *17*, *24–5*, 137–9, *213*–16, 220, 221
   Craig-y-cwm 268–71
   Glen Vale 172–6
   Largs Coast **7**, *127*, 188–92
   Siccar Point to Hawk's Heugh 184, 187

Famennian–Courceyan **7**, 127–8, 137–9, 181–7
Farlovian **17**, *212*–17, 220, 221
Farlow Sandstone Formation 221
fault-bounded basins **7**, 47, 112–17, 284–6, 287–9, 290–1, 302–7
faulting *152*, 158, 211, 250
   Albion Sands and Gateholm Island 284
   Dunnottar Coast Section *146*, 149–51
   Glenthorne *318*–19
   Little Castle Head 285
   Pantymaes Quarry *246*
   Red Point 74–8
   Rhynie 117–24
   The Cletts 47–8
   The Toutties *140*, 142, 144
   Tynet Burn 108
   West Angle Bay (North) 287–91
   see also named faults
Fenno–Scandian High 73
Ferryside 219, 252
Fersness 61–2
Ffynnon limestones 218
Fidler Geo to Wats Ness *see* Footabrough to Wick of Watsness
Fife *127*, 137–9, 168
Fidlar Stack Monocline *40*
Finavon Conglomerate *134–5*
Findon Fault *113*–14
Findon Group 29, 112–16
Fintona Beds 178
fish 13, 17–18, 29, 30–7, 83–5, 110–12, 173–6, 202–5, 217–18, 220–1, 259, 277–80, 288, 290–1, 295–6, 297
   Aberlemno Quarry **6**, **8**, 90–3, 161–3
   biozones *212–13*
   Achanarras Quarry **6**, **8**, 90
   Cusop Dingle 229–30
   Den of Findon 112–17
   Devil's Hole **7**, **9**, 228–9
   Duffryn Crawnon 265–8
   Greenan Nev Coast 59–61
   Heol Senni Quarry **7**, **9**, 251
   Llansteffan 308–*10*, 310
   Oak Dingle **7**, **9**, 229–31
   Pennyland **6**, **8**, 26, 78–83

   Portishead 312–18
   Red Point 74–8
   Siccar Point to Hawk's Heugh **7**, *181*, 184–7
   Taracliff Bay to Newark Bay 56–9
   The Cletts **10**, 26, 47–8, *50*–2
   The Toutties **6**, **8**, 140–5
   Tillywhandland Quarry **6**, **8**, 161, 163–6
   Tynet Burn **6**, **8**, 107–12
   Wilderness (Land Grove) Quarry 274–6
   Wolf's Hole Quarry **7–8**, 176
Flimston Bay Fault 292–4
Footabrough to Wick of Watsness **6**, *24–5*, 37–43
   Melby Formation *39*
   Sandness Formation *39*, 41
   Sulma Water Formation *39*
   Walls Formation **6**, *39–43*
Forbidden Cave 167, 169
Forest of Dean 214, 219, 271, 274, 312, 317
Forest of Dean Coalfield 211, 221
Forfarshire Fish Bed 164
Foyers Pluton 102, 104, 106
Frasnian *4–5*, **8**, **9**, *17*, *24–5*, 129–30, *212*–17, 221
Frasnian–Famennian *129–30*, 211–*16*, 219–21, 262–5
   Afon y Waen 262–5
   Freshwater East–Skrinkle Haven 220
   Freshwater West 220
   Portishead 220
   Tarbet Ness *29*–30
   West Angle Bay (North) 220
Freshwater East Anticline 301
Freshwater East Formation *212–13*, 293, *294*, 299, 301
Freshwater East–Skrinkle Haven **7**, **10**, **11**, *214*, 217, 220–1, *293*, 301–8
   Basement Group 302
   Gray Sandstone Group 301–2
   Lower Limestone Shale Group 301, 306
   Milford Haven Group *302*, *303*
      Freshwater West Formation 302, *303*, 305–6

Index

Rat Island Mudstone Member *303*, 305, 307
Conigar Pit Sandstone Member *303*, 305, 307
Moor Cliffs Formation 301–7
 Chapel Point Calcretes Member 305, *306*
 Pickard Bay Tuff Bed 301–4, 306, 308
 Rook's Cave Tuff 301–4, 308
 Townsend Tuff Bed 301–4, 306, 308
Pembroke Group *302*
Ridgeway Conglomerate Formation 301–3, 305–8
Skrinkle Sandstones Group 301–3, 306–8
Tenby Group *302*
Freshwater West 7, *214*, 217–18, 220–1, 291–301, 307
 Gray Sandstone Group 292–4, 299
 Lower Limestone Shale Group 291, *294*
 Milford Haven Group *294*
  Freshwater East Formation 293, *294*, 299
  Freshwater West Formation 294–5, 300
   Conigar Pit Sandstone Member *294*–6, 300
   Rat Island Mudstone Member *294*, 296, 300
   Moor Cliffs Formation 293–4, 299–300
   Townsend Tuff Bed 291, 293, *294*, 300
 Ridgeway Conglomerate Formation 291–2, *294*, 296–8, 300
 Skrinkle Sandstones Group 291–2, *294*, *297*, 298, 300
  West Angle Formation *294*, 298–301
   Heterolithic Member *294*
   Conglomerate Member *294*, 298, 300
   Red-Grey Member 301
   Gupton Formation *294*, 298, 300
   Lower Sandstone Member *294*, 298, 300
   Stackpole Sandstone Member *294*, 298, 300
 Freshwater West Formation *212–13*, 294–5, 300, 302, *303*, 305–6, 308–9
  Conigar Pit Sandstone Member *212–13*, 250, *294*–6, 300, *303*, 305, 307
  Rat Island Mudstone Member *212–13*, 294, 296, 300, *303*, 305, 307
fungi 117, 120

Gala Group *181*, 182
Gamrie Bay *see* Den of Findon
Gamrie Fish Bed 112, *113*, 116
Gannochy Bridge *155*, 157
Gannochy Conglomerate Formation *134–5*, 155–8
Gannochy Tower *155*–7
Gargunnock Sandstones *138*
Garn-gofen Formation *212–13*, 221, 268–71
Garthna Geo *64*–6
Garthna Geo Fault *64*
gastropods 241, 244
Gateholm Island *see* Albion Sands and Gateholm Island
Gateholm Stack 282
Gaza Formation 97–102
Gedinnian 10, 17, *212–13*
 Aberlemno Quarry 162–3
 Craig-y-Fro Quarry 7, **10**, 257–8
 Porth-y-Mor 221–8
 Sawdde Gorge *240*
Gelliswick Bay Formation *212–13*, 217–18, 250
Geminospora lemurata 72
Geo of Hellia *68*–9
Gerdipaddle Syncline *40*
Giants's Rib 318–20
Givetian 4–5, **6**, **8**, **10**, 12, 17, 21, 24–5, 27–8, *212–14*, 216
 Achanarras Quarry **6**, **8**, 90
 Bay of Berstane 71–4
 Easter Rova Head 43–7
 Footabrough to Wick of Watsness *26*, 37–43
 Greenan Nev Coast 59–61

John o'Groats **6**, 83–5
Old Man of Hoy Coast 68–71
Pennyland **6**, **8**, 78–83
Red Point **6**, 74–8
South Fersness Bay 61–3
Taracliff Bay to Newark Bay **6**, 56–9
Tarbet Ness *29*–30
The Cletts **6**, **8**, 47–52
Wick Quarries **6**, 85–90
Glen Burn *173*–6
Glen Coe 16, 128
Glen Vale 7, *127*, 137–8, 172–6
 Burnside Sandstone Formation 173–4
 Glenvale Sandstone Formation 173–4, 176
 Kinnesswood Formation 173, 175–6
 Knox Pulpit Sandstone Formation 7, 172–6
 Lower Limestone Formation *173*
  Blackhall Limestone *173*
  Pathhead Formation *173*
Glenvale Sandstone Formation *138*, 173–4, 176
Glenbervie Porphyry 154
Glenthorne 7, *214*, 318–23
 Hangman Sandstone Formation 318–23
  Hollowbrook Member 322
  Little Hangman Member 322
  Rawn's Member 322–3
  Sherrycombe Member 322
  Trentishoe Member 318–23
 Ilfracombe Slates 318
 Lynton Formation 318
Glewstone 272
Glomites rhyniensis 120
Glyptolepis **108**, 111
 G. leptopterus 110
 G. paucidens 110
Glyptopomus 174, 314
Golspie 28
Gondwana 5, 13, *14*, 16
Gosslingia breconensis 219, 258
Gourdon Sandstone Formation *134–5*
graben 127–30
Graemsay *26*, 53
Grampian Group 102–4

380

Index

Grampian Highlands 8–9, 105
Grampian Highland Terrane *130–3*, 135, 137, 153, 157–8
granites *64–7*, 102, 105–6
Gray Sandstone Group 7, 213
   Albion Sands and Gateholm Island 282, *283*
   Freshwater East–Skrinkle Haven 301–2
   Freshwater West 292–4, 299
Great Conglomerate *132–3*
Great Conglomerate Formation 199
Great Furzenip *292*, 298
Great Glen Fault *14*, 15–16, *22*, 29, 102–7, *132–3*
Great Scar Limestone Group 205
Green Beds 243
Greenan Nev Coast 6, *24–5*, 27–8, 59–61
   Eday Marl Formation 6, 59–61
   Middle Eday Sandstone Formation *60*
   Upper Eday Sandstone Formation 60
Greenheugh Sandstone Formation *181–2*, 184, 186
Grey Grits Formation 7, *212–13*, 220, 258–62, 265–6, 268, 271
Greystoke 206
Greywacke Conglomerate Formation *136*
Groenlandaspis 314, 317
Grossilepis brandi 184
Gupton Formation 7, *212–13*, 287–90, 291, *294*, 298, 300
   Lower Sandstone Member *294*, 298, 300
   Stackpole Sandstone Member *288*, 290, *294*, 298, 300
Gwna Melange 317
Gwynfe Formation *212–13*, 240–2, *243–4*
gypsum 54–6, 89–90
Gyroptychius 92, 113

Hagshaw Hills 135
Haist Pebbly Sandstone Member 69
half-graben 189–92
halite 59, 71–4, 82, 89

Ham Skarfskerry Subgroup *24–5*
Hamar Anticline *40*
Hangman Sandstone Formation 7, 211, 217, 318–23
   Hollowbrook Member 322
   Little Hangman Member 322
   Rawn's Member 322–3
   Sherrycombe Member 322
   Trentishoe Member 318–23
Harra Ebb Sandstone Formation *24–5*, 64–7
Hatton Conglomerate *134–5*
hawaiite *134–5*, *146*, 148–51
Hawk's Heugh *see* Siccar Point to Hawk's Heugh
Hazeldean Burn 184
heavy minerals 38, 39–43, 154, 215, 221, 227, 322
Hellia *68–9*
Helman Head Formation *24–5*, 94
Hemicylaspis 9
Hemiteleaspis heintzi 143
Heol Senni Quarry 7, 9, *214*, 219, 250–3
   Brownstones Formation *251*, 253
   St Maughans Formation *251*, 253
   Senni Formation 250, *251*, 253
Heterocrania rhyniensis 121
Heterolithic Member *287*, 289, *294*
heterostracan 143, 145, 277
hexapods 117, 121
Highland Border Complex 147–54
Highland Boundary Fault *4*, 16, *22*, *26*, 128–36, 155–8, *188–9*
Hill of Borwick 66
Hill Head Red Bed Subgroup *24–5*
Hillhead Group *24–5*
Hirst Rocks *181*, 183, 186
Holburn Head Quarry *8*
Hollowbrook Member 322
Holoptychius sp. 9, 174, 184–6, 201, 220, 264–7, 288, 291, 298, 314
   H. nobilissimus 201, 203, 221
Homosteus milleri 34

Horneophyton lignieri 120
Horse Shoe 168
Horse Tooth Stone 66
Horse Neck *281*, 282
Hostinella sp. *229*, 303
   H. beardii 219
hot spring 117–24
Hoy Sandstone Formation *24–5*, 27, 68–71
   Haist Pebbly Sandstone Member 69
   Hoy Volcanic Member 6, *24–5*, 68–71
   Lang Geo Sandstone Member 69
   Lyre Geo Sandstone Member 69
   Trowie Glen Sandstone Member 69
Hoy Volcanic Member 6, *24–5*, 68–71
Hughmilleria norvegica 143
Hutton's Unconformity 185, 199, *201*
   North Newton Shore 7, 185, 192–5
   Siccar Point to Hawk's Heugh 7, 181–7
Huxter 30–1

Iapetus Ocean 13–16, 128
Iapetus Suture *4*, 13–16, 128
ignimbrites 131, 137, 154
Ilfracombe Slates 318
Inchbervis Castle 172
Inchbonny 185
   see also Jedburgh
Inches and Holm Burn Flagstone Group *24–5*
Inchmurrin Conglomerate *134–5*
insects 117
Inverclyde Group 158, 173, *181*, 192–3, 199–200, 202
   Ballagan Formation *181*, 185–6, 199, 201, *203*
   Kinnesswood Formation 6, 7, *128*, 137, 158–9, 168, 173, 175–6, *181*, 184–7, 192–3, 199, 202–4
Inverness 28–9, 102, 104
Inverness Sandstone Group *103*
Ireland 64, 67, 131, 178, 217

381

Index

Irish Sea 127, 239, 255
Ischnacanthus 166
  I. gracilis 166
Isle of Arran 7, 134, 192–5
Isle Port Conglomerate Member *136*
Isopodichnus 114

Jedburgh 137, 186, 199–202
John o'Groats 6, **8**, 30, 83–5
  John o'Groats Sandstone Group 83–5
    Last House Formation 83
      John o'Groats Fish Bed 83–85
    Upper Caithness Flagstone Group 83, *84*
    Mey Subgroup 83–5
John o'Groats Fish Bed 83–5
John o'Groats Sandstone Group 24–5, 29–30, 79, 81–2, 83–5
  Last House Formation 83
    John o'Groats Fish Bed 83–5
Johnston Thrust *293*

Kaellan Hellier 65
Kame of Hoy *68*
Kampecaris 143, *229*
Kelly Burn Sandstone Formation *138*, 187–8
Kelso 203, 204
Kerrera *see* South Kerrera
Kerrera Sandstone Formation 139
Kidstoniella fritschii 120
Kidstonophyton discoides 120
Kidwelly 219, 254
Kilgour Craigs 174
Kilkenny Bay 311–13, *316*
Kilmorack Group *24–5*
King's Quoit 305
Kinnesswood Formation *128*, 137, 168, 199
  Glen Vale 173, 175–6
  Largs Coast 187
  Milton Ness 6, 158–9
  North Newton Shore **7**, 192–3
  Palmers Hill Rail Cutting 202–4
  Siccar Point to Hawk's Heugh *181*, 184–6
Kirk Rigging 183

Kirkbean 199–*200*, 202–5
Kirkwall *27*, 71
Knock Castle *188–90*
Knock Farril Conglomerate *24–5*
Knox Pulpit Sandstone Formation 7, 128, *138*, 172–6
Krispiromyces discoides 120
Krithodeophyton croftii 219

Laidlehope Burn 202–*3*
Laird's Burn Member *136*
Lake District 4–5, *132–3*, 211
Lake Distict Massif 199–202
Lake Forfar 131, 166
Lammermuir Fault 199, *200*
Lammermuir Hills *200*
Lanark Basin 15–16, *128*–38
Lanark Group *129*, 134–7
  Auchtitench Sandstone Formation 128, *136*
    Dungavel Hill Conglomerate Member *136*
    Laird's Burn Member *136*
    Wiston Grey Volcaniclastic Sandstone Member *136*
  Biggar Volcanic Formation *136*
  Carrick Volcanic Formation *136*
  Duneaton Volcanic Formation *136*
  Greywacke Conglomerate Formation *136*
  Pentland Hills Volcanic Formation 128, 135–*6*
  Swanshaw Sandstone Formation *136*
    Chapelton Burn Member *136*
    Isle Port Conglomerate Member *136*
Lang Geo Sandstone Member 69
Lang Rigg *see* Melby
Langiella scourfieldii 120
Langholm 199–201, 202–*3*
Langiophyton mackiei 120
Lappie Farm 174
Largs Coast 7, *127*, 187–91
  Kinnesswood Formation 187
  Skelmorlie Conglomerate Formation 187–8
  Stratheden Group 187

Kelly Burn Sandstone Formation 187–8
Last House Formation 83
Latheron Subgroup *24–5*, 75, 93
Laurasia 73
Laurencekirk lavas 171
Laurentia *4–5*, 13–16, 128, 217
Laurussia 13–16, 211
lavas 131, 205–8
Lawrenny Cliff Formation *212–13*
lead 54–5
Leanach and Dores Sandstone *24–5*
Leck Leenaun Fault *132–3*
Ledbury cutting 9
Ledbury Formation *212–13*
Leintwardine Formation 241
Leperditia sp. 244
Lepidocaris rhyniensis 121
Leptodesma cf. *lichas* 264
Lerwick 26, 43
Lerwick Sandstone Facies *24–5*
Leverhulmia 117
  L. mariae 121
lichens 117, 120
Liddesdale 199, 202–*3*
Lindsway Bay Formation *212–13*, 217, 282, 284
lingulids 259, 264, 267, 282, 293, 299, 301, 302
Lingula sp. 220, 264, 289, 306
  L. cornea 244
  L. minima 241, 244
Lintrathen Tuff Bed 153–4, *155*, 158
Little Castle Head 7, **11**, *214*, 217, 284–6, *293*
  Sandy Haven Formation 284–6
  Pickard Bay Tuff Bed 284–5
  Townsend Tuff Bed 7, **11**, 284–6
Little Furzenip 292–4
Little Hangman Member 322
Little Mell Fell *205*, 208
Llanddeusant Formation *212–13*, 240–1, 245
Llandeilo *4–5*
Llandovery *4–5*, *181*, 218, 211
Llandybie 253, 254

Index

Llandyfaelog Disturbance *308*
Llangefni 218
Llanishen Conglomerate Formation *212-13*, 217-18
Llanover Quarry **10**
Llanstadwell Formation *212-13*
Llansteffan 7, *214*, 217-18, 253, 308-11
   Chapel Point Calcretes Member 7, 308-11
   Freshwater West Formation 308-*9*
Llansteffan Castle 308-*9*
Llanvirn 7, **10**, *206*, *302-3*
Lligwy Bay 218
Llyn-y-Fan Fawr 219, 254, 256
Loch Duntelchaig see Dun Chia Hill (Loch Duntelchaig)
Loch Ness 102
Loch Ruthven 102-*3*
Lochkovian *4-5*, 9, *14-16*, 17, *128-30*, 199, *212-13*, 215
   Freshwater West 293-301
   Heol Senni Quarry 7, **9**, 252-3
   Oak Dingle **9**
   Pantymaes Quarry 246-50
   Porth-y-Mor 227
   Sawdde Gorge *240*
Long Mountain 211
Long Quarry Member *240*, *242*, 245
Long Quarry (Sandstone) Formation **10**, *212-13*, 217, 240, 241
   see also Tilestones Formation
Longcroft Fault *118-19*
Longcroft Tuffs *118-19*, 121, 124
Lorn Plateau Volcanic Formation **8**, **12**, 16, 128, *135-7*
Loups Bridge *155-7*
Lower Abdon Limestone *212-13*
Lower Caithness Flagstone Group *24-5*, 85-6, 90, *94*
   Badbea Breccia *24-5*
   Berriedale Flagstone Formation *24-5*
   Berriedale Sandstone Formation *24-5*
   Clyth Subgroup *24-5*
   Ellen's Geo Conglomerate *24-5*, *94*

Helman Head Formation *24-5*, *94*
Hill Head Red Bed Subgroup *24-5*
Lybster Subgroup *24-5*, 85
Robbery Head Subgroup *24-5*, 90-3
Lower Eday Sandstone Formation *24-5*, 56-8, 61-3, *72*
Lower Limestone Formation *173*
   Blackhall Limestone *173*
Lower Limestone Shale Group 221, 254, 258-62, 264, *265*, 286-7, 289, 291, *294*, 301, 306
   Castell Coch Limestone Formation 259, *265*, 266
   Cwmyniscoy Formation *265*
Lower Melby Fish Bed 31-5
Lower Mell Fell Conglomerate Formation 199-201, 205-8
Lower Nodule Bed 107-109, 111
Lower Roman Camp Formation 240, 241, *243*
Lower Sandstone Member *294*, 298, 300
Lower Shales *118*-20, 122
Lower Stromness Flagstone Formation, *24-5*, 52-6, 64, 66-7
Lower Wallop Quarry **11**
Loxonema conicum 244
   L. gregaria 241
Ludford Lane and Ludford Corner **9**, **11**
Ludlow *4-5*, 13-*17*, 256
   Freshwater East-Skrinkle Haven 7, **10**, *302-3*, 307
   Siccar Point to Hawk's Heugh *240-3*, 245-6
   Sawdde Gorge *240-3*, 245-6
Ludlow Anticline 241
Ludlow Bone Bed 4, **9**, **10**, **11**
Ludlow- Přídolí **9**, **11**, 217
lungfish 180
Lybster Subgroup *24-5*, 85
lycopsid 117
Lydney 7, **9**, *214*, 217, 277-80
   Raglan Mudstone Formation 277-8, 280
      Psammosteus Limestone 7, 277-80

St Maughans Formation 277-80
Lydney Harbour 277-*8*
Lynton Anticline 319
Lynton Formation 318
Lyonomyces pyriformis 120
Lyonophyton rhyniensis 120
Lyre Geo Sandstone Member 69

Mackiella rotunda 120
Maiden Kaim 148
Main Cliff, Tynet Burn *107-12*
malachite 289, 303
Malling Conglomerate *134-5*
Manorbier Bay *302-5*
Marloes **11**, 213, *281*
   see also Albion Sands and Gateholm Island
Marloes Anticline 281-2
Marloes Fault *293*
Marloes Sands 211, 281-2
Marly Brae 183
Marsett Formation 205, 207
Matta Taing see Melby
Melby **6**, **8**, *24-5*, *26*, 30-7
   Melby Formation 6, 30-1, 37
      Lower Melby Fish Bed 31-5
      Upper Melby Fish Bed 31, 33-7
   Melby Rhyolite *31*
   Sandness Formation *31*
Melby Fault 25, *26*, 30, *31*, *39*
Melby Fish Beds *24-5*
   see also Lower Melby Fish Bed; Upper Melby Fish Bed
Melby Formation *24-5*, 25-*6*, 30-7
   Footabrough to Wick of Watsness *39*
   Melby **6**, 30-1, 37
   Melby Rhyolite *31*
Melgund Sandstone Member 161
Mell Fell Conglomerate Formation 4, 7, 199-202, 204-8
Mell Fell Trough *4-5*, 205-8
Mellerstain Hill 203
Merbach Brook 235, 238
Merbach Ridge 232, 235
Mercia Mudstone Group *312-13*
Merse of Berwickshire *200*, 202

Index

Merthyr Tydfil 258
Mesacanthus sp. 34, 55, 66, *108*, 110, 166
  M. mitchelli 166
  M. pusillus 110
Mey Subgroup *24–5*, *79*, 83–5
Michrhystridium 72
Microbrachius dicki 83
Micrornatus newportensis Zone 139, 163
Microsphaeridiorhynchus nucula 241
Middle Eday Sandstone Formation *24–5*, *60*, 61–3
Middle Mell Fell Conglomerate Formation 199–201, 205–8
Middlewood Sandstone 235
Midland Valley of Scotland *4*, 8, 13–16, *127–95*, 227
Midland Microcraton *14–16*
Milford Haven Group *212–13*, 282, *294*, *302*, *303*
  Albion Sands Formation *212–13*, 217, *281–4*
  Freshwater East Formation *212–13*, 293, *294*, 299, 301
  Freshwater West Formation *212–13*, 294–5, 300, 302, *303*, 305–6, 308–9
    Conigar Pit Sandstone Member *212–13*, 250, *294–6*, 300, *303*, 305, 307
    Rat Island Mudstone Member *212–13*, 294, 296, 300, *303*, 305, 307
  Gelliswick Bay Formation *212–13*, 217–18, 250
  Lindsway Bay Formation 211, *212–13*, 282, 284
  Moor Cliffs Formation *212–13*, 215, 217, 250, 293–4, 299–300, 301–7
  Red Cliff Formation *212–13*, *281–4*
  Sandy Haven Formation *212–13*, *281–4*, 284–6
Mill Bay Formation *212–13*, 253
Millbuie Sandstone Group *24–5*
Millburn Geo 47
Milleromyces rhyniensis 120
Millerosteus minor 79

millipedes 140, 143, 145, 166
Millstone Grit *254*, *269*
  see also Tenby Group
Milton Ness 6, *127*, 137–8, 158–61, 167, 176, 199
  Arbuthnott–Garvock Group *158*
  Inverclyde Group 158
    Kinnesswood Formation 6, 158–9
Milton of Noth Andesite *119*
Minehead 319, 320, 322
miospores 17–18, 28, 73, 176, 180, 208, 251
  see also spores
Mitcheldean 274
mites 117, 121
Modiola sp. 306
  M. lata 290
modiolopsid bivalves 229–30, 244, 277
Modiolopsis sp. 230, 241, 244
  M. complanata 230, 244
  M. complanata var. *trimpleyensis* 277
Mona Complex *222*, 317
Monington-on-Wye *232*
Monkeys Fold Formation *212–13*, 273
Monograptus colonus 206
Monograptus ultimus Biozone 4
Montrose 128, 139, 158
Montrose Volcanic Formation 12, *134–5*, 139, *167*
Moor Cliffs Formation *212–13*, 215, 217, 250, 293–4, 299–300, 301–7
Moray 21–5, 28–30, 107–12
Moray Firth 28–9, 96–7
Mousavord Loch Anticline *39*
Muckle Castle 58
Mudstone Facies Association 247
Musselwick Fault *293*
myriapod *229*, 247, 249

Nairn Sandstone *24–5*
Nairnside 102
Nairnside Fish Bed *24–5*
Nant ddu *265*
nautiloids 241, 289
nematodes 120, 121
nematophytes 120

Nematophyton taiti 120
Nematoplexus rhyniensis 120
Neobeyrichia lauensis 241
Ness of Clousta to the Brigs 12
Ness of Duncansby 83–4
Nesting Fault *26*, *43*, 47
New Aberdour see Den of Findon
New Red Sandstone 3
New Shipping Formation *212–13*
Newark Bay see Taracliff Bay to Newark Bay
Newport 218, 271
Newton Marlstone 235
Niandt Limestone Member *24–5*, 66
North Bight of Rovahead 44
North Esk River 6, *127*, 134, 155–8
  Dalradian Supergroup *155*
  Dunnottar–Crawton Group *155*
    Lintrathen Tuff Bed *155*, 158
  Strathmore Group 6, 155, 157–8
    Cromlix Mudstone Formation 155–7
    Gannochy Conglomerate Formation 155–8
    Teith Sandstone Formation *155*, 157
North Minches Basin 21
North Newton Shore 7, *127*, 137, 185, 192–5
  Dalradian Supergroup 192–3, 195
    Southern Highland Group 193
      North Sannox Grits 193
  Inverclyde Group 192–3
    Kinnesswood Formation 192–3–195
North Sannox Grits 193
North Scapa Fault *27*
North-eastern Scottish Mainland 21–4, 28–30, 74–124
Northern Ireland 178
Northumberland 4, 16
Northumberland Basin 139, 199, *200*, 204
Nothia aphylla 120
Noust of Nethertown *53–5*

Index

Oak Dingle 7, 9, *213*–18, 229–31
  Raglan Mudstone Formation 230
  Bishop's Frome Limestone 230
  St Maughans Formation 229–30
Ochil Fault *132–3*
Ochil Hills 16, 128, 175
Ochil Volcanic Formation *134–5*, 139, 176–7
Old Castle Head *302–5*
Old Hall Bay 148
Old Man of Hoy Coast 6, *24–5*, 27, 28, 68–71
  Hoy Sandstone Formation 68–71
    Haist Pebbly Sandstone Member 69
    Hoy Volcanic Member 6, 68–71
    Lang Geo Sandstone Member 69
    Lyre Geo Sandstone Member 69
    Trowie Glen Sandstone Member 69
  Upper Stromness Flagstone Formation 68–71
Old Millstone Quarry 66
Oldhamstocks Basin 182
olivine diorite 57, 59
Onchus wheathillensis 284
opilionid 117, 121
Orcadian Basin 4, 6, 8, 10, 15–16, 21–124, 137, 159, 163, 166
Orkney *3*, 4–5, 6, 8, 12, 21, 24, 26–8, 52–74
Orlock Bridge (–Kingledores) Fault *4*, *132–3*
Orthis (*Dalmanella*) *elegantula* 254
Orthoceras 241, 289
osteolepid *108*–112
Osteolepis 110, 111
  O. macrolepidotus 58, 66, 110, 113, 221
  O. panderi 8
ostracoderm 244, 253, 275, 293, 302, 305, 311
ostracods 241, 244, 289
Ousdale Arkose *24–5*

Outer Isles Basin 21
Oxendean Burn 9

Pachytheca sp. 139, 162, *229*, 244, 279, 282, 293, 310
Palaeoblastocladia milleri 120
Palaeocharinus rhyniensis 121
  P. tuberculatus 121
  P. hornei 121
Palaeocteniza crassipes 121
palaeography 13–16, 22, 30, 67–8, 71, 73–4, 78–9, 82, 85, 90, 93, 93, 107, 128, *132–3*, 167, 176, 186, 199, 203–4, 208, *216*, 221–2, 227, 245, 290, 302–3
Palaeomyces agglomerata 120
  P. asteroxyli 120
  P. gordonii 120
  P. horneae 120
  P. simpsonii 120
  P. vestita 120
Palaeonitella cranii 120
Palaeospondylus gunni 91
Palaeotydeus devonicus 121
Palmers Hill Rail Cutting 7, 199, *200*, 202–4
  Inverclyde Group 202
    Ballagan Formation *203*
    Birrenswark Volcanic Formation 203–4
    Kinnesswood Formation 202–4
  Riccarton Group 202
  Stratheden Group 203
palynomorphs 38, 72, 145
Pantymaes Quarry 7, *214*, 246–50
  Mudstone Facies Association 247
  St Maughans Formation 246–7, 250
  Sandstone Facies Association 247–50
Papa Stour 8, *26*
Parahughmilleria sp. 166
Paraprotacarus hirsti 121
Parka decipiens 162, 229, 248
'Passage Group' 202
Pathhead Formation *173*
Pease Bay *see* Siccar Point to Hawk's Heugh
Pease Burn *181*, 183

Pebbly Beds *see* Caeras Conglomerate Member
Peerie Castle 57
Pembroke Group *265*, *302*
Pembroke Syncline *301*
Pembrokeshire *4–5*, 16, 211, *214*, 217–19, 221, 227, 262
Pen-Cerrig-Calch 271
Pengau (Pen-y-gau) Farm 219
Pennan *see* Den of Findon
Pennyland 6, 8, *24–5*, 26, *29*, 78–83
  John o'Groats Sandstone Group 79, 81, 82
  Upper Caithness Flagstone Group 6, 78, 79, 82
  Mey Subgroup 79
Penrith 199, 204, 206–8
Pentland Hills 16, 128, 135, *136*
Pentland Hills Volcanic Formation *136*
Pentlandia macroptera 57, 83
Perton Lane **10**
Pettico Wick to St Abb's Harbour **12**, 139
phytoplankton 23, 41, 52, 82, 93, 166
Pickard Bay Tuff Bed *240–2*, 244, 284–5, 301–4, 306, 308
Pinksey Gill Beds 201
placoderms 110, 112
Planolites 264, 276, 289, 321
plants 10, 11, 17, 139, 164–6, 178, 187, 219, 220, 251, 290, 295–6, 303, 315
  Aberlemno Quarry 6, 8, 161–3, *162–3*
  Afon y Waen *264–5*
  Auchensail Quarry 7, 10, 178–80
  Craig-y-Fro Quarry 7, 10, 257–8
  Cusop Dingle *236*, 238–9
  Duffryn Crawnon 7, 267
  Freshwater East–Skrinkle Haven 7, 10, 303–7
  Freshwater West 292–301
  Glen Vale 175
  Heol Senni Quarry 252–3
  Llansteffan 310
  Lydney 277–80
  Melby 32–7
  Milton Ness 159–61
  North Esk River 157

385

Index

Pantymaes Quarry 247–50
Rhynie **6**, **10**, 117, 120–4
Sawdde Gorge 241–4
Siccar Point to Hawk's Heugh 7, **9**, 187
Tillywhandland Quarry **8**, 164–6
West Angle Bay (North) 290
Yesnaby and Gaulton Coast Section 66
Plateau Beds Formation *212–13*, 219–20
   Abercriban Quarries *258*, 259–60, 262
   Afon y Waen 7, 262–5
   Duffryn Crawnon 7, 265–8
Pneumodesmus newmani 140, 143, 145
Pobie Skeo *31*, *35*
Point of Ayre **12**
Point of Blo-Geo *48*, 49
Point of Ness 54
polychaete worms 72, 121
Polygenetic Conglomerate 199, 206, 208
polygonalis-emsiensis biozone 258
Pont-ar-llechau bridge 240, 241, 243
Pont-ar-llechau Member *242–5*
Pontypool Limestone 218
Pooley Bridge 7, 199, *200*, 204–8
   Borrowdale Volcanic Group 205–7
   Marsett Formation 205, 207
   Mell Fell Conglomerate Formation 204–8
   Ravenstonedale Group 205
   Yoredale Group *205*
Poraspis polaris 229
Port Albion 98
Port Schuchan to Dunure Castle **12**, 139
Port Skerra 28, 74
Port Tarsuinn 96
Port Tarsuinn Fault *97*
Port Tarsuinn Member 96, 101
Porth-y-Mor 7, *214*, 218, 221–8
   Bodafon Formation 221–2, 226
   Porth-y-Mor Formation 221–7
   Traeth Bach Formation 221–5

Traeth Lligwy Formation *222*, 224–7
Porth-y-Mor Formation *212–13*, 218, 221–7
Portishead 7, **9**, *214*, 219, 220, 227, 268, 312–18
   Black Nore Sandstone Formation 312–18
   Mercia Mudstone Group *312*, 313
   Portishead Formation 312–18
     Woodhill Bay Conglomerate 312, 314–18
     Woodhill Fish Bed 312, 314, 315, 317–18
Portishead Formation *212–13*, 220, 221, 312–18
   Woodhill Bay Conglomerate 312, 314–18
   Woodhill Bay Fish Bed 312, 314, 315, 317–18
Pragian *4–5*, *17–18*, *24–5*, 116, *129*, 139, *212–13*, *213*–17
   Caeras Quarry *254–6*
   Craig-y-Fro Quarry 258
   Rhynie 117–24
Prague Basin 17
Pre-lava Sandstones *118–19*, 121, 124
Presipe 301–4
Pretannia 217, 245, 284, 285
Přídolí *4–5*, 7, **9**, **10**, *17*, *129*, 211–*14*, 215–18, 232–4, 234–9, 240–6, 284–6
   Cusop Dingle 7, 234–9
   Freshwater East **10**
   Little Castle Head 7, **11**, 284–6
   Lydney 277–80
   Sawdde Gorge 235, 240–6
   The Scar 7, 232–4
Priest's Nose 302–3
Primrose Hill Quarry 219, 252
Protacarus crani 121
Proto-Southern Upland Fault *132–3*
Protochonetes ludloviensis 241
Protopteraspis 178, 228
   P. gosseleti 219, *229*
Protopteridium thomsonii **10**, 66
Protospeleorchestes pseudoprotacarus 121

Prototaxites sp. 178, 180, *229*
   P. forfarensis 162
Psammosteus Limestone 7, 170, 215, 217, 218, 227, 231, 245, 246, 277–80, 294
   see also Bishop's Frome Limestone, Chapel Point Calcretes Member
pseudo-anticlines 115, 235, 279–80, 282, 293–4, 296, 299, 303–5, 308, 310
pseudonodule 34, 40, 81
Pseudoprotacarus scoticus 121
Pseudosauripterus anglicus 220, 264
Psilophyton 139, 157
Psilophyton Zone **10**
pteraspids **8**, 176, 178, 228, 253, 294
Pteraspis cf. *dunensis* 276
   P. dixoni 219, 252
   P. mitchelli 177–8
   P. rostrata 178, 230
   P. rostrata var. *trimpleyensis* 230
Pterichthyodes milleri 34, 110
Pterinopecten sp. 220, 264
pterygotid arthropods *229*
Pterygotus 143, 162, 166
Ptychomaletoechia omaliusi 220, 264, 267
Purple Sandstone Member *141*, 142, 144
Purton Passage *see* Tite's Point
Pwll-y-Wrach 235

Quantock Hills 319
Quarry Haven *see* Den of Findon
Quarry Hill Sandstone Formation *119*, 124
Quarter *188*, 189–91
Quartz Conglomerate Group 7, *213*–16, 221, 268–71, 316
   Craig-y-cwm Formation *212–13*, 221, 268–71
   Garn-gofen Formation *212–13*, 221, 268–71
   Wern Watkin Formation *212–13*, 221, 268, 269, 271
Qui Ayre Sandstone Formation *24–5*, 64–7

Index

Raglan Mudstone Formation 7, *212–13*, 215, 218–19, 228, 230, 232–3, *234–7*, *240–6*, 277–8, 280
Ramnageo *64*, 66
Ranns of Kutch 73
Ram's Head *38, 40*
Rat Island Mudstone Member *212–13*, 294, 296, 300, *303*, 305, 307
Ravenstonedale 204
Ravenstonedale Group *205*
 Marsett Formation 205, 207
Rawn's Member 322–3
Red Cliff Formation *212–13*, 213, 215, 217, 281–4
Red Head Sandstone Member *134–5, 167*
Red Marl Group 235, 247, 284, 301
 Middlewood Sandstone 235
 Newton Marlstone 235
Red Point 6, *24–5, 29–30*, 74–8, 111
 Upper Caithness Flagstone Group 75, 78
 Latheron Subgroup 75
Red Rock *181*, 184–6
Red Sandstone Member 140–2, 144
Red-Grey Member 220, 287–91, *294*, 299, 301
Redbank Cliff 232
Redheugh Mudstone Formation *181–2*, 184, 186
Redheugh Shore *181*, 183
Remigolepis 184, 185–7
Remyophyton delicatum 120
Reston Group *129*, 137–9, 199–202
 Cheviot Volcanic Formation 199
 Eyemouth Volcanic Formation 199
 Great Conglomerate Formation 199
Rhadinacanthus longispinus 111
Rhamphodopsis threiplandi 91
 R. trispinatus 111
Rheic Ocean 211, 239, 322
Rheic suture 16
Rhenish 13, 17–18

Rhinopteraspis crouchi 9
 R. dunensis (=*cornubica*) 219, 252, 253, 276
cf. *Rhinodipterus* 220
rhizocretions 187
rhizoliths 158–61, 176, 223
Rhynchertia punctata 120
Rhynia 117, 120
 R. gwynne-vaughanii 120
Rhynie 6, **10**, *24–5*, 28–9, 117–24
 Dryden Flags Formation 118–19, *122*, 124
  Lower Shales *118*–20, 122
  Rhynie Cherts Unit 117–120, 123
  Shales with Muddy Sandstones *118*–20, 122–3
  Upper Shales *118*–20, 122
  White Sandstones *118*–19, 122, 124
 Longcroft Tuffs *118–19*, 121, 124
 Pre-lava Sandstones *118*–19, 121, 124
 Tillybrachty Sandstone Formation 119, 124
 Windyfield Chert 117–*19*, 121, 124
 Windyfield Sandstones and Shales *118, 119*, 121
Rhynie Basin 29, 119, 124
Rhynie Cherts Unit 117–120, 123
Rhynicoccus uniformis 120
Rhyniella praecursor 120, 121
 R. vermiformis 120
Rhynimonstrum dunlopi 121
Rhyniognatha hirsti 117, 121
Riccarton Burn 202, *203*
Riccarton Group 202
Ridgeway Conglomerate Formation *4*, 211, *212–13*, 217–20
 Freshwater East–Skrinkle Haven 301–*3*, 305–8
 Freshwater West 291–2, *294*, 296–8, 300
 West Angle Bay (North) 7, 286–8, 290–1
Riera Geo 94
Ritec Fault *5*, 211, 215–18, 220, 285–6, 288–91, *293*, 300–1

River Wye 232, *234*, *271*
Robbery Head Subgroup *24–5*, 90–3
Robert's Linn 202–3
Rob's Cove Conglomerate *134–5*
Rock Hall 159
Rook's Cave 302
Rook's Cave Tuff 301–4, 308
Rosebrae Beds *24–5*
Rosneath Conglomerate Formation *138*
Ross-on-Wye, Royal Hotel 7, *214*, 219, 271–4
 Brownstones Formation 7, 271–4
Rotliegendes 11
Rotten Craig *31, 36*
Rouen Bay Sandstone *134–5*
Rousay Flagstone Member *24–5*, 27, 61–3
Rova Head Conglomerate Formation *24–5*, 43–7
Roxburghshire 199
Royal Hotel *see* Ross-on-Wye, Royal Hotel
River Usk *234*
Ruchill Flagstone Formation *134–5*
Ruperra Limestone 218
Rusophycus 114, 264
Ruthery Head 142

Sabrinacanthus 9
 S. arcuatus 277
Sacquoy Sandstone Member *24–5*, 27, 61–2
St Abb's Head *132–3*, 139, 199–*200*
St Abb's Volcanic Formation *132–3*, 139
St Anthony's Anticline 309
St Anthony's Cottage 308–10
St John's Knapp Sandstone *134–5*
St Maughans Formation 9, *212–14*, 217–18
 Cusop Dingle *234–5*
 Devil's Hole 228
 Heol Senni Quarry *251*, 253
 Lydney 277–80
 Oak Dingle 229–30
 Pantymaes Quarry 246–7, 250

387

Index

Sawdde Gorge *240–1*, 245–6
  The Scar 232
  Wilderness (Land Grove) Quarry 274, 276
Samson's Footmarks *79*
Sanday 57, 58
Sandness Formation *24–5*, *26*, *31*, *39*
Sandsting Plutonic Complex *24–5*, 38–9
Sandstone Facies Association 247–50
Sandwick Fish Bed *10*, *24–5*, 27, 53–6, *64*, 66, 116
Sandy Haven Formation *212–13*, *281–4*, 284–6
Sanguinolites sp. 220, 259, 268
Sarclet **6**, *24–5*, 28, *29*, 93–5
  Lower Caithness Flagstone Group *94*
    Ellen's Geo Conglomerate *94*
    Helman Head Formation *94*
  Sarclet Group **6**, 93–5
    Sarclet Conglomerate Formation 94
    Sarclet Sandstone Formation 94–5
    Ulbster Mudstone Formation 94–5
    Ulbster Sandstone Formation 94–5
Sarclet Conglomerate Formation *24–5*, 94
Sarclet Group **6**, *24–5*, 93–5
  Ousdale Arkose *24–5*
  Sarclet Conglomerate Formation *24–5*, 94
  Sarclet Sandstone Formation *24–5*, 94–5
  Ulbster Mudstone Formation *24–5*, 94–5
  Ulbster Sandstone Formation *24–5*, 94–5
Sarclet Head 93, *94*
Sarclet Sandstone Formation *24–5*, 94–5
sarcopterygians 111
Sauripterus 220, 264
Sawdde Gorge **7**, 217, 240–6
  Black Cock Formation *240*
    Carn Powell Member *240*
  Gwynfe Formation *240–2*, *243–4*
    Pont-ar-llechau Member *242–5*
  Llanddeusant Formation *240–1*, 245
    Lower Roman Camp Formation *240*, 241, *243*
    Raglan Mudstone Formation *240–6*
      Green Beds *243*
      Pickard Bay Tuff Bed *240–2*, 244
      Townsend Tuff Bed *240–1*, 244, 246
    St Maughans Formation *240–1*, 245–6
    Tilestones Formation *240–5*
      Capel Horeb Member *242*, *243*, 245
      Long Quarry Member *240*, *242*, 245
    Trichrûg Formation *240*
    Upper Cwm Clyd Formation *240*
    Upper Roman Camp Formation 241
cf. *Sawdonia* 178, 252
  S. ornata 180
Scaat Craig **8**
Scandian Orogeny *4*, 13–16, 128–30, 158, 168, 217
Scarvister Syncline *39*
Scolenaspis sp. *229*
  S. cf. *whitei* 229
Scone Sandstone Formation 128, *134–5*, 161, 163, 166–8, 170–2, 176, 178
  Arbroath Sandstone Member *134–5*, *167–9*
  Auchmithie Conglomerate Member *134–5*, 167
  Campsie Limestone Member **7**, *134–5*, *170–2*
    Stanley Limestone **7**, *170–2*
  Dunblane Sandstone *134–5*
  Melgund Sandstone Member 161
  Red Head Sandstone Member *134–5*, *167*
  Sheriffmuir Sandstone Member *134–5*, 176–8
  Tannadice Sandstone *134–5*
Scottish Border Basin *4*, *14*–16, 199, 199–201, 202–8
Scurdie Ness to Usan Harbour **12**, 139
scutigeromorph 121
Seamill Sandstone Formation *138*
Securiaspis 178
  S. caledonica 177
  S. waterstoni 177
Senni Formation *212–13*, 219
  Craig-y-cwm *269*
  Craig-y-Fro Quarry **7**, 257
  Heol Senni Quarry 250, *251*, 253
Sennicaulis hippocrepiformis 219
Shales with Muddy Sandstones *118–20*, *122–3*
Shannochill Conglomerate *134–5*
Shap Wells Conglomerate Formation 206, 208
Shap Wells Trough 206
Shapinsay *27*, 57–8
Sheriffmuir Road to Menstrie Burn **12**, 139
Sheriffmuir Sandstone Member *134–5*, 176–8
Sherrycombe Member 322
Shetland 3–4, **6**, **8**, **12**, 21–2, *24–5*, 30–52
Shingly Geo *48*, *49*, 51–2
Shirehampton Formation 317, 318
Siccar Point Conglomerate Member *181*, *182*, 184
Siccar Point to Hawk's Heugh **7**, *127*, 137, 181–7, 192
  Inverclyde Group *181*
    Ballagan Formation *181*, 185, 186
      Eastern Hole Conglomerate 185, 187
    Kinnesswood Formation *181*, 184–7
  Gala Group *181*, *182*
  Stratheden Group 181–2
    Greenheugh Sandstone Formation *181–2*, 184, 186
    Redheugh Mudstone Formation *181–2*, 184, 186

388

Index

Siccar Point Conglomerate Member *181*, 182, 184
Sidlaw Anticline 167
Sidlaw Hills 16, 128
Siegenian 9, **10**, 17, *212–13*, 213–17, 219
Silurian 4–11, 12, 13–17, *129–37*
Silurian–Devonian 9, 288
sinistral shear 130–1
Skelmorlie Conglomerate Formation *138*, 187–8
Skerry of Ness 52, 53
Skiddaw Group *205*
    Buttermere Formation *205*
    Tarn Moor Formation *205*
Skolithos 79, 110, 157, 175–6, 220, 235–6, 243, 249, 259, 288
Skomer Basin 282
Skomer Island *293*
Skrinkle Haven 301, *302–3*, *302*–7, 306
Skrinkle Sandstones Group *212–13*, 220, 262, 286–91, 291–2, *294*, *297*, 298, 300, 301–3, 306–8
    Gupton Formation 7, *212–13*, 287–90, 291, *294*, 298, 300
        Lower Sandstone Member *294*, 298, 300
        Stackpole Sandstone Member *288*, 290, *294*, 298, 300
    West Angle Formation *212–13*, 287–91, *294*–301
        Conglomerate Member 287–90, *294*, 298, 300
        Heterolithic Member *287*, 289, *294*
        Red-Grey Member 220, 287–91, *294*, 299, 301
Sloagar **10**
Smerwick Group 64
Smuggan Syncline 40
Solway Basin 16, 199–200, 202, 204
Somerset 311–18, 322–3
South Bight of Rovahead 45
Southern Highland Group 193
    North Sannox Grits 193

South Fersness Bay 6, *24–5*, 27, 28, 56, 61–3
    Eday Group **6**, 61
    Eday Flagstone Formation 61–3
    Lower Eday Sandstone Formation 61–3
    Middle Eday Sandstone Formation 61–3
    Upper Stromness Flagstone Formation 61–3
        Rousay Flagstone Member 61–3
        Sacquoy Sandstone Member 61–2
South Head *see* Wick Quarries
South Kerrera **12**, 139
South Ronaldsay 27, 58
South Shore of Moray Firth and Strath Nairn *24–5*
South Stromness Coast Section 6, *24–5*, 26–7, 52–6
    Caithness Flagstone Group 52–6
    Lower Stromness Flagstone Formation 52–6
    Upper Stromness Flagstone Formation 52–6
        Sandwick Fish Bed 53–6
South Wales Coalfield 4, 5, 211, 221, *254–6*, 256–62, 265–71
South Wilkhaven Fault 96–7
Southern Upland Fault *14*–16, 128–30, *132–3*, *136*, *188*–9, 199
Southern Uplands 4, **9**, **12**, 13–16, 128, 137–9, 199–201, 202–5
Southern Uplands Terrane *129–30*, 131–3, 137–9, 199–202
Southern Uplands–Longford Down High *132–3*
Spittal Beds *24–5*
Spittal Quarry **8**
?*Sporogonites* sp. 180
spores 116, 139, *213*–17, 221, 295
    Albion Sands and Gateholm Island 282, 284
    Craig-y-Fro Quarry 258

    Freshwater West–Skrinkle Haven 303
    Sarclet 94
    West Angle Bay (North) 287–8, 291
    zonation *212–13*
    see also miospores
Stackpole Sandstone Member *288*, 290, *294*, 298, 300
Stanley Limestone **7**, *170–2*
Steath Geo *48*–52
Steeple Rock 168
Stegotrachelus finlayi 51–2
Stensiöpelta sp. 230
Stirling 127, *134–5*, *138*
Stob Dearg and Cam Ghleann **12**, 139
Stockiemuir Sandstone Formation *138*
Stonehaven Basin 15–16, *127*–35, *140*, 142, 144, 145
Stonehaven Group *130*, 131, 140, 144, 145, 149–50
    Carron Sandstone Formation *134–5*, 140–5, 145–6, 149–50
    Cowie Sandstone Formation *134–5*, 140–6, 149
Storey Arms Quarry *see* Craig-y-Fro Quarry
Stourborough Hill Syncline 38–9, 43
Strath Rory Group 6, *24–5*, 96–102
Strathclyde Group *181*
Stratheden Group *127*, *129*–30, 137–9, 168, 181–2, 187, 199–201, 203
    Burnside Sandstone Formation *129*, *138*–9, *167*–9, 173–4
    Bute Conglomerate Formation *138*
    Clashbenny Formation *138*
    Dura Den Member *138*
    Fairlie Sandstone Formation *138*
    Gargunnock Sandstones *138*
    Glenvale Sandstone Formation *138*, 173–4, 176
    Greenheugh Sandstone Formation *181*–2, 184, 186
    Kelly Burn Sandstone Formation *138*, 187–8

Index

Knox Pulpit Sandstone Formation 7, 128, *138*, 172–6
Redheugh Mudstone Formation *181–2*, 184, 186
Rosneath Conglomerate Formation *138*
Seamill Sandstone Formation *138*
Skelmorlie Conglomerate Formation *138*, 187–8
Stockiemuir Sandstone Formation *138*
Wemyss Bay Sandstone Formation *138*
Strathfinella 158
Strathfinella Hill Conglomerate *134–5*
Strathlethan Bay *146–7*, 150
Strathlethan Sandstone Member *134–5*, *146–7*, 150
Strathmore Basin 4, 15–16, 128–35, 154, 166–9
Strathmore Group 6, *129–30*, *134–5*, 155, 157–8, 171–2, 179
  Bracklinn Falls Conglomerate *134–5*
  Cromlix Mudstone Formation 128, *134–5*, 155–7, *170–2*
  Dalmary Sandstone *134–5*
  Gannochy Conglomerate Formation *134–5*, 155–8
  Malling Conglomerate *134–5*
  Shannochill Conglomerate *134–5*
  Strathfinella Hill Conglomerate *134–5*
  Teith Sandstone Formation 7, *134–5*, 139, *155*, 157, *170*, 176–8, 179–80
    Buchlyvie Sandstone *134–5*
  Tom Dubh Conglomerate *134–5*
  Uamh Beag Conglomerate *134–5*
Strathmore Syncline 140, 145, 155, 157, 158
Strathpeffer Group *24–5*
strike slip movement 13–16, 21, *22*, 25, 128, 131–7, *140*, 142, 144–5, *188–9*, 250, 256, 305

stromatolites 6, 22, 54–6, *64*–7, 79, 81
Stromness *see* South Stromness Coast Section
Stronsay *27*, 57–8
Struie Group *24–5*
Sugar Loaf 271
Sulma Water Fault *39*, 41
Sulma Water Formation *39*
Sumburgh Head 8, 47–8
Sutherland Geo *48–9*, 51–2
Svalbardia 10
Swanlake Bay *302*
Swansea Valley Fault 215, 219, 256
Swanshaw Sandstone Formation *136*

Taghanic highstand 73
Talgarth 218, *234*, 235
Tannadice Sandstone *134–5*
Taracliff Bay to Newark Bay 6, *24–5*, *27–8*, *56–9*
  Caithness Flagstone Group 56–9
  Eday Group 6, 56, 58–9
    Eday Flagstone Formation 56–9
    Deerness Volcanic Member 57
    Eday Marl Formation 56–7, 59
    Lower Eday Sandstone Formation 56–8
Tarbet Ness 6, *24–5*, *29–30*, 96–102, 106
  Balnagown Group 6, 96–102
  Gaza Formation 97–102
  Tarbet Ness Formation 97–102
  Strath Rory Group 6, 96–7, 100–2
Tarbet Ness Formation *24–5*, 97–102
Tarella 257
  T. trowenii 219, 258
Targrove Quarry 10
Tarn Moor Formation *205*
Tautilicaulis 302
Tay Bank 7, *127*, 131, 170–2
  Arbuthnott–Garvock Group 171–2
  Scone Sandstone Formation 170–2

  Campsie Limestone Member 7, 170–2
  Strathmore Group 171–2
  Cromlix Mudstone Formation *170–2*
  Teith Sandstone Formation *170*
Tealing Fish Bed 139
Teith Sandstone Formation 7, *134–5*, 139, *155*, 157, *170*, 176–8, 179–80
  Buchlyvie Sandstone *134–5*
Temeside, Ludlow 9
Temeside Mudstone Formation 9, 217
Tenby Group 7, **10**, 302–7
  see also Millstone Grit
Tenby–Angle Basin *213*–17, 220, 289–291, 307
Tesseraspis tessellata 229, 277
tetrapod 100–1
Tewel 142
Thalassinoides 72
The Cletts 6, 8, *24–5*, 26, 47–52
  Brindister Flagstone Formation 47
  Exnaboe Fish Bed 51–2
The Flaes Syncline *40*
The Scar 7, *214*, 217, 232–4
  Raglan Mudstone Formation 232–3
  Bishop's Frome Limestone 232
  Townsend Tuff Bed 232
  St Maughans Formation 232
The Snook *113*, 114, 116
The Toutties 6, 8, *127*, 130, 140–5
  Stonehaven Group 140, 144
  Carron Sandstone Formation 140–5
  Cowie Sandstone Formation 140–5
  Basal Breccia Member 142
  Brown and Grey Sandstone Member 140–2, 144
  Castle of Cowie Member *141*, 142, 144
  Cowie Harbour Conglomerate Member 140–2, 144

Index

Cowie Harbour Siltstone Member 140–5
Purple Sandstone Member *141*, 142, 144
Red Sandstone Member 140–2, 144
thelodonts **9**, 25, *212–13*, *229*, 235–6, 237, 276, 286,
Three Steps Geo *48–50*, 52
Three Wells Sandstone *134–5*
Thurso *29*, 30, 78–83
Tilestones Formation 217, 240–5
    Capel Horeb Member *242*, *243*, 245
    Long Quarry Member *240*, *242*, 245
Tillybrachty Sandstone Formation 119, 124
Tillycoultrie **12**, 139
Tillywhandland Quarry **6**, **8**, *127*, 131, 161, 163–6
    Dundee Flagstone Formation **6**, **8**, 163–4, 166
    see also Turin Hill *and* Aberlemno Quarry
Tintern Sandstone Formation *212–13*, 221, 312, 317
Tite's Point **9**
Tom Dubh Conglomerate *134–5*
Tom Mor 102–3
Tomintoul 28
Tongwynlais 221
Tongwynlais Formation 221
Too of the Head **12**
Tornquist Sea 13–15, 73
Tournaisian *4–5*, **17**, 155–6, 189, 202–5, *212–17*, 221
    Largs Coast 189
    Siccar Point to Hawk's Heugh 187
Tower Burn *181–3*
Townsend Tuff Bed **7**, **11**, *212–13*, 217, *232*, *234–6*, 238–9, *240–1*, 244, 246, 272, 284–6, 291, 293, 294, 300, 301–4, 306, 308
Towy estuary 308
trace fossils 100, 218, 224, 302, 303, 304–5, 307
    Abercriban Quarries 220
    Afon y Waen 264
    Den of Findon 114
    Freshwater East–Skrinkle Haven 302, 304–5, 307
    Freshwater West *295–6*, *298*
    Glen Vale 175–6
    Greenan Nev Coast *59–61*, *60–1*
    Llansteffan 308, 310–11
    Milton Ness 159–61
    North Esk River 157
    Pantymaes Quarry 247, 249
    Porth-y-Mor 224
    Portishead 313, 315, 317
    Tarbet Ness 100–2
    The Scar 233–4
    Tillywhandland Quarry **6**, **8**, 163–6
    West Angle Bay (North) 288, 290
    Wilderness (Land Grove) Quarry 275–6
Trachyophytes 120
Traeth Bach Formation *212–13*, 218, 219, 221–5
Traeth Dulas 221, *224–5*
Traeth Lligwy 221, *224–5*
Traeth Lligwy Formation *212–13*, 219, *222*, 224–7
Traeth yr Ora *222*, 227
transpression *4*, 13–16, 128–30, 215–17
transtension 5, *14*–16, 211, 215–*16*, 284
Traquairaspis campbelli 143, 145
    T. (Phialaspis) symondsi 228–9
Tredomen Quarry 218
Tremadoc–Arenig *14*
Tremuda Bay Volcanic Formation *134–5*, 145, *146*, 148
Trentishoe Member 318–23
Trichopherophyton 117
    T. teuchansii 120
Trichrûg Formation 211, *240*
trigonotarbids 117, 120, 121
Trinkie Pool *85*, 87, 89
Tristichopterus alatus 83
Trollochy 151–3
Troup Head *113*, 114
Trowie Glen Sandstone Member 69
Tudhope Quarry 201
Tugford *see* Oak Dingle
Turbochelius helicites 241, 244
Turin Hill **8**, **10**, *161*, *163*
    see also Aberlemno Quarry *and* Tillywhandland Quarry
Turinia pagei 212–13, *229*, 230, 275
Turkey Cottage *240*, 244–5
Turnberry Lighthouse to Port Murray **12**, 139
Turriff Basin *4*, 21–2, *24–5*, 28–*9*, 112–17
Tweeddale 199
Tynet Burn **6**, **8**, *29*, 30, 107–12
    Coccosteus Bed *108*, 109, 111
    Lower Nodule Bed 107–9, 111
    Upper Nodule Bed *107–11*
Tywi Anticline 245

Uamh Beag Conglomerate *134–5*
Ulbster 94
Ulbster Mudstone Formation *24–5*, 94–5
Ulbster Sandstone Formation *24–5*, 94–5
Ullswater 204–7
unconformities **11**, 28–9, *113*, 114–*15*, 117, 142, 199–201, 315–17
    Abercriban Quarries 259
    Dun Chia Hill (Loch Duntelchaig) 102–7
    Freshwater East–Skrinkle Haven 220
    Freshwater West 220, 295
    Largs Coast 190–2
    North Newton Shore **7**, 192–5
    Palmers Hill Rail Cutting 202–5
    Pooley Bridge 205–8
    Siccar Point to Hawk's Heugh **7**, 192–5
    Whiting Ness **6**, 166–9
    Yesnaby and Gaulton Coast Section 65–7
    see also Hutton's Unconformity
Undichna 218

Index

Upper Abdon Limestone *212–13*
Upper Caithness Flagstone Group **6**, *24–5*, 75, 78, 79, 82, 83, *84*, 90–3
  Ham Skarfskerry Subgroup *24–5*
  Latheron Subgroup *24–5*, 75, 93
  Mey Subgroup *24–5*, 79, 83–5
  Spittal Beds *24–5*
Upper Cwm Clyd Formation *240*
Upper Eday Sandstone Formation *24–5*, 60, *72*
Upper Leintwardine Formation 241
Upper Melby Fish Bed 31, 33–7
Upper Mell Fell Conglomerate Formation 199–201, 205–8
Upper Nodule Bed *107–11*
Upper Roman Camp Formation 241
Upper Shales *118–20*, 122
Upper Stromness Flagstone Formation 52–6, 61–2, *64*, 66–7, 68–71
  Rousay Flagstone Member *24–5*, *27*, 61–3
  Sacquoy Sandstone Member *24–5*, *27*, 61–2
Uraniacanthus sp. 166
Uskiella spargens 219

Vaakel Craigs 47–8, *49*
Vale of Eden 199, *200*
Variscan Orogeny *4–5*, 182, 211, 268, 285
Ventarura lyonii 117, 120
vertebrates 17, 219–21, 235–6, 238–9, 251, 276, 277–80
Veryhachium 72
vitrinite 140
Voe of Footabrough 37–38
Volcanic rocks **8**, *12*, 15–16, 26, 34, 57–9, 68–71, 83, 105–6, *118*, 128, 131–6, 139, 140–5, 199–200, *234–6*, 238–9, 284–6
Volcanic Conglomerate Member 142
volcaniclastic *12*, *136*, 146–51
volcanoes *132–3*

Wales 11, 13, 211, 217
Wales–Brabant Massif 288, 319, 322
Walls Boundary Fault *14*, *22*, *26*, 37, *39*
Walls Formation **10**, *24–5*, 37–43
Walls Peninsula 25–6, *31*, *39–3*
Warebeth *24–5*, 53–5, 66
Wats Ness 31–*2*
Watsonosteus fletti 83
Wauchope Burn 201
Wayne Herbert Quarry **9**, 218
Welsh Borderland *4–5*, 11, 13, 176–8, 211–15, *213–15*, 217, 227, 239, 250, 256
Wemyss Bay Sandstone Formation *138*
Wenlock 3, *4*, **7**, **11**, 13–16, *14*, 17, 128–30, 140–5, 211, 281–4, *302*
  Albion Sands and Gateholm Island **7**, 281–4
  Freshwater West 299, 301
  The Toutties 140, 144–5
Wern Watkin Formation *212–13*, 221, 268, 269, 271
West Angle Bay (North) **7**, *214*, 217, 220, 221, 286–91, 301
  Lower Limestone Shale Group 286–7, 289, 291
  Ridgeway Conglomerate Formation 286–88, 290–1
  Skrinkle Sandstones Group 286–91
    Gupton Formation **7**, *287–90*
      Stackpole Sandstone Member 288, 290
    West Angle Formation 287–91
      Conglomerate Member *287–90*
      Red-Grey Member *287–91*
West Angle Formation *212–13*, 287–91, *294*, 298– 301
  Conglomerate Member *287–90*, *294*, 298, 300
  Heterolithic Member *287*, 289, *294*
  Red-Grey Member 220, 287–91, *294*, 299, 301
Wester Rova Head *44*

Weydale Quarry **8**
Wharley Point 309, 311
Wharley Point Anticline 309
Whitcliffe Formation 241
White Sandstones *118*–19, 122, 124
Whitehouse Conglomerate Formation *134–5*, 139, *146*, 152
Whitehouse Den **8**, 139
Whiting Ness **6**, *127*, 137, 166–9, 178
  Arbuthnott–Garvock Group 166–7
    Scone Sandstone Formation 166–8
      Arbroath Sandstone Member 167–9
      Auchmithie Conglomerate Member 167
      Red Head Sandstone Member *167*
    Burnside Sandstone Formation *167–9*
    Dundee Flagstone Formation *167*
    Montrose Volcanic Formation *167*
Wick Quarries **6**, *22–4*, *29*, 85–90
  Lower Caithness Flagstone Group 85–6, 90
  Lybster Subgroup 85
Wick of Watsness see Footabrough to Wick of Watsness
Wiegeltaspis n. sp. 230
Wilderness (Land Grove) Quarry **7**, *214*, 219, 274–6
  Brownstones Formation **7**, 274, 276
  St Maughans Formation 274, 276
Wilkhaven Jetty Fault *97*
Wilton Road 271
Windermere Supergroup 206
  Coniston Group 206
Windyfield Block *117–24*
Windyfield Chert *117–19*, 121, 124
Windyfield Sandstones and Shales *118*, *119*, 121
Winfrenatia reticulata 120
Winsle Anticline *293*

Index

Wiston Grey Volcaniclastic Sandstone Member *136*
Wolf's Hole Quarry 7, *127*, 176–8
  Arbuthnott–Garvock Group 176
    Scone Sandstone Formation 176, 178
      Sheriffmuir Sandstone Member 176–7
  Ochil Volcanic Formation 176–7
Woodbank Series 256
Woodhill Bay Conglomerate 312, 314–18
Woodhill Bay Fish Bed 312, 314, *315*, 317–18

Yellow Hammer Rock *319*–22
Yelting Geo 69
Yesnaby and Gaulton Coast Section 6, *24–5*, *26–7*, 53, 64–7
  Caithness Flagstone Group *64*, 67
    Lower Stromness Flagstone Formation 64, 66–7
    Sandwick Fish Bed *64*, 66
    Upper Stromness Flagstone Formation *64*, 66–7
  Yesnaby Sandstone Group 6, 64–7
    Qui Ayre Sandstone Formation 64–7
    Harra Ebb Sandstone Formation 64–7
Yesnaby Sandstone Group 6, 64–7
  Harra Ebb Sandstone Formation *24–5*, 64–7
  Qui Ayre Sandstone Formation *24–5*, 64–7
Yoredale Group *205*

Zosterophyllum 162, 180, *229*, 257–8
  Z. llanoveranum 219
  Z. myretonianum 162–3
Zosterophyllum Zone **10**, 162